# The Videodisc Book
**A Guide and Directory**

# The Videodisc Book
## A GUIDE AND DIRECTORY

**1984 EDITION**

Rod Daynes, *Editor*

Beverly Butler, *Associate Editor*

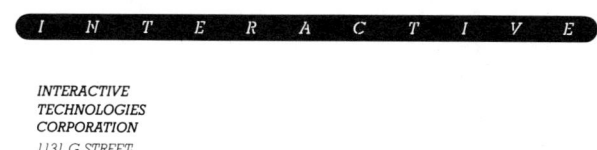

*INTERACTIVE
TECHNOLOGIES
CORPORATION
1131 G STREET
SAN DIEGO, CA 92101*

John Wiley & Sons, Inc.
New York

Copyright © 1984, by John Wiley & Sons, Inc.

All rights reserved. Published simultaneously in Canada.

Reproduction or translation of any part of this work beyond that permitted by Sections 107 or 108 of the 1976 United States Copyright Act without the permission of the copyright owner is unlawful. Requests for permission or further information should be addressed to the Permissions Department, John Wiley & Sons, Inc.

**Library of Congress Cataloging in Publication Data**
Main entry under title:

The Videodisc book.

   Bibliography: p.
   Discography: p.
   Includes index.
   1. Video discs—Handbooks, manuals, etc.   I. Daynes, Rod.  II. Butler, Beverly, 1947-
TK6685.V53   1984        621.388'332       84-5210
ISBN 0-471-80342-1

# Contents

Preface vii
How to Use this Book ix

## PART ONE: THE GUIDE

Production Flowchart 2

Who, What, Where, Why and How Much of Videodisc Technology 7
ROD DAYNES

An Interactive Videodisc Glossary 23
ROD DAYNES

**ANALYSIS**

The Front End: Systems Analysis and Media Selection 27
RICHARD WALKER and LAURALEE BUTLER

**DESIGN**

A Few Principles of Interactive Videodisc Design 35
JIM HOEKEMA

**PREPRODUCTION**

Scripting for Videodisc 45
CASEY GARHART

**PRODUCTION**

Producing a Videodisc: An Old Filmmaker's Perspective 51
CHAD WORCESTER

The Videodisc Production Worksheet 59
CHAD WORCESTER and JIM SMELOFF

Interchangeability 61
JEFF KEMPH

Production Tools and Techniques 67
RAY FEENEY

Videodisc Interfaces and Software 73
DEAN ZOLLMAN

**POSTPRODUCTION**

Videodisc Premastering     77
RONALD W. NUGENT

**MASTERING AND REPLICATION**

DRAW Technology The Instant Videodisc     83
JOHN WINSLOW

**IMPLEMENTATION**

The Tacoma Narrows Bridge Videodisc: a Personal History     87
ROBERT G. FULLER

What Can You Do with a Disc?: a Classroom Teacher's Perspective     93
PHIL KESSINGER

Future Developments     95
   *A Technological Perspective*     95
   Gershon Weltman

   *Educational Perspectives*     97
   Frank B. Withrow and Linda G. Roberts

   *The Entertainment Videodisc (Chapter Two)*     100
   Ken Christie

   *Archival Perspectives*     101
   Henry A. Bertram

   *Trends in the Integration of Computer and Video Technology*     102
   Mark Heyer

   *Industrial Perspectives*     105
   Perry Reeves

   *Game Technology: An Instructional Designer's Wish*     106
   James J. L'Allier

   *Publishing Perspectives*     107
   Gary Carlson

A Selected Bibliography of Videodisc Articles     111

## PART TWO: THE DIRECTORY

Production and Development Resources     119
Hardware Manufacturers/Software Developers/Integrated Systems     179
Mastering and Replication Resources     211
Videodiscography     217
Indexes     455
   *Individual Name*     457
   *Hardware/Software/Integrated Systems Companies by Type of Product*     463
   *Organizations by Geographic Location*     467
   *Levels 1, 2, and 3 Videodiscs*     472
   *Videodiscs by Type*     477

# Preface

*The Videodisc Book: A Guide and Directory*, 1984 Edition, is an invaluable resource for anyone needing to keep up with the ever- expanding field of videodisc technology. It is the first authoritative source of information for the vital new videodisc industry and will be published annually to help those involved in the industry keep up with this rapidly changing technology. The Videodisc Book consists of two major parts, as follows.

*Part One: the Guide* is a compilation of articles by acknowledged authorities in the field on:

- the who, what, where, why, and how much of videodisc;
- front-end analysis as a means to effectively plan and organize an interactive videodisc project;
- how interactive videodisc capabilities affect program design;
- scripting for nonlinear programs;
- the cardinal rules of interactive videodisc production;
- strategies for making videodisc programs compatible with the greatest number of players;
- production tools that aid visualization of a finished interactive videodisc program;
- interfacing arrangements between computers and videodisc players;
- how to assemble a successful videodisc premaster;
- the emerging DRAW technology as a means to make "instant" videodisc masters;
- suggestions for implementing finished videodisc programs;
- predictions for the future of major videodisc applications.

Part One also contains such aids as a production flowchart, a glossary of videodisc terminology, a scripting worksheet for interactive videodisc programs, a worksheet for projecting the cost of producing a videodisc program, and a bibliography for further reading.

*Part Two: the Directory* is a resource list for anyone interested in interactive videodisc production. It includes:

- the names, addresses, telephone numbers, contacts, services offered, and descriptions of over 200 individuals and organizations experienced in all phases of videodisc production and development;
- the names, addresses, telephone numbers, contacts, types of products offered, and descriptions of over 100 companies who offer videodisc-related hardware, software, or integrated systems;
- the names, addresses, telephone numbers, contacts, services offered, and price information for all the mastering and replication facilities in the U.S.;
- complete descriptions of over 1500 laser and CED consumer, educational, research, and industrial videodisc titles, including contact persons and availability information for industrial titles;
- indexes for fast and convenient access to all the directory entries.

Listings for Part Two of this edition were obtained by a mass mailing of questionnaires in the fall of 1983. The inclusion of any individual or organization in this book is not intended as an endorsement of their products or services by the editors or by John Wiley & Sons.

The questionnaires stated that involvement in the production of a videodisc or videodisc-related product was the criterion for inclusion in the book, but time did not permit verification of each questionnaire. The editors attempted to include all relevant information provided by each respondent, and they feel that each entry is an accurate reflection of the material provided.

## ACKNOWLEDGMENTS

The editors want to express their gratitude to all those persons who contributed to the development of this first edition of *The Videodisc Book:* Mary Breach, who sorted and entered names, organizations, titles, and articles; Jack Spiegelberg, who helped decipher manuals and write computer programs to manage all of the data; Phil Smith and Paul Zarins of the University of California Library, who offered advice on how to format a reference book and helped with research; Dick Palmer, who designed the book and served as publishing liasion; Elliott and Bennett Derman and all the others at GTS Typesetting, who provided invaluable computer-based typesetting services; our friends at John Wiley & Sons in New York, Gary Carlson, Bill Rosen, and Lorie Rothstein, who supported us throughout the project; all those listed in Part Two, who took time out from their busy schedules to respond to our questionnaires; and finally, the authors of the articles in Part One, who tolerated our ravings throughout the process.

# How to Use This Book

Part One is an excellent guide for anyone who needs to become familiar with the process for developing and producing interactive videodisc programs. Authors were selected for their expertise in the particular phase of the process for which they have contributed articles. The articles are organized according to the phases in the videodisc production and development process, which gives a cohesive overview for novices in the field and allows those with more experience to easily find the articles that will supplement gaps in their knowledge. The editors recommend, however, that even experienced readers peruse all the articles to gain fresh insights into interactive videodisc development and production.

The entries in Part Two are designed to be self-explanatory. All of the information in the entries was taken from questionnaires returned by the organizations listed. A description of the information asked for on the questionnaires follows. If a listing does not include a particular piece of information, such as a contact person, a phone number, or a description, that piece of information was not provided by the respondent. Each questionnaire provided ample opportunity for reponses that did not fit within categories provided.

The questionnaire for Production and Development Resources produced the following:

1. *Organization.* The name of the organization that responded is given here. Most of the listings are for companies engaged in various aspects of videodisc development, but some entries are for groups within large companies, university departments involved in supporting campus videodisc production, or individuals providing independent consulting.

2. *Type.* Respondents were asked to classify themselves according to one or more of these categories: company, research and development, not-for-profit, association, independent consultant, or "other." Some of the "other" categories offered by the respondents were educational institution or newsletter publisher.

3. *Address.* For large organizations, addresses for offices in more than one location are often given.

4. *Contact.* More than one contact person is sometimes given.

5. *Phone.* Toll free numbers are listed when they were provided.

6. *Services.* Respondents categorized their services within the overall headings of preproduction, production, and postproduction. Under preproduction they were asked to indicate which of the following services they offered: analysis, design, scripting, or funding. Within production they were asked about: producing, directing, special effects, and computer programming. Postproduction services surveyed were: premastering, program simulation, and system design and configuration. Seminars, workshops, and "other" were included as categories in all three headings.

7. *Description.* This section includes any additional important information, such as research and development work, company specialities, important projects, etc. provided by the respondent.

The questionnaire for Hardware Manufacturers/Software Developers/ Integrated Systems/Mastering and Replication Resources produced the following:

1. *Organization.* The name of the organization, mostly companies, that responded is given here.

2. *Address.* For large companies, addresses for offices in more than one location are often given.

3. *Contact.* More than one contact person is sometimes given.

4. *Phone.* Toll free numbers are listed when they were provided.

5. *Hardware Manufactured.* Respondents were asked which of the following categories applied to the products they manufactured: videodisc players, overlays/superimpose units, still-frame audio, digital audio, input devices, interfaces, or "other."

6. *Videodisc Related Software Produced.* Respondents were asked to categorize their software products according to the following: drivers/video utilities, authoring systems, graphics editors/titlers, disc simulators, "other."

7. *Integrated Systems Offered.* Respondents were asked which of the following types of systems they offer: training work stations/carrels, information displays, point-of-purchase systems, video games, and "other."

8. *Mastering and Replication Services.* If the company provides mastering and replication services, they were asked to send a rate card detailing their services and prices. This information appears in the "Mastering and Replication Resources" section of Part Two.

9. *Description.* This section includes any additional important information, such as research and development work, company specialities, or important products provided by the respondent. Respondents were encouraged to send photographs of their products for inclusion in their entry.

The questionnaire for the Videodiscography produced the following:

1. *Videodisc Title.* Some videodiscs are proprietary and the respondents could not furnish their titles. In those cases, the name of the organization that submitted the form is given followed by "Confidential" or "Not Given" in parentheses. For some large programs consisting of many videodiscs, often an overall title was given to cover all the discs.

2. *Description.* Respondents were asked to briefly describe the content and purpose of their videodisc. Again, for large programs one overall description may be given for all the component videodiscs.

3. *Type.* Respondents were asked to classify the subject matter of their videodiscs using as many of the following categories as apply: Motion Picture, Comedy, Music, Dance, Art, Game, Children, Adolescent, Adult, Medical, Research and Development, Technical, Instructional, In-house Train-

# How To Use This Book

ing, Archival, Catalog, Point-of-Purchase, Advertising, Promotional, or "Other." Many of the consumer videodisc titles were entered from distributor's catalogs and the editors used their judgment to classify the titles.

4. *MPAA Rating.* This only applies to motion pictures. Most motion picture titles were entered from catalogs, so a rating is included in the entry if it was given in the catalog. MPAA ratings are given in parentheses at the end of the "Type" heading.

5. *Release Date.* This is the date when a videodisc was released into distribution or delivered to a client.

6. *Format.* Respondents were asked which of the following formats apply to their videodisc: CED, Laser CAV, Laser CLV, or "Other." Some of the "others" given were VHD and Thomson CSF-Transmissive. Some discs are available in several formats. In the case of consumer discs distributed by both Pioneer and RCA, Pioneer always distributes the Laser version and RCA always the CED version.

7. *Level.* This heading describes the level of interactivity of the videodisc: 0 indicates no interactivity, 1 indicates the disc runs on a consumer level player with some interactivity, 2 indicates the disc runs on an industrial player, and 3 indicates a computer-controlled videodisc. When more than one level is given for a single videodisc, the disc is interchangeable and designed to run on more than one level of player. Most motion picture discs are level 0.

8. *System Description.* For level 3 discs, respondents were asked to provide a description of the system necessary to run the videodisc program, the purpose of the program, any relevant technical information, and any special features.   9. Developer. The individual or organization who developed the videodisc.

10. *Contact for more information.* Sometimes more than one individual or organization is given.

11. *Features.* Respondents were asked to indicate which of the following were part of their videodisc: Chapters, Picture Stops, Digital Dump-Sony, Digital Dump-Pioneer, Stereo, Dual Audio, CX Encoded, Sound Over Still, and "Other." For motion picture entries, the Laser and CED versions of the same title may differ in the features offered. In those cases, the format is given in parentheses preceding the features that apply to it.

12. *Running Time.* Does not apply to interactive discs. Sometimes a typical range of hours for completion of the program is given for interactive titles.

13. *Producer.* Self-explanatory.

14. *Distributor.* The individual or organization who can be contacted to obtain a copy of the videodisc.

15. *Retail Outlet.* Usually applies to consumer discs.

16. *Availability.* Many industrial or training videodiscs are not publicly available. If an internal-use videodisc can be viewed on-site or seen as a demonstration of a company's videodisc capabilities, that information is included here. Any special conditions, such as a disc only being available to health care professionals, is also included here.

17. *Order Number.* For consumer discs, this number is listed directly after the name of the distributor or retail outlet.

18. *Suggested Price.* For consumer discs, the price is given directly after the order number.

# PART ONE

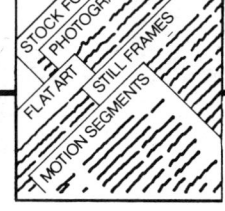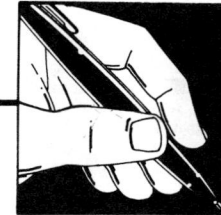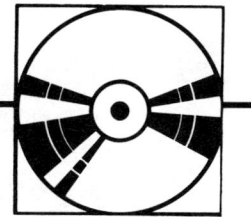

## THE GUIDE

# ANALYSIS                                                    DESIGN

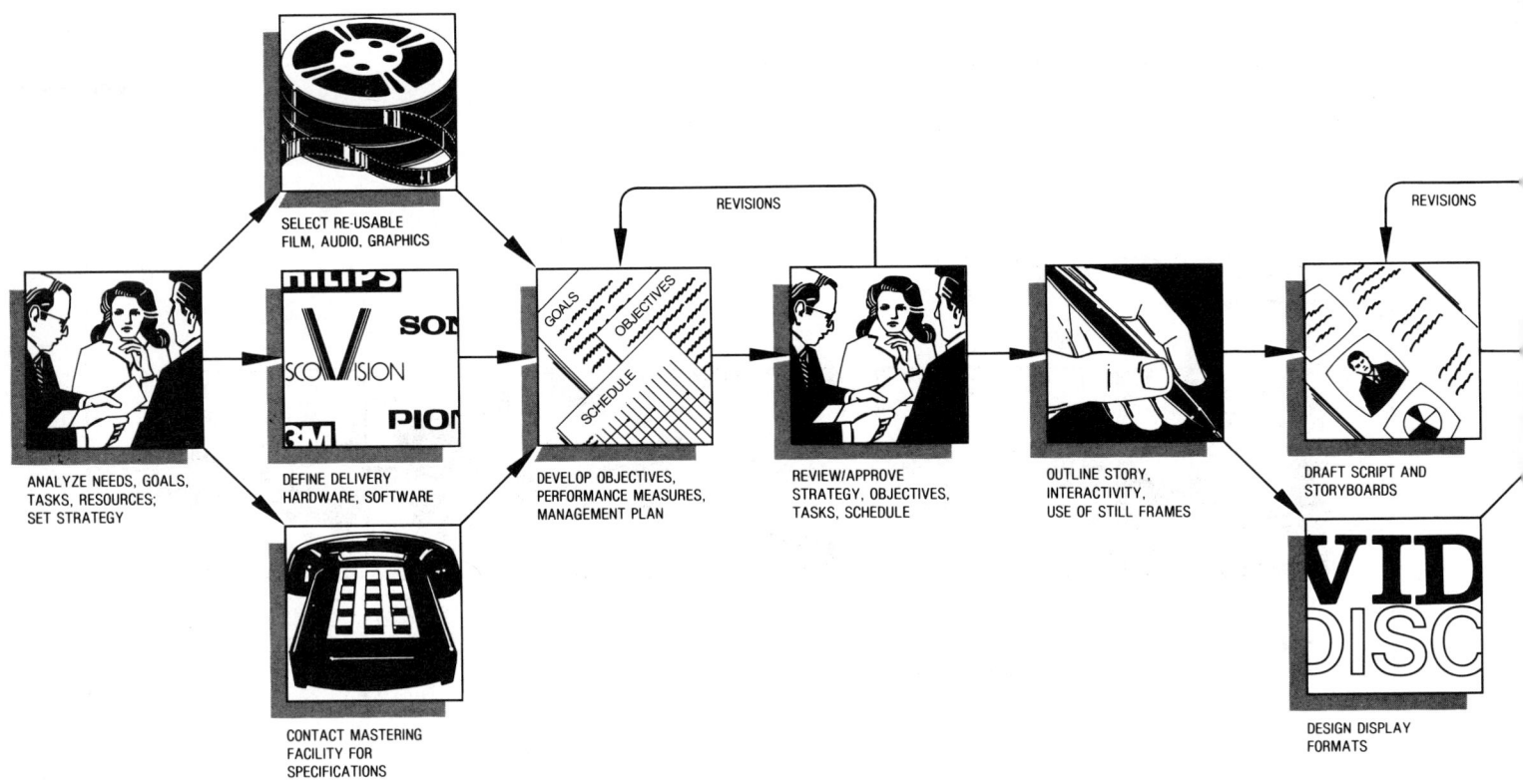

Written and designed by Jim Hoekma for WICAT Systems. Copyright 1981, 1984 WICAT Systems, Inc. Used with permission.

**WEEK**      1           2           3           4           5           6

Most of the graphics in this edition of *The Videodisc Book: A Guide and Directory*, have been taken from an "Interactive Videodisc Design" flowchart, originally developed by Jim Hoekema and published by WICAT systems in 1981. The flowchart is included in its entirety on the following pages. Seen as a whole, the articles and information in *The Videodisc Book: A Guide and Directory* amplify on the various processes shown in this flowchart.

The purpose of the chart is to give those who are new to this field some idea of the number of tasks involved in making a videodisc program. Although the medium of interactive videodisc is so powerful and flexible that it is impossible to describe a "typical" project, this flowchart is included as an example of the requirement for a systematic approach to developing interactive videodisc.

The chart itself is based primarily on several "Level 2" videodiscs that WICAT produced for a client's customer marketing presentations at about 200 national sales offices. These discs relied on a professional crew and high production values; they generally used about

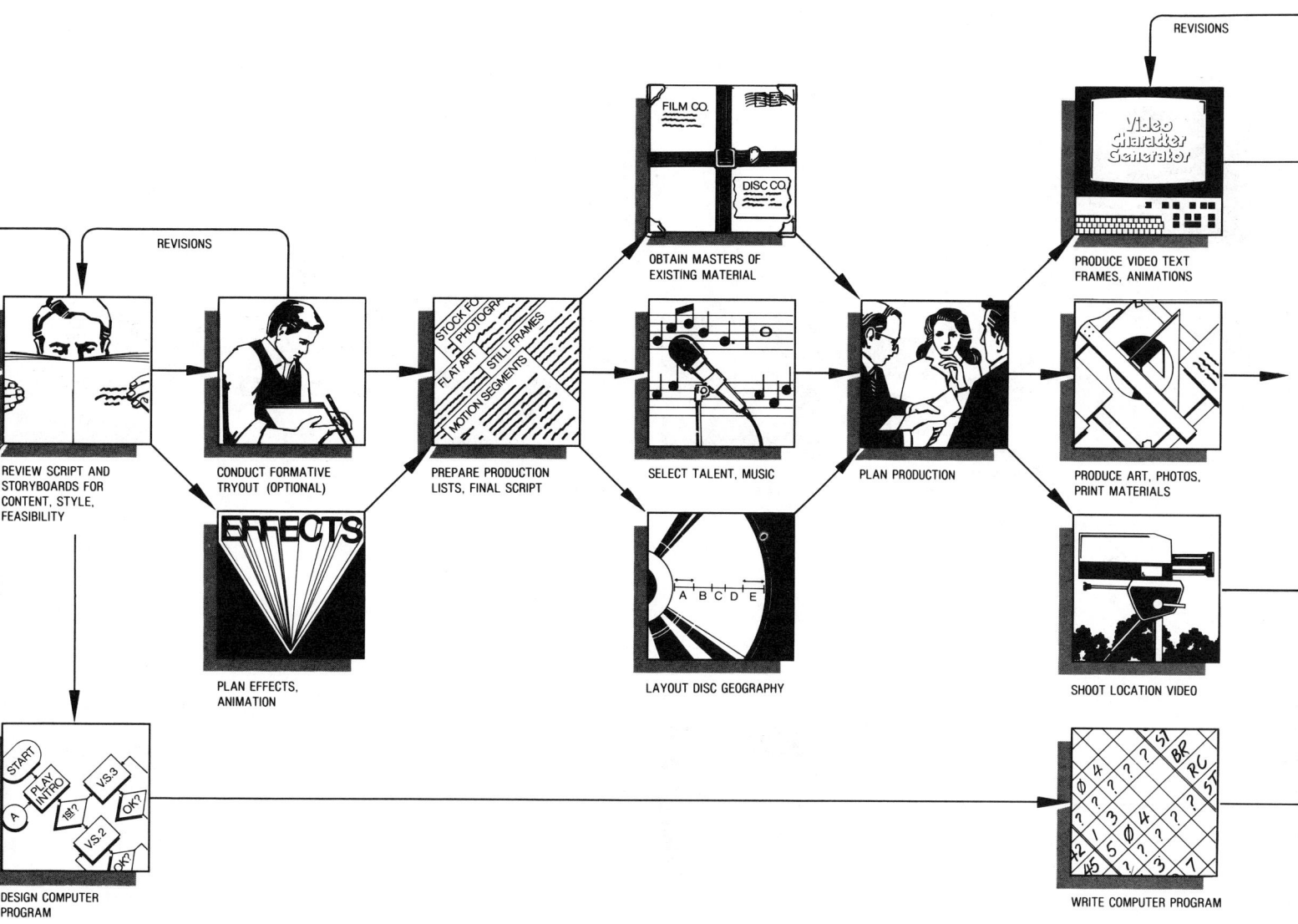

400-600 stillframes; and the programming component was limited to the very small 1000-byte capacity of the now defunct PR-7820 "industrial" player. A typical schedule for such projects is 16-20 weeks (without mastering).

More sophisticated videodisc programs using external computers and perhaps some special peripheral devices take much longer, and there are many more review points than the few stages shown here. On the other hand, some programs can be created more quickly and cheaper if existing images are used and there is little programming.

In short, given the wide variety possible in subject matter, production values, programming complexity, audiences, and delivery systems, interactive videodisc programs can differ greatly in cost, schedule, and final quality.

# POST-PRODUCTION/PRE-MASTERING

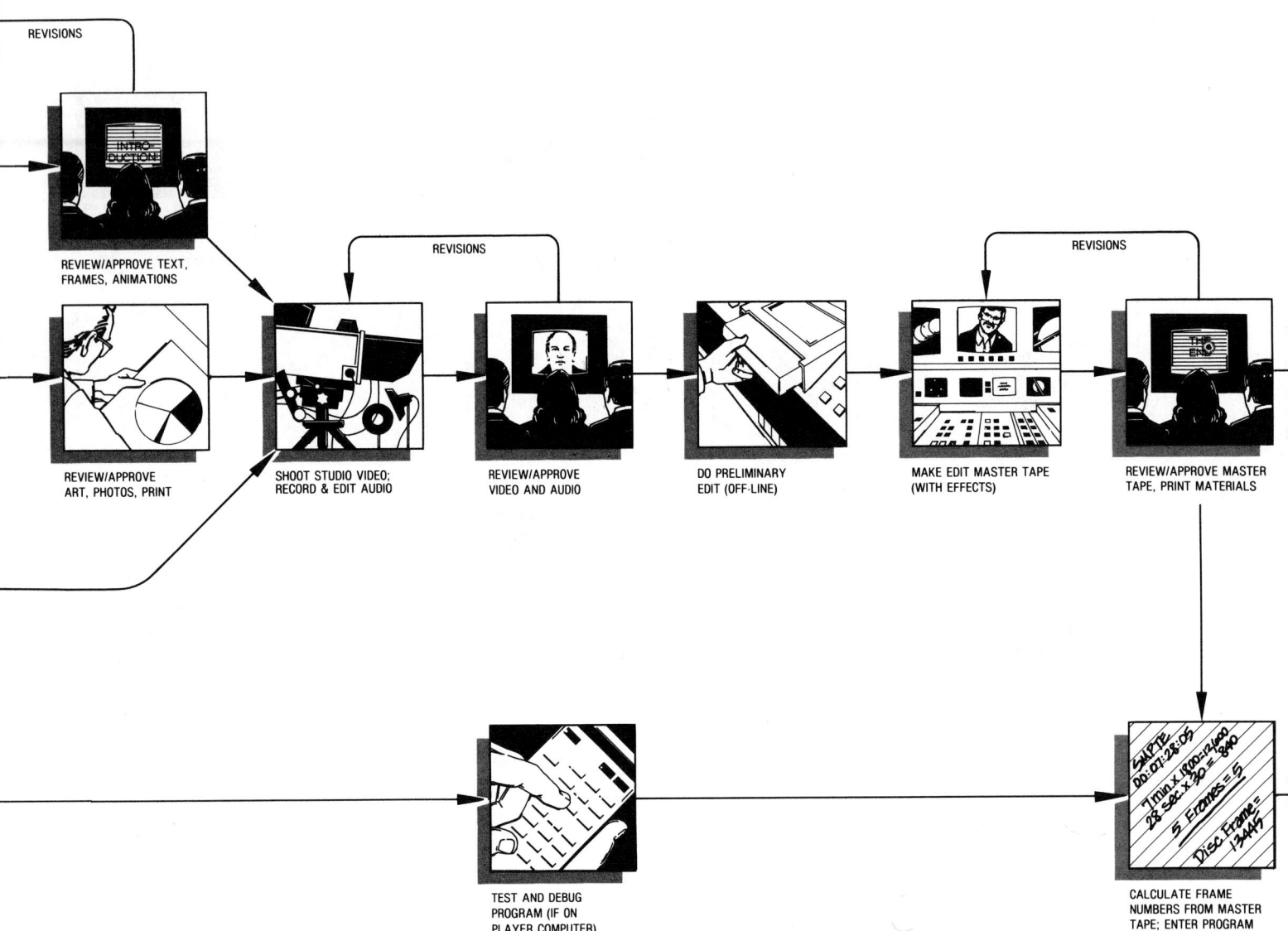

Written and designed by Jim Hoekma for WICAT Systems. Copyright 1981, 1984 WICAT Systems, Inc. Used with permission.

# MASTERING

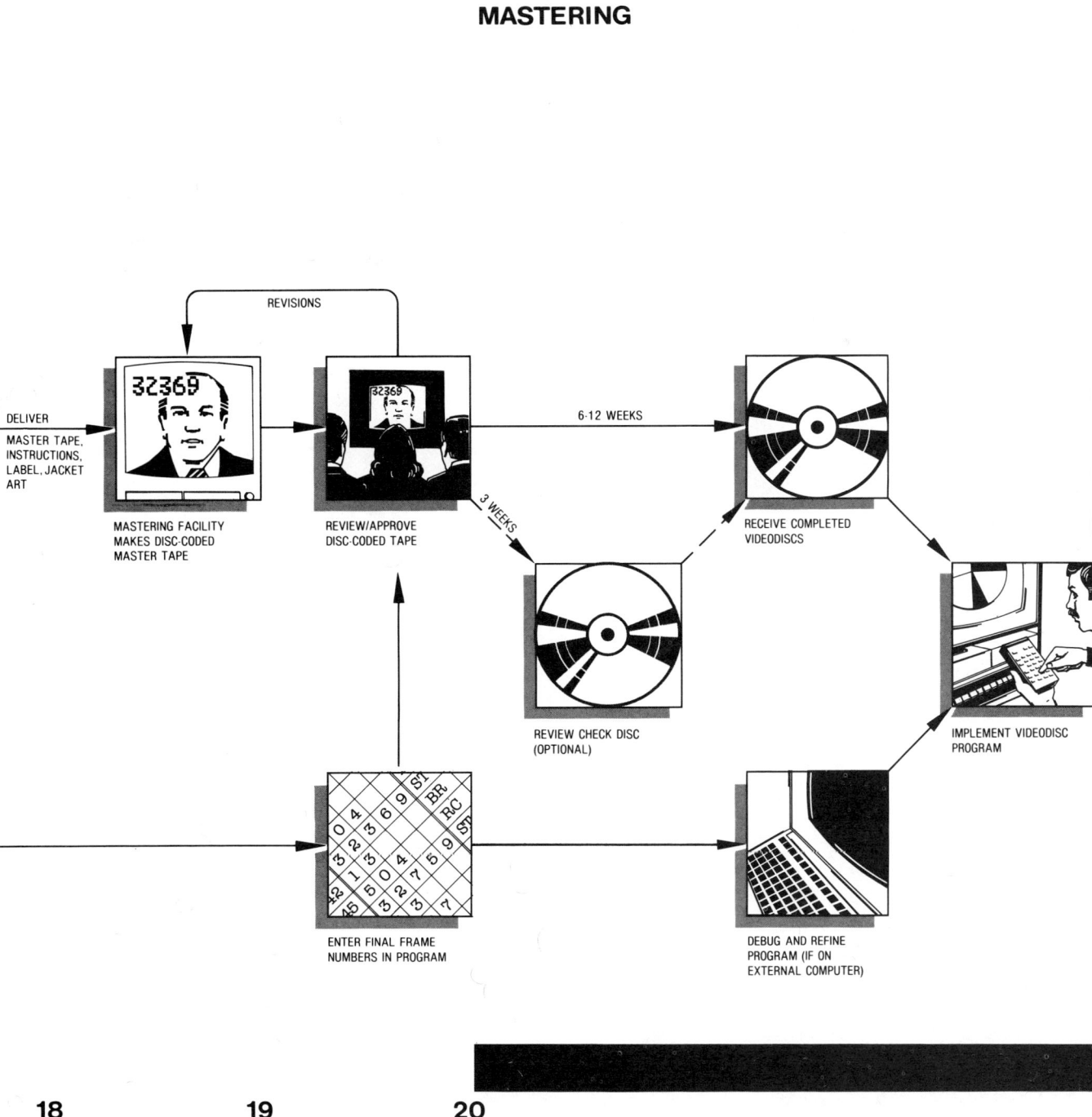

# Who, What, Where, Why, and How Much of Videodisc Technology

*Rod Daynes*

Several years ago, Mr. Michael Dann, former chief of network television programming for CBS, said in a speech that the success of any new information medium, be it the telephone, movies, radio, or television, depends on creative applications most appropriately suited to the attributes of the particular medium. These media, he said, might not have been so ubiquitous today were it not for a few innovators, as opposed to inventors, who at the outset saw commercial value in the unique capabilities of these otherwise useless gadgets. The telephone, for example, was thought by its inventor to be a device that would benefit the hearing-impaired.

With all due respect for the genius of Bell, Edison, De Forest, and the rest, this article will acknowledge a larger group of "geniuses," those of much lesser fame, who have seen potential (such an American word!) in a new medium, the videodisc, and have set to work on their own brand of experimentation.

## WHO INVENTED IT

**For Starters**

By now, millions of people have at least heard of the videodisc. Historians of science and technology generally credit James Logie Baird with its invention in 1926. He called it *phonovision* (see Photo 1). These first discs, made for early television transmissions by the British Broadcasting Corporation, had only 30 lines of resolution and played back at a frame repetition rate of 12.5 frames per second (fps). Compare this with today's National Television Standards Committee (NTSC) standard resolution of 525 lines and 30 fps.

It was not until the early 1960s that 3M engineers David Paul Gregg, Wayne Johnson, and Dean DeMoss came up with the first videodisc that was capable of recording and playing back full bandwidth images on an "optical" (lasers were not yet available) videodisc formatted in the same manner as today's Constant Angular Velocity (CAV) discs.

Then, in the early 1970s, engineers in both the United States and Holland almost simultaneously came up with remarkably similar optical disc development and playback strategies, using lasers. Mutual agreements between MCA and N.V. Philips in 1975 led to what is today's consumer "laser disc." Meanwhile, Keiser, Clemmons and others at RCA in Princeton, New Jersey were at work developing a videodisc that did not require highly focused light, but rather used a stylus, much like an ordinary LP. This, of course, resulted in what is now called the capacitance electronic disc (CED). There are other interesting derivatives of both optical and CED discs, which will be discussed later,

**Photo 1.** This is the world's first videodisc player, the hand built Baird "Televisor" (photo taken, November 1928). The television played back a large disc which displayed picture through the "window" on the right. (Photo, courtesy of Dr. Philip Rice, Menlo Park, Ca., and Mr. M. Hallett, Manager of the Broadcasting Gallery, Independent Broadcasting Authority, London.)

but all things being relative, these two formats, laser and capacitance, define videodisc technology in most people's minds today.

### Repackaging: Then and Now

In his talk, mentioned earlier, Michael Dann noted an interesting phenomenom: according to him, the first incarnations of the most significant new media of this century were nothing more than attempts at repackaging existing media and did not come into their own until someone discovered what was truly unique about them. For example, there was no movie industry to speak of until a few entrepreneurs saw past the notion of the nickelodeon-style traveling vaudeville theater. They realized that large numbers of people would be willing to pay money to watch the same thing if it was projected on a screen, any screen, anywhere. And, in keeping with the unique capabilities of the film medium, these entrepreneurs also realized that they could exhibit the same picture show to different audiences simply by reproducing the film and sending the copies to as many places as possible. Taking things further, these same entrepreneurs learned that picture shows were not like stage plays; they could manipulate time and space in ways no live theater ever had, and thus, the movie industry was born.

Later in the twentieth century, according to Mr. Dann, the first attempts at marketing color television again revealed the mistaken notion that one could use a new medium to repackage the attributes of an established medium. RCA and a few others had had color tv sets on the market for several years, but could not figure out why the public remained lukewarm to this exciting new development. RCA's idea was simply to broadcast color movies. With the advantage of being able to watch a color movie right in one's own living room, millions of people, RCA assumed, would be persuaded to buy these new tv sets. But the plan just was not working. The public, it seemed, did not particularly need the convenience of watching color reruns at home. By the mid 1950s, color television was nearing the edge of oblivion. It was only when NBC, after a major programming rethink, began broadcasting shows that were specifically made for color television (remember Mary Martin in "Peter Pan"?) that the sales of color tv sets finally began a corresponding and rather dramatic increase.

Until recently, repackaging was also the situation with the videodisc. The original conception of the technology's owners, the principals in the movie and broadcasting industries, was that the videodisc could serve as a dandy means of moving all of those titles that were not currently being shown in theaters. In 1973, MCA/Universal Pictures Inc., the original purveyor of movies on videodisc, had over 11,000 titles gathering dust in its vaults, and by the looks of the original MCA videodisc catalog (see Photo 2), they had every intention of placing ALL of them on disc. As might be expected, old movie buffs, film critics, and media watchers were immediately taken in by all of this and enthusiasm for the videodisc was enormous.

### A Period of Anticipation

There was another group of people, considerably less visible, who had been reading the technical papers from MCA Labs, Philips, RCA, and the rest, and had discovered that the videodisc could be something much more than just a device for movie collectors. The first group to start talking about the disc's real potential consisted of several individuals in the fledgling discipline now known as instructional science. A loose confederation of educators, research psychologists, systems analysts, and computer programmers, they had been involved in studies of perception, learning theory, computer graphics, and large memory systems, and had been developing computer-assisted instruction for the few nationally funded projects that were under way at the time. In 1974, Schneider and Bennion at BYU published a series of papers on the usefulness of videodisc technology for interactive computer-based instruction. This touched

**Photos 2a and b.** An early (1973) MCA Disco-Vision Video-disc Catalog, front and back. Note the suggested prices for discs.

off a flurry of activity in research organizations around the United States, including various Department of Defense organizations, other universities, and the private sector. The list included the computer companies, who were impressed with the disc's theoretical mass data storage capabilities. Grudgingly, MCA admitted that markets other than movies might actually exist for the videodisc, so MCA Discovision formed an "industrial" marketing division. But RCA stuck to its original course: movies only.

All of this enthusiasm was over a device that would not be launched in the consumer market for three more years and in the industrial market a year after that. Those were magic times though, the mid 1970s: it was the birthing period for the new technologies, including first incarnations of the personal computer, alternative energy systems; and for the videodisc, a period of heightened expectations. Everyone, it seemed, was making predictions about profound cultural and economic effects that would result from the advent of the videodisc. One educator, who would probably prefer to remain nameless, was heard to predict at a 1975 videodisc conference (minus the discs and players) that paper, pencils, perhaps even language as we know it could now be declared obsolete with the imminence of the videodisc on the horizon. Several more realistic educators and entrepreneurs, however, did see a tremendous potential for the disc, especially if it were interfaced to one of these new personal computers. In 1976, more than a year before most people would get a chance to actually use one, an authoring manual for interactive videodiscs was actually published. Although it was dated and somewhat limited in scope, it was practical, and some of it is useful even today. (Editor's note: see the *Videodisc Production Worksheet*, by Worcester and Smeloff.)

**Launch and the Competition**

It was not until 1978 that the first quantities of consumer videodisc players built by Philips and sold under the Magnavox label went on sale in test cities around the United States. MCA was to manufacture the discs. There were few titles available at the launch, mostly movies, and most of them did not work very well. In addition, these players were basically useless for anything but consumer playback of discs. There was no easy way to interface this player to an external computer. Owing to the novelty factor, however, and the fact that the zealots had been waiting for over five years, both players and discs were sold out within hours.

Not long after that, the RCA disc system was also launched. RCA's claim was that since they used off-the-shelf technology, their system would be more reliable and cheaper than the laser types. At $499.95, the RCA player's suggested price was $275 less than the Magnavox system.

From the outset, both systems were viewed with skepticism by the public at large, whose perception was (and still is) that "you can't record with videodiscs like you can with videotape." To overcome this "handicap," both the laser and the CED camps looked to the fact that videocassette systems were at the time considerably more expensive than disc systems. The disc manufacturers hoped that if enough titles were available, people would see an economic advantage of owning videodisc systems over videocassette systems.

To come to terms with the videocassette competition, the disc manufacturers had to overcome a major hurdle, the problem of limited disc manufacturing sources. Videodisc technology was (and still is) a "top down" manufacturing system owned and operated by a few very large corporations. By contrast, videocassette technology was (and still is) largely a "bottom up" manufacturing system. In other words, anyone with a duplicating machine could manufacture videocassette copies, but the technology to mass-produce videodiscs was not widely available.

Discs could be reproduced in a laboratory, but it was not clear if they could be reproduced with the same quality in a factory situation. Until this question could be answered, the licensing of mastering and replication facilities would not be possible. And since the disc manufacturers had deep concerns about their public image, pornography and other similar categories, which have been a major source of income for videocassettes, had to be avoided on videodisc. Owing to this and the limited amount of disc manufacturing sources, virtually 100 percent of the initial titles offered on videodisc (both laser and CED) were repackaged media (mostly hit movies) with some repackaged "educational" and "documentary" films added to see how they would sell. This continued for the next two years, with predictable results.

### The Benchmark System

In 1979, MCA Discovision announced the availability of the first mass-produced "industrial" videodisc player, the PR-7820, accompanied by the first custom mastering and replication service. Although far from the ideal, the PR-7820 was nonetheless considered a benchmark, representing the end of seven years of anticipation, false starts, wrong directions, and frustrations.

Unlike the consumer models, the PR-7820 provided user access to freeze frame, rapid frame access, variable motion, and discrete and stereo audio features. And most importantly, the player contained a primitive, but functional microprocessor with a user-oriented "language," plus an 8-bit parallel interface. The PR-7820 was seen by the instructional scientists and others in the research community as a computer peripheral: a disk drive with a gigantic capacity to retrieve pictures and sound as well as data. Curiously, several MCA Discovision executives seemed to resent this characterization. It was seen as an attempt at upstaging. They preferred to consider their player a "stand-alone" system.

### WHAT IT IS

Perhaps at this point we should define our terms. Normally, this sort of thing is done at the beginning. As I had suggested, most of us have at least heard of the videodisc by now, but there are so many aspects to the new videodisc technology, including what it is and how it works, that we might find ourselves in a bind for the remainder of this article (and the other articles in Section One) if we do not do something about it now. Since this article is, for the most part, about interactivity, this might serve as a "branch point" for those of you who are slightly more knowledgeable than the rest of us.

#### A Few Facts

Videodiscs are really quite simple to understand from a user's point of view. First of all, a simple convention: "disc" is the manufacturer-sanctioned and generally accepted spelling for the videodisc. On the other hand, "disk" seems to refer to magnetic media, such as floppy disks, hard disks, etc. In any case, this is the accepted differentiation between the two.

The following paragraphs summarize the other key points to remember about videodisc technology. Let's start with a few basics.

*They Can Play Back Nice Pictures.* Depending on the quality of the source material, most videodiscs are capable of reproducing images at or near "broadcast quality" (42-48 db); a general improvement over half-inch videocassettes.

*They Can Reproduce Both Stereo and Discrete Audio.* Currently, using noise-reduction equipment (standard with some players), an encoded disc can reproduce sound at approximately 70 db, much better than a standard LP, which reproduces sound at 55 db under good circumstances. With discrete (separate) audio tracks, videodiscs can be used for bilingual applications, for example, with English on one track and Spanish on the other.

*They Are Durable.* Laser discs are coated with a thin layer of clear plastic. Since nothing actually touches the disc except a light beam, and since the focal length of the laser is so small, scratches on this layer are ignored to a large extent. Further, since the coating protects the actual disc material, fingerprints and smudges need only be wiped away with a damp cloth (window cleaner works fine). This means that image quality does not gradually decay with repeated plays. Capacitance discs, on the other hand, are protected by a caddy; one never needs to touch them. And to the surprise of the advocates of laser discs, the image quality of the CED format does not appear to decay with repeated use.

*They Are Capable of Fast Random Access.* A videodisc is round and flat like an ordinary LP. Unlike tape or film, which require winding or unwinding in various lengths to find a given piece of information, all of the information on a videodisc is accessible merely by moving the laser (or stylus) from one point to another. Some videodisc players can do this with remarkable speed and accuracy (more about this later). Moving from one point to another over the surface of a disc, rather than getting from one point to another by winding or unwinding a reel of tape or film, means that a disc is inherently capable of faster access than is tape or film.

*They Can Be Reproduced At Low Cost.* This roundness and flatness has the potential for mass duplication, using techniques that are similar in principle to the duplication techniques used for standard LPs. This makes the videodisc very attractive in economies of scale.

*Not All of Them Are Compatible With Each Other.* Although videodiscs share a basic resemblance, that does not make them compatible. There are two basic divisions: laser and capacitance. One type uses a low-power laser to read information from the disc, and the other uses a stylus.

Videodiscs are further divided based on differences in the speeds at which they revolve. Since today's National Television Standards Committee (NTSC) video rate is 30 fps, all videodiscs must play back at 30 fps to display images on standard tv sets. Laser disc systems do this in two ways: first, at a CAV rate of 30 disc revolutions per second, or one frame per revolution (1800 rpm); and second, at a CLV rate, wherein the display rate remains constant at 30 fps but the rpms begin to decrease from 1800 to 600 rpm to compensate for changes in disc area (there is more area to cover on the outside edge of a disc as opposed to the inside edge).

In most cases, the same laser disc player can play back both laser CAV and CLV discs. On the other hand, all capacitance systems are CAV. However, the CED system revolves at four frames per revolution (450 rpm) and the JVC system, called video high density (VHD), revolves at two frames per revolution (900 rpm), making the two capacitance systems incompatible. Altogether, there are currently five disc formats: laser, CAV and CLV, CED, and VHD; and three incompatible systems: laser, CED, and VHD.

*Playing Back Movies and Other "Linear" Media is not Their Sole Purpose in Life.* Linear playback refers to programming that a viewer watches from beginning to end, like a movie or television program. All linear programs have a beginning, a middle, and an ending. Nonlinear programming, on the other hand, may have a beginning and an ending, but does not necessarily proceed in a straight line. Nonlinear programs are designed with "branch" points, which, like branches emerging from a tree trunk, provide several options for continuing the program. A feature like fast random access to any combination of still pictures or text, motion sequences, and audio, or even digital data stored on the disc, make the videodisc into something much more than a device for merely playing back movies.

## The Cardinal Rule

Instructional scientists profess that there is a difference between facts and rules. Facts have a demonstrable existence. Rules are statements based on facts that describe what is true in all or most cases. Based on the functional capabilities of videodisc technology, if one wishes to develop interactive programs, there is one cardinal rule:

*Interactive videodiscs must revolve at a constant rate of speed.* Again, we refer to the CAV format. CAV means that revolutions of the disc do not change throughout playback; i.e., the velocity remains constant even though there is more area to cover on the outside edge than on the inside edge. Why is this important

**Photo 3.** Left to right, a CAV laser disc and a CLV laser disc. Outwardly similar, CAV and CLV discs differ significantly when it comes to functional capabilities.

**Photo 4.** The same two discs close up. The CAV disc (left) shows lines fanning from a center point. These lines define the NTSC (525 lines) standard at a constant angular velocity. The CLV disc (right), on the other hand, shows no such lines. The velocity changes from the inside to the outside edge on the CLV disc.

for interactive programming? Because if the velocity remains constant, there will always be the same number of frames per revolution of the disc. If the number of frames per revolution is constant, then each revolution of the disc can be given an identification number or numbers. In the case of the laser CAV discs, each revolution of the disc equals one frame. Technically, this is the most suitable CAV strategy because each frame can contain an individual image, like a photo or text.

The way a CAV laser disc captures and displays an individual image is directly related to the one frame per revolution strategy. The laser merely finds any given frame and stays put. The disc keeps revolving.

As mentioned previously, the capacitance discs are also CAV formats. The CED discs revolve at four frames per revolution, and the VHD discs revolve at a constant two frames per revolution. While the advantage to adding more than one frame per revolution means that more frames can be stored on one side, the problem is that it is technically difficult, but not impossible, for capacitance discs to access and hold any single frame.

### CAV and CLV: a Closer Look

On the surface, CAV and CLV discs look almost identical, that is, they have the same dimensions and the same lovely silver appearance (see Photo 3). But if one looks more closely, one can see lines on the CAV disc that appear to fan outward like spokes emanating from the hub in a wheel. These lines (525 of them) are related to the horizontal scanning lines on an NTSC standard television set, but these lines are not the actual tracks. The tracks (where all of the pictures, sound, and data are stored) are not visible to the naked eye. Each track is less than six-millionths of an inch apart from the next one. If one were to attempt to count each of these tracks on a laser disc, the total would be approximately 54,000. The "fanning" lines tell you that while the track spacing is tighter on the inside than on the outside, one revolution of the disc still equals one frame.

Look closely at Photo 4. On the laser CLV disc the fanning lines are not at all evident. This is because CLV discs are not proportionally spaced from inside to outside; the spacing remains uniform from track to track. Since the motor that controls the velocity of the disc gradually slows down to about 600 rpms as the laser compensates for the increased number of tracks on the outside edge, CLV makes it impossible for the laser to find and hold a single frame.

CLV is a great thing for the laser disc manufacturers interested in repackaging movies, and similar material, because up to one hour's worth of information can be stored on each side of a laser disc. With CLV, movies and such do not require as many discs. With fewer disc sides to press per movie, more titles can be pressed in less time, and the cost to press each title can be cheaper. Unfortunately, CLV is not such a great thing for users who may be interested in anything other than real-time playback. CLV eliminates almost all of the capabilities of the laser CAV discs; i.e., one has no fast frame accurate access, no freeze framing, no frame stepping forward or reverse, no variable slow motion forward or reverse, and no picture stops. What is left for the CLV user is a "pause" function that blanks the video and holds the laser in the general vicinity, a scan function with the option for searching to chapters if the codes are available on the disc, stereo audio, and a time counter that displays elapsed time in minutes. CLV creates an unfortunate situation for movie buffs, film critics, educators, or hobbyists who might want to create their own videodisc games with or without external computers.

There is also some degree of irony in the CLV format., which was largely adopted by the laser disc manufacturers for real-time-only discs in 1981. All of the movies released prior to 1981 for the laser videodisc players were in the CAV format. These CAV movies are now becoming collector's items. It is nice to know, however, that the manufacturers are beginning to respond to the increasing demand for optional CAV versions of movies. "Raiders of the Lost Ark," for example, is now available in both laser CLV and CAV versions.

### Classifying Player Intelligence

Since there are several types of videodisc players now on the market and an emerging catalog of made-for-disc titles (well over 300 in this book), it becomes important to attempt some form of classification. In 1979, the Nebraska Videodisc Design/ Production Group came up with a fairly useful classification based on "intelligence levels" of various videodisc systems. The classification, which has been widely adopted, is as follows.

Level 0   A videodisc player designed for linear playback only. This also includes laser videodisc players that accept both CLV and CAV discs.

Level 1   Level 0 plus such built-in capabilities as fast frame accurate access, variable motion, stereo and discrete audio, freeze frame, chapter search, picture stop, and scanning functions; plus the capability, however difficult, to be interfaced to an external computer. These capabilities are inherent in most consumer videodisc players. A sampling of level 1 players includes the Pioneer VP-1000, LD-1100, LD-770, 8210, Magnavox 8010, Sylvania 7200, and RCA 400. (Editor's note: see Jeff Kemph's article on "interchangeability" for a complete listing.)

Level 2   All of the capabilities of level 1 plus faster access, increased durability, built-in interface port, and an on-board microprocessor with a user-programmable memory. Level 2 players include the Pioneer (formerly MCA Discovision) PR-7820 series, the Pioneer LD-V6000, and the Sony LDP-1000 series.

Level 3   Either level 1 or 2 interfaced to an external computer.

Level 3 adds the idea of management to the program. If it is an instructional program, for example, record keeping becomes possible. Teachers can write their lesson plans on the computer, and determine the required motion and stillframe sequences in advance, obtaining hard copy printouts of student performances at the conclusion of each lesson. With the addition of such devices as touch screens, joysticks, and "overlays," (the ability to superimpose computer text and graphics onto the videodisc-generated images), level 3 offers increased potential for interaction between user and system. With level 3, increased memory and processing power provide the videodisc with substantially more versatility than with levels 1 and 2.

Lately, there has been increasing talk of a "level 4." Level 4 is a theoretical domain in which all things are possible.

### The Idea Takes Hold

In 1980, Discovision Associates (DVA) was formed by joint agreement between MCA and IBM, and while DVA was short-lived, the partnership maintained the largest single commitment to videodisc technology to that point, which was made by General Motors (GM). GM purchased approximately 11,000 PR-7820 players. These players were placed in GM dealerships around the country and used as point-of-purchase devices and sales training and maintenance aids for the dealers. A year later, Sony announced a similar sale of several thousand of its LDP-1000 (level 2) videodisc players to the Ford Motor Company.

By 1981, not quite three years after the first Magnavox players went on sale, the idea of using videodisc technology as a device other than to play back movies had caught on. Conferences on the topic were extremely popular with a wide variety of audiences. The University of Nebraska, for example, had already held two sellout "National Symposia on Videodisc Programming," with presenters from such diverse organizations as academia, medical organizations, aerospace companies, the movie business, and the military. People came from around the world to see and hear what videodisc technology was really all about, and they saw and heard some remarkable things. For example, a few of the applications they saw at the conferences were:

- videodisc-based systems that could diagnose a simple or complex maintenance problem, automatically order the necessary parts, and then guide the technician through the repair procedure based on the technician's own skill level and experience;

- "surrogate travel" devices giving the user full control to walk, fly, or drive through the streets of a town and other places with further options to stop and look around at buildings, instantly switching from fall to winter, present to past;
- "electronic textbooks" on such diverse topics as college-level physics, whale watching for third graders, problem solving skills for hearing-impaired children, interactive fiction for high school students, tumbling, metrics, psychology, art history, language instruction, evidence and objections for law students, etc.;
- a videodisc interfaced to an exercycle, which displayed point-of-view sequences of bike paths or roller coasters, capable of speeding up or slowing down the pictures based on the level of exertion;
- several medical programs including patient education on insulin, various pulmonary problems, hematology, and a CPR course complete with a mannequin to practice on;
- an interactive videodisc shopping catalog for a major retail chain;
- level 1 discs on cooking, exercise, "rainy day" activities for kids, etc.;
- several videodisc-oriented computer-based tools such as interfaces, a generic controller, an authoring system, touch screens, and a low-cost "overlay" that superimposes computer text and graphics with pictures from the videodisc;
- an inexpensive training system consisting of printed textbooks with bar codes interspersed among the pages, and a bar code reader/computer interfaced to a videodisc, which provides motion and audio enhancement for the text;
- a low-cost "part task" trainer for flight training;
- an information center and exhibition system for a major theme park.

Significantly, the videodisc manufacturers themselves attended these conferences, and by 1983 practically all of them, including Pioneer, which had assumed DVA's responsibilities in 1982, were making announcements and unveiling exciting new services and product lines designed to enhance the interactive capabilities of videodisc technology.

In addition, large sums of money were starting to be invested by the manufacturers for the creation of new programming specifically meant for the videodisc, and many of these new discs met with immediate success, in particular "The First National Kidisc" released by Optical Programming Associates (a consortium of laser videodisc manufacturers).

**The Benchmark Project**

At a videodisc conference at Columbia University, Nicholas Negroponte, former director of MIT's Architecture Machine Group, was heard to remark that applications of a new technology many times do not even closely resemble the intentions of its inventors; a theme echoed in this article. It was Negroponte's group that developed the idea of "surrogate travel." The Architecture Machine Group, a computer graphics laboratory, originally saw the videodisc as an enhancement to the computer-based environments they were creating. Naturally, things began to grow and evolve as videodiscs were used at MIT. Before long, they had created the "world's largest slide tray," 54,000 individual pictures from MIT's architecture slide library. A sampling of other disc projects included: a bicycle repair system that actually pointed to a new approach for using technical manuals (the Office of Naval Research funded this one); a picture retrieval system called "news peek" (not to be confused with Orwell's "newspeak"), which combined picture data bases stored on videodisc with newspaper wire services; and the "interactive movie map," better known as the "Aspen" project.

If the PR-7820 was the benchmark videodisc player, then "Aspen" had to be the benchmark videodisc project. The ideas for using discs that went into Aspen have opened up an entirely new way of using videodisc technology. Basically a mapping project, the experiment focused on organizing a large picture data base to simulate a drive through the streets of a town. The goal was to give the user complete control of where he/she might wish to go. To produce the data base, members of the Architecture Machine Group drove down every street in Aspen, taking pictures with a specially developed group of four cameras mounted on a truck, with one camera aimed to the front, one to the back, one to the right, and one to the left. Shutters were synchronized with a fifth wheel at the rear of the truck. They cataloged every straight road and every turn from point-of-view positions, looking straight ahead and sideways.

After the pictures had been placed onto a laser CAV videodisc format, two identical discs were placed on two synchronized videodisc players that were interfaced to a computer. Users were invited to interact with the system by using a variety of input devices including joysticks, touch screens, voice actuation systems, etc. The videodisc would "drive" down a street and, based upon decisions made by the user, it would either turn right, left, stop, or continue forward.

By using two videodisc players instead of one, the Architecture Machine Group could eliminate "screen blanking." Screen blanking is what occurs when the video is temporarily turned off while the disc player searches to another location (worst case access time on the PR7820 was approximately five seconds). So when the user came to an intersection and would choose a right turn, for example, the second videodisc player, which had been waiting at the intersection, would find the appropriate turn footage and play it. Meanwhile, the first disc player would search to the appropriate straight street footage and play that directly after the turn. The second disc player would then search ahead to the next intersection and wait. And so on. The user could drive all over town as if he/she were actually there. Further information could also be requested on a particular building merely by touching the screen, and in some cases, the users "walk around" inside.

The system also had a "season knob" that would give the user precisely the same view at different times of the year and a "history knob" that matched the present day appearance of older buildings with how they looked one hundred years previous. In a sense, the Architecture Machine Group had created a time machine using videodiscs.

Aspen awakened the potential for "high fidelity" simulations using videodisc technology. The aerospace industry and the military began to see discs for flight simulation, etc., the education and training communities began work on instructional strategies that went far beyond the traditional tutorial modes of computer-assisted instruction. For example, proposals for surrogate travel through the human circulatory system, along the path of a circuit card, or to the planets and back, to name just a few, were heard. And the entertainment and exhibition industries began working on games, games, games.

### Some Other Influential Projects

The following description of influential projects is, for lack of space, woefully incomplete, but it does include some of the finest early applications of videodisc technology; projects that have inspired further explorations and developments.

*WICAT and "The Development of Living Things".* In May of 1978, the World Institute for Computer-Assisted Teaching (WICAT) developed the first interactive videodisc, based on a film by McGraw-Hill entitled, "The Development of Living Things." A disc on developmental biology for high school and college students, the first version was barely level 1; i.e., it was designed to take advantage of the capabilities of the first Magnavox player. To WICAT's credit, this disc was well designed and worked well within the limitations imposed by the Magnavox player.

The importance of this project was not fully realized, however, until the National Science Foundation awarded WICAT a contract to evaluate the disc with college students, then upgrade the concept to level 3. The report that was issued by WICAT (Bunderson, Olsen & Baillio, 1981) as a result of the evaluation, entitled "proof of concept demonstration and comparative evaluation of a prototype intelligent videodisc system," provided promising data on the effectiveness of videodisc technology "to produce significant (learning) gains, positive attitudes, and more productive use of time among science students" in relation to traditional classroom techniques. Thus, WICAT demonstrated the claim that interactive videodisc technology could be an instructionally effective teaching tool in the classroom, more so than the "lecture" style of teaching.

*Nebraska.* In 1978, Mr. Philip Rubin, then director of engineering research (later the Office of Science and Technology) at the Corporation for Public Broadcasting, developed what turned into a multiyear contract with the University of Nebraska to develop "the full potential" of videodisc technology. A tall order, but the University of Nebraska, which, through its licensee, KUON-TV, is home of one of the nation's finest teleproduction facilities, went right to work.

Different in scope from the projects at MIT, Nebraska's efforts were focused on improving the state of televised instruction. Thanks to the foresight and leadership of Mr. Rubin, Nebraska had the freedom to concentrate on developing a wide variety of topics; discs that would point the way to practical implementation of videodisc technology in the public and private sectors. In a period of five years, Nebraska produced dozens of interactive discs, including the first closed-captioned videodisc for the hearing impaired. Nebraska was the first to establish replicable standards for "premastering." (Editor's note: see Ron Nugent's article on "Videodisc Premastering.") They were also the first to develop a low-cost "overlay," which was publicly demonstrated in April 1981.

Having been among the first to be granted the privilege of designing and producing interactive videodiscs, Nebraska was obliged to share its knowledge via workshops, symposia, technical reports, and newsletters. Nebraska's most significant contributions, therefore, had to be these early demonstrations of widely diverse but entirely feasible disc applications, and its dissemination of information designed to make people aware of the potential of interactive videodisc technology.

*The American Heart Association's "CPR Computer/Videodisc Learning System".* Designed in 1981 by Mr. David Hon, the CPR videodisc course was created to be a self-contained, certified course in cardio-pulmonary resuscitation. Using a personal computer, videodisc, and a Laerdal mannequin wired with an array of sensors, a CPR trainee could practice developing the correct depth and placement of CPR compressions. The computer would monitor the student's progress throughout the course, and the videodisc made it possible for the student to receive immediate feedback and further assistance from a doctor on the correctness of the procedures. The system provided graphics, audio tones, multiple choice evaluations, and a dictionary. But even more importantly, it provided individualized, self-paced interactive instruction. Results of field evaluations of the "CPR Computer/Videodisc Learning System" not only indicated that the system was able to properly train students in CPR, but that it exceeded all training requirements. For obvious reasons, this outstanding use of technology to train people in the life-saving CPR technique gained international recognition, and the system is widely used today.

*The GM Network.* The GM Network was the result of the first large scale videodisc purchase (11,000 players) finalized in 1980 by MCA Discovision (later Discovision Associates, and still later Pioneer Video). Subsequent to the hardware sales came the development of disc programming, which was done under subcontract to GM by Sandy Corporation, Detroit. Although all of the programs were level 2 and did not vary widely in design and programming strategies, the significance of this project was that it was the first large scale interactive videodisc network. It served to legitimatize videodisc technology in industry, and established a precedent upon which other large corporations could build.

*Army Communicative Technology Office.* ACTO was begun in the mid 1970s and first commanded by Colonel Hank Gibbons, followed by Colonel John Goetz.

ACTO was formed to investigate practical applications of new audio and visual technologies as they might be applied to Army training. The first projects that ACTO became involved with focused upon what was called, in the mid 1970s, Integrated Technical Documentation and Training (ITDT), and later changed to Skill Performance Aids (SPA), an acronymn that was less of a mouthful. SPA was based on the idea that recruits could be better utilized if they could begin their military occupational training as soon as possible, circumventing as much classroom and theory as deemed advisable by their superiors. A system of SPA manuals was developed that included on the job training as well as maintenance information for Army systems. SPA was an effective concept with one exception: the manuals were voluminous. Videodisc technology's media and data storage capabilities were seen as a potential solution to this ever-mounting problem of paper. The Concept Evaluation of Soldier Instructional Delivery Equipment (CESIDE) project was contracted to Hughes Aircraft Company in 1977 to test soldier effectiveness by comparing a videodisc version of one of these SPA manuals with a conventional printed version. In terms of time to complete each task, the videodisc SPA manuals greatly outscored their paper counterparts. Meanwhile, ACTO and a sister organization, the Army Research Institute for the Behavioral Sciences, were conducting other tests of videodisc technology in relation to the instructional effectiveness and cost-effectiveness of videodisc technology. Still other projects for the Army included Perceptronics' Tank Gunnery Trainer (1980), which used a cleverly designed molded plastic gunnery station that could be "bolted" onto the screen of a television monitor. Controlled by a microcomputer and videodisc, the student could view an oncoming tank through a wide-angle reticle lens and could fire at the tank in a simulated battle. ACTO has remained to this day a leader in the field of videodisc applications.

## WHERE IT IS

### New Hardware

Where is videodisc technology in 1984? From a hardware and systems viewpoint, 1984 could be considered the year of the second generation (or third, in some cases) of videodisc technology (see Photo 5). First of all, the players themselves have improved. They are lighter, faster, cheaper, and more easily interfaceable to external systems. Some of them contain noteworthy features, such as solid state lasers

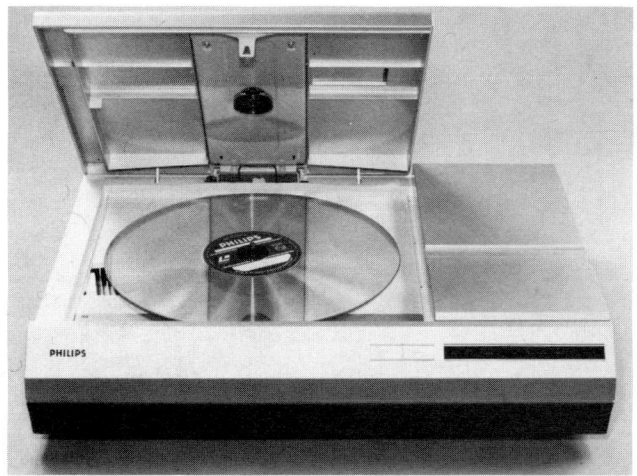

**Photo 5.** The new Philips 832 player, typical of the next generation of videodisc technology. These new players are smaller, lighter, faster, and cheaper than their predecessors.

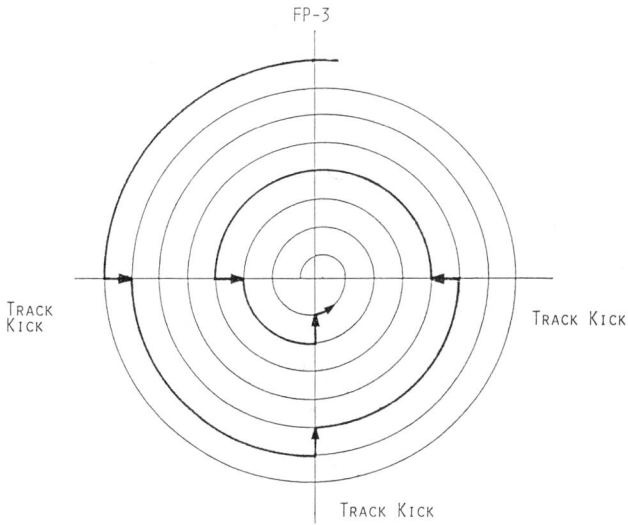

TRACK KICK AT EVERY QUARTER OF THE ROTATION EXCEPT FP-3

X4 MODE

**Figure 1.** The VHD "field kicking" strategy, wherein the video from camera 1 is recorded at 1/4 of the rotation, and the successive fields are recorded at another 1/4, and so on. The video information from camera 2 is recorded at the next 1/4 of the track following into the orbit of camera 1 in playback, a sensor traces the recorded program path. An external computer then directs the stylus servo to shift the sensor to a different part of the same track.

(laser diodes), "field kicking," and "instant jump." Moreover, in response to user demand, many laser disc player lines include players that are front-loaded rather than top-loaded. Stillframe audio (sometimes called compressed audio) is at long last widely available. Finally, 1984 marks the year of introduction of some significant new industrial players from JVC, and the McDonnell Douglas Electronics Company (MDEC). The following is a brief review of these hardware developments.

*Weight.* A typical first generation industrial player weighed an average of 43 pounds. A typical second generation player weighs on the average of 24 pounds.

*Speed.* The speed of a disc player refers to "worst case access," or how long it takes the player to search from frame 1 to 54,000 (a standard 30 minutes). The early consumer disc players, the Pioneer VP-1000 for example, searched this distance in approximately 18 seconds, the PR-7820 industrial player searched in 4.5 seconds. The new industrial players, such as the Philips 832 and the Pioneer LD-V1000, have improved upon the PR-7820 by almost a full second and the new Hitachi VIP-9500 can search in approximately 2 seconds.

*Price.* Suggested retail price of the PR-7820 was $2,200, with quantity discounts. The Sony LDP-1000 suggested retail price was $3000 with quantity discounts. In 1984, the second generation players are much less expensive. To sample a few of them, the Philips 832 sells for $1385, the Pioneer LD-V1000 is quoted at $1200,

the Hitachi VIP-9500 sells for $1640, and all of these players also offer quantity discounts. These figures represent better than a fifty percent decrease in average price for the new players.

*Laser Diodes.* One of the most expensive components inside a laser videodisc player is the laser. All laser disc players, both the level 1 and level 2 varieties, have until now used low-power helium neon (gas) lasers to read information from the surface of the disc. Laser diodes, on the other hand, are quite small (the "lens" is about the size of a tie tack) and, if produced in quantities, could bring the price of a laser disc player down considerably. The Hitachi "VIP" series players are the first industrial players to use laser diodes. The Pioneer LD-700 is the first consumer player to use laser diodes. I have seen both players and was impressed with the visual quality. There appeared to be no difference between images produced by gas lasers and laser diodes.

*Computer Interfaces.* Virtually all of the new industrial videodisc players and some of the new consumer players are designed for level 3 applications and contain built-in computer interfaces; mainly RS-232c serial interfaces, although parallel interfaces are available as options on most of them.

*Field Kicking and Instant Jump.* One of the most intriguing new features of the new generation of players is what has been called "field kicking," mainly in relation to the capacitance systems, and "instant jump" in relation to the laser systems. These innovations have sprung from the surrogate travel concept for videodiscs. Field kicking, for example, takes advantage of the fact that capacitance CAV discs revolve at more than one frame per revolution. As shown in figure 1, the VHD system can be made to "kick" any one of four fields. This makes it possible to record up to four parallel motion sequences on a single disc, which can be switched back and forth instantaneously. Used for simulations, field kicking could be effectively used to change lanes on a freeway, or to land on an aircraft carrier, for example.

Similar in anticipated effect to field kicking, the instant jump feature, which is presently available on some of the new industrial laser disc players, enables the laser to instantly locate any frame within a 200-frame range. This gives the player the ability to "step" up to 200 frames without any screen blanking. In real time, 200 frames are more than 6 1/2 seconds; thus, one can step to and from motion sequences (with audio) in addition to still frames.

*Front Loading.* Front-loaded players, which have been available from the start from the capacitance systems, are now available in some laser disc systems. Compared to top-loaded players, where the user inserts the disc by opening the lid of the player, front-loaded players are considered by many industrial users as an improvement in terms of convenience. Front-loaded systems are space savers and can be rack-mounted.

*Still-Frame Audio.* Still-frame audio, a means of digitally loading several seconds of audio, storing the audio in a buffer, and then playing it back while the screen displays a single frame, is now available as an option for videodisc players. Long awaited, several such systems are now available from which to choose. In addition, a few of the disc mastering and replication facilities (3M and Sony) are now capable of reproducing discs from premaster tapes that are still-frame audio encoded.

*Entirely New Players.* A new laser disc player from Pioneer Video, designed to replace the LD-1100 consumer player is the LD-700 (see Photo 6). It is much

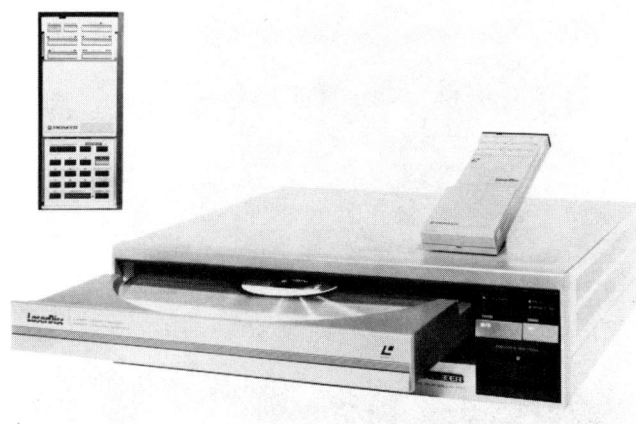

**Photo 6.** The Pioneer LD-700, A consumer-level laser videodisc player that is front loading, like its more rugged, faster brother, the LDV-4000. The LD-700 uses a laser diode and has a built-in RS-232 interface port.

smaller (16 1/2 inches wide by 4 3/4 inches high by 16 3/8 inches deep), uses a laser diode, has better motion control and can be controlled by an external computer.

The long-awaited MDEC system (see Photo 7) from McDonnell Douglas Electronics Company will also make its debut in 1984. The Model 5000 will offer most of the capabilities of the other new videodisc players, including fast search times, motion forward and reverse with audio, computer-controlled variable motion, discrete audio, computer interfaces, and front loading. In addition, the MDEC player will contain built-in still-frame audio. The most unique feature about the MDEC system is that it uses standard photographic film as disc material. This means that ownership of disc mastering and replication is possible, in fact encouraged by MDEC. This could be extremely useful for large scale in-house networks requiring security.

## New Concepts

*Games.* In terms of concepts, 1984 also marks a major turning point for videodisc technology. As an executive at Pioneer Video stated at a recent videodisc conference, "It feels good to finally be operating in the black." And the major reason for this turning point has to be the success of videodisc-based games, a significant new videodisc concept.

Actually, the idea of videodisc based games is not really all that new. People have been talking about using videodiscs to enhance arcade games ever since the first version of "Pong," and instructional scien-

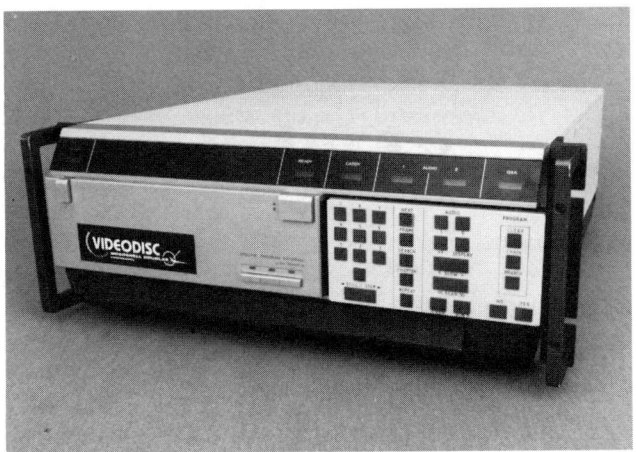

**Photo 7.** The McDonnell Douglas Electronics Company's "MDEC" player, model 5000. The MDEC system offers unique, built-in compressed (still frame) audio. Most unique, however, is the "photographic" disc strategy employed with the MDEC system.

tists have long known the power of games as instructional strategies. The pieces have been in place since 1979 with the release of the PR-7820, the first demonstrations of MIT's "Aspen" project, and the development of low cost computer/NTSC overlays. So what is so new about videodisc games? Other than the fact that hardware manufacturers and a few entrepreneurs are making a lot of money from them, the answer has something to do with an appropriate application of the basic capabilities of the disc and much more to do with (for the lack of a better word) the "transparency" of the medium. That is, the videodisc in 1984 is no longer an icon.

By now, most of us have marveled at the crowds in the arcade parlors craning their necks to see "Dirk the Daring" accomplish his heroic feats, etc. But if one were to inquire within, one would learn a most amazing thing: with the exception of all but a few "video junkies" and "techies," these game players do not know or even care whether or not these new games use videodisc technology. They are much more interested in the heightened fantasy facilitated by the imagery, coupled with the challenge to win the game. Putting the player away in a box or treating it like a peripheral may be irksome to some, but by doing so the hype is separated from the actual utility of videodisc technology.

One of the most closely watched videodisc developments for 1984 will be the entry of level 3 systems into the home market spearheaded by the success of videodisc based arcade games.

*Integrated Systems.* In line with featuring the effect and not the means of gaining it, another important new concept for the industrial community is the integrated system. An integrated system is a level 3 system that has been coordinated by a single manufacturer. Essentially, one manufacturer sells the entire system, and is likewise responsible for service contracts, etc. Such systems usually include extensive software support, including authoring systems and graphics packages. Significant entries include: Digital Equipment Corporation's Interactive Video Information System (IVIS); Wang's new system; IBM's system, which is designed to support their personal computer dealer network; and Sony's SMC-70, the first videodisc/personal computer system with "superimpose" (overlay) made by a single manufacturer.

## Developing Markets

*Point of Purchase.* One of the most comprehensive, economically self-regenerating ideas for videodisc technology is the implementation of information networks. In 1984, the first vestiges of these systems are reaching the public. The so-called point-of-purchase marketing systems, for example, have improved a great deal; most companies involved in their development having learned from earlier mistakes. There are a few credible level 1 point-of-purchase marketing devices, but most systems range from level 2 to level 3 complete with autodial modems, credit card readers, touch screens, and even sonic sensors that begin pitching their wares within earshot of a prospective customer. The systems can either be designed as stand-alone kiosks or as interlinked terminals in a network.

A typical example of a level 2 point-of-purchase system is a product presentation kiosk that might be placed in a catalog showroom. The kiosk invites the user with a graphic on the kiosk or menu on the screen to learn more about a blender, hair dryer, etc., merely by touching the appropriate key on an easy-to-use key pad. Beyond that, nothing else happens except, perhaps a pitch at the end of the presentation that the appliance is on sale and can be found on aisle 17, etc. This type of system is an outgrowth of the cartridge film loops that we have all seen in the hardware stores and department stores. The advantages of using videodiscs instead of film for this type of application should be clear. Discs offer faster, more user-directed access and the systems are more durable. From a marketing viewpoint, the systems themselves are supported by space sold on the discs for product presentations.

*Electronic Video Retailing.* Level 3 point-of-purchase systems are much more versatile. With the addition

**Photo 8.** A "point-of-purchase" system used by J.P. Stevens. A level three configuration, it is based on the Sony SMC-70 microcomputer interfaced to a Sony LDP-1000A player.

of credit card readers, modems, and graphic overlays to update prices and other timely information, we enter into what Nolan Bushnell, founder of Atari and now chairman of a point-of-purchase company called ByVideo, refers to as Electronic Video Retailing (EVR).

Using the same basic idea as a level 2 point-of-purchase system, an EVR system would "know" when the customer was passing by and would cleverly entice him or her over to the machine, which might be in a hotel lobby, airport, train station, etc. Using voice actuation, a touch screen, or key pad, or a combination of all of these (bells and whistles), the customer could request information on hotel and airline reservations, preview theater teasers, floral arrangements from florists, etc. By inserting a credit card, the customer could actually make a purchase, and in the case of tickets and confirmations, receive the goods right there via a printer (see Photo 8). (Editor's note: Perry Reeves' article "Future Developments: Industrial Perspectives," sights a few specific examples of point-of-purchase systems.)

**Europe and Japan**

Having observed the gestation period of videodisc technology in the United States, studying the successes and learning from the failures, both Japan and Europe are now beginning to leapfrog directly into an implementation phase.

*Europe: Great Britain.* Most of the European implementation of videodisc technology is gaining momentum sooner in the United Kingdom than in other parts of the continent. A major reason for this has to be due to the various incarnations of the venerable British Broadcasting Corporation (BBC). First of all, the BBC has developed its own personal computer, the Acorn, and is just now beginning to release videodisc projects designed for the Acorn and the Apple IIe. One of the first to be released will be "Conundrum," a mystery series. Computer-assisted exercises and activities (suitable for both individual students and small group interaction) are also being developed as extensions of existing video- and audio-based courses.

BBC English by Radio and Television, the world's largest producer of video materials for English as a second/foreign language, is developing videodisc-based courseware within the context of its product line as a whole. And the BBC Open University, which has long been interested in videodisc technology for delivery of credit courses, is currently working on computer-based videodisc materials.

Other major developments in the United Kingdom include an IBM Personal Computer-based integrated system designed for computer dealers who feature IBM products. The system includes a PC with a 16-inch touch screen and a Philips VP-835 industrial player (discussed in the following section, "Europe: The Continent"). Reportedly, this point-of-purchase system will find its way into as many as 1,000 European IBM Personal Computer dealerships.

Thorn EMI, one of the original partners in the VHD consortium, which included General Electric and JVC, has finally launched the VHD system in Great Britain, providing competition for CED and laser formats that were already available.

*Europe: The Continent.* N.V. Philips of Holland, parent company of North American Philips, has released an industrial player, the VP-835, which is well-suited for the current video climate in Europe. A PAL player (625 lines of resolution), the VP-835 is modular and includes optional features not available on any other player. These features include a facility for a teletext-style overlay and mixer, which can be useful for the European market. The Europeans have developed teletext and videotex networks to a much greater extent than the Americans. Also included with the VP-835 is an on-board microprocessor with 48K RAM and an erasable programmable read-only memory (EPROM). In essence, the VP-835 is a small personal computer with a built-in videodisc player. In fact, a keyboard and touch screen are available as options.

*Japan.* In May 1983, Tokyo held its first "International Videodisc Symposium." In addition to various presentations from American representatives seen in

Tokyo at that time was the first videodisc "jukebox" capable of playing up to 200 laser discs. Plans were also revealed for a science information center that will rely heavily on videodisc technology for exhibits and information dissemination.

In Japan, educational publishing occupies a larger market segment than it does in America. Japanese families are more willing to spend a larger proportion of their incomes for continuing education materials. English language education is particularly popular for both home and industry. For example, the BBC and BYU have just completed a co-production called "Flight 505," which will be used to teach English to Japanese businessmen. Designed as a level 3 interactive course, the discs are designed to operate on a Sony SMC-70 interactive video computer system. In addition to the discs (five sides), the course comes with workbooks and audio cassettes. The course is targeted for the large industrial organizations in Japan where the series will be used to train executives who must do business in English-speaking countries. "Flight 505" is a simulated journey to America by such a businessman. On the plane he is handed his immigration cards, etc., and as culture shock looms on the horizon, he realizes that his business contacts and decisions in America are going to depend on his language skills. A frightful experience perhaps, but the series contains ample instructional "helps" for the protagonist, and, based on the decisions that the student makes on his behalf, the businessman's first experience in America can have one of many outcomes, some more productive than others.

## WHERE IT IS GOING

### In the Works

A few of the new products and services to watch for include inexpensive direct-read-after-write (DRAW) disc records that do not require a clean-room environment. (Editor's note: see John Winslow's article, "DRAW Technology: The Instant Videodisc.") In addition, we can expect to see multihead machines providing even faster search times plus numerous other benefits, and "one-stop" premastering services that can dramatically cut the price of postproduction, treating the process much like film processing. These new service companies will also offer editing services for filmmakers and videodisc producers requiring the ability to control several disc players at once; the disc players being used to play back "check disc dailies" for evaluation. And from an educational viewpoint, the proliferation of more powerful, multi-tasking 16-bit and 32-bit personal computers will permit further development of "concurrent" videodisc instructional strategies, wherein students will be able to learn popular word processing and data base management programs by building real files, not simulated ones, under the constant supervision of an observant videodisc tutor.

### Surprise, Mr. Baird!

One wonders what John Logie Baird would think of his invention if he were alive today. My guess is that he would be impressed with its progress as a technology, what with the addition of precision lasers, high-resolution images, massive data storage capabilities, and all. But I think he would be much more impressed with the wide variety of creative applications that his "phonovision" system has inspired. Surely, he would be surprised to learn that his invention eventually facilitated a simulated "you-drive-it" tour through the streets of a town, a fierce, highly-pitched battle to the death with a fantastic dragon, and an exercycle that lets you ride your bike over a roller coaster.

Do you think that he ever considered, as most of us do today, his invention to be analogous to a book with tables of contents and chapters that talk and move at the touch of a finger? I do not think so, but I do think that he would be pleased with all of the vitality of this new industry that is finally finding a home.

## REFERENCES

1. Dann, Michael, "Speech on New Technologies." VIDCA, Brussells, (1972).

2. Bennion, Junius L. *Possible Applications of Optical Videodiscs to Individualized Instruction Systems.* Provo: Division of Instructional Research, Development and Evaluation, BYU, (1974).

3. Bennion, Junius L. and Schneider, Edward W. Interactive *Videodisc Systems for Education.* Provo: Division of Instructional Research, Development and Evaluation, BYU, (1975).

Rod Daynes is a videodisc producer and designer. Now with Interactive Technologies Corporation, Mr. Daynes has been directly involved with the design, production, and premastering of more discs on a wider variety of topics than anyone else in the world. Mr. Daynes organized the Nebraska Videodisc Design/Production Group under a grant from the Corporation for Public Broadcasting to the University of Nebraska, and was its director for four years.

# An Interactive Videodisc Glossary

*Rod Daynes*

Reprinted from BYTE Magazine, June 1982

**authoring**  A structured approach to developing all elements of an interactive videodisc program, with emphasis on preproduction.

**blanking**  During the time it takes for a videodisc to search from one sequence to another, the video image is turned off. This results in a "blank" screen; thus the search interval between sequences is referred to as "blanking."

**branch**  An instruction to diverge from one sequence in a program to another.

**capacitance disc**  A videodisc system that uses capacitance signals embedded on the disc and a "stylus" that touches the surface of the disc to read encoded information.

**CED**  Capacitance Electronic Disc developed by RCA.

**check disc**  A disc that is used to evaluate video material prior to the replication of release copies; similar in concept to an answer print in the motion picture industry.

**compressed audio**  *See* STILL FRAME AUDIO.

**constant angular velocity (CAV)**  A CAV disc revolves continuously at 1800 rpm, one revolution per frame, making each frame of the disc addressable. This feature is a basic requirement for interactive videodiscs.

**constant linear velocity (CLV)**  A CLV or "extended play" disc maintains a consistent length for each frame, thus enabling longer playing time per side, but sacrificing individual frame addressability. Reference to locations on CLV discs is limited to time in minutes and seconds. CLV discs are basically useless for interactive videodisc applications.

**chapter**  A consecutive sequence of frames.

**chapter stop**  A code embedded in the vertical interval of the videodisc that enables certain videodisc players (mostly Level 1 Reflective) to locate the beginnings of chapters.

**cue**  A pulse entered onto one of the lines in the vertical blanking interval (VBI) that results in frame numbers, picture codes, chapter codes, closed captions, white flags, etc. on the disc.

**digital dump**  A computer program that has been encoded onto a videodisc as a read-only memory (ROM), which in turn is loaded in a random access memory (RAM) resident in the videodisc player. The term "digital dump" is normally used in reference to level 2 videodisc players. *See* LEVEL 2.

From the "Videodisc Interfacing Primer" by Rod Daynes appearing in the June 1982 issue of BYTE magazine. Copyright 1982 Byte Publications, Inc. Used with the permission of McGraw-Hill Publications, Inc.

**direct-read-after-write (DRAW)** A record-once optical disc technology primarily used for mass storage of digital data, archival data, in-house, or confidential information, and daily checks of film footage and interactive video material.

**field** A scan of 262.5 lines on the screen at 1/60 of a second, constituting one-half of a complete video frame. Each field scans every other line; i.e., field one ("up" field) scans odd numbered lines, field two ("down" field) scans even numbered lines. *See* FRAME; INTERLACE.

**field dominance** The determination of a consistent edit point (either field one or field two) on the videotape record and playback machines prior to beginning of a premastering session.

**field kicking** Sometimes called "field search," "field ID," "field selection," or "field monitoring," field kicking is a method used by some capacitance videodisc systems that enables the stylus sensor to randomly access any one of four fields or four frames instaneously. *See* INSTANT JUMP.

**flicker** Sometimes known as "interfield jitter," or "jitter," flicker is a phenomenon that occurs in a videodisc freeze frame or still frame when both fields are not identically matched, thus creating two different pictures alternating every 1/60 of a second.

**frame** Two complete scans of the video screen at 1/30 of a second. A frame is composed of two fields (at 262.5 lines) and a retrace; a single frame is a standard CAV videodisc reference point. There can be as many as 54,000 addressable frames on one side of a CAV videodisc.

**freeze frame** A single frame from a motion sequence that is stopped.

**full frame time code** Otherwise known as non-drop frame time code, full frame time code (non-drop frame) is a standardized method (from the Society of Motion Picture and Television Engineers—SMPTE) of address coding a videotape. It gives an accurate frame count rather than an accurate clock time. (The latter is sometimes referred to as "skip frame" or "drop frame.") Full frame time code is required for videodisc premastering.

**generic disc** A videodisc that serves as a generic data base, usually on a single subject, such as hematology or NASA footage.

**instant jump** Similar in effect, but not technique. to a capacitance "field kicking" system, instant jump is the ability of a laser disc player to search up to 200 frames instantly, without any screen blanking. *See* FIELD KICKING; SCREEN BLANKING.

**interaction** A reciprocal dialogue between the user and the system; Interactivity (adj., interactive).

**interchangeability** A videodisc design strategy that includes information readable on consumer, industrial, and computer- controlled systems.

**interlace** In NTSC video, half of the horizontal scanning lines are laid down, and after retrace, the other half are laid down in such a manner as to fall in between the first lines.

**intermediate materials** All media selected for assembly onto the videodisc premaster, i.e., 16 mm film, videotape, 35 mm slides.

**jaggies** Jaggies are tearing phenomena around the edges of NTSC images. Research at MIT has effectively solved this problem, and by so doing has created a new way of thinking about displays, computer graphics, and NTSC video. *See* SOFT FONTS.

**keyer** A keyer cuts a hole in the background video and fills in the hole from a different video source, i.e., computer-generated text and graphics keyed over NTSC video. *See* OVERLAY; VIDEO REPROCESSING.

**landing pad** A range of frames within which a player can locate a frame or frame sequence. Landing Pad (LPD) is also a command that modifies the number of times a player attempts to locate a frame following an unsuccessful search.

**laser disc** A designation of an optically read disc that uses laser light to read information from the disc. *See* OPTICAL DISC.

**level of interactivity** The potential for interaction prescribed by the capabilities of videodisc hardware.

**level 1** Usually a consumer model videodisc player with still-freeze frame, picture stop, chapter stop, frame addressability, and dual-channel audio, but with limited memory and less processing power.

**level 2** An "industrial" model videodisc player with the capabilities of level 1 plus on-board programmable memory, and improved access times.

**level 3** Level 1 or level 2 player(s) interfaced to an external computer.

**level 4** A theoretical configuration wherein all things are possible.

# An Interactive Videodisc Glossary

**mastering**  A real-time process in which the pre-master videotape is used to modulate a laser beam onto a photosensitive glass master disc.

**NTSC video**  The American television standard for video set at 525 lines by the National Television Standards Committee (NTSC).

**optical disc**  A videodisc that uses a light beam to read information from the surface of the disc.

**optical memory**  Digital data encoded on an optical disc used for mass data storage. It is estimated that one side of an optical disc could store up to 10 billion bits.

**overlay**  A term used to describe the keying of computer-generated text/graphics onto NTSC video.

**picture stop**  An instruction encoded in the vertical interval on the videodisc to stop the videodisc player on a predetermined frame.

**postproduction/premastering**  Sometimes called "video processing," this is the process of editing, assembly, evaluation, revision, and coding of intermediate materials. A premaster is a fully coded videotape.

**preproduction**  All design tasks, such as flowcharting, storyboarding, script writing, software design, etc., prior to where media production starts—with sets, lighting, acting, sound recording, etc.

**reflective (optical) disc**  A designation by which the laser beam reads data encoded on an optical videodisc. In the case of a reflective disc, the laser beam is reflected off the shiny surface on the disc.

**scan**  To traverse the surface of the disc with video displayed.

**screen blanking**  A variable interval of time in which video is turned off while the disc player searches from one point to another on the disc.

**search**  To rapidly access a single frame or a sequence of frames on a disc with video off.

**sequence**  Two or more frames forming one unit, e.g., motion sequence and still frame sequence.

**slow motion**  In videodisc technology, the controlled movement of the laser from one frame to another frame at a variable rate less than 30 frames per second.

**SMPTE**  Society of Motion Picture and Television Engineers.

**SOFT FONTS**  A grey level scheme developed by MIT for high quality fonts in NTSC video that adds legibility, removes scintillation, enhances encodability, and which results in over eighty characters per line on a color television receiver.

**step**  To advance one frame forward or reverse.

**still frame**  Still material, including photographs, line drawings, and pages, designed and presented as a single videodisc frame.

**still-frame audio (SFA)**  Sometimes called "compressed audio," still-frame audio describes a method of digitally encoding and decoding several seconds of voice quality audio per individual disc frame, resulting in a potential for several hours of audio per disc.

**three-two (3-2) pulldown**  A means of compensating film shot at 24 fps with video (30 fps). The first film frame is actually exposed on 3 video fields, and the next film frame is exposed on 2 fields.

**transmissive disc**  A designation of the means by which the laser beam reads data encoded on an optical videodisc. In the case of the transmissive disc, the laser beam passes through the transparent surface of the disc.

**vertical blanking interval (VBI)**  21 blanked lines during field 1 and 21 blanked lines during field 2, wherein frame numbers, picture stops, chapter stops, white flags, closed captions, etc. are encoded.

**vertical interval time code (VITC)**  A derivative of the Society of Motion Picture and Television Engineers (SMPTE) time code that is stored in the vertical blanking interval.

**VHD**  Video High Density. *See* CAPACITANCE DISC.

**video reprocessing**  The process of keying video from the computer over NTSC video.

**white flag**  A code that identifies a new film frame (shot at 24 fps). During premastering, the code is inserted on the first video field, which in turn enables the videodisc player to jump back to the previous two fields, resulting in a fully matched freeze frame (without flicker).

# ANALYSIS

# The Front End: Systems Analysis and Media Selection

*Richard Walker and Lauralee Butler*

### INTRODUCTION

Creating an effective videodisc program is a complex and involved process. Content, personnel, media, and cost must all be choreographed to create a product that satisfies the original objectives of the program. From the onset, good planning is critical. A front-end analysis of a project can provide a large part of this necessary planning and facilitate the organization and completion of preproduction, production and post-production. Front-end analysis is not limited to developing the "big picture," however. It is also essential to the completion of production tasks such as scripting, designing special effects, developing computer programs, and all of the other components involved in producing a videodisc.

The purpose of front-end analysis is to provide a clearly documented description of the character and scope of the project to be completed. An effective way to complete this analysis is to break it up into five phases, 1) goals analysis, 2) preliminary specification, 3) task or message definition, 4) media selection, and 5) final specification. Each of these phases examines specific questions concerning a project. The answers to those questions contribute to a well-rounded, detailed final specification for the project.

*Goals analysis* describes the overall program objectives, budgetary requirements, and deadlines. The first step towards succesful completion of a program is made when these aspects of the project are clearly outlined and agreed upon.

The *preliminary specification* is a comprehensive listing of all tasks or facts to be presented in the program. The content of this specification reflects decisions made during goals analysis concerning program objectives. Extraneous or obsolete information is easier to spot and eliminate when the purpose of the program is kept firmly in mind.

*Task or message definition* expands upon the preliminary specification. Each task or message is described in detail to clearly state what the user will encounter in each section of the program.

Decisions about the final presentation of program content are made during the process of *media selection*. Each task or message has been explicitly outlined during previous steps and now the most effective means of presentation must be determined.

The last step in front-end analysis is *final specification*, a culmination of results from all previous steps. A program flowchart is derived from an established hierarchy of all tasks or messages and describes program content, methods of presentation, and flow. The

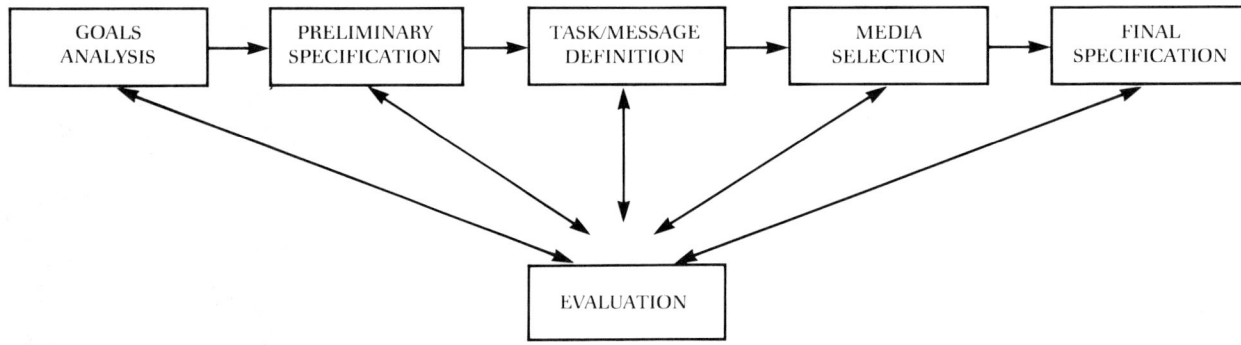

**Figure 1**

flowchart is also used to determine final project costs and time requirements. This information, along with a description of the overall goals and constraints of the project, is then used to produce the final project specification.

Figure 1 is a schematic representation of the different stages of front-end analysis. Notice that evaluation of the conclusions reached during each phase is a continuously important part of the process.

The five steps of front-end analysis are progressive in nature, each one elaborating on the conclusions of the last. The remainder of this chapter describes each of the five phases of the analysis in greater detail and demonstrates methods of application.

## GOALS ANALYSIS

As indicated in the above diagram, the first step in planning a project is to determine the project goals and constraints. This involves an evaluation of what you are trying to accomplish and the possible restrictions. Experience has shown that the success of a project, whether it be an educational, informational, or point-of-sales presentation, depends upon careful definition of goals at the outset. Clearly defined objectives lead to sound design and smooth production.

Together with your client (or internal management), evaluate the overall scope of the project. What are the program objectives and how can they be met? Who is the audience to be reached? What are their special requirements and objectives? How much time do you have to complete the project? What kind of budget do you have at your disposal? How does the budget compete with the time line and objectives?

The answers to these and related questions are then written up in a formal report and submitted to the client for approval. This report describes the product, and estimates a delivery date and cost. Alternatives or possible modifications to the product are also outlined and accompanied by cost and time estimates. In terms of a videodisc production, the report states whether the program is to be interactive, computer-assisted, bilingual, etc. Production variables are far more numerous for videodisc than for printed matter. A report of this kind is useful to explain what is expected of the product, and can reduce the number of surprises during and after production.

With goals and constraints agreed upon and approved, you can begin thinking about what it's going to take to complete the project. Determine whether or not a project of this kind has been done before. If so, was it effective? Can portions of it be incorporated into your project? You might be able to make use of existing videotape, film, slides, or film strips. Using existing resources can save time and effort, reduce cost, and can also remove an element of risk in production. Be careful, however, not to assume that just because film, videotape, or slides on your topic already exist, these media will necessarily save you time and costs. Weigh the benefits against any possible liabilities.

From the very beginning of a project, probe your clients to determine their needs, wants, and capabilities. Ferret out any hidden or even unknown preconceptions on the part of the client and others represented. All too often something is left out of a program, or needlessly included because one party or the other assumed it would be taken care of. Compare the client's requirements with the program objectives. Are they compatible? Does the program provide what is needed? Reviews should be frequent, either at the end of each phase or monthly, whichever is most appropriate. In this way, unpleasant surprises are kept to a minimum and a rich exchange of ideas and skills is encouraged.

Good planning is critical to smooth production and effective presentation. Forethought and frequent communication can significantly diminish wasted effort and frustration.

# The Front End: Systems Analysis and Media Selection

## PRELIMINARY SPECIFICATION

The next step in the project analysis is to compile a list of the facts or tasks that need to be communicated. For example, suppose the overall project goal is to teach business-oriented typing skills. Mastery of machine operation, keyboarding, and letter formatting are necessary parts of learning to type. But in order to gain proficiency in those areas, the student first needs to acquire more fundamental skills. Machine operation requires learning the keyboard layout, how to replace the ribbon, how to set margins, etc. Breaking a complex task objective down into its supporting elements is a very effective way to see exactly what needs to be presented to the user.[1]

A strictly informational program would contain a list of categories to be explained rather than tasks to be learned. Take, for example, a program about going to the zoo; information about everything found there, exhibits, animal shows, administration and research, and plant life might be included. Those categories could then be broken down into smaller groups of information. "Exhibits" would be divided into sections concerning mammals, reptiles, and birds, just as machine operation was divided into supporting skills in the previous example. The mammals section would no doubt be even further divided into sections containing information about bears, monkeys, giraffes, and so on. Figure 2 shows how a portion of this information listing might appear.

An analysis of a program providing sales information would generate a list very similar to the zoo list. The method of presentation, however, would be more promotional in nature. A program designed to display what a furniture store offers might be organized according to the rooms of a typical house. The user could request to see furniture for a living room, bedroom, kitchen, playroom, or perhaps even a patio. The available furniture styles as well as inventory would be shown for each room.

Once the tasks or concepts list is assembled, you and the client may need to delete those items that are not applicable or supportive of the project objectives. Certain tasks in instructional programs may be of too low a level for the target audience to spend time on. Obsolete or extraneous information should be deleted from programs designed to convey a message. In addition, a sales program should probably refrain from displaying a product that provides your client's competitor with free advertising.

This list is the preliminary specification, and when approved, it is a fairly comprehensive description of what ground is covered in the program to be produced.

A Program About the Zoo

1.0 Animal Exhibits
   1.1 Mammals
      1.1.1 Bears
      1.1.2 Monkeys
      1.1.3 Giraffes
      1.1.4 Lions
      1.1.5 Tigers
      1.1.6 Antelope
   1.2 Birds
      1.2.1 Pelicans
      1.2.2 Flamingoes
   1.3 Reptiles
      1.3.1 Snakes
      1.3.2 Crocodiles

2.0 Animal Shows
   2.1 Elephant Show
   2.2 Bird Show

3.0 Administration and Research
   3.1 Zoo Operations
   3.2 Research
   3.3 Animal Health

4.0 Zoo Plant Life
   4.1 Conifers
   4.2 Water Plants

**Figure 2**

## TASK OR MESSAGE DEFINITION

The purpose of this step is to refine the preliminary specification by clearly describing everything the user may encounter in each task or message presented in the program. As mentioned before, it is necessary to identify those low level skills that lead to a greater understanding of higher level ideas.

The tasks that make up an instructional program are written with eventual student performance in mind. What are the conditions that dictate when, how, and why a student will perform the different tasks learned? What determines successful completion?[2]

Supporting information in a sales-oriented program would consist of product descriptions. These descriptions could include product features and any warranties or guarantees offered. For example, a lounge chair may be part of a three-piece set, and available in a variety of upholstery fabrics and wood finishes. It is a good product because it's kid- and dog-proof and competitively priced. It also has a lifetime guarantee except in the case of fire, flood, meteor showers, etc.

Strictly informative programs have a similar flow, progressing from general facts and figures to the more specific. Motivational statements to convey why this knowledge is important and how it relates to the observer's life are also appropriate here. The observer is then able to build a conceptual framework and make the facts and other information presented more personally meaningful.

Once these ideas have been developed, they are ready for customer review. It can't be stressed often enough how important it is to have frequent and comprehensive review cycles. Working closely with a client or management contributes greatly to mutual understanding and enhanced productivity.

## MEDIA SELECTION

The next step in project analysis is to decide upon the most effective method of presentation for each task or message. Videodisc technology provides a number of options that may at first seem somewhat overwhelming. We have devised a method of assessment to easily and systematically categorize each program unit in terms of how a videodisc could most effectively present information. In this phase, as with all others, program objectives on the part of the user and producer remain the strongest influence on descisions made. A model airplane can be displayed in a box, but actually seeing the assembled plane flying is what really captivates the potential consumer and may lead the way to a purchase.

The matrix in Figure 3 has been developed as a useful tool to help with the decision-making process when selecting the most appropriate mode of presentation. The values assigned to each box designate how effectively videodisc, computer, or printed mat-

MEDIA

| FUNCTION | videodisc still | videodisc motion | computer | printed matter | videodisc & computer | videodisc, computer & printed matter |
|---|---|---|---|---|---|---|
| high quality visuals | 2 | 2 | 1 | 2 | 2 | 2 |
| color | 2 | 2 | 1 | 1 | 2 | 2 |
| sound | 2 | 2 | 1 | 0 | 2 | 2 |
| motion film | na | 2 | 0 | 0 | 2 | 2 |
| motion video | na | 2 | 0 | 0 | 2 | 2 |
| motion animation | na | 2 | 1 | 0 | 2 | 2 |
| still | 2 | na | 1 | 2 | 2 | 2 |
| text | 1 | 1 | 2 | 2 | 2 | 2 |
| system portability * | 1 | 1 | 1 | 2 | 1 | 1 |
| content easy to index? | 1 | 1 | 2 | 1 | 2 | 2 |
| graphics | 2 | 2 | 2 | 2 | 2 | 2 |
| content easy to change? | 0 | 0 | 2 | 0 | 2 † | 2 |
| interactive | 1 | 1 | 2 | 1 | 2 | 2 |

Performance Level: 0 = None
                                1 = Fair
                                2 = Excellent

* In general, videodisc players and computers do not travel. There is a disproportionate risk of damage when moving this equipment for distances greater than several yards. The 1 indicates that, for example, a school purchasing a videodisc can expect to transport it from classroom to classroom only *with great caution*.

†Of course the images and narration remain impossible to change once a videodisc is mastered. Data stored in a computer, however, is very easy to adapt to changing prices, population numbers and so on.

**Figure 3**

# The Front End: Systems Analysis and Media Selection

| | |
|---|---|
| high quality visuals | x |
| color | x |
| sound | x |
| film | x |
| motion video | x |
| animation | |
| still | |
| text | |
| system portability | |
| content easy to index? | |
| graphics | |
| content easy to change? | |
| interactive | |

|  |  |
|---|---|
| PROJECT: | About the Zoo |
| EXAMPLE OR TASK: | The Brown Bear |
| SEGMENT DESCRIPTION: | This segment shows The bear's natural habitat |
| APPROXIMATE RUN TIME: | 90 seconds |
| MESSAGE OR TASK #: | 1.1.1.1 |

**Figure 4**

| | |
|---|---|
| high quality visuals | x |
| color | x |
| sound | |
| film | |
| motion video | |
| animation | |
| still | x |
| text | x |
| system portability | |
| content easy to index? | |
| graphics | |
| content easy to change? | |
| interactive | |

|  |  |
|---|---|
| PROJECT: | About the Zoo |
| EXAMPLE OR TASK: | The Brown Bear |
| SEGMENT DESCRIPTION: | This segment shows what the bear looks like at different stages of growth. |
| APPROXIMATE RUN TIME: | 120 seconds |
| MESSAGE OR TASK #: | 1.1.1.2 |

**Figure 5**

ter uses graphics, sound, etc. In this chart, a 0 indicates that a function, listed on the left axis, can not be performed by the medium shown on the top axis. A 1 indicates an adequate but perhaps limited performance, and a 2 means the medium is excellent for that particular purpose. In some cases, a function that could actually have been given a 2 for performance has been downgraded to a 1 because of feasibility constraints. For example, computer-generated illustrations can be of high resolution and quality, but are very costly in terms of computer memory space and actual dollars at this time.

Media selection is begun by completing a worksheet for each task or message. The worksheet contains a check list to designate what media functions are necessary; and includes space for a description of the task or message to be presented. Following is an example of a completed worksheet.

According to this worksheet it has been decided that color videotape or film and sound would best show the bear's natural habitat. Are those options feasible in terms of the capabilities of the medium? A look at the matrix (Figure 3) shows that, yes, use of film and sound on videodisc can be very effective. Another example of a completed worksheet is shown following.

This worksheet indicates that good-quality color photos or artwork along with text matter is thought to be the best way to show the brown bear's appearance. According to the matrix, these options are all possible with videodisc; but videodisc is not ideal for presenting extensive text. Consideration should be given to using some sort of supplemental printed matter in addition to videodisc.

With a clear outline of program objectives, supporting tasks, and information, the appropriate media for presentation becomes more apparent. The worksheet helps to uncover the character of each message or task, and points to one medium over another. Does the task require student participation? Computers are most effective for interactive lessons. Is the program about national parks? Motion and stills via videodisc would be most effective in presenting that information.

Once the presentation media has been decided upon, the next step is to group tasks according to method of presentation. An estimate of the amount of artwork, film footage, or even computer programming necessary can be obtained from this grouping. The number of personnel hours required and production costs can also be derived.

With this media selection model, program units are not only evaluated for best possible presentation, but information concerning overall project appearance and more specific cost estimates can be easily extracted as well. A summary of the data gathered from the worksheets, the worksheets themselves, a task listing, and any other information describing the scope of the project thus far are now gathered together to produce a high level product specification for customer approval.

## FINAL SPECIFICATION

By the time you have reached this step in front-end analysis, the scope of the project should be quite apparent. The increasingly detailed specifications clearly show objectives, underlying tasks and concepts, and methods of presentation. Before preliminary scripting and storyboarding can begin, however, the order of program events must be determined and a final project specification drawn up.

The organization of an educational program is quite different from that of an informational or sales presentation. For educational material, appropriate task sequencing is most important to ensure that a student isn't asked to complete an assignment he or she isn't yet prepared for. Informational or advertising material, on the other hand, allows a considerable amount of artistic leeway in presentation, but organization and planning are certainly still as critical to deliver effective and relevant messages and at the same time fulfill the program objectives.

The worksheets completed during media selection provide a convenient way to organize content. Laid out in the proper order, like storyboard pages, they represent an extensive program flowchart. A schematization of this flowchart, as shown in Figure 6, is then included as part of the final product specification.

This specification is a complete guide to project development and production, now that program objectives, tasks, and concepts have been clearly defined and prioritized. Methods of presentation are described along with an evaluation of the personnel hours needed for production and the overall cost of the project.

In summary, the result of a front-end analysis is a comprehensive and refined description of the project under consideration. Each step, beginning with goals analysis and ending with the final specification, contributes to a complete framework that guides later production. The advantages of front-end analysis are many: good planning leads to smooth and efficient production, frustrations are reduced and work fulfillment enhanced, program objectives remain clear. One of the most important results of the front-end analysis, however, is the establishment of continuous

# The Front End: Systems Analysis and Media Selection

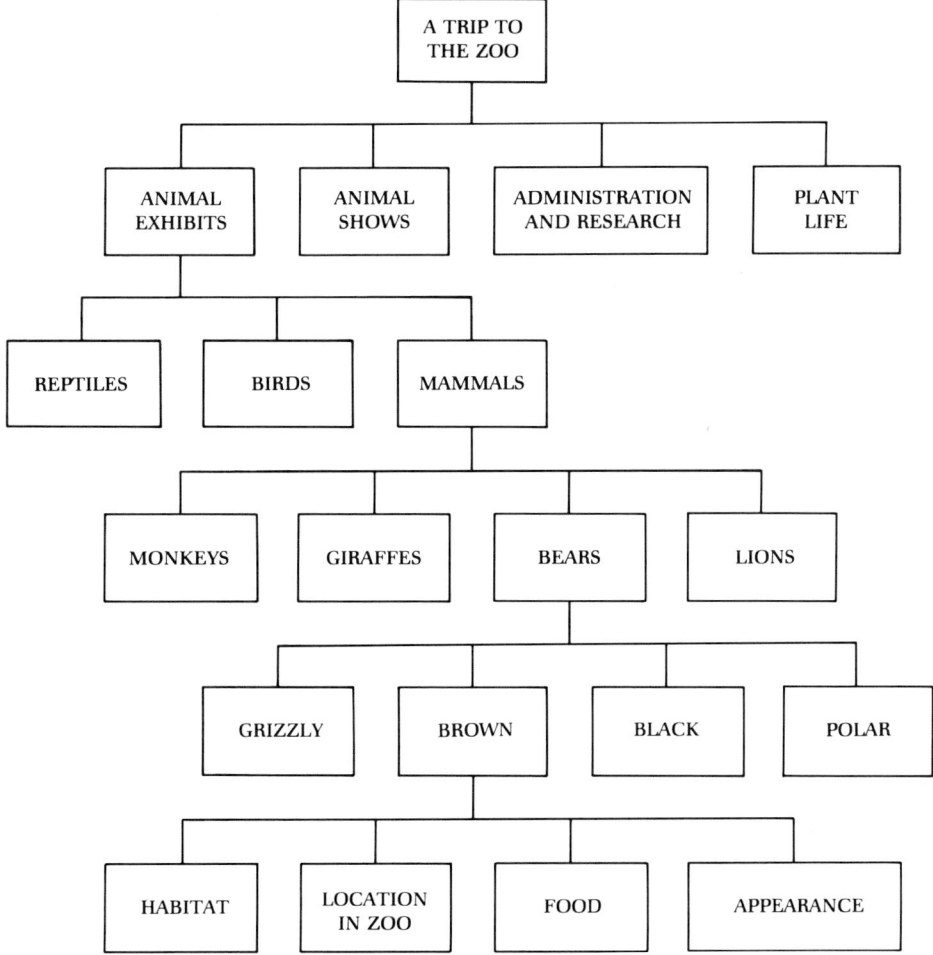

**Figure 6**

and open communication of goals and expectations between client and producer. This mutual understanding is the basis of a sound working relationship leading to a mutually satisfying project conclusion and the possible initiation of future creative undertakings.

## RECOMMENDED READINGS

1. R.M. Gagne, *Principles of Instructional Design*, (2d Ed. New York: Holt, Rinehart, and Winston, 1979).

2. R.M. Gagne, *Conditions of Learning*, [3d Ed. New York: Holt, Rinehart, and Winston, 1977(a)].

Dr. Richard A. Walker holds a Ph.D. in Instructional Science from Brigham Young University where he conducted CAI research for the National Science Foundation with Dr. M. David Merrill. Dr. Walker has been involved in numerous CAI training projects for the military and participated in the evaluation of the TICCIT CAI System at NAS North Island. Dr. Walker was manager of Electronic Publishing at Courseware, Inc., and is currently in charge of Instructional Development at Interactive Technologies Corporation in San Diego, California.

Lauralee Butler holds a B.A. degree in Psychology from the University of California at Los Angeles. As part of a team of programmers and instructional designers, she worked to develop the production process to create interactive programs for the computerized Interactive Video System at Primarius, Inc. and produced over 20 CAI programs. Ms. Butler is currently working as an Instructional Developer at Interactive Technologies Corporation in San Diego, California.

# DESIGN

# A Few Principles of Interactive Videodisc Design

*Jim Hoekema*

One of the curious aspects about making interactive videodiscs is that no one in this field has been formally qualified for it. The technology is so new that, aside from a short workshop here and there (such as the excellent ones given by the University of Nebraska), the only way to learn how to produce an interactive videodisc is to produce one! Since we all come from some other background, it means that everyone involved in interactive video is learning one or more new disciplines—filmmaking, video technology, computer programming, instructional science, or the hodgepodge of human factors that link them all together.

Not only does interactive videodisc thrust these previously unrelated disciplines together, but it also changes each one of them in small but significant ways. If a viewer can see any one of ten motion sequences in any order, this fact certainly changes (or should change) how a filmmaker would write, shoot, and edit each scene. And a traditional background in graphic design would offer little direct preparation for the special problems of designing, say, text and pictures for five hundred videodisc still frames.

So a lot of people have been learning from experience what does and doesn't work with interactive videodisc, acquiring new disciplines and practical wisdom on the fly. Lacking a prescriptive science, we have had to rely on the faster but less well-documented powers of art and common sense to integrate the diverse components of this complex new technology. Based on such experience, the following observations are intended as a group of general but essentially practical hints to others involved in designing and producing interactive video for educational or other purposes.

## DESIGNING FOR THE MEDIUM

The process of designing an interactive videodisc program is a lot like the process an architect follows in designing a building. Starting from the client's guidelines, the purpose of the building, the population who will inhabit the structure, and constraints of budget and schedule, the architect must create an environment that accommodates not only the stated

The first half of this article originally appeared as "Interactive Videodisc: A New Architecture" in *Performance and Instruction*, XXII:9 (1983), 6ff.

goals but also the unstated goals and the unanticipated purposes that the real users may demand of the building after its completion.

We would do well, therefore, to recall the widespread preoccupation of architects (popularized especially by Frank Lloyd Wright) with doing what is natural with the materials being used. In this "organic" approach to design, the architect treats wood as wood, steel as steel, and does not try to make one material behave unnaturally like another. (Philadelphia architect Louis Kahn told audiences that he tried to "ask the bricks what they wanted to become" in designing his buildings.)

If videodiscs could speak, and we were silly enough to ask them, what would they tell us they want to become? In other words, are there any intrinsic qualities of the computer-controlled optical videodisc that demand certain design approaches from the outset, quite aside from the particular application? Are there any "generic" principles of interactive videodisc design? I believe there are a few of them, as follows.

### User Control

Given the rapid random access by which the videodisc player is able to search from any one point on a videodisc to any other (about 3 seconds maximum on industrial players), it follows that one of the "natural" or "organic" features of any application of this technology is to provide the user with some measure of control over the pace and sequence of the material. Such control is provided on a basic level by the functions built into the player (SCAN, STOP, FAST, SLOW, etc.). Instructional developers generally agree that user control is a good thing, yet a surprising number of sophisticated videodisc training programs (controlled by external computers) provide users with a lower level of control over program pace and sequence than that afforded by purely "manual" or "level 1" (no computer) videodiscs.

### Rapid Pace

Given the feature of user control, it follows that motion sequences shot for videodiscs must earn the viewer's interest continually, since he or she may elect to skip the rest of a given segment at any time. A slow or boring experience is easily remedied by the viewer, while the opposite extreme—a motion segment so packed with information that it seems to go by much too fast—can always be repeated if necessary. Thus a natural style of filmmaking for videodisc would be rapidly paced and visually rich.

### High Quality

An interactive videodisc program on any subject demands a design and a production of higher quality than would be required for the same subject in any other medium. There are several reasons for this, as follows.

1. The argument above for a rapid pace also applies to the quality of motion sequences: videodisc is a medium that begs for scrutiny. Imagine, for example, that you are shooting a sequence on the operation of a computer terminal and, lacking the correct close-up of a particular screen or key press, you are forced to substitute another shot. In a videotape or film, the shot is gone before the student has a chance to notice the anachronism. On videodisc, the student can find those few frames and study them. If the shot had been correct, it might have provided a useful added level of detail for a student who needed it. A student inspecting an incorrect shot, on the other hand, might go marching down a path of misleading assumptions, or perhaps recognize the error and lose all faith in the instruction. In short, a videodisc program must be designed to bear the scrutiny it invites.

2. Because a videodisc is delivered on a television screen, viewers' expectations are set by the standards of broadcast television. This is not a problem when it comes to plot and characterization in narrative: there, thanks to the networks, we can assume that a typical American viewer brings very low expectations to the experience of watching television. I am thinking more of the technical areas of motion picture production, such as lighting, sets, camera work, graphics, animations, and special digital effects. In these areas, news programs, commercials, and station logos have set high expectations, which are roundly frustrated in most educational and industrial production. This is particularly true of the crude and amateurish graphics found on the still frames of most videodiscs—even on such an otherwise well-produced program as the first "Mystery Disc."

3. Another argument in favor of high-quality videodisc productions is that, in most cases, we have no control over the quality of the television monitor on which our program will be viewed. Yet we do know, if the program is at all interactive, that users will spend a considerable amount of time with it—more, say, than with a videotape. Therefore, we should prepare for the worst, and design a program that will communicate effectively even on a badly tuned, poor-quality monitor. The message needs to be well composed and strongly beamed in order to survive such a "noisy" channel of communication.

## A Few Principles of Interactive Videodisc Design

4. Perhaps the most persuasive argument for quality in videodisc productions is the most practical one. Because they are so complex, combining as they do most of the strengths of all other media of communication, interactive videodisc programs tend to take a long time to develop: almost invariably longer than originally anticipated, and certainly longer than other media (videotape, a CAI program of equal complexity, or a programmed text). A common tendency, when scripts and storyboards are finally complete, is to rush into production and try to save time and money by cutting a few corners here and there. When this happens, the result is usually a slovenly product whose poor "surface" qualities create such negative impressions on users that many subtleties built into the deeper structure of the program never have a chance to see the light of day. In short, the care and attention necessary for design and development should be matched by an equal level of professionalism in production.

### One Frame, One Message

Information presented on still frames will of necessity appear as a sequence of whole frames, advanced by some form of user input. This suggests that the most natural or organic way to present text on videodisc is *not* to fill each screen with as much text as can be fit (as if it were important to minimize the total number of keypresses for getting through a given number of words). Such a design, in fact, would guarantee that almost every keypress would interrupt a thought. A more organic method is to make each frame communicate one message or thought, however short it may be. Then each keypress becomes a part of the communication: it is the signal by which the user tells the system that he or she has absorbed the last piece of information and is ready for the next one.

### Self-Evident Structure

The organization or structure of an interactive videodisc program is *not* immediately apparent to the viewer by any quality inherent to the system. As Nicholas Negroponte of MIT has often said, a book provides more information about where we are in relation to the whole through the simple tactile evidence of how much text we hold in our left hand versus how much remains in our right hand! Lacking such automatic signs about where we are in relation to an overall structure, interactive videodisc programs require a conscious effort by the designers to provide such clues. Useful devices to this end include

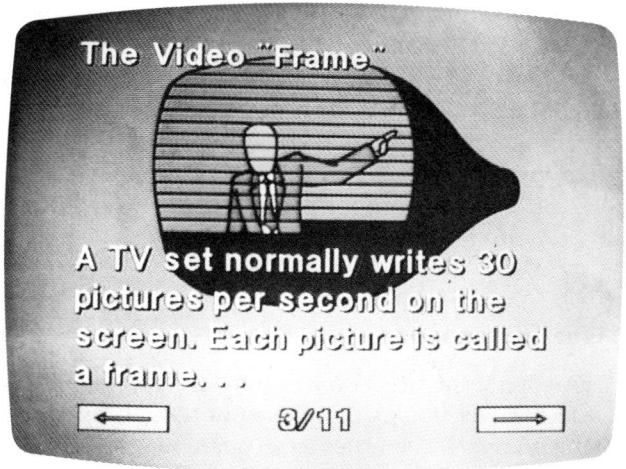

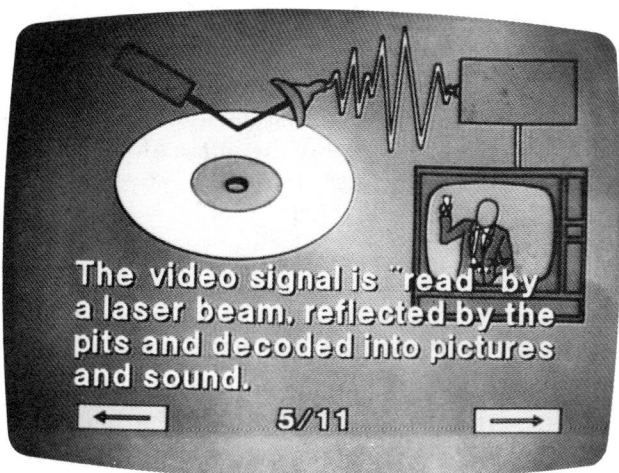

Examples of the principle of "one frame, one message." Each frame presents a single idea with verbal and visual elements reinforcing each other in a strong, concise statement.

nested menu structures,* numbered lessons, section titles with running "headers" or "footers" on every frame, and even less explicit "signposts" such as color codes.

### Self-Contained Components

A corollary to the need for self-evident organization is the need for self-contained components. If a user can determine the sequence in which a number of

---

\* The menu-driven structure is by far the more popular. A menu is a choice frame, where a user selects which of several units or sections to see next; when the selected activity has been completed, the user should always be returned to the same menu. A menu scheme is not only the easiest way to provide user control; the menu itself is also a valuable "map" of the whole program, providing a sense of the overall structure of the content.

different segments can be viewed, it follows that each segment must make sense to the viewer regardless of which of the other segments had been viewed immediately beforehand. In such situations (which are actually rather rare in both training and marketing applications, when a prescribed sequence usually prevails), it is important to avoid saying things like, "As we learned in a previous lesson . . ."

**Visual Information**

Finally, it seems obvious that an application of videodisc technology that does not take advantage of visual imagery—whether the realism of photography or the clarity of intelligent graphics—would waste a major strength of the medium. By the same token, if a given idea could be communicated by words alone or by a combination of words and pictures, the latter would be a more effective method.

So much for design principles that seem to flow naturally from the nature of the interactive videodisc medium. The guidelines that follow are more personal, but no less firmly held!

## THE MACHINE AS A GOOD SERVANT

All interactive programs have personalities—primitive ones, to be sure, but sufficient to leave impressions based, as we base our impressions of people, on what they say to us and how they respond to what we say. A program can have a clear, distinct personality (especially if written by a single individual), or it can have a weak, disorganized or even schizoid personality (especially when developed by multiple authors). Whatever its degree of coherence, a program's behavior inevitably communicates a set of attitudes toward both the subject matter and the viewer.

As designers of interactive videodisc programs, we have the opportunity, indeed the responsibility, to create coherent programs imbued with a "personality"—a set of attitudes and predictable behavior—that best serves the needs of users. But what is the nature of this ideal personality?

"User-friendliness" is a quality that most computer users can easily recognize, though more often in the breach than in the observing. User-friendliness, or the lack of it, is most conspicuous in the messages by which the computer asks for user input, and in the ways it responds to that input. For example, most people would agree that the following message is not very user-friendly:

> YOU MUST ANSWER THE FOLLOWING QUESTIONS.
> IF YOU DON'T, YOU WILL NOT BE ALLOWED
> TO SEE ANY MORE OF THE PROGRAM.
> PRESS THE "5" KEY NOW.

One could argue that the following is a lot friendlier:

> HIYA JIM! YOU'RE DOING GREAT!
> HEY, WE'VE GOT SOME QUESTIONS FOR YOU,
> AND WE KNOW YOU'RE GOING TO LOVE
> ANSWERING EACH AND EVERY ONE OF THEM.
> PRESS "5," AND STAY COOL.

Those who enjoy being accosted by drunken strangers will probably find such messages admirably friendly, but for most of us, such a message makes up in obnoxiousness for what it lacks in sincerity. A more appropriate way of communicating the content of these two admittedly exaggerated examples would be:

> PLEASE ANSWER
> THE FOLLOWING QUESTIONS.
> TO BEGIN, PRESS "5."

## A Few Principles of Interactive Videodisc Design

Not only is this message courteous, it is also brief. By not cluttering up the essential with extraneous information, it shows consideration for the viewer's time.

Like people, machines reveal their personalities more strongly in what they do than in what they say. A program that sends out friendly greetings now and then may still be a rigid taskmaster at heart, forcing users into unnatural patterns of behavior that happen to suit the convenience of the programmer, as follows.

A feedback that looks "friendly"—unless it is used every time the user completes the simplest task.

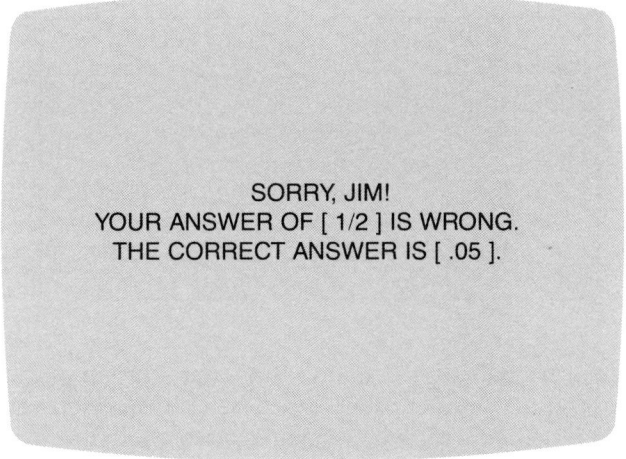

Particularly irksome is a program that attempts to simulate human emotions, as if *it* could somehow be made "happy" or "sad" by the user's performance, or that it "knows" the user once his or her name has been entered. Also, the excessive display of emotion often conveys the opposite of what was intended. For example, a feedback like "GREAT! YOU DID IT!" seems very friendly on the face of it—until you discover that the same feedback is used every time the user completes even the most rudimentary task! In that case, such excessive enthusiasm conveys the system's (or rather the author's) obviously low opinion of the user's intelligence.

No, this kind of friendliness is not what is needed, either. What is really needed is not an artificial friendliness but rather a sense of good service. A good role model would be the butler Hudson on the PBS television series of a few years ago, "Upstairs, Downstairs." An interactive program (or any intelligent machine for that matter) should behave like that kind of good servant, with the following traits:

- capable, unobtrusive
- considerate, respectful
- presenting choices when necessary with a minimum of fuss and bother
- responding immediately and appropriately, and
- anticipating the master's desires when possible.

In practical terms, this means that programs should address the viewer in a tone that is courteous, economical, and to the point; the program structure should be designed so that it is easy for the user to know where he or she is at any time, and how to move to another part of the program; and both the writing and the visual design of everything in the program should provide utmost clarity of communication.

### PROGRAM STRUCTURE

The single most important ingredient in a truly user-serving interactive program is a clear and flexible program structure. "Program structure" includes the arrangement of the content into component parts, the relationships among those parts as presented to and perceived by the user, and the inputs by which the user moves through or around the material.

The analysis and segmentation of the content make up a design process. The resulting organization of information must relate to the most significant "natural" distinctions in the content as well as to the artificial structures imposed by program control and the need for user inputs. Often, the organization used in books on a given subject may not be the most appropriate for an interactive program. Many of the most successful interactive programs are those for which the program designer and the subject-matter expert are the same person. If that is not possible, subject experts and program designers need to learn each other's disciplines as thoroughly as possible: the closer the collaboration the better.

A program structure is generally not perceived all at once; rather, a user builds up a mental image of the structure over time, starting from the first few frames. A well-designed structure must therefore be coherent and consistent throughout, but above all it should establish a clear set of expectations at the beginning, which will not be frustrated as the program progresses. There should never be something in the opening sequence that "works a little differently" from the rest of the program.

No interactive program (videodisc or otherwise) can serve the first-time user and the experienced user equally; one or the other will always be favored. The case can be made that certain types of computer programs, such as word processors and electronic spreadsheets, should be optimized for experienced users: once they have learned how the program works they want speed and procedural shortcuts to create and edit their own files.

By contrast, most interactive videodisc programs are information-intensive and are typically viewed only a few times by the same viewer. Therefore, most interactive videodisc programs should be *optimized* for the first-time user. This means that the program should be entirely understandable to a novice user without any special preparation or the need to consult any offline documentation. Additional aids and shortcuts can be made available to experienced users, but they should not distract.

## Names and Titles

Regardless of the type of structure used, one of the most important tasks of program design is to name the parts, and to use those names with rigid consistency in title frames, menu selections, and cross-references.

Titles should be short, descriptive, unique, and inviting (in that order of priority). A common mistake is to devise snappy titles for dull subjects, a strategy that turns the interactive experience into a string of disappointments.

Numbers or letters are often helpful in a series of lessons or units, but "Lesson 5" (by itself) is not a very informative title, especially if a user is expected to select such an item from a menu. It is best to add a content-related title to the rubric and number, for example, "Lesson 5: Ignition Systems."

When there is a hierarchical organization, such as several units each composed of lessons, which in turn comprise activities or exercises, it is very important to remind users of each element's place within the hierarchy. Units and Lessons might require end frames ("End of Unit 3, Lesson 5") as well as title frames.

All major divisions should begin with a title frame, containing only the title information. In hierarchical structures, the main title should be accompanied by a smaller "superheading" at the top of the frame, reminding the user of the larger division of which this is a part:

```
┌─────────────────────────────────────┐
│         Part II: Automobiles        │
│         Unit C: Maintenance         │
│ ─ ─ ─ ─ ─ ─ ─ ─ ─ ─ ─ ─ ─ ─ ─ ─ ─ ─ │
│                                     │
│              Lesson 5               │
│          CHANGING THE OIL           │
│                                     │
└─────────────────────────────────────┘
```

Such small signposts do not demand much of the user's attention, but they are there if needed to help users maintain their orientation as they move through the material.

In menu-driven programs, it is essential that the titles on the title frames correspond exactly with those listed as selections on a menu. One important purpose of the title frame is to confirm that the system has recieved the user's instruction correctly.

## Types of Structure

Most interactive videodisc programs are either menu-driven or linear in their overall structure. A given program may include other types of structures within it, such as games, simulations, or self-adjusting tests; but the other layer of a program must generally follow one of two paths from the opening frames to the various parts: users are either guided automatically from one portion to another (linear structure), or they are asked to choose among options at one or more menu frames. The key to designing a user-serving environment lies in taking one or the other path consistently.

One disadvantage of the menu-driven program is that it constantly forces users to make decisions, which may distract and interrupt them from the content. If the choices are unclear, the requirement to choose one may seem arbitrary and irritating.

# A Few Principles of Interactive Videodisc Design

An example of the use of headings and superheadings. This is the menu for a section entitled "Course Development," which in turn is a subset of a larger unit identified by the superheading "Management Issues." (In this example, taken from a videodisc program using touch screens, the user touches one of the bars below to make a selection.)

A "Master Index" (or main menu) and the title frame for one of the sections selected from the menu. The lesson title "Solving a Problem" is identical on the two displays, and the title frame repeats the lesson number ("4"), which is also the number pressed by the user to make the selection from the menu.

Linear programs, which begin at the beginning and progress through the content in a (hopefully) logical order, have the advantage of avoiding the periodic intrusion of mechanics on the viewer's concentration on content. Many applications, especially in education and marketing, are strongly linear in nature, with clear prerequisite relationships among the parts.

On the other hand, users often lose sight of the larger organization of the content in a linear program, especially if there is more information than can be absorbed in one sitting. Resuming a linear program where it had been left off previously is awkward and difficult without a means of skimming and moving back and forth (or an online "manager" to keep track of student records). Also, most users sooner or later want to go back to only certain portions, or view material in some order other than the normal one. A useful device to handle such needs is an optional or hidden menu structure in addition to the normal linear path.

### User Inputs and Screen Prompts

User inputs are the commands by which users advance the program or make choices; screen prompts are the messages from the system to the user inviting such inputs. If inputs are in the form of labeled function keys, the number of functions should be kept to a minimum. Prompts and function keys should match exactly. Functions should behave consistently and be clearly distinct from one another. Very few programs require more than five or six distinct control functions.

User-input functions can be divided into three classes, each with different rules for prompting, as follows.

1. Inputs required to advance a program from one state to the next should be prompted explicitly on the screen. Users should never be left wondering whether the next move is theirs or the computer's. At any time, the screen should display either some unfolding event or action (showing computer control), or a prompt indicating that the program will remain as is until some user input is received.

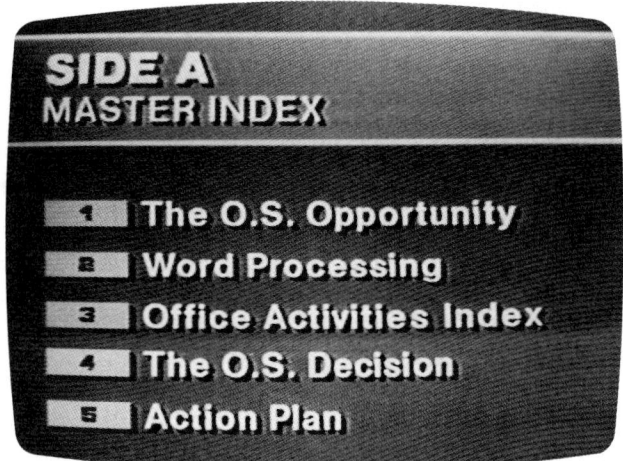

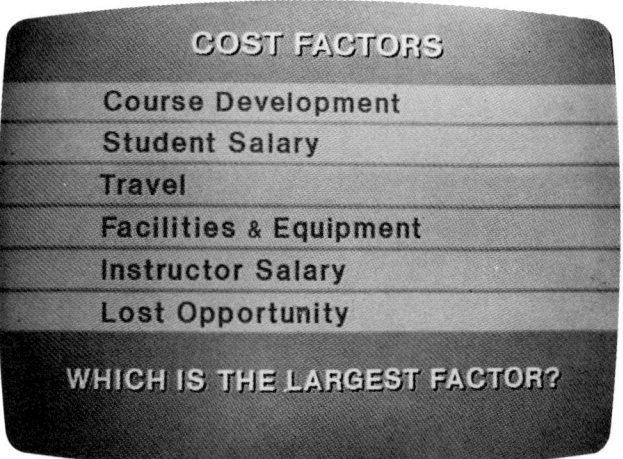

A user-input frame from a program using a touch screen.

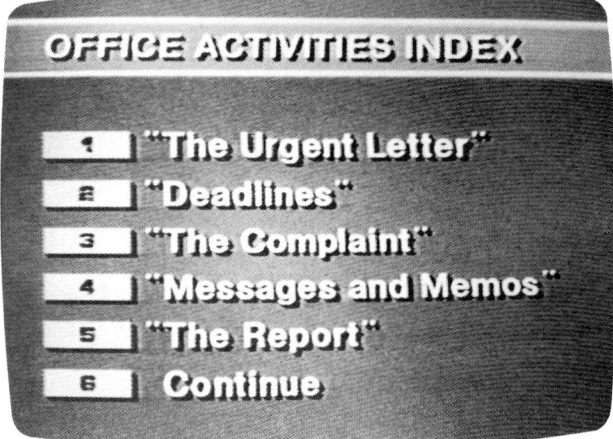

The "Office Activities Index" is one of the choices on the "Master Index" that users see only if they ask for it—an example of an optional menu structure in an otherwise linear program.

A user-input frame from a program in which the input device was the keypad of the now defunct PR-7820 player. The number symbols are blue, recalling in shape and color the keys on the keypad.

2. Functions that are active only some of the time should be prompted when they are available, such as "HELP" during an instruction sequence, or the capacity to "REPEAT" a short motion sequence or audio cue. The prompts should disappear as soon as the functions become inactive.

3. A third category consists of functions or controls by which the user can *interrupt* the normal path of a program. For example, a key marked "MENU" might always take a user to the applicable (next highest) menu. Or a key marked "SKIP" might allow users to skip a problem or go immediately to the next activity. Since these are user-interrupt functions, not a part of the normal sequence of interactions, they should generally not be prompted. Such functions work well, however, only if they are always active, and always work in the same predictable way.

Designers do not always have control over the names printed on the keys of the input device, in which case users must remember certain less-than-obvious associations between name and function. In this situation, it is often a good idea to make greater use of explicitly prompted choices (i.e., menus).

Systems using a touchpanel, mouse, or tablet input, offer an intoxicating freedom; user choices can be named and configured differently for a variety of displays. Designers should resist the temptation, however, of using a new set of "functions" on every frame. Users will develop a better sense of control through the consistent use of a limited number of functions and screen configurations. (Your programmers will also be happier).

# A Few Principles of Interactive Videodisc Design

## SCREEN DESIGN

Screen design for videodisc programs is as important as it is for television and film production, and then some. The biggest issue of screen design for videodisc is the effort and cost of producing large numbers of still frames.

Because the still frames on a videodisc carry valuable information which the producers want communicated, it is important that each of these frames be composed with the same care as, say, a magazine spread. Sometimes pictures must be taken expressly to go with certain words, and often the two elements must be rearranged before the desired effect is achieved. On a magazine, this sort of thing is done by an art director earning $25-40 per hour; but for a videodisc, this work often can only be done in a postproduction studio costing $350-500 per hour! Also, a typical videodisc program with, say, 600 still frames involves considerably more work than a typical 100-page issue of *Newsweek* or even a 200-page art book!

Most videodisc producers with at least one project's experience have learned to anticipate these problems by designing a system of graphic formats and doing as much design work as possible before going into postproduction.

Like any magazine, an interactive videodisc needs an overall approach to visual and graphic design, just as it needs an overall structure. This approach is simply a set of general rules which are applied with variations to each particular situation on the disc.

### Distinctive Frame Types

Probably the single most important task of this visual design program is to distinguish among different classes of messages presented to users. For example, a distinctive treatment or symbol for all prompts will help users know that *some* input is required even before they read the frame to see *which* input is involved. In terms of classic communication theory, the delimitation of the class of possible messages allows the user to receive the intended message faster, and with less possibility of error.

Visually distinctive frame types provide a kind of advance warning of the kind of information being presented. All videodisc programs can benefit from clear visual differences in the treatment of at least the following three distinct types of frames.

1. Title frames, announcing a new section (and usually confirming a menu choice).
2. Information frames, describing content (but with prompts indicating how to advance).
3. User-input frames, such as menus, where the whole point is to present and accept user inputs.

Other frame types requiring distinctive treatment might include right-answer and wrong-answer feedbacks, "help" frames, and score displays.

### Less is More

Given the enormous still-frame capacity of a videodisc, there is no excuse for a frame filled with text. As a general rule, one frame should present one message. Careful use of blank space around the words, short line lengths, and the use of lines or rules will help carry the message to the user with speed and impact. If each screen contains one message, the user's input to advance to the next frame will never interrupt a thought but will in effect mean: "Message received; ready for next message."

### Type Styles and Sizes

The video character generators available in most postproduction studios (Compositor, Chyron, Vidifont) allow an impressive variety of different type styles and sizes (although with varying limits on the number of different combinations available at one time). In general, it is not worth the bother of using more than two or three *styles* of type, because the subtle differences among them lose effectiveness when spread across many frames that are never seen simultaneously. Also, differences in color, composition, and other graphic devices tend to overpower type style distinctions. On the other hand, a range of type *sizes* is very useful, especially if each size is used for a separate purpose.

### Color

Color is by far the most powerful graphic device for underscoring differences in frame types. A single color used consistently for prompts, for example, can be a very effective cue. Different chapters or levels might also have distinctive colors.

In the use of color, however, there are a few common temptations to avoid. First, a color-coding scheme will work only with a relatively small number of colors (say, four or five). Although the human eye can distinguish hundreds of colors in juxtaposition, most of us have a poor memory for recognizing precise hues without direct comparison.

Also, it is a good idea to avoid bright colors, especially in backgrounds. Poor monitors and televisions will present a noisy and unstable picture if strong, bright colors are juxtaposed, especially with text. As background, highly saturated colors tend to advance and swallow up the text, undermining the intended relationship between foreground and background.

At the risk of being overly formulaic, I would recommend a neutral gray value as a background for all videodisc still frames. It recedes optically and provides an ideal middle value, against which any other color can be used as a "foreground." After a half-dozen projects, the gray background today seems equivalent to the standard off-white pages used in almost every book ever published.

Jim Hoekema began designing and producing interactive videodisc programs in 1977, while he was a developer in the exhibit-design, graphics, and filmmaking office of Charles and Ray Eames in Venice, California. As Design Director for the training division of WICAT Systems, he produced several interactive marketing videodiscs for IBM and a prototype videodisc training system for AT&T. He is currently Design Director for software development in WICAT's Basic Education Division.

**PREPRODUCTION**

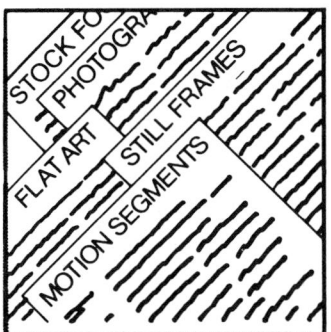

# Scripting for Videodisc

*Casey Garhart*

In an examination of the scripting process for videodiscs, the first question that arises is, "What's the difference?" The videodisc is just a new way to present audio and video information, so how does writing for disc differ from writing for film or television? Certainly, there are still frames, but that's just a matter of integrating a slide presentation with tape. Is there really a need for information on how to write for videodiscs?

The answer is "Yes." Not so much because motion and still frames can be combined into a single format, and not because there is the potential for two audio tracks. The real difference is that while film, videotape, and slide presentations are basically linear in nature, the videodisc is nonlinear. It is an interactive medium where the user makes the final edit. This interactivity becomes a major strength of the medium if it is reflected in the program design and production; if it is not taken into account, it can become a liability.

Everyone is familiar with linear video. There is usually a soft introduction and then the opening credits come up. Music establishes the mood and setting and then we go into Scene I. Scene I fades into Scene II and the music builds. There is a flow to the program as established conventions of filmmaking carry the viewer through transitions. Certain types of cuts indicate the passage of time or change of setting. The editing establishes the pace of the program as ideas build and character development occurs. And under it all the music becomes a through line that connects the various segments. The program builds to a peak or climax and then relaxes into the closing scenes. Finally, we reach the end, and with the closing credits the program is over. Whether it was a full-length movie, an instructional television program, or a training film, the general structure remains the same. There is a beginning, a middle, and an end, in that order, which have been carefully constructed into a unified whole according to certain concepts and conventions.

Not so with interactive programs. Instead of sitting passively, absorbing information, the viewer takes an active part in the presentation. In fact, the user is ultimately the controlling force behind the presentation. If the program is truly interactive (and meaningless interactivity is a waste of both time and money) the final sequence cannot be known until after each viewing. And it can change each time through. This ability to restructure the program means that the flow and movement traditionally designed into video productions is no longer applicable. In fact, many of the established conventions are not applicable with interactive videodiscs. Where to put the credits, for example, is suddenly a matter for discussion. Inter-

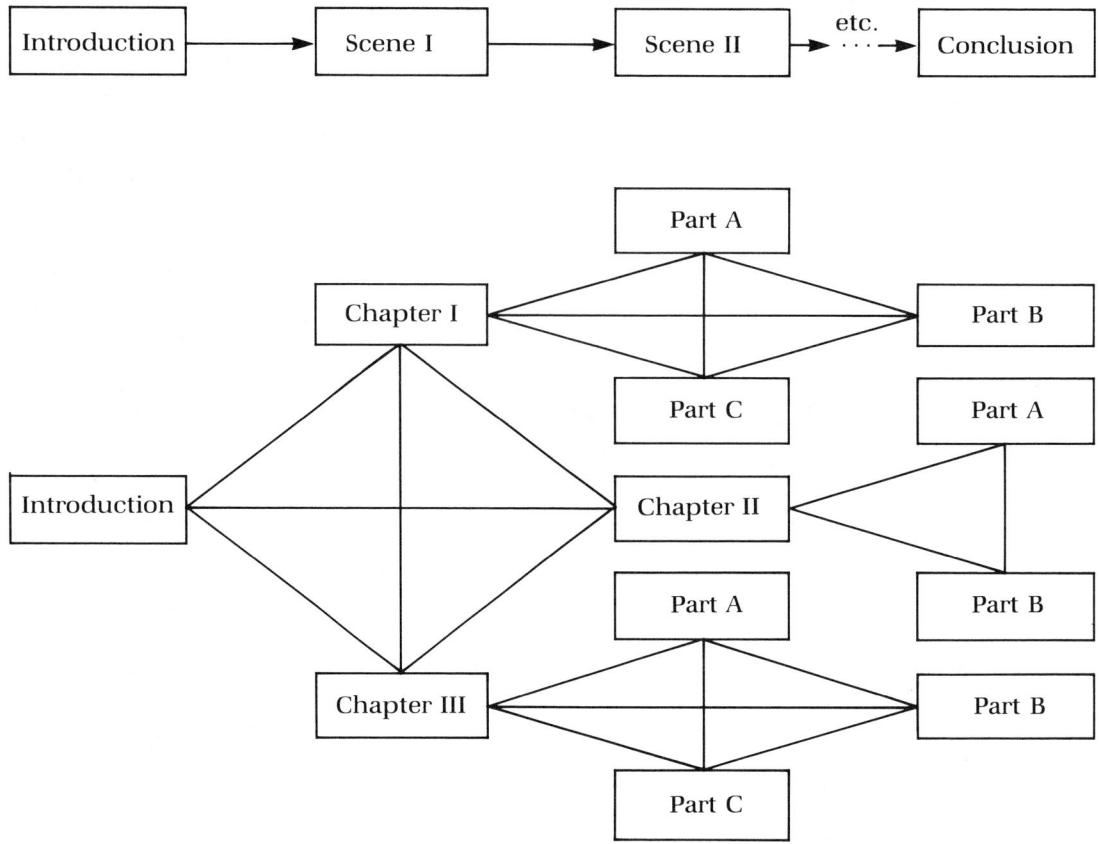

**Figure 1.** Linear model: The introduction is at the beginning, the conclusion is at the end, and all the pieces in between follow in a fixed sequence.
Interactive model: The introduction is at the beginning, but after that the user determines the sequence by selecting Chapter I, II, or III. Within each chapter the user can further select various parts. Although the sequence isn't fixed, the relationships among the parts are. For instance, this model does not allow the user to go directly from Chapter I, Part B to Chapter III, Part C.

active programs are segmented. Although the beginning may remain constant, there is no way to know what part will come second or third. And although the program may contain a set ending, there is no way to insure that everyone will get to it. This segmentation and reordering mean that transitions need to be more generic. A nice dissolve between parts one and two won't be effective if the user goes from part one to part four. And every time the user has to make a choice the flow of the program is interrupted. These continual interruptions have the potential to make the program seem choppy, especially if a program designed and produced for linear videotape or film is simply being reformatted. It is therefore important to build continuity into the program in new ways.

In order to create this continuity, the writer must have a clear vision of how the various segments relate to one another and to the program as a whole. In a linear program this relationship among parts is sequential and often causal in nature: B occurs because A has just occurred. The relationships in a linear program are also static. B will always follow A. In a sense, the structure of the program could be considered two-dimensional. An interactive program, on the other hand, is three-dimensional [figure 1]. This can make the structure somewhat more difficult to conceptualize and maintain, but it also makes the maintenance of the conceptual framework crucial. The pieces no longer fit into a train with one following another; instead they form a network where one piece can be connected to, and can influence, several pieces at one time. Although the structure that the user creates is dynamic in the sense that different people will put the pieces together in different ways, the overall structure with its multiple potential interactions remains the same and it is this structure that the writer must not lose sight of. To some extent, this structure should also be made clear to the user so that he or she knows what kind of choices will be available and what kind of movement is possible. Without a general conceptual understanding of the structure, it is possible for a user to feel lost in a maze of decisions. With the framework, users can bring their own sense of continuity to the program.

Continuity can also be fostered through repetition with the overall structure. In figure 1, the program is divided into chapters and each chapter is further divided into parts. In order to provide a sense of unity, each chapter introduction and menu should have a similar format. This enables the user to easily recognize chapter beginnings. Methods for moving around within and among chapters and their subparts should remain consistent to provide for fluidity of use. Too much repetition, however, can lead to monotony. If there are many segments and they are all structurally the same, the user will soon become bored. Within the structure it is necessary to provide some variety in order to differentiate one segment from the next. As with so many things in this world, just because some degree of sameness is good doesn't mean that more will be better. There should be enough similarity to keep the program from seeming disjointed, but not so much that it becomes monotonous.

In designing a videodisc the first thing that must be done is to develop the general program goals and a broad, overall structure to achieve these goals. Then the level of interactivity of the disc must be determined. A level 1 videodisc has only a minimum of interactivity and utilizes primarily chapter stops and picture stops to provide user input. A level 2 videodisc uses program dumps encoded onto the disc to control the program through the player's resident microprocessor. The user interacts through the keypad. Finally, a level 3 videodisc interfaces an external computer with the player for maximum interactivity. The user's input can be through the computer keyboard, or it can utilize a touchscreen, light pen, joy stick, or any other computer input device. In some cases, the level of interactivity may be predetermined by the hardware available; in other cases the nature of the program or the intended audience may make the decision. But however the decision is made, the hardware configuration and level of interactivity incumbent with it will influence all further development.

Once the level of interactivity is known, the structural components of the disc should be identified. Will it have chapters? Will there be multiple levels and how will the user's level be determined? Will there be testing? If so, will each chapter have its own test, or will there be one test for the entire program? What kind of feedback and/or remediation will the user receive? And when? How will the chapters be subdivided? And how much opportunity will the user have to control the presentation?

Once this structure has been determined it is a good idea to map it with some sort of a flowchart to show visually what the major components are and how they interrelate with one another. At this stage the flowchart must necessarily be fairly general without much detail, but it will help to keep the overall structure in mind as details evolve. It also helps to point out any major conceptual errors, such as branches that dead-end, leaving the user with no place to go. The flowchart also presents a holistic view of the program, which will be necessary as work begins to focus on the individual parts. This flowchart, and the more detailed ones that will follow as the process continues, are invaluable not only to the writer but also to others involved in the production as a starting place for communication.

At this point the structure can be refined and the various components fleshed out more. Forms that will provide the program with some degree of continuity should be developed, while ways to provide variety among segments must also be considered. Details of interaction and conventions for moving around within the program must be decided. This would include whether or not the user will always select from menus or whether other forms of input will be accepted. It would also include whether or not the user can exit from the program at any point and, if so, what sort of input will accomplish this. Will the user only be able to make decisions at specific, predetermined points, such as menus or test questions; or will he or she also be able to jump around at will through the use of function keys or other established procedures?

If the videodisc is to be a level 3 system with an external computer interface, it is possible to have some information computer generated and therefore the role of the computer must be determined. Is it only to provide control, or will it also present textual and/or graphics content? How will the two media be interwoven to create an integrated whole? These are the types of questions that must be answered as the actual scripting begins. And as they are answered the flowchart should be revised to reflect the depth and detail of the developing structure. It would be convenient in many ways if a single flowchart could be developed and then a final script written from it, but the reality of the situation is that there is usually an interactive process wherein the developing script influences changes in the flowchart; and subsequent changes in the flowchart mandate changes in the script. Together, the flowchart and the script create a dynamic

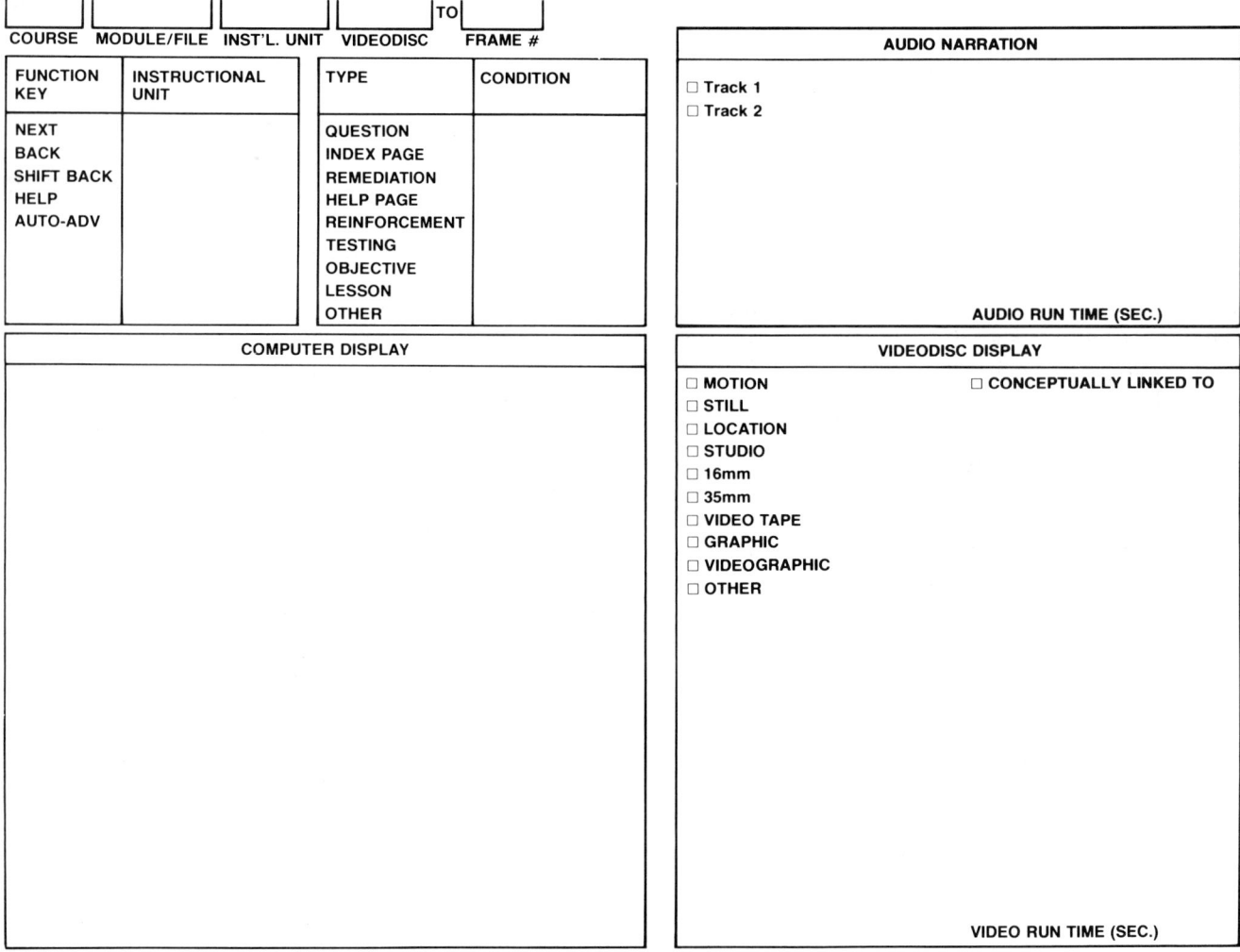

Figure 2.

former tries to encompass the whole, while the latter focuses on the component parts.

The actual scripting of these many pieces also requires some adjustments. The traditional two-column script with one side for audio and the other for video is no longer adequate. The videodisc involves so many components that must be accounted for that new forms have been developed [figure 2].

Although there is no single, standardized videodisc script form, most interactive videodisc scripts resemble the one in figure 2. These boxes enable the writer to clearly specify the nature of the video—whether it is motion or still frame; the audio track(s) being used—audio 1, audio 2, or stereo; the textual material that may be electronically generated on the videodisc or computer generated; the branching instructions for both the producer and the programmer—what preceded this segment and what will follow it; and finally, page or segment numbers so that it is feasible to keep track of the reams of paper that comprise an interactive videodisc script. Each page should represent either a single still frame or a single motion segment.

In general, user input occurs at the end of a page, although there are times when progress from one segment to another is automatic, such as when a motion sequence culminates in a menu from which the user must select the next option. The motion sequence would occupy one page, or however many pages are required to describe it, while the still frame menu would occupy a page of its own. This script is intended for both the producer and the programmer and should include information relevant to both; but it is probably not the script that the talent will receive. A traditional, two-column acting script is generally created from the interactive script, since much of the information on the initial script is potentially confusing and irrelevant for the actors.

Videodisc scripting is still new and very few rules have evolved. The way one writer deals with a problem may vary from the solutions discovered by another. But one rule has surfaced: a videodisc script must be explicit. Nothing should be assumed because there are too many interacting factors and too few mutually understood conventions. The only admonition for a writer of interactive videodiscs is to remember the whole while working on the parts, and the parts while working on the whole.

Casey Garhart began writing scripts for interactive videodiscs in 1979 for the Media Development Project for the Hearing Impaired (MDPHI) at the University of Nebraska-Lincoln. In association with the Nebraska Videodisc Group, she developed ten projects for MDPHI, including the first computer-controlled "sleuth game" for deaf children, "Think It Through"; two discs on finger spelling (Casey is fluent in American Sign Language); and other projects. Casey also wrote the manual for 3M's "Producing Interactive Videodiscs." Other writing credits include interactive videodiscs for World Book International and a major project for IBM Personal Computers, Boca Raton, Florida. Currently finishing her Ph.D. in Instructional Design at Penn State, Casey is, among other things, teaching a Computer Science course in Pascal.

## PRODUCTION

# Producing a Videodisc: An Old Filmmaker's Perspective

*Chad Worcester*

I remember well that day in September in upstate New York when I shot my first roll of 16 mm film with the used Bolex I had just bought on Long Island for $160. With visions of Truffaut and Goddard dancing in my head, I filmed, from every conceivable angle, my roommate and his girl friend romping around on the third green of the college golf course, playfully throwing leaves. Thus, my first (and quite forgettable) cinematic statement. Humble as it may have been, it was the start of my filmmaking career.

Fifteen years and many documentaries and dramatic films later I find myself a producer of videodiscs with the Nebraska Videodisc Design/Production Group. Why? What's a filmmaker doing in the world of videotape, nonlinearity and microcomputers?

I made the transition willingly and enthusiastically. The growing interactive videodisc industry is very promising and offers a host of opportunities. Also, I feel industrial training and marketing (two popular applications of interactive videodiscs) can use a heavy infusion of people with my kind of background, experience, and attitude. After producing numerous discs I have learned many things to do and not to do to ensure an effective and reliable outcome. For those producers actively making discs now, and those contemplating it, I offer the following personal suggestions, tips and opinions.

*Cardinal Rule #1 of Interactive Videodisc Production.* All producers should make every effort to get involved in the project as early as possible. If the project director knows who the producer is going to be early on, then that person should be included in all the major developmental steps of disc design, budgeting and scripting. This does not mean that the producer should do the job of the instructional designer, the disc designer, the project director and/or the writer. What it does mean is that the producer can be an invaluable aid in helping to determine what can and can't be done for the money allotted and the time projected.

If you're fortunate enough to be included in the early meetings, the following are some things to be mindful of.

1. Be sure the client is informed, in general, of the quality of production he or she is going to get for the money committed. There are many levels of production and each has its place. Clients should be aware that an interactive videodisc is no weekend project. It normally requires months to produce and the cooperation of many different people. A project of this size will cost a considerable amount—anywhere from $1,500 to $5,000 per finished linear minute of disc time, according to the complexity of interactivity

---

Portions of this article originally appeared as "Interactive Video - A New Video: An Ex-Filmmaker's Perspective" in *Performance and Instruction*, XXII:9 (1983), 14ff.

Documentary techniques can be employed with the videodisc. By using generously the still frame mode and the new compressed audio capability you can take regular documentary footage shot in the classic documentary style with a minimum and mobile crew and distill it down into short realtime segments on the disc. This way you can include more information in the limited space available with optical/laser interactive disc formats. You can also include interactive questions and prompts to help the user better appreciate the content of the documentary.

and production values desired. Even on the low end ($1,500/finished disc minute), for a 30-minute, one-sided disc the cost will be $45,000. A general rule of thumb is that about one-third of the entire budget will be used for line production (the actual shooting and editing of the disc). The other two-thirds will pay for design, writing, programming, coding, evaluation, administration, and overhead. If the client wants a $45,000 disc, there will then be $15,000 allotted for production. The client has to know that $15,000 for 30 minutes will not buy very much more than simple sets, local talent, and basic premastering.

2. It's nice to be around when the designer, the project director, and the writer are brainstorming about how they plan to inject all sorts of Hollywood production values into their design and script. Your role should not be that of the spoiler, but rather a steady and rational voice amid the strong winds of creativity. And of course at the appropriate time you may add to the wind—if, that is, it's blowing in the right direction.

Unfortunately, most designers and writers of interactive videodiscs today have little practical production experience. More often than not their barometer for video production standards is prime time television. This is not bad in itself because prime time television exhibits some of the highest video production values you'll ever see, but also at a very high price.

The producer must always keep the projected budget (if it hasn't been clearly defined yet) and the time line clearly in mind. Designers and writers are generally strong willed people and are not likely to change their minds easily. Compromise, then, is the only solution. Some areas may have very high production values with high-priced talent, out-of-town locations, and expensive sets. To balance these out, other areas of the disc will have to be quite basic.

3. The producer must impress upon the writer the importance of writing the disc to time. That is, making sure that the linear time for both audio channels 1 and 2 do not exceed the linear video/audio limits of the disc (generally 30 minutes per side). I would

# Producing a Videodisc: An Old Filmmaker's Perspective

When you are shooting single camera video take full advantage of the instant video image. Make sure the director has a monitor on the set with what the camera is seeing. It's also a good idea to set up some kind of communication system so the director can talk to the video engineer (who is generally in another room) and the sound mixer. This can be done with simple public address equipment, or if your budget warrants it personal battery powered headsets.

recommend that a writer also leave a 2-minute cushion for each 30-minute side. In other words, if a writer is writing a script for a 30-minute disc, the script should time out (by reading and timing all audio 1 and 2 script pages, as well as calculating time for stillframes and video expositions) to 28 minutes.

4. Determine early if the client wants any named talent and if so how much of the talent's normal entourage will be included in the deal. More than once my talent budget has been prematurely spent when, after the contracts have been signed and pre-production started, the client demands that a name talent be signed.

5. And speaking of contracts, it's the project director's responsibility, first and foremost, to get as much as possible into the contract regarding time lines, budgets, talent, delivery dates, delays, rewriting, reediting, remastering, and purchase order adjustments. Producers should definitely participate in the contract development process!

6. Finally, if the project director neglects to get the producer's input on the formulation of the budget, then he or she can surely expect problems later on. Only a person who knows what it costs to hire production personnel, to build sets, to put together a crew, to edit a sequence, etc., can estimate accurately what it will cost to produce a particular script treatment. Budgets should not be thrown together in a frenzy to meet a bid deadline. They should be carefully constructed from lengthy lists of all possible areas where money can be spent. The most embarrassing part of producing an interactive videodisc is to promise a client a specific product and then, after completing two-thirds of the production, be forced to ask the client for additional funds. This can be avoided if a good contract is written and an accurate budget calculated.

*Cardinal Rule #2 of Interactive Videodisc Production.* The early production decisions you make can be the most important ones for the entire project.

Before you begin hiring lots of people and spending lots of money you will be expected to make some important determinations, such as the following. What kind of format will be used for the action sequences (film or videotape)? What kind of camera(s) will you use? What kind of recorder will you use? What location(s) will you use? How elaborate will your sound recording be?

Most interactive videodiscs are shot entirely with videotape. But if you have sequences where the user will be expected to analyze the motion of an object, animal, or person, for example, you should shoot those segments on film (16 mm, color negative) at 30 frames per second (fps) and transfer them to videotape at 30 fps. This will insure your videodisc user of rock-steady still frames anywhere in the motion analysis sequence. You could do the same thing with videotape, but it would require a special video processing technique to eliminate the inherent "flicker" of video. A video frame (the equivalent of one image of film) is made up of two video fields, which interlace continuously on the television screen at 1/60th of a second per field. When you create a still frame of two fields together, they constitute a bit of action 1/60th of a second apart. Any motion that occurred in that bit of action fast enough to be noticed in 1/60th of a second will result in a repeating flicker. It is recommended, therefore, that you always use videotape for discs and use film only when you need it for a specific analytical purpose. Besides, all your critical editing will be done on tape, so why not start with it?

What kind of videotape format should you use: 2", 1", 3/4", or 1/2"? At present, 2" recording is available for studio sessions only: 2" will give you excellent results, as it has done for the broadcast industry for decades. It is, however, the format of the past. Many postproduction houses have only one or two 2" machines, which they use generally for dubbing commercials and public service announcements; they aren't equipped to edit in 2".

The 1" format is the accepted professional standard for most videotape productions and postproduction editing. Any reputable 1" recorder will give you very high quality video quite suitable for any interactive videodisc application. There are two main standards of 1" videotape: "C" format and "B" format. In the United States the "C" format is by far the most prevalent. In Europe and elsewhere in the world there is wide acceptance of the "B" format. Both produce the highest quality video

The differences lie primarily in the construction of the video heads ("C" format has one head, while "B" format has two) and also in the threading pattern. It's a no-win situation to find yourself between "C" format engineers and "B" format engineers debating about which is the superior system. Because the "B" format has two heads it will provide a denser video picture but it also puts more wear on the tape.

There are very few postproduction houses in the United States that have "B" format editing suites; so if you are planning to record on "B" format be sure to plan for dubbing all your field and studio recordings to "C" before you go into premastering.

Very good quality video can be produced with 3/4" videotape if you take care to use a good camera (with a good lens!) and a good professional, broadcast grade recorder. Because most interactive videodiscs are intended for discrete learning systems, kiosks, or games, there is no pressing need to conform to FCC regulations regarding blanking limits for broadcast. You are more likely to run into those kinds of problems when you use 3/4" equipment. If you're working with an industrial in-house A/V unit that has been using 3/4" equipment and turning out a good-looking product, there's no reason why you can't stay with the same equipment to make your videodiscs. But problems can arise when you get into editing your premaster, and your project requires a number of still frame edits. If you have access to a reliable computer interface to professional, broadcast grade 3/4" playback and record VCRs, then you might not encounter any problems.

But proceed with caution, because if any of your field frame relationships are inaccurate or you are making bad color frame edits, the videodisc mastering machines will not accept them. Then one of two things will happen: the mastering house will reject your tape, or you'll get back a check disc with jittery still frames and/or jumbled frame numbers. The best bet if you want to originate on 3/4" is to include a dub up to 1" for the online premaster editing.

One-half inch is supposed to be the broadcast standard of the near future, at least for electronic news gathering (ENG) purposes. This might be true, but to convert to 1/2" will be expensive. Moreover, there will always be some kind of dubbing to a final edit format because the 1/2" format is not NTSC compatible. My feeling is that unless someone plops a state-of-the-art 1/2" system in my lap, there would be no need for me to use it for standard videodisc production: 1" and 3/4" will do just fine.

Just as for any other type of electronic gear, you can choose from a wide variety of videotape cameras. One consideration should always remain in the videodisc producer's mind when he or she is selecting camera equipment: videodiscs are generally not broadcast material. Essentially what this means is, don't let some fast-talking rental house salesman sell you on broadcast-quality equipment when a cheaper camera will give you all the resolution and video signal you'll need. All your major camera/recorder manufacturers offer relatively low-cost ($10,000-$15,000) systems that produce perfectly acceptable pictures. These cameras/recorders are also readily available for rental.

Producers should be cognizant of the standards that constitute good video camera/recorder performance: lines of resolution; signal to noise; auto-registration; and auto-centering.

And if you want a tip for producing good-looking video, follow this rule: spend your money on a good lens and a good lighting videographer. Someone who has a good eye and knows how to light for it can give a low-end camera/recorder an excellent look.

The selection of locations—be they in a sound studio or in a residence or building—can be a very significant part of your early production tasks. If you pick these places too quickly or without adequate site surveys, it's a certainty that one or more of them will prove disastrous in the middle of shooting. I feel the best location is the one where you completely "own" the period of time and where you've made very clear to the landlord(s) what you plan to do. There is nothing worse during a production than to be at the mercy of someone's goodwill when you've used up your welcome, you're five pages behind, and you know the only way to finish is to stay very late. Even if someone offers you a location for free, be sure to pay them something for it. And be very careful about using your friends' places to shoot in. If they are not aware of what a production crew can do to a home, then spare them the agony and don't ask. If you want to shoot in your own place, you'd be advised to time it just before a major redecorating. In large cities you can usually rent homes and other locations that are used exclusively for filming and taping. This might be the best answer, because you don't have to worry about marking up the walls or putting staples into the woodwork.

Most of the locations for the videodiscs I've worked on have been outside of sound studios. Occasionally, however, I've needed a studio with a full three-camera crew. If you have a series of discs to do with a central host and a set you use over and over again, it's advisable to do these scenes in a studio. Working in a good one with an experienced crew can be a very enjoyable experience. If you are well prepared and you don't come up with a lot of new things the crew wasn't briefed on beforehand, you can tape many script pages in a day.

Following is a brief checklist of things to consider when selecting locations.

1. Be sure you have full control of the space you'll be using and will have access to it at any time.

2. Make sure your electrician or videographer/cinematographer has determined if there is enough electricity for the lights, and can tie in and distribute it without problems.

3. Make sure the location of the sun will not create any problems, with harsh light coming in through windows and/or doors. If that might be the case, be sure to have enough neutral density gel to reduce the brightness.

4. Think about how you are going to handle the waste that the production crew will produce. You may want to hire a garbage collection service for the time you're there.

5. Be very sure that you have phone service and that you are able to make and receive calls during shooting.

6. Be sure that access to the location is adequate, especially if you will be using any sizable equipment like a heavy dolly, large set pieces, or large lights. If you are not on the ground floor it's important to find out where the elevators are (freight elevators are preferable).

7. If you will be in a congested area, be sure to provide parking for the production vehicles and, if possible, for the crew and talent.

8. Determine with your sound person if there will be any serious problems with noise or sound.

If you will be shooting in and around large buildings, you should take the time to see if there are any strong RF (radio frequencies) affecting the location. If so and if you are using tape, you will find that these radio signals will ruin your recordings. I have a friend who had this happen to him. He lost a half-day finding a new location, and of course all his preparation for the scrapped location was worthless.

An unfortunate development with a lot of low-budget industrial video projects is that the sound recording often gets cursory attention. Perhaps this happens because it is true that in a number of situations if the sound is not acceptable it can be "fixed in the mix," and producers will rely on this if the production is behind schedule. It's my opinion, however, that low-quality sound will *always* sound low-quality, and "fixing it in the mix" is only a partial solution.

Because the premastering demands of a videodisc with all its still frames and extra sound tracks are often more time consuming than those for a linear project of the same length, one of the last things you want to deal with in the middle of premastering is trying to improve poorly recorded sound.

I would suggest that there be a sound crew of two, and also that as much of the sound as possible be recorded with an omnidirectional microphone in the hands of an experienced boom operator. If you have the budget to afford radio mikes, then all the better. Radio mikes can be wonderful, but they can also pick up police radios and ham operators. Also make sure your sound people bring along lots of sound blankets and use them liberally. A lot of videodiscs are shot in very "boomy" offices. Some well-placed sound blankets can warm up your actors' voices considerably.

Although it will seem like a major "extra" cost when you contract with two sound people as opposed to one, it will definitely pay off. I have spent, on more than one occasion, as much or more money to fix up bad sound in postproduction than it would have cost me to have a second experienced sound person on the set.

Location selection can sometimes seem insignificant when so many other problems beset the producer early on. But don't let this be the case. Give special attention to this most important task.

*Cardinal Rule #3 of Interactive Videodisc Production.* Think interactively, not linearly. A common mistake for traditionally trained (linear) producers and directors when working with videodiscs for the first time is that in the midst of production they tend to handle actors, camera moves, and lens moves (zooming and how much) as though they were shooting a dramatic film. Although the differences between videodisc productions and linear productions are subtle, if they are overlooked or ignored, the videodisc can appear disjointed and choppy when it is completed.

For the rest of this personal perspective on producing interactive videodiscs, I will list a number of things I keep in mind while shooting and premastering that help me to keep thinking interactively.

1. *Actors:* Always use professional actors for any part that requires the recitation of more than one or two lines. For very small parts and extras I've found I can use nonprofessional friends, acquaintances, and/or the employees at the place I am working in.

As for the professional actors, make sure they have done film or videotape acting before. If they've had a lot of theater experience, but none or very little film/tape experience, use them only if you can't find anybody else to do the part. Find out if they can read teleprompters without glasses (if they will be required to act without glasses). Incidentally, camera people hate to work with actors with glasses.

If an actor says that he or she can memorize a lot of lines quickly, don't be too impressed. Videodiscs, especially those for the corporate marketplace, tend to have a lot of technical, difficult-to-remember copy. Without a doubt the best way to make sure that the script is read exactly right is to make generous use of a teleprompter (more on teleprompters later).

Always be on the lookout for a wide variety of dependable actors with different ethnic backgrounds. Most corporate clients want representatives of different ethnic groups prominent throughout the disc. During the preliminary brainstorming script sessions, try to encourage the writer and instructional designer not to consider kids and animals seriously. If there's one sure way to prolong the length of your production time and shorten the fuse of your director, it's to make a child or an animal one of the principal actors.

2. *Directing actors:* A videodisc is forever and, because of the nature of the medium, your eventual audience can inspect literally every frame of the production, if they want to. Consequently, don't let your actors get sloppy. Be very mindful of continuity, pacing, and characterization. If you are able to afford an experienced continuity person, it's money well spent to hire one. Go through all the motion sequences very carefully before production and determine how they will be accessed once the disc is done.

I have gotten into the habit of putting fade-ins and fade-outs at the beginning and end of most of them. I also direct the actors to put in slight pauses at the beginning of the opening shot of a sequence and at the end of the last shot of the sequence. This allows a programmer the option of starting or ending with a fade-in or fade-out, or accessing the sequence directly by searching to just after the fade-in or just before the fade-out.

If you adopt a consistent approach to the beginnings and endings of all the short motion sequences that most interactive videodiscs are made up of, you'll find that even though the user is watching one short segment after another, there is an identifiable style to the disc that makes it more cohesive and enjoyable. Remember, there's nothing that says a videodisc has to be boring. Although most videodiscs are a far cry from dramatic productions, the casting, acting and attention to detail can be just as professional and challenging as a good linear production.

3. *Teleprompters:* Most interactive videodiscs take good advantage of their ability to store enormous amounts of information. A good deal of that information is usually communicated by someone talking about it on camera. (A long discussion could ensue

now on how to handle the problem of the "talking head"—but that's another chapter at least!) Because so much of the information is generally technical in nature, I find it dangerous and naive to expect an actor to memorize thirty or forty pages—or even five for that matter—and recite them accurately on camera. I have solved my problems with the regular use of teleprompters—those marvelous machines that use a partially silvered mirror in front of the lens to present the text of the script to the actor as he or she looks into the lens.

Don't use cue cards or teleprompters that require the actor to look off-screen. These methods are suitable if you've got a dramatic scene, but they are sloppy and amateurish if your talent is expected to be talking directly to the camera.

If you will be using a sound studio with videotape, chances are the studio cameras already have teleprompters mounted on them.

Be sure you check ahead to make sure what kind of paper the teleprompter requires and what are the proper margins for your copy. The other kind of teleprompters are the location type that are made to mount on most popular video and film cameras. These are readily available from major rental houses. They generally require you to handwrite the copy on special paper. But if you have access to a word processor with a font compose software option plus a matrix printer with a pressure roll feed, then you can compose large type and print it onto the special paper with the matrix printer. Secretaries absolutely hate to hand write location teleprompter copy. Be sure to have on the set Scotchtape, white correction fluid, black pens, and extra rolls of prompter paper. Videodiscs have a tendency to change constantly, which means the prompter copy changes too.

4. *Shooting slides for still frames:* A common mistake made by a linear producer new to videodiscs is to drastically underestimate the time and money necessary to produce the still frames required by the videodisc script. If a two-sided disc (30 minutes per side) requires in the neighborhood of 1,200 still frames (all of which are original production), it's recommended to budget approximately the same amount of time to shoot the slides as it will take to shoot the motion. The cost of the slide production will be about one-third of the motion sequence expense. Use a competent photographer who preferably has lots of experience shooting technical slides.

Most technical slides are shot against a seamless background. Survey the color of the equipment you'll be shooting. Pick a color for the background that is different. Off-white may seem like the perfect color but you'd be surprised how many machines, gadgets and gizmos are colored off-white. Light grey or light blue are usually good choices. Shoot all of your pictures with the exact same color of seamless background and, if at all possible, with the same photographer, same lighting, same camera, same lenses, and same processing house.

When shooting inside of machines (as most technical videodiscs eventually require) be very careful of the contrast ranges you'll encounter. It is very easy to overlook providing enough light for the nooks and crannies of the machine. If these areas are not lit correctly they'll end up as dark blobs on the tv screen. One way that works well for lighting the deep recesses of machines is to use a circular, lens-mounted flash, the kind that dentists regularly use for photographing inside mouths.

Be sure to bracket your exposures at least one over and one under, and try to get all your slides processed before you pack up the lights and equipment and release the photographer. Your photographer should have a good macro lens capable of shooting close enough to fill the screen with one-half of a postage stamp.

And for heaven's sake don't forget tv cutoff. Be sure your photographer has a ground glass in his or her camera with tv cutoff inscribed. For relevant technical information, make sure it's all within the "safe title" tv cutoff parameters. If your photographer doesn't know what this means, find another one.

5. *Laying down still frames:* If you are producing a disc designed for level 3 (computer-controlled) interactivity, you are advised to edit in three frames of video for every still frame, instead of one. When moving from one still frame to the next on the disc, the computer can tell the disc player to step ahead three times in succession, a move virtually imperceptible to the user. The advantage of having three frames instead of one is to protect against a "flicker frame" resulting from a bad edit. If your premaster isn't perfect from the standpoint of the stability of its video and colorframe edits, a mastering machine might misread one or more of them and produce a still frame where the two fields comprising it are slightly displaced, producing a frame that jitters or jumps. If this were the case, the computer could search to the middle of the three frames, which is almost always steady, unaffected by the bad edit(s). And, with the new generations of videodisc players on the market now, a new feature is the instant jump within a few hundred frames without the traditional screen blank and/or roll. If bad frames are on the disc, as long as there is at least one good one, the user will never

know the difference. However, if you are preparing a disc for level 1 or level 2 interactivity, you will have to use single frame still frames at any time. Before sending it to the mastering house, be very careful to give your premaster tape as much scrutiny as possible to guard against disc jitter frames.

6. *Video Playback:* As a producer trained in the techniques of film, I never had a chance to see what was happening in the camera. I always trusted my judgment and my cinematographer's word. Now with video, however, I can see everything, as it will be, right there during production. This can be both a great boon and a great time-waster.

I would recommend that the privilege of the video playback be used mainly for those takes that have a special need to be viewed (continuity, script content, etc.) and for those takes whose production values are so high that you want to make very sure that everything is correct before you go on to another setup. To playback practically every take, to reassure yourself that it's usable, will chew up lots of time and lots of money. Also, with a good engineer at the recorder, you should be able to reuse the tape after busted or bad takes. Ask him or her to save the slates when you recue. If you are able to arrange for the equipment, it's recommended that there be a monitor on the set of what the camera is shooting. Instead of watching the actor(s) on the set (as you would if shooting film), be sure to watch the monitor. You can make a lot of important decisions quickly that way. But don't be too harsh on your videographers if they don't have the identical framing eyes that you do. Take them aside and talk to them privately about what you want. They are artists too and they don't like to be reprimanded in front of the entire crew. In addition, their talents should be used completely. It is also possible that they are seeing the scene in a better way than you are.

7. *Note-taking:* Videodiscs will generally use up a lot of tape because most discs are comprised of many short scenes and each scene requires four or five takes. It's extremely important that a person be assigned to maintain a complete shooting log of all of the material you're taping. These notes should include scene and take number, day, location, tape roll, description of the scene, script page number, time code, and assessment of the take. If you are able to set up the time code to reflect which number of tape you're using (generally in the hour portion of the time code) this can sometimes help speed up premastering. This copious note-taking goes for all sound effects, as well as for music and narration voice-overs. Be sure to use frequent slates identifying the script page and take number, and make sure the slate voice is up front and clear. This will help minimize search time for the audio editor. Another technique I have adopted is to keep a note pad handy at all times during production and premastering. When ideas, problems, questions, tasks, or reminders crop up I write them down on the pad. Over the course of the production I am constantly shuffling through these notes. When I have done what the note specifies, I throw it away. Ideally, at the end of the disc there are no more notes left!

8. *Special effects:* The combination of the character generator and the digital video effects (DVE) generator has made videodisc premastering both a joy and a curse. It's a joy because the character generator allows you a wide range of freedom to design and produce a great variety of character fonts and graphics for still frames and supers, and the DVE allows you the freedom to shrink, stretch, flip, and flop any or all of the video on the screen. If you've got access to a sufficiently powerful character generator and DVE, you can produce the same video gymnastics you regularly see on network news and sports productions.

But it's also a curse because too many writers and designers know its capabilities and they gleefully stuff the script with all sorts of fancy effects. I have nothing against fancy effects, because once they are completed and included in the disc the "oohs and aahs" they were designed to elicit invariably occur. The problem with them is that the writer and designer normally have only a vague idea of how much time a particular effect will take and how much it will cost. It's not unusual for a few still frames on a disc to take two or three hours to prepare. A word to the wise is to use the character generator and DVE generously, but be aware that your enthusiasm to "dazzle 'em with brilliance" could result in a production going over its budget and past its deadline.

*Cardinal Rule #4 of Interactive Videodisc Production*
Producing an interactive videodisc is a complete team effort. The normal videodisc production team is made up of five main characters: the client, the project director, the instructional designer (and writer), the producer, and the programmer. Each element of the team is equal to the others and the successful interactive videodisc depends on all the elements performing completely and dependably. The producer works very closely with the writer at first, later on with the programmer, and throughout the production with the project director. It's very important then that the producer learn as much as he or she can about script writing, instructional design, computers, and administration. A really good team effort is the result of the individuals doing the best they can with

# THE VIDEODISC PRODUCTION WORKSHEET

## Chad Worcester and Jim Smeloff

The Videodisc Production Worksheet is a guideline by which an interactive videodisc project director can estimate the cost of a project.

To use the worksheet you must first determine how many hours of instruction/use you plan to provide the student/customer. Once that has been established you can approximate the number of stills (both videodisc and computer generated) needed by subtracting the amount of realtime video you plan to produce (one-sided disc, two-sided disc, etc.) from the hours of instruction/use you expect to provide. Be sure to account for any realtime video you know will be accessed more than once. Take the remaining time and divide by 15 seconds. This will give you the number of still frames you'll need to produce.

In 1976, Junius Bennion determined that the average time for the use of a still frame in an interactive video program was 10 seconds. Because of the increasing complexity of interactive video content, we have revised that figure upward. However, the 15 seconds per still frame is not a hardset rule. You can change it according to your application.

Next, determine how many of these still frames will be generated solely from the videodisc, solely from the computer, or a combination of the two (computer overlay). With these calculations completed you will be able to determine how much realtime video you'll need to produce.

Plug these figures into the worksheet. Take notice of the footnote information. Before you find yourself spending wildly, the Videodisc Production Worksheet will give you a general idea of how much your proposed interactive videodisc project will cost to produce, exclusive of delivery system hardware.

It is a good idea to contract for a complete development package. A typical package includes interactive design (based on accepted instructional design models), the selection of a delivery system, the development of a script treatment and the preparation of a budget. These are the important elements of a proposed interactive videodisc project and should be defined before any major production spending begins.

Remember, the Videodisc Production Worksheet is a guide. If you feel any of the dollar figures need to be adjusted according to your production circumstances feel free to do so. Also, these figures should be reviewed every six months and altered, if necessary, according to the fluctuations in the market.

### VIDEODISC PRODUCTION WORKSHEET

1. How many hours of instruction desired for interactive disc(s) __A__ ?
2. A minus realtime motion[1] times 3600 = __B__ .
3. B divided by $15^2$ = C.
4. C = number of still frames (videodisc & computer).
5. C times $4^3$ = D.
6. D = number of discframes needed for still frames.
7. D divided by 1800 = E.
8. E = realtime disc space needed for still frames.[4]
9. Initial contacts/consultation = (1%) of budget #11 through 17.
10. Development Package[5] = 6% of budget #11 through 17.

|     |                          | Level I       | Level II      | Level III     |
|-----|--------------------------|---------------|---------------|---------------|
| 11. | Final Design/Scripting[6] | $200/lm[7]   | $300/lm       | $400/lm       |
| 12. | Production/Premastering[8] Video Still Frames[9] | $16/still | $15/still | $13/still |
|     | Realtime motion          | $2,100/lm[10] | $2,100/lm    | $2,100/lm     |
|     |                          | $1,000/lm[11] | $1,000/lm    | $1,000/lm     |
| 13. | Programming              | -0-           | $5/still      | $15/still (computer/video)[12] |
| 14. | Mastering (aver. price)[13] | $2,000     | $2,000        | $2,000        |
| 15. | Replicating (aver. price)[14] | $18/disc  | $18/disc      | $18/disc      |
| 16. | Project Evaluation (optional) | $150/lm  | $150/lm       | $200/lm       |
| 17. | Packaging Design/Packaging | $5-20/unit  | $5-20/unit    | $5-20/unit    |

18. Contingency = 10% of budget items #11 through 17.
19. Project Administration = 15% of budget items #11 through 18.
20. TOTAL BUDGET

### NOTES

(1) Be sure to include any video segments that will be accessed more than once. Add that time to the real time motion amount. Also, take into account the disc space needed for still frames. See footnote #3.
(2) Each still frame will be viewed by the user for an average of 15 seconds.
(3) Each single frame will be laid down 4 times. This figure will vary according to the delivery system. This is a conservative figure for optical laser systems and a generous figure for capacitance systems.
(4) If you will be using more than 300 disc frames for still frames, you must plan carefully for the space they will take up on the disc.
(5) See the explanation about the development package in the accompanying comments.
(6) Includes time for instructional designer and writer.
(7) "lm" stands for linear minute of disc time. If your disc program is to be 25 minutes long (real time motion and still frames) then use this figure for multiplication.
(8) Includes costs for all production personnel.
(9) The production costs per still include design, photography, character generation and digital video effects (DVE's).
(10) This figure represents a high level of production, generally original production on 1" videotape with full 1" computer-controlled postproduction.
(11) This figure represents acceptable production elements, perhaps original production on 3/4" or professional 1/2" videotape.
(12) Multiply this figure by the total number of still frames (both videodisc and computer generated).
(13) These are average prices. You will need to contact the mastering house of your choice for the specific prices.
(14) See footnote #13.

their specialty, knowing full well how their individual task will fit into and affect the other elements of the videodisc.

Producing interactive videodiscs could conceivably be called a new art form. It's the blending of a number of different disciplines (writing, instructional design, programming, and video production) into a new product that no audio/visual format can presently match. Because the interactive videodisc team members are, by design, weighted towards the academic and the technical side, it's very important for the producer to assume the essential role of making the videodisc lively, fresh, and stimulating. I feel my background in documentary and dramatic filmmaking has been an important influence on the videodiscs I've worked on. As grating as "edutainment" is to my ears, I have to admit that the philosophy behind it is good: an effective interactive videodisc should have excellent production values, good casting, good characterizations, suspense, conflict, resolution, and a healthy dose of "pizzazz."

Of the five main members of the videodisc team, the producer is the one in the best position to achieve these results. The more that interactive videodisc production can attract and benefit by the skills of traditionally trained producers, the better the discs of the future will be. I hope the trend continues.

Chad Worcester earned an MS in film production at Boston University's School of Public Communication and spent two years at the American Film Institute to become a Conservatory Fellow in Cinematography. He has shot and directed educational films, commercials, and industrials. He has been with the Nebraska Videodisc Design/Production Group for almost four years and has worked on the design and production of over twenty videodiscs. He is presently the Group's Senior Producer in charge of all disc production.

# Interchangeability

*Jeff Kemph*

Logic structure—this is what separates the interactive videodisc program from its linear counterpart. Interactive programs all have a logic structure, which is a system of options or branching points that enable the user to affect the nature of the presentation. This program structure, if well-conceived, imparts to the presentation a rhythm and a character that act on the individual user in individual ways.

Like the corridors of the East Wing of the National Gallery of Art, the program structure of an interactive disc program leads the user on an exploratory journey; entertaining and rewarding the user for having chosen a particular journey through the program. Yet, during the course of the journey the logic structure of the program maintains a consistency that keeps the traveler oriented within the maze. Like the corridors of the East Wing, the program structure relates all of the various rooms to the whole.

The advent, with interactive videodisc, of participatory audio/video programs, which are woven together by the user under the guidance of the program's logic structure, has brought about new opportunities and new limitations. On the one hand, the interactive videodisc has presented producers, directors, and designers with the opportunity to explore the new world of interactive narrative. But the potential number of videodisc systems on which the program may be experienced may be limited by the very logic structure in which the interactive narrative is woven.

The effort to make interactive videodisc programs compatible with the greatest number of players is called "interchangeability." These players encode and decode the information in different manners, have different interactive capabilities, label the same function with different names and operate at different levels of interactivity.

Currently there are two videodisc systems in the marketplace, the LaserVision (LV) and the Capacitance Electronic Disc (CED). The LV camp is made up of players marketed by Pioneer, Magnavox, Sylvania, Hitachi, and Sony. The CED format is represented by the new interactive player from RCA.

There is yet a third noncompatible disc format, the video high density (VHD) system. Under the flag of JVC, VHD is currently being demonstrated to interested parties.

Playback systems are also divided, much like the rest of the world, into different intelligence levels. The more intelligent a player system is, the more sophisticated the quality of interaction can be. Level 1 programs rely on codes embedded in the vertical interval of the videodisc picture to define its logic structure. Level 2 programs are given logic structures by the inclusion of digital dumps, small one-kilobyte computer programs located on audio channel 2. Level 3 programs rely on external computers to provide the logic structure from which the different elements of the program are accessed.

In order to design any interactive videodisc program you must be cognizant of the functional capabilities of the player systems. What opportunities for interaction with the user do they offer? How are the interactive functions of the player achieved? How does the player communicate these functions to the user? What labels have the equipment manufacturers used to describe specific player actions? If the answers to these questions are placed in a design matrix, the different player system capabilities can be, to some degree, rationalized. The result will be an interactive program that will entertain, and perhaps inform, the greatest number of potential users at the lowest possible cost. This is the objective of interchangeability.

Let's look at the LV players first. What techniques do they use to allow the user to interact with the program?

The active LV player population currently includes:

| Pioneer | Magnavox | Sony | Sylvania | Hitachi | Philips |
|---|---|---|---|---|---|
| LD-600 | 8000 | LDP-1000 series | VP7200 | VIP8500 | 832 |
| VP-1000 | 8005 | LDP-1000A | | VIP9500 | |
| LD-1100 | VP8010GY | | | | |
| LD-V7000 | VP832 | | | | |
| PR-8210 | | | | | |
| LD-V4000 | | | | | |
| PR-7820 series | | | | | |
| LD-V1000 | | | | | |
| LD-770 | | | | | |

The LV player population is made up of players that read two different specification code groups: the 24-bit Manchester-L or Philips code, and the 40-bit or DVA code.

Both of the code groups, the 24-bit and the 40-bit, are divided into two code subgroups. The Constant Angular Velocity (CAV) codes enable each side of the disc to contain up to 54,000 addressable frames, or 30 minutes of playing time. One-hour disc sides are possible with Constant Linear Velocity (CLV) coding.

## A LITTLE BACKGROUND INFORMATION

The NTSC television picture is made up of 30 video frames per second (fps). Each of the 30 frames of video is in turn composed of two interlaced fields of video, field one and field two. Each field contains 262 1/2 lines of video. Most of these lines contain information that produces the active video, or picture, that is reproduced by the monitor.

Some of the lines, however, do not contain picture information. These lines occur before the active, or picture portion of the video. They contain sync information and a variety of other data signals. These lines are known as the vertical interval. The vertical interval occurs before the start of each field of video. Vertical interval lines 10 to 21 occur at the beginning of field one, while the vertical interval containing disc codes on field two are 273 to 284. Different codes are stored on different lines in the vertical interval.

In interactive program design, the CLV format is too limiting. Many of the features that make the structuring of interactive videodisc programs possible are not available when using the CLV format. The CLV disc contains a variable number of fields, from two fields per track, or rotation, for the innermost part of the disc, to six fields per track at the outer edge of the disc. Because there are a varying number of fields per track, the vertical intervals are not in alignment. This makes some interactive features impossible in the CLV format. The CLV disc can be chapterized and indexed by time, but that's about it. In fact, one of the functions of the CLV code is to lock out variable speed play functions. Very simple logic structures are dictated by the limitations of this format, e.g., banded video music albums.

The second disc format, CAV, is capable of yielding a more complex program structure. The CAV disc contains two fields per track, from the innermost track of the disc to the outermost track. This results in the alignment of vertical intervals and allows for variable speed playback, from fast forward to still-frame. The material on the disc can be randomly accessed by either frame number or chapter number, and the disc can be automatically stopped on a preselected frame. Like the CLV disc codes, there are two versions of this code, the 24-bit and the 40-bit. These codes are stored on different lines in the vertical interval.

## 24-BIT CODES

All laser vision players, with the exception of the Pioneer 7820 series and the LD-600, read some or all of the 24-bit Philips codes. This code grouping includes codes for lead-in, lead-out, picture numbers, chapter numbers, and picture stops. When using the codes to structure a program logic system, we are mainly concerned with those codes that are present during the active video program; i.e., the codes from frame one to the last frame of the program. Three of these codes, picture numbers, chapter numbers, and picture stops, occur during active program video and give the disc designer some tools with which an inter-

**Interchangeability**

active disc structure can be created. Lead-in and lead-out codes occur, respectively, before and after active program video. Consequently, they do not significantly affect the logic structure of the program and will not be discussed here. (Editor's note: see Ron Nugent's article on "Premastering.")

The interactive structure in the level 1 program is built by placing various codes in the vertical interval of the video. These codes—picture numbers, chapter numbers, and picture stops—enable the player to read the videodisc in a variety of modes, such as normal motion (30 frames per second), variable speed motion, or still-frame. Program material can be indexed by chapter or frame number.

**Picture Numbers**

Every frame on a CAV disc has a picture number. These numbers begin with 00001 at the first frame of the active program and increase in a normal count sequence to the end of the active program. The picture numbers are inserted into lines 17 and 18, or into lines 280 and 281, depending on which field is the first field of the videodisc frame. The maximum number of picture numbers in the 24-bit coding scheme is 79,999.

**Chapter Numbers**

Chapter numbers are optional. They can be used to provide easy access to larger program segments. They are unique and occur in a normal count sequence, beginning with a selectable chapter number at the beginning of the active program. On CAV discs, the chapter numbers are inserted into lines 18 and 281 in fields of the active program that do not contain picture numbers. In addition, the first 400 tracks of each chapter are encoded with a stop bit, enabling the player to locate the beginning portion of each chapter while in scan mode. The remainder of the chapter does not contain the stop bit. Chapters can be as short as one frame in duration, although picture numbers are normally used as a branch address for single frame information. And, they of course, can be as long as the length of the active program video. The maximum number of chapters per disc side is 79.

**Picture Stop**

Picture stop codes are also optional. They are, normally, available only in the CAV format. The picture stop codes automatically switch the videodisc player from play or slow motion to the still-frame mode. They make the intermixing of motion and still-frame segments possible. Picture stop codes are inserted into lines 16 and 17 or 279 and 280 on the field immediately following the field that contains the picture number.

## 40-BIT CAV CODES

Only the Pioneer 7820 series players and the LD-600 player read some or all of the 40-bit code. The CAV codes available in this group are picture numbers and first field "white flags". Since the LD-600 reads only white flags, it is too limited to be used for the playback of interactive videodiscs.

**Picture Numbers**

Like 24-bit picture numbers, 40-bit picture numbers are always present on CAV discs. They are unique and in a normal count sequence, beginning at 00001 at the first frame of active program video. The maximum number of picture numbers is 99,999. They are inserted on lines 10 and 273 of the video signal.

**First Field White Flag**

The white flag, a 100IRE (white) signal, defines the first field of a videodisc frame. This signal tells the videodisc player which two fields make up a frame when the player is in a variable speed play mode. It is located on either line 11 or 274 during the active program video.

## THE LEVEL 1 PROGRAM

Level 1 programs rely on picture numbers, chapter numbers, and picture stop codes to define the program's structure. This means that the players playing back a level 1 program will have to be able to read and process these 24-bit codes in order for the disc to play back the various sequences as defined by the program's logic structure.

Level 1 discs are primarily designed for the home market. One of the most well-known level 1 discs is "Murder, Anyone?" In this program the user takes on the role of the private eye, helping Stew Cavanaugh, the disc's on-screen private eye, solve one of sixteen murder mysteries. The disc employs frame and chapter searches and picture stop codes to give the program a logic structure. In that logic structure the participant can view short motion scenes, examine single frame clue files, and accuse the guilty party after the mystery has been solved. Since the program is non-linear, instructions as to valid branching options must be given the user at appropriate times. These instructions are usually presented in a still-frame format. They consist of either chapter or frame searches and/or audio channel selection. In order to play back the information as the program designer intended, the player must be able to read and act on level 1 codes. Otherwise, the program will play linearly, turning the nonlinear program design into a stream of meaningless scenes.

## THE LEVEL 2 PROGRAM

Videodisc history is rich in examples of videodisc programs distributed to a controlled network of players. The automobile companies, as well as other companies in the point-of-sale arena, have adopted this form of videodisc network. The General Motors videodisc network is designed around the capabilities of the Pioneer 7820. Ford's network is designed around the Sony LDP-1000.

Both the GM and Ford videodisc networks are built around level 2 programs. Level 2 programs have the interactive logic structure of the program contained in one or more program dumps, which is a small computer program stored on audio channel 2 of the videodisc. When the player spins up to speed, it encounters the first program dump. The video and audio portion of the disc are both blanked, or not displayed, and that portion of the control program is loaded. The player then plays back the disc material, consisting of variable length segments of audio and video material, according to the design of the control program and the inputs made by the user.

The segments of the interactive level 2 program are defined by picture numbers; the start of a motion segment, the end of a motion segment, the still-frame visual, and the start and end of an audio segment. These picture numbers, in conjunction with the various programmable commands, e.g., Audio Channel 1/2 On/Off, Search, Autostop, Decrement Register, Timed Wait, Input, Branch, etc., form the logic structure of the video program.

The programs of the GM Disc Network are designed to be played back only on the 7820 player. The program's logic structure is defined by 40-bit picture numbers and the computer program, which is written for an F8 microprocessor.

Similarly, the Ford Disc Network programs are designed to play back only on the Sony LDP-1000 player. The logic structure of the Ford discs is defined by the 24-bit picture numbers and the computer program, which is written for a Z-80 microprocessor.

Although the Ford and GM Disc programs both use CAV discs, they are not interchangeable, for the computer programming languages that control the programs differ. Admittedly, I'm sure that no one at either Ford or GM is losing any sleep over this incompatibility problem. In fact, most of the level 2 programs produced to date have been "special application" programs, programs intended for playback in a controlled network of players. For these disc applications, interchangeability is not an issue, although the technical issues that determine interchangeability are important.

## LEVEL 3 PROGRAMS

Level 3 programs are similar to level 2 programs in that the two variables that constitute the program logic structure are the picture numbers and the computer programming language. In level 3 programs, however, many more variables exist. These are the high-end disc programs. The amount of information accessible in a single level 3 program can be staggering. Likewise, an endless combination of player, player interface, input and output devices exist. These programs may include in their design such features as voice recognition, branching, graphic overlay devices, or movement recognition devices. The list is endless.

These programs are typically produced for very specialized applications, such as education or training. Here, the programs are designed around the capabilities of a single player system. And, like the Ford and GM networks, the users of sophisticated level 3 systems care little if the logic system on which the program is constructed is transportable to another playback system. "If you want to use our disc, use our playback system."

## CED

The RCA SJT 400 player introduces interactive programming to the CED player line. The CED technology represents a radical departure from the older LV disc format. In marketing this new interactive player, RCA has introduced a new set of parameters for design and production. The CED disc contains four times the number of frames per revolution as the LV format, codes which automatically control more player functions, and a new jargon for many of the functions.

The CED disc format contains four video frames, or eight fields, per revolution, and can hold up to one hour of video on a single disc side. This eight-field format presents some special production problems for the videodisc designer, but it also allows for some

# Interchangeability

new program design features not previously possible with the LV format. Still-frame animation, for example, is possible with eight fields per revolution.

But before we get too deeply into a discussion of the interactive CED disc, let's look at the CED's coding structure and parameters.

The CED format contains a Digital Auxilliary Information (DAXI) code on lines 17 and 280 of each video field. This code is a 77-bit code. In the DAXI coding system, each field is given a unique field number. The field numbers begin at 00:00:00 and run to 59:59:59.

## Field Numbers

Remember that the RCA system plays back eight fields per revolution. These eight fields are analogous to the single frame still frame in the LV format. Each field number in the CED format provides an address for a "page." When accessed, the page consists of the field with a predetermined field number desired, plus the seven previous fields. If the page is to contain coherent information, the designer needs to ensure that all eight fields contain essentially identical information. The "frame search" of the LV disc becomes a "page seek" in the CED system.

## Band Numbers

The structure of the interactive program can also be divided into larger program elements, called *bands*. Bands are equivalent to LV chapters in the way that they function. There are 62 bands available for program use. Bands 0 and 63 are reserved for the player; they are similar in function to the lead-in and lead-out codes of the LV format. Bands must be at least 10 seconds in length. The DAXI code for bands occurs only on the first field of the tv frame, fields 0, 2, 4 and 6.

## Flags

The second field of each television frame, fields 1, 3, 5, and 7, is used for DAXI codes, which control the stop flag and determine whether the player is in a stereo or independent audio display mode.

The stop flag acts like the picture stop code of the LV format. When encountered, the stop flag signals the player to repeat an eight-field sequence. According to the RCA program production specification, the stop flag is inserted on the 12th frame of a still and continues until the last field of the 24th frame. RCA encourages redundancy in the reproduction of still-frame material, 24 frames per still visual. Hardly a still frame.

Flags can also be used to control the audio display status of the player. The DAXI codes can dictate that a user listen to a certain videodisc segment in stereo, with no option for the user to override the command. Or, the DAXI code can let the user select which audio channel to listen to when an independent audio code is encoded.

As mentioned earlier, along with this new format of disc comes a new jargon for the labeling of playback functions. Some terms are time seek, band seek, page seek, previous, next, repeat, memory and program.

## Video High Density (VHD) Discs

The third disc format is the VHD format. This format encodes audio, video, and control information on a grooveless capacitance disc. Each track on the VHD disc contains four fields, and, like the CAV and CED discs, the vertical intervals are in alignment, enabling specialized playback, such as, variable speed playback, chapterization, automatic stop, field numbering, and 2-channel audio. The discs contain up to one hour of information per side.

## Interchangeability—Level of Interaction

Now, let's discuss a few strategies for the creation of an interchangeable program. Designing videodisc programs that are interchangeable between systems offering different levels of interaction is easily possible within the same disc format, LV or CED. Many of the interactive programs being produced today are composed of short sequences strung together by a branching structure. These programs do not take full use of the capabilities offered by the technology, such as random playback of program segments, random beginning points for the program logic, or automation monitoring of user inputs to continually change the program sequence.

Be sure to design for the player system with the lowest level of interactivity first. Since the consumer players search for information at a slower speed than the industrial players do, program segments should be configured for level 1 interaction first. Industrial players, typically used in level 2 and 3 programs, offer much shorter search times. Hence, the program material unique to level 2 and 3 programs can be located at the end of the disc, and rapidly searched for as needed.

The interchangeable disc is typically one in which the same disc contains different versions of the same program. The goal is to make the program logically and functionally identical. To do this, the motion segments that all of the program levels will access are placed on the disc first. Interwoven within the motion segments are the still-frame visuals, which contain program information as well as branching directions for level 1 players. Intermeshed with the level 1 program are special frames that will be used to overlay control information for level 3 programs. Taking advantage of the persistence of vision phenomena, on which all video and film programs are based, it is very easy to hide visuals that relate to level 3 programs operating on systems that provide for graphic overlay. At the end of the motion sequence, insert a single frame on video that is identical to the level 1 frame except for one feature. The level 3 frame must not carry the level 1 instruction that defines the next user action or actions. During the playback of the level 1 program the level 3 frame will not be noticed. For the level 3 user, the control program will branch the player over the frames that contain level 1 directions.

The new industrial players have the capability to jump to another visual without blanking the screen. When the player jumps tracks, the result is really real-time editing. One visual is cut directly against another, with no visible clue that any material was jumped over. The degree to which the disc designer can make use of the jump feature is dependent on the player model. Different players have different jump parameters. Check to determine the jump parameters of the players available. The LD-V1000, for example, can execute jumps forward in user-definable 10-frame increments up to 100 frames. The Philips VP832, on the other hand, offers user definable jumps in frame increments over a much larger window. If the nature of the program material is such that a series of level 3 frames must be included, these should be stored at the end of the level 1 program.

Frames that are unique to level 2 programs should also be stored at the end of the level 1 program. Since the branching instructions for level 2 and level 3 programs differ from the level 1 program, picture stops, branching instructions, and black can all serve to keep the level 1 participant from accessing these frames.

**Mastering and Replication**

Interchangeability, in the final analysis, refers to the ability of discs of the same format to be played on player systems offering different levels of intelligence. It does not refer to the transportability of a single premaster videotape between edit houses. This is because the physical nature and the control information between the formats vary so greatly.

Mastering and replication facilities today are capable of encoding the interchangeable disc. It is standard practice today for all LV discs to contain both 24-bit and 40-bit codes. Pioneer, 3M, and Technidisc offer encoding and program dumps in the formats read by all LV players. A single disc can be mastered today that contains a full complement of 24-bit and 40-bit codes as well as program dumps for the Sony LDP-1000 and Pioneer 7820 series players.

Sony, although it does not offer Pioneer 7820 specification program dumps, does offer all other LV encoding formats. In fact, the Sony LDP-1000A is the interchangeable player. It can read the logic structure of level 1 programs, store Sony format level 2 program dumps, and serves as the player for level 3 program systems.

The technology exists today to produce the interchangeable disc. Although there are additional costs for editing and coding the interchangeable disc, the result is a single videodisc, one mastering and replication charge, one disc to package and inventory, and a single program to ship. The resulting program can be marketed to the largest player system population. So when designing your next videodisc, ask yourself, "Is this program valuable to more than just one market segment?" If the answer is "yes," and the market segments use different playback systems, consider interchangeability and open up new marketing possibilities for your program.

Jeff Kemph, Vice President of Video Production at Digital Controls, directs all facits of developing and producing videodisc materials. Mr. Kemph served as producer director for the Nebraska Videodisc Design/Production group during its videodisc development period and worked as an independent videodisc designer. He has produced videodiscs for World Book, Inc., the National Geographic Society, 3M Company, Media Development for the Hearing Impaired, the National Library of Medicine, and the Corporation for Public Broadcasting. Mr. Kemph has received numerous awards for his work.

# Production Tools and Techniques

*Ray Feeney*

Videodisc production techniques are different from normal television production techniques. They challenge existing procedures for bringing creative ideas to the public. The current process of visualizing non-videodisc products for television and film works reasonably well with existing script styles, storyboards, and the special vocabulary now used in normal entertainment production. Tools such as word-processed scripts and electronic identification slates allow for considerable flexibility once production is under way. Alterations can be made with little effort to accommodate for changes that occur during the production. The postproduction process can then be used to meet the few precision requirements such as overall running time and the positioning of commercial breaks. The immediate visual and aural feedback of videotape technology allows for the editing of loose ends into a smoothly running production.

But producing videodiscs requires more complicated procedures than those used in present (linear) production. While many of the same tools are used, the preproduction phase is much more extensive. Conceivably, there could be up to 50,000 storyboards per half hour of video. Each storyboard could represent a different picture, which would have to stand the scrutiny of from five seconds to five hours of examination by the end user. Each of these single frame storyboards could have from 25 to 250 words to describe the shooting procedure, as well as one or two diagrams. The individual frames would most likely have exact frame numbers assigned prior to production; there would be no allowance for missing or extra frames. Then, adding further confusion, it may be necessary to scramble the order of still frames and motion sequences for optimum access and search time, which would require proper sorting by computer-controlled disc players during playback.

In the past, many proof-of-concept discs were created using traditional production techniques that, though they may have been suited to present entertainment production needs, were sadly lacking when faced with the interactive videodisc requirement for seamless branching (no screen blanking) or large picture data bases comprised of single unique frames. Large picture data base discs refer to those that contain thousands of single frames of information and are presented in still-frame mode, often with computer generated overlays on top. This obviously presents several production difficulties. Such discs might contain as many as 50,000 individual edits—a considerable number more than the 300 edits in an average network half hour show. These 50,000 individual edits certainly put a strain on the traditional methods of script notes and identification slates.

There are few production techniques at present which require that frame counts must be assigned before production begins. One such example would be special effects work, where there is often the need to coordinate between two or more teams or production facilities. In postproduction there are several areas where frame accuracy is commonplace, including negative cutting, timing, and videotape editing. None of these systems, however, can effectively handle single frame precision editing or timing (color correction) for up to 50,000 individual frames.

In dealing with the recording of single frames which, by definition, require more careful attention than the recording of motion sequences, it is generally considered that productions shot in film and transfered to tape have the best look. Many of the data base discs have been shot on film in order to use the single frame stop motion animation capability (see Photo 1). By recording directly on tape, however, it is possible to present more information and detail in the final disc. The electronic camera can be used more efficiently than can the film process to compensate for the deficiencies of the videodisc and monitor. One simplified explanation for this is that the gamma of the film is fixed by the manufacturing process at a pleasing compromise, and the electronic camera can be adjusted for optimum contour, gamma, and enhancement for the kinds of materials being shot, be it pictures, documents, maps, artwork, models, etc. In all fairness, the film technique is still used by two-thirds of the media producers because an "acceptable" quality level can be assured by taking some care and with a minimum of customer hassle. In addition, the sheer number of suppliers of computer-controlled effects cameras make for competitive pricing.

So, despite the potential for increased quality inherent with electronic recording, the current market is still dominated by the broadcast industry. For example, most slow-motion disc storage or frame store technology that could be effectively used in videodisc production is aimed at the sports replay market, where the demands are not nearly as stringent as for videodisc recording.

"Broadcast quality does not necessarily mean optimum quality." This is a common complaint at postproduction facilities, whose engineers find themselves caught in the middle with, on one hand, existing equipment designed for an established (mature) broadcasting industry; and on the other hand, an ever-increasing challenge presented by the new computerized videodisc technology. Videodiscs are capable of such high levels of quality that the makers of video equipment are just now seeing the need to improve: to build faster, more accurate, and more flexible electronic tools.

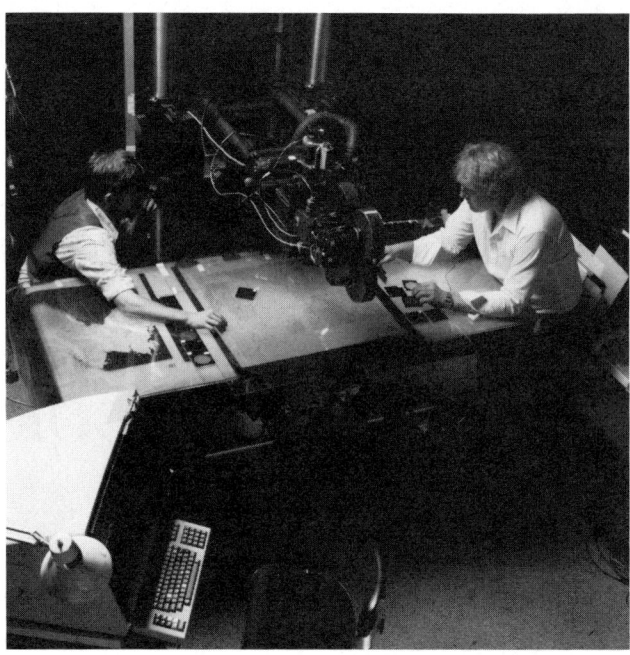

Using an animation stand to develop a computer-controlled terrain map for the military.

Another example of the inadequacies of current production tools can be shown by one of the most significant advances in the latest family of videodisc players: a feature called "instant jump," (sometimes called "field kicking," depending upon the system you are using). This feature allows a controlling computer to direct the videodisc player to skip over several fields/frames without any tell-tale blanking of the picture. In essence, the instant jump feature can seamlessly edit entire scenes. Conceivably, a videodisc game producer could develop several separate motion sequences, each running in parallel so that the game player could in turn have several decision points at any given moment.

Think for a moment, though, how one might go about these motion sequences on tape. Using something simple, for example, say four parallel (motion) images, one red, one yellow, one green, and one blue, the postproduction facility would need at least five "broadcast quality" machines (one for each image and a master machine), a first-rate switcher, and a computerized editing system with highly elaborate software. Given this, a "simple" real time command to the machines might be: machine 1: play back four frames, beginning at track one; machine 2: play back four frames, beginning at track five; machine 3: play back four frames, beginning at track nine; and, machine 4: play back four frames, beginning at track thirteen; and then repeat the sequences with machine 1, starting up again at track seventeen, all in real time. The

# Production Tools and Techniques

Taping a "travel" sequence through a model castle for an interactive videodisc game. Note the overhead grid. The camera can be mounted overhead and precisely stepped through the corridors, using a computer and "snorkel" lens.

problem here is that each tape machine can only play back four frames and then it has to "jump" to the beginning of its next series. Since, in this case, frame one to frame seventeen represents slightly more than one half-second (there are 30 frames per second), the current generation of machine cannot "rerack" fast enough. Thus, the whole recording process would come to an abrupt halt in less than one second!

Of course, there are solutions. Currently, the best one is to record all images on special digital storage media and then request random playback. The other, less desirable, solution would be to record the parallel images in a manner similar to conventional animation (very time-consuming and expensive).

Besides specialized equipment, interactive videodisc production requires unique applications of many of the conventional production tools. Special effects equipment can be adapted to the needs of videodiscs; however, some changes in philosophy, procedure, and the configuration of equipment are required for a special effects company to be able to handle a videodisc production cost effectively. Since the budget for a half-hour training disc usually is lower than for a typical thirty-second television commercial, the key items for videodisc production are automation, volume, and throughput in excess of that which is normal to animation facilities.

The specialized equipment necessary for this has evolved from the motion-controlled animation stands and model articulators invented for big budget commercials and motion pictures, which use effects heavily. A procedure that is used once or twice in a commercial might be required hundreds of times with an interactive disc. For example, by using computer-

A "knight's-eye" view of the model castle.

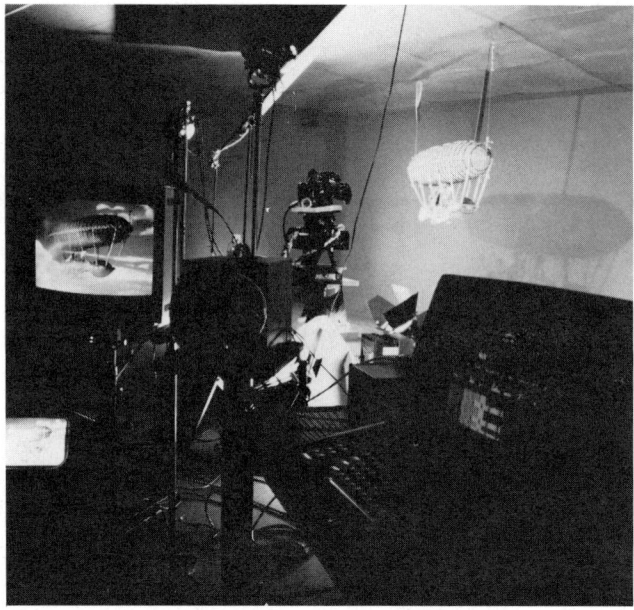

Filming an Ultimatte sequence. Note the clouds on the monitor matted with the imaginary dirigible. The Ultimatte strategy is relatively inexpensive and is far more versatile than a standard video keyer.

controlled cameras that can exactly repeat the same move, it is possible to make seamless branches. A seamless branch is created by filming or recording a particular scene two or more times while the camera makes precisely the same move and photographs an alternate take or a different response. The overlapping parts fit exactly together so that switches can be pulled from one to the other.

By way of illustration, suppose the point of view on the screen moves towards an intersection and turns right. The same computer-controlled move could be repeated with a left turn. The viewer could then choose to move forward and turn either right or left without any noticeable bump or transition in the action. Computer-controlled hookups and software can provide developers with motion and branches as indiscernible as the time and budget allow.

This leaves the most important consideration—the content of the imagery itself. Since traditional entertainment plays sequentially, visual continuity is more important than visual consistency. The individual scenes only have to match at the cut points and can be somewhat different 10 or 20 seconds later after several cutaways. On the other hand, the interactive disc is used in a random access mode. Visual consistency throughout the disc is critical because it may be possible to cut to a particular sequence from several different portions of the disc, or to branch to a considerable number of choices. There are several production techniques and tools that can enhance these characteristics. In the case of a "surrogate travel" disc (like MIT's "Aspen" project), or even a simulated flight over the countryside, the greatest problem is the uncontrollable changes that occur during the production. These changes show up as shifts in clouds, sun angle and shadow changes created by the sun rising and setting, wind shift in trees, car movement, etc.

The solution to the visual consistency problem is to build models, miniatures, and terrain boards to replace exterior shooting. The computer-controlled cameras, using a periscope-type ("snorkel") lens, can exactly step and repeat across a terrain board to create a flight simulator background. Once the model is built, the lighting grid hung, and the sky background painted, nothing need change for hours or even days, if necessary. Any path can be created, altered, shot, and processed to match exactly with any other. Speeds can be changed, pickups can be shot, and all with intercutability. Practically speaking, time is frozen on the miniature set (see Photos 2 and 3).

There is another very large advantage to miniature work: the object need never exist in full scale. A proposed real estate development could be shot from a model for planning or even sales purposes. The Rhine Valley in 1943 could be created for a World War II bombing game for arcade use. Places where access is denied could cost-effectively be recreated in miniature. This might include a hostile environment such as the interior of a nuclear reactor, oil drilling platforms, undersea habitats, military targeting and recognition systems, etc.

Another production tool is the Ultimatte (see Photo 4). This device can be used for electronic "cutting and

## Production Tools and Techniques

pasting." Since the final display medium is an electronic image on the disc, there is no quality loss when two or more pictures are electronically matted. The immediacy of electronic matting allows the client, who may not have the experience to visualize the end product during preproduction, a chance to zero in on a specific result. The ability to lock two or more cameras together in scale can produce exciting and meaningful matte effects, or be a means of providing comic relief.

Little-by-little, existing entertainment television and film resources are being adapted and modified to produce interactive videodiscs. Experience will generate the proper supervision techniques for selecting, educating, training, and interfacing with the Hollywood community. It is clear that videodisc technology will continue to have a profound impact on the established film and video production tools and techniques.

Ray Feeney is Vice President at Spectra-Image, Inc., a postproduction facility targeted for videodisc production. Ray has nine years of experience in the computer-controlled special effects business. He has a B.S. in electrical engineering from Caltech.

# Videodisc Interfaces and Software

*Dean Zollman*

To create a level 3 videodisc delivery system one must establish communication between a videodisc player and a computer. This type of communication requires two additional components—an interface and software. The interface must translate data "understood" by the computer into information recognized by the player, and vice versa. The software talks to the interface so that it knows when and how to communicate with the player. These components now exist in many different varieties.

## INTERFACES

Basically three types of interfacing situations exist. The easiest interfacing arrangement is encountered with the videodisc players that use a computer industry standard for communication. All Sony, Phillips, and Hitachi players, and also recent industrial/educational players manufactured by Pioneer (LDV-1000) use such a standard, the RS-232 port. With this port one needs only a standard serial (RS-232) interface in the computer. Then two-way communication between the computer and the videodisc player can be completed.

The early DiscoVision and Pioneer model PR-7820 Series players and the Pioneer PR-8210 player had a nonstandard computer port. Thus, an interface that communicates specifically with these players is necessary. Several such interfaces are being marketed.

An alternative for the PR series players is to use an RS-232 output from the computer, and then translate it to signals understood by the videodisc player. To accomplish this task one needs an electronic component that will complete the translation. Pioneer sells such boxes for each of the PR series of players. One box will work with the PR-7820 Models 1 and 2, but a different box is needed for the PR-7820-3 and PR-8210.

Interfacing computers with the consumer model players such as the Pioneer VP-1000, Pioneer LDP-1100, Magnavox VC-8010, and Sylvania VP7200 is somewhat more difficult. These players are not designed to be controlled by a computer, although they do have the random access features of other videodisc players. The interface must be accomplished by using the computer to act as if it were "pushing the buttons" on the remote control keypad used in the home. Such interfaces can be purchased for several computers and, if you want to do it yourself, you can find complete instructions in the June 1983 issue of *Byte* magazine (pp 60-74). Even though this method of driving a less expensive videodisc player is quite effective, it does have one major drawback. The communication is always one-way. The computer can send data to the videodisc player, but no

The Kansas State University Physics Department has built eight inexpensive level-3 systems by utilizing a consumer player (Pioneer VP-1000), an Apple II computer, and the interface sold by Anthro-Digital.

information can be sent from the player back to the computer.

Another approach to interfacing with the consumer player is to translate the output from the RS-232 port of the computer into signals that look to the disc player as if they are coming from the consumer's keypad. Again, one can buy a translation box to do the job.

Pioneer has created one additional difficulty in connecting a computer to the LD-1100. The disc player does not have a jack to connect the keypad to the player with wire. Instead, the communication is meant to be done through infrared signals. To overcome this problem one must purchase an add-on device, which enables the computer to be plugged into the player. No such difficulty exists with the other consumer players listed in earlier paragraphs.

## SOFTWARE

Software for the computer is necessary to tell the interface which commands to send to the videodisc player and, for industrial/educational players, to interpret the information received from the player by the interface. The software is usually available in two forms: (1) a set of procedures or subroutines for a programming language such as BASIC or Pascal, or (2) as part of an authoring language.

The procedures or subroutines are particularly useful for people who have some experience with computer programming. Many of the interface manufacturers have developed procedures that enable a programmer to write statements reasonably close to English and that control the videodisc player. For example, many manufacturers of interfaces supply

### VIDEODISC-COMPUTER INTERFACES

| Interface | Source |
|---|---|
| RS-232 to industrial/educational player | 1*, 4*, 5*, 8*, 12* |
| RS-232 to consumer player | 1, 10 |
| Apple II or IIe to industrial/educational player | 1*, 3*, 4*, 5*, 7*, 13*, 21* |
| Apple II or IIe to consumer | 1*, 2*, 3*, 4* |
| Atari 400 to consumer | 10 |
| Commodore PET 2001 to consumer | 23 |
| Commodore VIC-20 to consumer | 23 |
| Radio Shack TRS-80 Model 3 to industrial/educational player | 9 |
| Sony SMC-70 to industrial/educational player | 16* |
| Terak 8510 to industrial/educational player | 19* |
| Texas Instruments TI 99/4 to consumer | 18* |
| WICAT System 100 to 150 to any player | 22* |
| Audiotape to consumer player | 20, 21 |

* Authoring language available

### LIST OF INTERFACE DEVELOPERS

1. Allen Communications
2. Anthro-Digital, Inc.
3. Blue Lakes Computing
4. CAVRI Systems, Inc.
5. Digital Controls Inc.
6. Digital Equipment Corp.
7. GenTech
8. Interactive Research Corp.
9. Nebraska Videodisc Design/Production Group
10. New Media Graphic
11. Nuvatec, Inc.
12. Online Computer Systems
13. Owl Micro-Communications
14. Pergamon International Information Cooperation
15. Positron
16. Sony Communications Products Co.
17. Symtec, Inc.
18. Texas Instruments
19. University of Utah Video-Computer Learning Project
20. Video Vision
21. Whitney Educational Systems
22. WICAT Systems
23. Windelott Co.

### COMPLETE LEVEL-3 SYSTEMS

| Name | Source |
|---|---|
| Interactive Video Instructional Systems (IVIS) | 6** |
| POP-I | 5 |
| Positron | 15 |
| VIDEO PATSEARCH | 14 |
| Video Spond | 11 |

*Authoring Language must be used (requires a VAX computer for authoring only)

BASIC and Pascal procedures, which allow one to use commands such as "VIDEOSEARCH" and "PLAY." These commands enable one to control the videodisc player as an integral part of the program.

Authoring languages provide similar types of commands to the user. The authoring languages, however, are usually developed for specific purposes, such as education and training, and are designed to be used by nonprogrammers. These languages generally allow individuals to place information on the screen and analyze responses with a minimum of knowledge of computers. Perhaps the most well-known authoring language is PILOT, which has been used extensively in education for a number of years. Some authoring languages have been modified or extended to include addressing a videodisc player. Several interface manufacturers list authoring languages for sale. Both Apple Computers and Sony market authoring languages that have the capabilities of addressing videodisc or videotape players. (Texas Instruments also sold an authoring language with commands for the discontinued TI 99/4.) These languages can be very useful for many applications involving level-3 systems, but some developers find that they do not provide the same flexibility as a programming language.

Dean Zollman is Professor of Physics at Kansas State University. In addition to developing the first commercially available college-level videodisc, "The Puzzle of the Tacoma Narrows Bridge Collapse," with Robert Fuller and Thomas Campbell, Professor Zollman has developed three computer-based discs on physics. Currently, he is in charge of a program to develop computer software for already available videodiscs. He recently developed a Physics course for elementary education majors at KSU. He is the former editor of the American Association of Physics Teachers' film repository and is Chairperson of the Association's Committee on Computers in Physics Education. He received his BS in Physics in 1964 and his MS in 1965 from the University of Indiana-Bloomington. His Ph.D. in Theoretical Nuclear Physics was received in 1970 from the University of Maryland-College Park.

POSTPRODUCTION

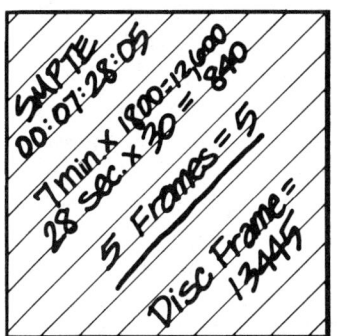

# Videodisc Premastering

*Ronald W. Nugent*

Premastering is the process of assembling onto a single medium all of the intermediate media, such as slides, motion picture film, videotape, etc., that will comprise the finished videodisc program. The product of this process is called a premaster. A master videodisc is recorded from the premaster. Premastering consists of four distinct steps:

1. editing
2. coding
3. evaluation
4. revision

Each step is necessary to ensure that the premaster will produce a satisfactory master videodisc. Anyone who has produced a traditional linear videotape program should be familiar with all of this process except for coding, which is unique to videodisc. One needs to be aware, however, of significant differences in terms of the care that must be observed, as well as certain unique technical requirements that must be followed when producing a master videodisc. Each step in the premastering process will be described in detail in this article.

## WHAT IS VIDEO AND HOW DOES A VIDEODISC PLAYER WORK?

To understand some of the unique characteristics of videodisc premastering, it helps to understand how video images are displayed on a television screen, and how a videodisc works. Let's begin with an explanation of National Television Standards Committee (NTSC) video, which is the North American standard.

NTSC video is created by what is called "interlace" scanning. Each image, or frame, is electronically scanned twice and divided into 525 lines of information. Each of the two scans contains one-half of the total information. When the video is displayed on a screen, the two scans are then recombined to make one frame. The individual scans are called fields; two fields equal one frame. The two fields are called field one and field two. An individual frame can start with either field. The starting point is called "field dominance." A frame starting on field one is referred to as first field or field one dominant; a frame starting on field two is second field or field two dominant. Field dominance is determined during video editing by the point at which the recorder makes its edit; that is, whether the edit starts at the head of field one or field two.

A videodisc player will read a disc that is mastered from either a field one or field two dominant premaster, but the disc must be correctly coded to match the field dominance of the premaster. During a motion sequence (the play mode) this is unimportant because the player is reading a series of fields in succession. But when the player plays each field more than once, as it does in the still frame or slow-motion mode, it becomes critically important because the player must be reading the two fields that correctly constitute one frame. If the two fields are part of two different images, the result is a totally unreadable frame.

There is another problem inherent in NTSC interlace scanning that causes an additional undesirable effect. The two scans are actually one-sixtieth of a second apart in time, one-half of the thirty-frame-per-second frame rate of NTSC video. When the player "freezes" on a frame from a motion segment shot on video, it is actually reading two images one-sixtieth of a second apart in time. If the motion is fairly rapid, the still image will appear to oscillate back and forth. This phenomenon is called "frame flicker" or simply "flicker." There are a couple of methods for overcoming this problem and they will be discussed later. What follows is a prescriptive, albeit somewhat technical, treatment of the premastering process.

## EDITING

Before editing begins, contact the videodisc mastering facility you will be using. (Editor's note: Part Two of the *Videodisc Book* lists videodisc mastering and replication facilities under the heading "Mastering and Replication Resources.")

The mastering facility will supply you with all of their technical requirements for a premaster submitted to them. They will also provide you with specifications for coding and the formatting requirements for such features as digital dumps (i.e., computer programs loaded from the disc) and compressed (still frame) audio. You should also meet with the engineering staff at the premastering facility that will be assembling the premaster. A good premaster requires that all production and engineering staff be acutely aware of the technical requirements.

Virtually all premastering is done on videotape. A few early discs were made from film premasters but film has been almost universally abandoned as a premastering medium because of the problem of integrating slides and text. Videotape can integrate film, slides, and video; and the creation of text and graphics is easily accomplished by using an electronic character generator. The advent of DRAW technology (direct read-after-write or user-recordable videodiscs) may make it possible to assemble all materials directly onto a master videodisc depending on how the DRAW system records. (Editor's note: See John Winslow's article on DRAW technology.)

If the system can record a series of segments from various sources and can be controlled by a synchronized editing system, then all materials can be assembled directly on the disc. If, however, the disc can only be recorded in real time from a source containing all elements of the program, a premaster will still be required.

Frame accuracy is required for the assembly of single frame materials such as a series of slides or text frames. The editing system you use must be field accurate as well; i.e., the field dominance of the recorder must be consistent, to ensure that the final videodisc will always read the correct two fields in the still frame or slow motion mode. Field/frame accuracy requires the use of an industry standard time code established by the Society for Motion Picture and Television Engineers, the SMPTE time code. Incidentally, there are two kinds of SMPTE time codes, one that is used for precision work and another that is used for many broadcast applications. The code used for precision work is called "full frame" or "non-drop frame" time code because it counts every single frame. The other time code is called "skip frame" or "drop frame," and, as the names imply, they do not count every frame. Since it is critical for videodisc applications (still frames in particular) that all frames be counted, you *must* use the full frame time code.

There is yet another time code called Vertical Interval Time Code (VITC). Similar to the SMPTE time code, VITC is used by Sony for all constant angular velocity (CAV) discs. If you intend to use Sony for mastering and replication, the VITC will have to be employed during editing.

Before editing begins, the field dominance of the record machine must be determined. There are two ways to do this. The first is to use a wave form monitor to look at certain lines in the vertical interval of the video fields. The vertical interval is the normally unseen portion of the video that contains synchronization and control signals. Determining field dominance by looking at the vertical interval is a highly technical process that requires the assistance of a skilled video engineer. The procedure is outlined in an excellent manual from 3M entitled "Premastering/Post Production Procedures for Scotch Videodiscs."

The second and simpler method for determining field dominance requires a videotape player that can be advanced one field at a time. In this method the video is dubbed in such a manner as to show the time code numbers from the master tape. At the edit points, note where the picture change occurs in relation to the time code. If the picture changes exactly as the frame number changes, the tape is first field dominant. If the picture changes between frame number changes, the tape is second field dominant. Examine a number of edits to see if the dominance is consistent. Any videotape material that is to be inserted into the premaster should also be examined for field dominance. If the field dominance is different from the record machine on which the premaster is to be assembled, it may be possible to change the dominance of the recorder or to have the final videodisc master coded to reflect changing field dominance between segments. We will say more about the coding process later in the coding discussion.

Most two-inch quadraplex recorders are field two dominant and cannot be changed. Most one-inch recorders are switchable. Three-quarter-inch U-Matic recorders may change back and forth depending on the quality of the editing system. Ampex one-inch recorders with the automatic scan tracking (AST) feature will produce inconsistent dominance, depending on how the playback machine locks up. This has a simple solution: turn off the AST whenever you are recording a premaster tape.

Premasters may be assembled on three-quarter-inch systems if the system is capable of the required field/frame accuracy and if the system meets the necessary technical standards. But mastering houses will not necessarily accept all three-quarter-inch premasters and you may have to dub the three-quarter-inch tape to another format, usually one-inch, before release to the mastering facility. This is one of the things that should be determined in consulting with a mastering facility before editing begins. One-inch videotape is the preferred medium of all of the mastering houses, and a one-inch premaster will produce a videodisc of significantly greater quality.

All of the mastering facilities have specified requirements for blank leader, color bars, audio tone, and color black that must precede and follow active program material on the premaster tape. There are also exact specifications for audio levels, chroma levels, and luminance. Chroma levels may not exceed 100% modulation for a premaster submitted to any of the houses. Luminance may not exceed 100-110 IRE units; the exact standard varies between houses. It is recommended that a peak signal clipper be used to limit video levels to the maximum acceptable level. Unfortunately, not all such clippers do an acceptable job and the video may be degraded by excessive clipping. It is a better idea to make sure that all video elements are recorded at acceptable levels in the production stage and also that the record machine is carefully set up during editing. Controlling video levels is a matter of carefully watching the scopes that are monitoring video levels in the editing process. Engineering staff must be made aware of the standards and follow them closely. Again, consult with the mastering facility of choice to obtain their exact requirements before the editing process begins.

Another critical standard for premasters is technically perfect edits. It may sound strange to call this a standard, but the simple fact is that marginal edits are very common. Unless the problem is severe they quite frequently go undetected. Bad edits in a premaster can result in unusable frames on the final disc. The loss may be limited to one frame immediately following the edit on the premaster, or extend to dozens of frames if the error is great enough to cause instability when the premaster is played back in the mastering process. Color framing error is a particularly difficult problem as it is difficult to detect during editing. Great care must be taken in the editing process; and every edit, particularly edits in which just one frame of material is recorded, must be checked.

The recording of single frames on a premaster may be done directly on the premaster, or first recorded on an intermediate medium and transferred to the premaster in strings. It is generally practical to record up to 300 or so single frames directly on the premaster. Beyond that number it is usually better to use an intermediate medium. A significant exception to this rule is to use one of the new videotape recorders that is designed for single frames without the ten- to fifteen-second preroll that is required of standard recorders. This speeds up the process considerably and minimizes videotape wear. Regardless of the number of single frames recorded directly on the premaster, each edit must be checked carefully to ensure that all of the stills have been recorded correctly with no marginal edits and that no frames have been left out.

Either film or a large capacity electronic still store may be used to generate long strings of single frames for real time transfer to the premaster. Film is most practical for stills that originate from 35 mm slides. There are facilities that specialize in the high speed copying of 35 mm slides to 16 mm film. They generally charge about $1 per slide. The resulting filmstrip is then transferred to videotape for editing into a premaster. Copystand photography onto film may also be the most practical way to generate large strings of stills from photographs or documents. The field dominance of the videotape to which the filmstrip is transferred must match the dominance of the premastering recorder.

Large capacity electronic still stores are frequently used to build up long strings of stills. Still stores have the advantage of being able to store video images; they can be used to store text pages from an electronic character generator or composite video from a digital effects device. The strings can be directly recorded onto videotape from the still store. Still stores will be largely replaced by DRAW discs in the future. Strings of stills will be recorded on a DRAW disc and transferred directly to the premaster. The Panasonic OMDR (optical memory disc recorder) is the first such DRAW system commercially available.

There is a continuing debate about how many frames of each still should be edited onto the premaster. For stills that will be stepped through by the user, there must be only one frame for each still, or the user will have to step more than once to get to the next still. For level 2 or level 3 discs under programmed control, it is generally recommended that at least three frames per still be used. Then if there is a minor problem in editing, there will be at least one usable frame of each still. The same considerations apply to text frames.

The editing of motion segments that are to be played from the videodisc in real time only requires no special considerations beyond careful control of video levels. If, however, they are to be played in slow motion, or if individual frames from motion segments are to be displayed in the still frame mode, special techniques will be required. The reason for this has to do with the field/frame problem discussed earlier and the way NTSC video is displayed.

Standard motion picture film is shot at twenty-four frames per second. The frame rate for NTSC video is thirty frames per second. When film is transferred to video, a process known as 3/2 pulldown is used to solve the discrepancy in the two frame rates. Every other frame of film is scanned three times instead of two, resulting in three fields of video for that frame. If the disc player tries to read the extra field and the first field of the next film frame, the result will be frame flicker. Without special coding of the disc, two frames out of five will flicker. If the still frame point is preprogrammed, one can simply select a frame that is read by the player as the two correct fields. If the disc is to be stopped at random or slow motion is to be used, then special coding is required. The coding consists of electronic "flags" that tell the disc player which two fields to read as a frame. The process is called "white flagging."

The flagging process varies according to which videodisc mastering house is chosen. For 3M, the flags are inserted during the initial film-to-tape transfer, using a code inserter available from 3M and a specially modified telecine. No additional editing of the sequences should be done after the transfer because of the possibility of disturbing the field frame relationship during videotape editing. Pioneer will insert the flags at the time of mastering, and they can move the flags to correctly encode field/frame changes that occur during editing. Changing the flags within a premaster is a complex process, however, and again it is recommended that all film segments be edited as film and a single transfer to video be made. Sony does not offer a flagging process.

One way to overcome the 3/2 pulldown is to shoot and transfer all film segments at thirty frames per second. The film must be transferred with the correct field dominance, matching the dominance of the editing system. No special coding will be required. When film segments are transferred at thirty frames per second, the segments can be edited as video with no concern for shifts in the field/frame relationships. Film shot at thirty frames per second will produce excellent videodisc images and rock-solid stills, and slow motion at any point. There is an additional bonus in that the higher film frame rate will produce images superior to the standard frame rate and less strobing of the image in camera pans. The only complications are a more expensive film-to-tape transfer at one of the few postproduction facilities that offer such transfers, and the problems of sync sound. Most sync sound film cameras and editing systems are designed for twenty-four frames per second. For sync sound at thirty frames per second, the sound will have to be recorded by a free-rolling audio recorder and the audio matched with the picture in a postproduction process, or special equipment will have to be used.

For motion materials shot on video, the only solution to the flicker problem is electronic. The differences in motion between the two scans can be eliminated by processing the video with a field correlator that examines each field and, by repeating certain lines from each field, effectively eliminates the differences between the two.

Interactive videodiscs often involve a large number of text frames. These are generally produced on an electronic character generator and edited directly onto the premaster. In the selection of text fonts, readability should be the number one criterion. Font sizes and styles should be pleasant and easily readable at normal viewing distances. Pioneer, for example, recommends 15 lines of text with 40 characters per line for individual viewing and 10 lines of text with 30 characters per line for group viewing.

Background colors should also be chosen for readability. Strong background colors behind text can be very unpleasant to read. A light blue-grey is most often used as the background for text. If it is desirable to provide color coding to identify text by section or function, such as HELP or REVIEW, color bands or colored symbols should be used over the blue-grey background. In general, strong, saturated color backgrounds will not reproduce well on videodisc. Sony, for example, recommends that levels for solid color backgrounds not exceed burst level (40 IRE).

Some of the newer character generators have the capability of storing and releasing long strings of text pages at up to real time (thirty frames per second). Use of one of these generators will greatly speed up the editing process. They eliminate the need to record each page separately onto the premaster.

Audio recording on the premaster should be carefully controlled. Videodiscs are capable of very high quality audio reproduction, but they will sound no better than the audio on the premaster. Peak levels should be consistent throughout and short term peaks should not exceed +3 db. Stereo and dual independent audio channels should be in phase and within 2 db of each other. The mastering facility requires that you specify which channel on the premaster (1 or 2) is to go on the videodisc's channel one or two. If you wish to use noise reduction systems such as Dolby, you will have to consult with the mastering house to see what kinds of coding they will accept and how it must be formatted. Pioneer prefers an independent audio tape (synchronized with the video), which they incorporate during mastering. This is an extra-cost service but might be very desirable for very high quality audio that would otherwise be degraded by recording on the premaster.

Still frame audio is available from at least one mastering facility, Sony, and may be eventually available from others. The Sony system will allow up to 15.4 hours of audio accompanying 1384 single frames of video. The audio is encoded digitally on the disc and recovered by a peripheral device. The disc may contain both motion segments with standard audio and still frame audio. The digital audio is encoded in the video portion of the signal in frames preceding the accompanying still frame. There is thus a trade-off between the length and duration of the still frame audio and the amount of video information that can be put on one side of the disc. Preparation of a premaster is a complex process for still frame audio, and during the procedure close coordination with Sony is necessary.

# CODING

The coding process involves the insertion of certain codes into the program material that trigger automatic functions on certain videodisc players. There are six such codes currently used, as follows:

1. Picture codes that identify the start of each frame and the field dominance of the frame;

2. Chapter codes that identify sections of the disc by chapter number for level 1 players;

3. Picture stop codes that trigger automatic stops on level 1 players;

4. 3/2 pull down codes that identify the correct field/frame relationship for twenty-four frames per second film segments;

5. Programming dumps that load the internal memory of Level 2 players; and,

6. Digital audio codes for still frame audio.

Picture codes, chapter codes, picture stop codes, and 3/2 pull down codes are inserted at the time of mastering by Pioneer; the point of insertion is designated, using the time code on the premaster as a reference. Pioneer will also correct mixed field dominance premasters by adjusting the picture code to reflect the field dominance of each segment on the premaster. 3M will insert picture codes for a premaster with a single field dominance during mastering, but the other codes must be inserted during editing, using a cue code inserter available from 3M. The cue code inserter also adds picture codes and can be used to correctly code mixed field dominance premasters. Sony provides picture codes only, which are inserted during mastering.

Codes for digital dumps are also inserted during mastering, but they are prepared during the premastering stage. There are two types of digital dumps; one for the Pioneer player and another for the Sony player. Pioneer will provide Pioneer dumps only; Sony only Sony dumps. 3M will encode dumps for either player or both. The dumps are encoded in the second audio channel. For the Pioneer dumps, the premaster must have a five- second window with no audio on channel 2 and video black for each dump. The actual authoring process varies greatly at each facility and requires close coordination with it. Again, specifications can be obtained from the chosen facility.

The process for still frame audio coding was covered in the discussion on audio. At present it is available only from Sony.

## EVALUATION

There are two kinds of evaluation: first, a traditional evaluation with sample users, and second, a technical evaluation of the completed premaster videotape. The latter might be more correctly referred to as "checking." Evaluation with sample users is complicated by the complex nature of interactive videodiscs. A high fidelity evaluation pretty much requires a check disc (a check disc is a one-off replica after mastering has occurred but before the desired number of replicas are made). Videodisc simulators have been developed utilizing random access videotape players, but these are generally too slow in response time. Sony has such a simulator available at several sites, which is used primarily for authoring and checking digital dumps. DRAW technology will provide some relief in the future because a DRAW system will produce a disc very quickly that can then be used for evaluation.

Technical evaluation of the completed premaster is an often tedious but vital process. The premaster tape itself can be checked at every edit point, but generally this is not recommended because of the possibility of tape damage. The premaster should, however, be spot-checked during editing if there is any reason to doubt the quality of an edit or field dominance. After premastering is complete, a dub should be made of the premaster to a videotape player with frame or field step capability. A list of what to look for follows.

- Are the edits good? Very bad edits will be obvious. Marginal edits can be spotted by looking for breakup of the image in the top left-hand corner.
- Is the field dominance consistent?
- Are video and audio levels within specifications?
- If the 3M cue code inserter has been used, are the cues where they should be?
- Are the text frames all there, and are the spelling and grammar correct?
- Are all the single frame materials there and in the correct order?

## REVISION

The revision process will depend on the results of the evaluation process. All technical errors discovered in the evaluation of the premaster must be corrected. The results of the evaluation with sample users may require the production and premastering of new material. If the new material will fit within the space of the old, then it can be placed directly on the original premaster. If it is longer, a whole new premaster tape will have to be created. Great care must be taken in dubbing the material from the old premaster to the new. If the premaster has any codes on it, all codes must be carried over. There is also a potential problem of loss of quality. Solid colors will tend to degrade and video levels may become higher. If the premaster is three-quarter-inch or two-inch quadraplex, the degradation may be unacceptable and a whole new premaster may have to be assembled from the original source material. When revisions are completed, the new premaster must receive the same careful technical evaluation as the original premaster.

## A FINAL WORD

Premastering is a complex process that requires great care and a knowledge of the required technical standards. Close coordination with the mastering facility of choice is absolutely necessary. Excellent manuals are available from the mastering houses and these should be studied in detail before production and premastering begin.

Ron Nugent is Director of the Nebraska Videodisc Design/ Production Group, located at the University of Nebraska. He assumed his current responsibility in September 1983. Prior to that he was Assistant Director and Senior Producer with the Group. With the videodisc project since 1978, Ron has been involved in the production of over 80 interactive videodiscs. Ron graduated with a BS in Radio and Television Broadcasting from the University of Oregon in 1969. He spent six years as a television producer for Georgia Public Television, after which he joined the Instructional Television Unit at the Nebraska Educational Television Network.

## MASTERING AND REPLICATION

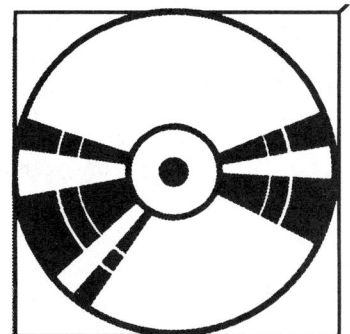

# DRAW Technology —the Instant Videodisc

*John Winslow*

### INTRODUCTION

The optical videodisc system was designed to deliver prerecorded video entertainment to the home. During the last decade the disc and player technologies have matured and manufacturers who can provide competent players and high-quality discs at reasonable cost have become established. Because the disc contains so much information in a small area, the system's per- formance as a large, rapidly accessed database is unmatched: 54,000 video frames or 5 billion error-corrected bits that can be randomly accessed in under 3 seconds on average.

When these properties were published they produced responses, such as the following, that had nothing to do with prerecorded entertainment.

"Can you make a system for recording all of my daily bank transactions?"

"Can I archive genealogical data? Production drawings? Photographs of a museum collection?"

"Can the daily footage of a film production be put on a disc? Can we edit from the disc rather than from film?"

"My material is confidential. Can I make discs in-house?"

"I have assembled a disc program on tape. Can I get a disc in a hurry to check the program?"

These questions could be generalized to, "Can I make my own discs and play them back right away?"

The producers of optical discs were not responsive to these requests because they considered their business to be the mass replication of discs rather than producing a variety of special disc-making equipment.

The number and content of these requests make it clear that the availability of instantly playable discs in-house is essential to the practical development of a great many proposed applications of the optical disc; hence, the development of Direct Read After Write (DRAW) technology. Simply put, a DRAW disc is a recordable laser videodisc.

### THE STATUS QUO

At present, laser videodisc players are available from Hitachi, Philips, Pioneer, and Sony. Prices range from under $500 to over $2,000. 3M, Pioneer, Sony, Sonopress, and Technidisc all produce discs from a client's tapes. 3M, Pioneer, and Sonopress also produce entertainment discs that typically retail for $30 for a two-sided disc or $15 per side.

A typical price schedule (Pioneer) for single-sided interactive discs made from a client's tape is given in the following table.

| Quantity | Price per Disc |
| --- | --- |
| 5-9 | $525 |
| 10-24 | $270 |
| 25-49 | $120 |
| 50-99 | $65 |
| 100-199 | $38 |
| 200-299 | $25 |
| 300-399 | $23 |
| 400-499 | $20 |
| 500-599 | $19 |
| 1000-1499 | $18 |
| 1500-1999 | $16 |
| 2000-up | $14 |

This schedule implies a tooling charge of $2,500 and a disc charge, exclusive of tooling, of about $14. The client who needs five discs or less will have to pay $2,625 to cover the tooling cost. This cost reflects the time and labor required to convert the video program into a nickel stamper and mount it in the disc mold. This conversion requires at least 24 hours. The normal turnaround time from submission of an acceptable tape to delivery of finished discs is several weeks unless a premium is paid for expedited delivery.

## THE DISC ARCHIVE

Optical videodiscs are generally regarded as the best available archival storage medium because of their durability, noncritical storage requirements, nonvolatility, compactness, and fast access. In spite of these qualities, few archival discs have been made because of the prohibitive cost of a few mass-replicated discs. A DRAW disc that would play on a standard player would make the optical disc strongly competitive with existing archival media on a cost basis and so make the optical disc, with all its advantages, available to the archivist.

The longevity of the DRAW disc is, of course, a central issue. There is reason for optimism: replicated discs made from poly methyl methacrylate (PMMA), an acrylic, have suffered no apparent degradation after ten years of storage in an office environment; and several DRAW media are claimed to exhibit archival stability. Therefore it is reasonable to suppose that a PMMA substrate bearing one of these stable DRAW media could be an archival videodisc.

## THE INTERACTIVE DISC: AN EDITING PROBLEM

Although entertainment programs account for a major portion of the discs produced, other types of programs are being made in increasing numbers: industrial communications, sales seminars, maintenance and repair instruction, catalogs, point-of-sale pitches, and video game image stores. Most of these programs are interactive. They are assembled from pieces of material that are logically related by branching and looping paths rather than by a straight line. The optical disc is the only medium that can support programs of this kind. The spiral form of the recording permits jumping from one point of the spiral to another in a distance less than 10 millionths of the length of the spiral path connecting the points. Instead of playing straight through from beginning to end, interactive discs are programmed to stop at branch points and present the viewer with alternatives. When the viewer makes his choice, the player shortcuts to the appropriate frame and resumes play or asks for another choice. Generally a computer spares the viewer from having to read and enter five-digit frame numbers. It presents choices labeled "1, 2, 3 ---" and supplies the frame number. In some programs the computer examines the viewer's choice logically, in order to give an appropriate response.

An interactive disc program is created as follows. First the logical structure of the disc program is defined. Then the program images are assembled on the tape in accordance with the logical structure and a computer program is written to supply commands to the player.

The interactive disc program must perform perfectly. In conventional programming there is rapid interaction with the computer, which allows the programmer to spot errors and repair them quickly in order to get a perfect result. The disc programmer's job is much harder, for he cannot test his work until it is a disc running on a player. The cost and delay in obtaining a mass-replicated test disc effectively prohibit functional testing of the program.

## THE CHECK DISC

If a DRAW recorder was included in the assembly process so that the programmer could check his work from time to time, he could easily find and correct logical errors in the program structure, incorrect frame numbers, and awkward delays between branch points. At present, the frame numbers of branch points are reckoned by counting frames on the assembled tape. This count is liable to be in error because tape frames may not be in one-to-one correspondence with disc frames. The check disc would supply the correct disc frame numbers directly during playback; and they could then be inserted into the computer program that controls the player.

Several proposals for editing with the videodisc have been made; and the following simplified version includes elements that are common to all of them.

Film editing consists, primarily, of discarding what is unwanted and arranging what is left. The optical videodisc can be commanded to do the same thing: selectively play back only what is wanted and present it in the desired order. The fast access time of the system enables it to skip the unwanted portions of the disc and play back the selected portions without undue delay. By changing the commands, different selections can be made and played back in the order desired until a satisfactory version is obtained. All this is done without spooling, threading, cutting or splicing film. The list of commands that has directed the player to skip here and play there is, in effect, the edit list, the end result of the edit process.

This description of editing with the videodisc has been simplified to present the concept more clearly; the reality is more complicated. In general, several discs and players would be required in a typical edit session. It would be impractical to command these players by having the harassed editor punching frame numbers into a bunch of keypads. A computer would be required to manage the system. It would enable the editor to use simple commands such as "play from frame—to frame —; skip to frame—; play—", and have the players respond correctly. The computer would also produce an edit list with film frame numbers that the film cutter could understand.

## WHERE IS IT TODAY

Optical Disc Corporation (ODC) of Cerritos, California was founded in 1982 for the purpose of developing and manufacturing DRAW discs and disc recorders that conform to the LaserDisc Standard. They will enable the user to produce his own discs in-house in an office environment. The turnaround time will equal the play (or record) time of the discs themselves. At this writing, the development work is substantially complete. ODC has produced blank discs containing a proprietary DRAW medium, has recorded them directly with a laser beam recorder, and has played them back on standard laser videodisc players. The discs are made from materials that withstand harsh environments, and will receive further testing to insure that the production disc will survive indefinitely in an office environment.

## SUMMARY

Because of its unique combination of storage capacity and speed of access the optical videodisc has been proposed for tasks that cannot be accomplished by any other means. So far there has been relatively little performance to match the proposals. By removing the constraints imposed by the mass replication system, the DRAW recorder and the DRAW disc will enable the optical videodisc system to actually do the jobs it has been expected to do for so long. The applications discussed in this article, archival storage, disc program editing, and film editing, were chosen from among many possible uses for the DRAW system.

John Winslow, Senior Member of the Technical Staff at Optical Disc Corporation (ODC), is a graduate of Cal Tech with a BS in Physics and an MSEE. In 1970, while at MCA Labs, John was instrumental in the development of mastering and replication procedures for early optical laser videodiscs. He worked with engineers and managers of Philips in the Netherlands to develop the present optical disc standard adopted by the IEG.

# IMPLEMENTATION

# The Tacoma Narrows Bridge Videodisc: A Personal History

*Robert G. Fuller*

## INTRODUCTION

In the mid 1950s, Dr. Franklin Miller of Kenyon College put together a single concept physics film loop from the archival newsreel film of the collapse of the Tacoma Narrows Bridge. It became far and away the most popular single concept physics film ever made. Our students at Nebraska have worn out many copies of the film loop from multiple showings.

In the later part of the 1970s, Dr. Alfred Bork of the University of California, Irvine, and others, organized a conference on the intelligent videodisc. The conference brought this new technology to the attention of some of us educators who were largely ignorant of its potentialities until then. Little could I have guessed that my participation in that conference would lead to the first commercially available interactive educational videodisc.

In 1978, the National Science Foundation put out a call for proof of concept projects for intelligent videodiscs for science education, planning to give out a few hundreds of thousands of dollars. To three of us plain physics professors in the Midwest, Thomas C. Campbell of Illinois Central College, Dean A. Zollman of Kansas State University, and me, it seemed that the production of a dumb videodisc was in order first. So we put together a Low Cost Approach To Videodisc Education (LCAVE) proposal to the NSF in 1979. We asked for a few tens of thousands of dollars to make a plain videodisc. Our proposal promised to make a videodisc as cheaply as possible by using existing film. We intended to use the film of the Tacoma Narrows Bridge collapse to introduce the physics of standing waves. Our videodisc would try to make maximum use of the capabilities of the existing stand-alone videodisc hardware. We came to discover that meant making what was called a level 2 disc. The final result is the present, not-so-dumb videodisc, *The Puzzle of the Tacoma Narrows Bridge Collapse*, John Wiley & Sons, Inc. (1982).

It is my intention in this article to discuss some of the personal aspects of the history of this videodisc, its production, and its public distribution. My hope is that you will find something in this story that will enlighten your activities with videodiscs.

## THE IMPORTANT INGREDIENTS

The first important ingredient of a videodisc is a good idea. We all had seen the Tacoma Narrows Bridge film loop and we knew it to be compelling film footage that leads to interesting physics. It was a logical choice for a first physics videodisc. Our commitment

The first Tacoma Narrows Bridge, November 8, 1940.

to the use of existing film footage and of limiting the content of our disc to the physics related to that film footage made it easy for us to decide what to put on the disc and what to leave off. Nowadays one sees educational discs that seem to have no focus or story line. We are not believers in the concept of the "generic" videodisc. We see a good videodisc as one that is developed with a fairly clear and specific educational goal in mind.

A second important ingredient is a real commitment to interactive education. Are you really convinced about the necessity of interactive education? When you go somewhere to share your professional knowledge, do you primarily tell people what you know, or do you engage your hosts in a discussion to attempt to bring them to new understandings on their own? In my experience, if you are of the latter type, which happens to be a very small percentage of the people engaged in education, then the videodisc is a likely medium for you to explore. The Tacoma Narrows Bridge team of physicists were already practitioners of active learning based on the constructivist research of Jean Piaget before the arrival of the videodisc. We were, in one sense, educators waiting for our telecommunications medium to arrive, and it did with the appearance of the interactive videodisc. Our experiences have convinced us that the videodisc can play a valuable role in education by letting learners manipulate the visual images of real, everyday events.

A third ingredient in the success of our videodisc is the fact that all three of us were experienced producers of single concept films for physics teaching. We had first met at a film loop instructional course (FLIC) at the University of Nebraska in 1972. Together in the summer of 1972 we learned about film production and storyboarding. We each developed a good

# The Tacoma Narrows Bridge Videodisc: A Personal History

Academics and videodisc producer worked together in the control room to produce the studio footage for the videodisc. From left to right, Dean Zollman and Robert Fuller, physicists, with Ron Nugent, NETV videodisc producer.

Putting academics into a tv studio causes cultural shock. Here the sound expert tries to show Thomas Campbell (center) and Robert Fuller (right) how to get the proper sound out of the fan for the model bridge demonstration on the disc.

beginner's sense of what an instructional film should be. Subsequently, we worked together on a number of physics film projects. Moreover, the three of us ran a workshop for physicists on how to produce single concept physics films and Tom Campbell and I had edited twelve single concept physics films based on the NASA Skylab missions. The point being that film is much closer to the videodisc medium than is television. Film editing is a frame-by-frame process, very similar to postproduction editing of a videodisc. The three of us, then, were especially well prepared for working with videodiscs.

A fourth essential ingredient in the story of our project is money. Our LCAVE proposal received a grant of $60,000 from the National Science Foundation. The production costs for the Tacoma Narrows disc were about $40,000, equipment costs about $10,000, and mastering costs about $10,000.

A fifth key aspect of our project was the availability of the Nebraska Videodisc Design/Production Group, also located at the University of Nebraska. By the time we started the production of our disc, the Nebraska group had already produced nine videodiscs for education. We had access to a lovely teleproduction facility (Nebraska Educational Television Network—NETN), experienced, professional film and television personnel as well as people who knew about videodiscs.

A wide variety of styles of cooperation and control is possible for joint projects between academics and television or videodisc producers. We had control of the content and purposes of our videodisc. The Nebraska group provided us with advice and technical production support. They did not try to tell us what physics to include nor how that content should be presented. When we had an idea of how we wanted to show a particular physics concept, whether in an inquiry, or discovery mode, the Nebraska Group tried to help us achieve our ends in the best video production way possible at the time. We worked in the postproduction editing room along with the television people. We were involved at all times in the decisions about edits, cuts, audio track matching, etc. It was a very satisfying learning experience for us. The final videodisc was a product of which we were proud.

## THE PAINS OF PRODUCTION

The project began with the making of a large overall storyboard for the videodisc. It included the opening "hook" consisting of the archival Tacoma Narrows Bridge collapse footage, a sequence on the effects of the wind on the bridge, a sequence on the physical properties of the bridge, and an ending that would enable the students to make sense out of the bridge collapse for themselves. Both the wind and physical properties sections of the disc required us to make new film footage of some physics experiments in the NETN studio. There is a good bit of cultural shock experienced by physicists when working in a television studio with lighting, sound, and photographic experts. Unlike our own laboratories, when things did not go right in the tv studio, we had to wait for someone else to correct it. One sequence had to be completely redone because of a malfunctioning camera.

We operated with shooting scripts for the physics experiments and created the desired effects in discussion with the videodisc design/production staff as we did the work in the studio. After all of the new film and the other source materials had been acquired, we were given time-coded copies from which to make our postproduction scripts.

Because we were seeking to make a level 2 disc, we included many branch points and multiple choice questions on the disc. In an attempt to use our understanding of the best ideas from the reinforcement theory of learning we decided to show our positive responses, such as "Good," or "Correct," on beautiful pictures from the magazine *Arizona Highways*. We also used lightning or cactus pictures as the background for the negative responses. Since the archival film occurred in 1940 we used a little Glenn Miller music at one place on the original disc. Of course, we did all of these things without worrying about the possible difficulties they could cause us if we were ever to go commercial. Remember, this was in 1979-1980 and the Nebraska group had not yet produced a videodisc with a digital dump on it. In fact, we discussed the possibility of putting a digital dump on our disc to control student use of it on a Pioneer (then DVA) PR 7820 player. We rejected the idea because at the time digital dumps were unreliable and experimental. We did not want to make the first disc from Nebraska to use a digital dump.

As we began to put together the postproduction scripts for the various parts of the disc, we found that the disc began to take on a character and personality of its own. It seemed to be forcing us to consider physical concepts and ideas that we did not have in mind when we first started the project. In fact, the explanation of the collapse of the bridge was more complex than we had been led to believe. We found ourselves forced to examine the historical records and studies of suspension bridges and their collapses. Not only did these new facts make their way onto the disc, but they also provided us with some hidden humor frames that we put on the disc.

Even at this early stage of videodisc development it was clear that eventually all discs would be used on systems that were controlled by advanced computer software. So we tried to develop a multilevel disc. While we were designing for a stand-alone PR 7820 player, we knew that the use of a videodisc player with a computer was only around the corner. So we added extra frames to our disc so that it could to be used with a computer without the PR 7820 commands appearing on the tv monitor. (Now, some four years since the origins of this disc, it has not yet been completely used in a computer-controlled interactive mode.)

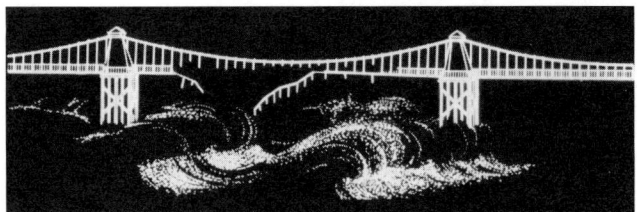

Several bridges in England designed by Sir Samual Brown collapsed in the 1830s. Eye witness sketches show behavior much like the Tacoma Narrows Bridge on the day of its collapse.

We were able to complete the whole disc by the late fall of 1980 and had received permission to make 50 copies of the disc. So we proudly took our disc to the January 1981 national meeting of the American Association of Physics Teachers to show it off. Of course, we had to carry our own player with us. There were no players easily available for rent. The fact that we had made a complex level 2 disc without a digital dump meant that we had to spend a couple of hours programming a player before we could properly demonstrate our videodisc. Those early days were marked by some frustrating experiences. For instance, there was the time Tom Campbell had spent a few hours programming his player in preparation for showing off our disc at a state physics meeting. Then he put on a programmed disc from a General Motors car dealer and wiped out his hours of effort.

We sold our first fifty copies at $50 each and they went like hot cakes! One physics department bought one. The rest were snapped up by videodisc aficionados. Those early versions of the disc may now be collector's items.

At that point our NSF project was just about out of money. We had a product that seemed pretty good, but there were NO commercial distributors and NO educational institutions with videodisc players.

**GOING COMMERCIAL**

By June 1981 we had received positive evaluation results for our disc. We had discovered the necessity of having a digital dump on the disc so we could easily use its complex logic with students. We were sold out of discs and we did not have the money to add a digital dump to the disc and make another pressing. With the encouragement of NSF we began to search for a commercial distributor for our videodisc. We placed ads in various media and library journals. We had a few conversations with some of the new videodisc software companies, but it was going to take an initial investment of a few thousand dollars to see the disc through the digital dump and mastering process. None of them was interested.

Meanwhile, Dean Zollman, who had done the original programming of the player for our disc, had been in touch with the DVA people who understood the process of putting a digital dump onto a disc. He found that if we converted our program to hexadecimal code we could get all of our programming and frame numbers stored on the disc with one program dump. So he proceeded to convert our program to hexadecimal code and we ordered a proof disc.

By means of an informal contact, John Wiley & Sons, a well-known publisher of physics textbooks at the college and university level, found out about the existence of our disc and our interest in finding a commercial distributor. Wiley representatives came to Lincoln to see the disc and discuss their possible interest in its commercial distribution.

At the national meeting of the physics teachers in January 1982, we verbally agreed to have the Tacoma Narrows videodisc commercially distributed by John Wiley & Sons. An official distribution contract was drawn up and submitted to the University of Nebraska-Lincoln legal department.

While the various lawyers were trying to agree on the contract language, I began to get all of the necessary commercial distribution agreements signed. I wrote to twenty-four of the photographers for *Arizona Highways* whose pictures we had used for reinforcers. Some refused to give us permission to use their photographs, even for the $25 per picture stipend we offered. That meant that we would have to reedit the master videotape and remove their pictures. At that point we decided to remove the Glenn Miller music rather than seek permission to use it. We asked John Wiley & Sons to negotiate with Ella Fitzgerald's agent and Memorex for permission to use the famous commercial on the disc. After all the difficulties I had had with the photographers for *Arizona Highways*, I was afraid to try to negotiate with a famous star. Barney Elliott, the photographer in Tacoma who took the original film of the bridge, signed a royalty agreement.

In our conversations with Wiley, we had decided that it would be best for the videodisc if the copyright were transferred from the University of Nebraska-Lincoln to John Wiley & Sons. Our reasoning was simple. The opening archival footage of our videodisc is seven and a half minutes of very interesting historic film with a documentary narrative. It could easily be copied onto a videotape and used as straight lecture material. We thought it would be extremely difficult to get a university legal staff to bring suit against another educational institution for such a violation of copyright. We thought that the legal staff of John

Barney Elliott, the person who took the original movie footage of the bridge, accepts a copy of the videodisc while standing in front of the present, still standing, Tacoma Narrows Bridge.

Wiley & Sons would be more likely to take the proper actions to keep such a black market videotape from being circulated. After all of the details in the contract had been accepted by both UN-L and Wiley, a process which took several months, the contract had to be sent to the National Science Foundation for approval. One of our initial goals was to have the commercial videodisc ready to demonstrate at the summer meeting of physics teachers in June. Here it was May and we were still trying to get the approval from NSF to bring out the product!

Meanwhile, back on the campus, we were working with Wiley on the packaging of the disc for distribution and preparation of the teacher's manual to accompany it. Since this was to be the very first interactive educational videodisc, we wanted it to set a standard of excellence that subsequent videodiscs in the education market would have to match. It was in these aspects of the project that the help given to us by the editorial staff at Wiley was superb. We discussed a double-fold jacket that would have enough information printed inside it so that the instructor could use the disc with just the jacket information.

Because we had already used a first version of the Tacoma Narrows disc, we had developed a variety of print materials for student use as well as the instructions we used to program a player for this disc. How much of this should be made available to the buyer of the videodisc? Of course, added material in a teacher's manual meant increased costs for Wiley. Once again we had lengthy discussions about these issues with Wiley's editorial staff about what John Wiley & Sons would be prepared to do for a teacher's manual. I think we benefited from their long experience in the science publishing field. They realized that a high-quality teacher's manual is an important part of a good product. They encouraged us to include in the teacher's manual all of the information that would be necessary for an instructor to completely and adequately use this videodisc. We knew that the videodisc represented a new technology and that the programming language of the PR 7820 player was crude. We decided that a complete listing of the program commands, as well as all of the frame numbers, stored on the disc and read into various memory locations in the microprocessor of the 7820 player, would be very helpful to the sophisticated user and would enable such a person to reprogram the player to use the disc in a variety of ways. While we thought it was unlikely that many instructors would actually want to program their player, we decided to include such technical information in the teacher's manual. In addition, we wanted to make our student laboratory pages available for use.

Wiley editors also suggested the tear-out pages at the back of the teacher's manual that are now a regular feature of the product. The *Teacher's Manual to accompany The Puzzle of the Tacoma Narrows Bridge Collapse Videodisc* includes sections on instructional guidelines, how to use the disc with various players, instructional strategies and goals, enrichment materials, technical and programming information, and acknowledgments along with the student laboratory pages.

The Wiley student manual is superior to any that we had developed and I use the Wiley materials with my own students. In fact, last spring I met an educator from France who had purchased both a PR-7820 player (which arrived from the USA with no manual), and also a copy of the Tacoma Narrows Bridge disc with the Instructor's Manual. He actually taught himself how to program the 7820 player from the Teacher's Manual published by Wiley.

By the time all of this was done it was near the end of May and the physics teachers' meeting was only a few weeks away. We still did not have the OK from the NSF. It finally came, but they insisted that UN-L keep the copyright. So with that last-minute change and pushing on both sets of lawyers for a quick agreement we managed to get the videodiscs mastered in California, the jackets and Teacher's Manual published in New York; and I put a few of them together in Nebraska as I headed for the airplane to leave for the summer physics teachers' meeting.

During these long discussions with Wiley I feel that I convinced them that videodiscs were really the educational concept of the future. While hardware sales are still slow in the educational marketplace, nonetheless at a price of $150, I consider the Tacoma Narrows Bridge disc one of the outstanding bargains in the videodisc industry. In addition, I think the total package sets a standard to be matched by every new educational videodisc that comes onto the market.

In closing, I hope that all of you producers of educational videodiscs will feel as positive about your final product as we do. We do not think that the Tacoma Narrows Bridge disc is perfect. In fact, we have discussed remaking it sometime in the future. But remember, it was conceived in 1979, a lifetime ago in a high technology industry. In fact, not one model of videodisc player that was on the market in 1979 is still available. The Tacoma Narrows Bridge disc stretched the technology of its day and presents a challenge to educational designers of today.

Finally, the total package as put together for us, and for you, by John Wiley & Sons is of the highest quality. All of us will be fortunate indeed if future educational videodisc projects are published and marketed with such expertise.

Robert Fuller is Professor of Physics at the University of Nebraska-Lincoln. In addition to developing the first commercially available college-level interactive videodisc with Dean Zollman and Thomas Campbell, Professor Fuller has for several years been involved in the development of experimental courseware using computers and videodiscs. He serves as the Director of UNL's ADAPT program, a multi-disciplinary Piagetian-based program for college freshmen. He is past-president of the American Association of Physics Teachers (1980). In 1973, he received UNL's distinguished teaching award. He obtained his Ph.D. from the University of Illinois in 1965. His undergraduate degree, also in physics, was from the University of Missouri-Rolla in 1957.

# What Can You Do With A Disc?: A Classroom Teacher's Perspective

*Phil Kessinger*

**WHAT CAN YOU DO WITH A DISC?**

Elting E. Morrison, a noted American historian, recently wrote about Disney's EPCOT, "The real problem is not to find out what the machinery can do—it can do almost anything; it is what will we do with the machinery?" Teachers, parents, and consumers in general are finding this is also the central problem with new technology. What can one do with a videodisc? We know that it has massive storage, quick access to key information, and excellent sound capabilities yet how do we use it as this heralded powerful new process of instruction and learning?

What follows are some ways teachers and parents can use videodiscs with excellent educational results.

**USING A SINGULAR VIDEODISC PLAYER**

Assuming that the budgets of many schools would not permit the purchase of any but the least expensive videodisc player, there is still a great deal one can do with older DiscoVision feature films (CAV); for example, "Tom Sawyer," or easily available interactive consumer discs on music (all kinds), cooking, football, jazzercise, photography, running, painting, twentieth-century history, etc. One of the most powerful features is the still frame capability. Following is a strategy for using it.

1. Preview the videodisc once in a general sense, noting how it would be related to teaching objectives.

2. View the videodisc a second time with frame display on, making an index of important frame numbers related to teaching objectives. This can be easily done with still frame or 3× (three times normal speed) forward or reverse to find a specific location.

3. The videodisc could first be presented to students as series of selected still frames (for example, key teaching objectives), and then have the students take notes either with or without directions as to what to take notes about. Or else a teacher could make a simple worksheet, to insure time on task, for all students to complete while viewing these still frames.

4. The videodisc could then be shown from beginning to end either with or without students asking questions as the videodisc progressed. This is quite easy to manage with still frame or pause functions on any player. The simple capability to stop or start an audiovisual presentation without losing educational momentum is quite effective.

5. Review/test: after showing the entire videodisc, important teaching objectives might be relocated again and discussed in class in terms of the context of the entire presentation. Students could ask their own follow-up questions for further study; and relevant parts of a videodisc could be quickly searched for precise class review and discussion. In addition to traditional

written tests, a test could be made using still or short segments from a videodisc, much like a "lab particum" test.

I have used this approach with a 21-minute videodisc, "Amish: A People of Preservation," in a United States history class, with surprising results. By being so specific in presenting this information, students were able to recall information in a detailed fashion without vague generalities. Further, students were able to conceptualize at a higher level of thinking in comparing specific information from the Amish videodisc to previously studied textbook information.

Combining the previously mentioned still frame approach with many of the consumer videodiscs that are now chapterized, one can easily deal with the information in a specific and controlled manner. Using a photography disc, "Creative Camera," one can study certain areas at a level that is appropriate for a certain learner. For instance, if a parent wanted to teach a young person how to take a photograph, information on a videodisc could be played without the audio, while the parent made explanations his or her child could understand. Or if one simply wanted to learn how to watch professional football, become a better gardener, create chocolate mousse, jog two miles, or become an expert on Van Gogh, such learning is fun and quite easy to do at home with a videodisc.

## USING A VIDEODISC/MICROCOMPUTER COMBINATION

It has often been said that a good film, or any good medium, does not need an introduction; it presents itself. But in a classroom, it is poor judgment to take students for granted and assume that they can "see what a teacher sees." With a videodisc player interfaced with a microcomputer, however, a teacher could present media to students as it stands and then follow it up in a review manner that was never possible previously with a 16 mm film presented in class.

Using the authoring system from the "Villa Alegre: Videodiscs-Microcomputer Learning Kit" (produced by the Videodisc Design/Production Group, University of Nebraska-Lincoln), I approached this problem of students really understanding a film shown in class in a new way, using a chapter authoring program (CAP) included in this kit. Basically it is a clear and easy way to establish an annotated menu for any videodisc. In this authoring system, each videodisc is considered a *file*. Each *file* may have up to nine *chapters* and each *chapter* may have up to nine discrete retrievable video *segments*. Consequently, with the use of this authoring system any videodisc could have an overall menu of 81 video segments (9 *chapters* × 9 *segments* = 81 video selections).

In teaching social studies (social science), a teacher is always trying to assist students to develop concepts to use as building blocks for greater understanding and insights into society. Three large concepts that regularly occur in any social science class are politics, economics, and culture. Using the film on videodisc "Lonely Are The Brave" (with Kirk Douglas and Walter Matthau) and the authoring system from the "Villa Alegre: Videodiscs-Microcomputer Learning Kit," it was quite simple for each side of a videodisc (*file*) to be made into three *chapters:* one on politics, another on economics, and a third on cultural/social issues. Thus it was possible to retrieve specific portions of the film for class review and accurate discussion. A fourth chapter was created for each side of the videodisc to divide it up into nine distinct segments, so that the user could search for specific unidentified parts of the videodisc under computer control. One of the nine segments would play a videodisc side through from start to finish.

The hardest part of creating this videodisc menu to review a film shown to class was studying the frame numbers. But this effort also turned out to be quite creative, because it was similar to re-editing the film. One had the power to stop and start a scene exactly where it seemed a teacher would want to identify political, economic, and cultural concepts in American society for students to learn.

A great fringe benefit is that in once producing the best model for complex videodisc/microcomputer review of a film for students, each class then receives the same quality of educational effort in a teaching day. Teachers using a videodisc/microcomputer film review, and who have 170 students divided among six classes a day, will teach a class at the end of the day with the same high-quality results as when they teach in the morning and both teachers and students are sharper.

As a guest on one of Johnny Carson's "Tonight Shows," Kirk Douglas commented that he thought the film "Lonely Are The Brave" was one of his better films, commenting that he felt he "did everything right."

Teachers using a film review process such as the one described here would probably feel the same way—that they were using videodisc/microcomputer technology in a precise and creative way to help students build a better understanding of society.

Philip C. Kessinger has been a teacher of secondary social studies at Eugene Public Schools, Eugene, Oregon since 1969. He has a BA in history and has done graduate work in history, education, filmmaking, and school psychology. He has produced film, video, and numerous slide/tape productions. He has been awarded a grant for Multi-Image slide/tape production and recently completed a grant from the National, Oregon, and Eugene Education Associations for videodisc in-service for teachers. He is currently working on interactive videodisc projects for educators.

# Future Developments

*The following eight articles are not related to the videodisc development process per se. That is, unlike the previous articles under the categories of analysis, design, preproduction, production, etc., these articles are speculations about the future of videodisc technology. Experienced videodisc developers and users from such diverse professions as engineering and computer science, educational administration, show business, library science, video consulting, hardware manufacturing, instructional design, and publishing were asked to take a brief look at what the future of videodisc technology will mean from their own individual perspectives. Subjective by necessity, the following viewpoints offer a multitude of authoritative opinions on the future of videodisc technology from individuals who are uniquely positioned to help shape the future of this vital new medium.*

## A TECHNOLOGICAL PERSPECTIVE

*Gershon Weltman*

The path of technology is strewn with the remains of people who have tried unsuccessfully to predict its course. Videodisc technology will no doubt prove no easier to fathom. The only hope we have is to study the past, try to discern the present, and at best establish logical directions for the future.

The idea of videodisc begins with television. In the public mind, television brought movies, plays, and live entertainment from afar, by unknown means, into the home. Once in the home, interaction with this novel medium was minimal. Turning it on and off and selecting the channel were about the limit of user involvement.

Videotape introduced the concept of controllable, interactive video to the public. A videotape cassette could be held in one's hand and played at will. Through its use, video programs could be displaced in time; and the video produced by the cassette could be shown faster, slower, forward, backward, and even one frame at a time. Thus video was identified as a medium that could be compactly contained and radically manipulated.

Videodisc is the logical extension of this identification. On a disc, containment is even more compact, and the potential for interactive manipulation is even greater. The question in people's minds is, "To what purpose should this great interactivity be put?" There is no one answer, but perhaps the most interesting and most important purposes will involve the simulation of visual experience. Consequently, the major technical innovations in the immediate future will go to make such simulation more natural, more realistic, more entertaining, and more economical.

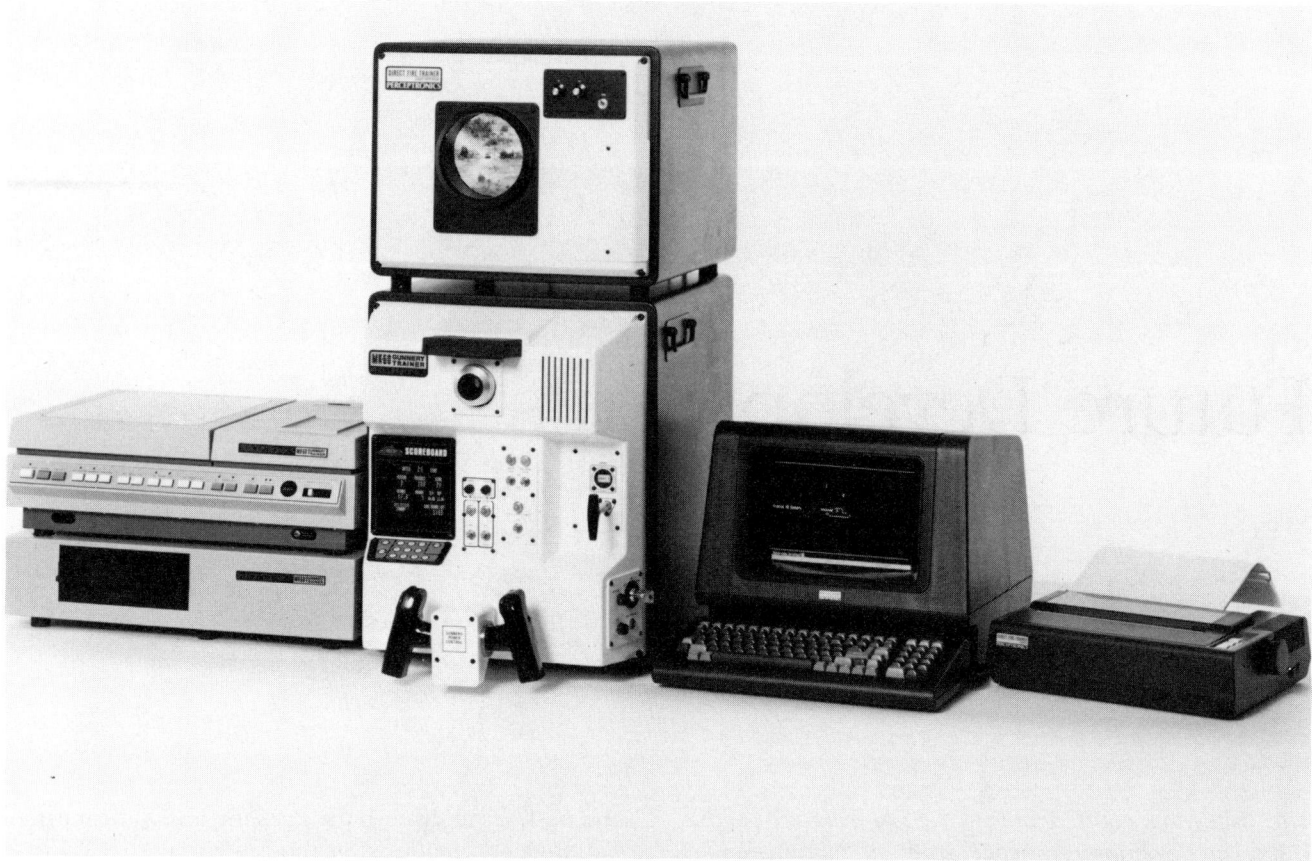

The Perceptronics MK Series Direct First Training System

The current use of videodisc for visual simulation is illustrated by products like the MK-60 Tank Gunnery Trainer, an M60A-1 tank; the LaserTour Exercise System, in which the disc provides a realistic bicycling scene; and the ActionCode Instructional System, in which the disc substitutes for direct views of an oscilloscope and other electronic equipment. Such simulations are made possible by the unique characteristics of the videodisc, coupled with the power of the modern microcomputer.

Since our own company has a stake in the future of videodisc technology, here are a few of the things we are working on. We are developing new techniques for enhancing videodisc-based simulation, including ways to "paste" many videodisc pictures together, so that one can view a large video scene. We are also developing new ways to insert one video picture into another, and to combine pictures with sound and motion effects. Video that reacts to the viewer's movements is already here; video that reacts to the viewer's thoughts may not be far away. So although it's difficult to predict the future, it is well worth waiting for.

Illustrated are the currently available products within the MK series. These are, left to right: a videodisc player, which provides the target problems, and below it the floppy disc drive, which stores the simulation software; the gunner's console, which simulates the actual gunner's station, and above it is the sight repeater, which shows the instructor the gunner's aim point; the instructor's console, which permits control of the training program; and a printer, which provides a permanent record of performance.

Dr. Gershon Weltman, Ph.D., has served as principal investigator and program director for a wide variety of research and development programs at Perceptronics. In general, the programs for which he has been responsible involve innovative applications of modern computer and display technology to man/machine systems; to these programs he brings an extensive background in human factors design, computer science and experimental investigation. Dr. Weltman received his BS in engineering, his MS in engineering and his Ph.D. in engineering from UCLA. He completed a post doctoral fellowship at Weizmann Institute of Science in Rehovoth, Israel.

Teaching using interactive videodisc for a group lesson. Project VIM classroom site.

## EDUCATIONAL PERSPECTIVES

*Frank B. Withrow and Linda G. Roberts*

The applications of videodisc technology in educational settings have been much slower to develop than those in training and in the military. Contrary to assumptions, however, there are videodisc players in elementary and secondary schools. A telephone survey made during the summer of 1983, of 330 randomly selected media directors, indicated that 59 of their schools had laser videodiscs which were being used primarily as audiovisual devices. Teachers were able to index and retrieve portions of the videodisc material on demand, unlike their experiences with slides, filmstrips or movies. Only a few, however, reported use of the full capabilities of computer control of the videodisc.

These findings by the Great Plains National Instructional Television Library parallel our experiences with the Videodisc Interactive Microcomputer (VIM) Project coordinated by the U.S. Department of Education, Division of Educational Technology. The VIM project involved forty-five elementary schools across the country, who had both videodiscs and microcomputers on site. The project provided experimental videodiscs, an interface between the videodisc and computer, an electronic mail system, and an authoring system. Sites that had more than one computer were more likely to try out the interactive videodisc format for individualized student instruction. But a single learning station was not considered as useful as was small group or whole class instruction. The primary use of the videodisc in traditional audiovisual format is significant. Teachers report that videodisc control now makes possible far greater use (in a variety of ways) of "Music Is," one of the programs made available on videodisc by the U.S. Department of Education.

Example of closed captions on videodisc.

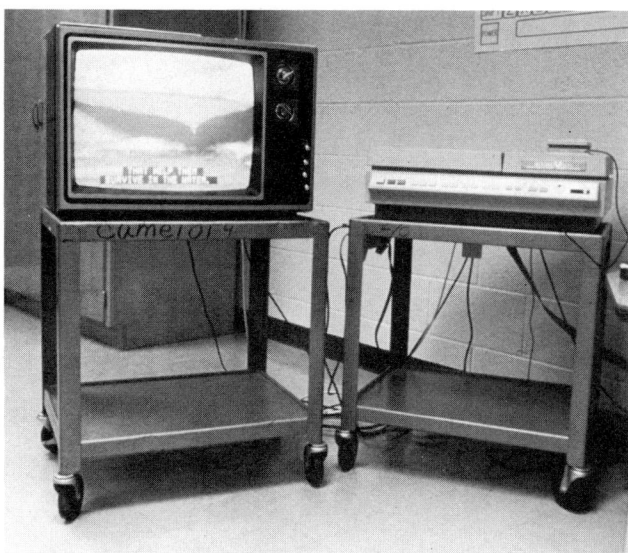
Using VIM script, an authoring language developed for demonstration sites, a teacher and student create an individual learning sequence.

The VIM Project was specifically designed to allow for maximum independence and local control. The teachers liked the flexibility of videodiscs, but they had neither the time nor the expertise to program interactive sequences. They needed well-developed lesson materials that made use of the videodisc formats and at the same time allowed them the option of individualizing lessons to suit individual or class needs. In general, the VIM Project indicates that there is a place for this new technology in the classroom. It was clear, however, that extensive teacher training and staff development, continuing support for operation and maintenance of equipment, and opportunities to try out and evaluate new approaches to teaching and learning processes were absolutely necessary.

## WHERE DO WE GO FROM HERE?

From these experiences and those in industry and the military, we can make some educated guesses about how to design educational videodiscs and how to use them effectively. Technology in general, and videodiscs in particular, require a clear definition of educational goals and a systematic strategy for achieving those goals. It implies a new instructional strategy that manages both human (teacher/tutor/trainer) and material (videodiscs, computer software, textbooks) resources with an understanding of the unique capabilities of both. In addition to the uses in group and individual instruction, there are opportunities to provide stand-alone instruction where other resources are simply not available. For example, it becomes possible to provide instruction in physics or economics in a school site with a remote teacher, and it becomes possible to provide drill in basic reading vocabulary to non-reading adults in a library, community center, or at home. What then are the barriers to be overcome?

### Cost

The cost of videodisc technology is considered a barrier to use by the education community. In its most elaborate configuration, i.e., high-resolution monitors, touch screen, and expanded computer memory, each learning station can reach a cost of $14,000. A learning station without the bells and whistles can be put together for $1,800. If we amortize the capitalization costs over a five-year period and assume 900 hours of use per year, the per-hour instructional cost is $3.10 for the most expensive and 40 cents for the least expensive configuration. If we raise the utilization rate to 1,500 hours per year, then the costs drop to $2.15 and 27 cents respectively. With even further reductions in the manufacture and delivery of videodisc systems, our projections demonstrate that videodisc educational tools will soon be affordable by schools.

## Software

As with other technologies, software is a major barrier to be overcome. Several things need to happen. One is to develop software through a comprehensive effort that incorporates leadership from the federal government, involvement of the private sector, input from instructional designers and researchers, and active coordination and implementation at the state and local level. The crisis in education mandates that the federal government develop mechanisms that allow for comprehensive programs such as those demonstrated by the development of "Sesame Street," NASA's landing of a man on the moon, or, as Education Secretary Bell stated in testimony before Congress, a "Manhattan Project."

Based on recent conversations with private, public, and government sources, we are encouraged by the promise of a number of such significant development efforts. It looks likely that several projects will be under way shortly, which will include the following:

- High school mathematics and science programs (70-100 discs);
- Literacy training for beginning readers and nonreading adults, including materials in English as a second language (50-75 discs); and,
- Vocational and technical education skills training, which builds on the base of materials already in development in the military and industry (100-150 discs).

## Training and Staff Development

At the same time, there is the need to deal with the urgent problem of how teachers and schools will use these new tools. There is no question that inservice staff development is a critical factor in helping teachers to use not only this technology, but other technologies as well. Many teachers are skeptical of this new wave; many remember the unmet promises of early teaching machines; others point to the need for more high quality software. As in software development, a cooperative effort is needed, involving the private developers, the schools of education, the local districts, the state departments of education, and the federal government. New approaches, hands-on experiences, and team efforts must be employed to encourage new uses and facilitate new and expanded instructional skills. An example of how teachers can be trained and at the same time become part of the team effort to develop videodisc courseware is presently under way in a joint project between the Lynfield and Lexington Public Schools in Massachusetts and the Digital Equipment Corporation.

## Summary

Over the last three years we have seen a growing interest in education and videodisc technology, particularly for applications at the secondary level and, more recently, for applications in elementary levels and in nontraditional and distant education settings. The growing demands for additional mathematics, science, technology, and foreign language instruction, coupled with critical teacher shortages and limited educational resources, increase the educational need for computer/videodisc systems. Reaching the future promise will require new and imaginative cooperation between all of the educational stakeholders. There must be a means to bring together the knowledge gained from research and prototype development with the experience and practical working expertise in our schools and educational institutions without sacrificing content or production quality, good teaching, and positive human interaction. The technology for future education rests in part with the marvelous technological advances we have made. To realize the promise, we can learn from the past.

## THE PAST IS PROLOGUE

The advances in civilization and intellectual development have been marked by the search for better and more efficient symbolic communication formats and systems. The development of speech and language was a quantum jump forward for the human species. The phonetic alphabet and the Gutenberg printing press revolutionized learning and thinking. Telephonic, radio, and broadcast television communications changed the social, political, and economic nature of the world. The present informational technologies built on silicon chips seem likely to advance these changes even further, and provide new opportunities, new demands, and new dreams.

Dr. Frank B. Withrow is director, Division of Technology, Resource Assessment and Development, in the Center for Libraries and Education Improvement, U.S. Department of Education. He is an experienced elementary school teacher, supervising teacher, researcher, and administrator. He has produced films and television, microcomputer courseware, and has contributed significantly to the development of technology programs for the handicapped. His research efforts have been directed towards basic studies of perception and memory of both auditory and visual stimuli. He has also taught at the university level. He was a Battelle Memorial Fellow and studied the applications of interactive television to the teaching of language and the education of handicapped students.

Dr. Linda G. Roberts is an OERI associate in Technology, Center for Libraries and Education Improvement, U.S. Department of Education, conducting research, analyzing trends, and projecting future uses of computer, interactive video, and telecommunications applications in education. She came to the Department in 1980 as a National Policy Fellow, and has served as a consultant to the U.S. Congress, Office of Technology Assessment. Her case studies of technology implementation at the local district and state agency level are part of the OTA Report, "Informational Technology and Its Impact on American Education" (1982). Before coming to Washington, Roberts was a classroom teacher, university professor, and academic dean, as well as a consultant to many school districts, publishers, and television producers, including the Children's Television Workshop. She received a B.S. in child development from Cornell University, an Ed.M. in elementary education from Harvard University, and an Ed.D. in curriculum and instruction from the University of Tennessee.

# THE ENTERTAINMENT VIDEODISC

## (CHAPTER TWO)

*Ken Christie*

It's 1984 and the videodisc industry has finally achieved the legitimacy it has been seeking ever since those first Magnavox players went on sale in Atlanta six years ago. Video games provided the necessary shot in the arm; and now the hardware manufacturers are turning a profit and the software explosion has begun. Millions are interacting with videodiscs at Epcot Center and world's fairs, a variety of videodisc-based games populates the arcades, and education and training practitioners continue to expand into videodisc courseware. For the first time, you don't have to explain to people what a videodisc is. The word is out!

So what's next? What does the future hold for videodiscs? What's possible in the area of entertainment? What can we reasonably expect to see? Will the medium even survive?

Virtually anything is possible. Consider the potential entertainment value of the following ideas.

- Multiple display arcade games where groups of players compete on the playing field simultaneously.
- Full-fledged interactive simulators, using videodisc and wrap-around projection techniques.
- Three-dimensional experiences. It would be hard to find a better way than the disc to insure rock-solid, headache-free imagery.
- Hologram storage on videodisc.

A really exciting possibility that is already being explored is interactive theatre, where audience members dictate the action from voting consoles built into the armrests of their seats. Imagine being able to determine Indiana Jones' fate or chart the course of the *Starship Enterprise* yourself!

Someone may also develop a way to produce a high-quality videodisc outside of a clean room environment. (Editor's note: see John Winslow's article: "DRAW Technology.") Videodisc recorders could become as commonplace as audio tape recorders are today, and Fotomat stores could add mastering and replication to their list of services!

There is no limit to what's possible in the field of entertainment, but the success of any of the above ideas will be determined by market factors too complex to speculate on at this time. What's likely in the near term is easier to predict.

The most immediate development should be more and varied interactivity. New types of interfaces—touch screen, voice recognition/synthesis, eye tracking, pressure and moisture sensing—will elevate action games to more than just a hand-eye coordination experience. Computer graphic overlays will make possible much more realistic and effective simulations and mini-movie arcade games.

If Coleco's planned videodisc computer peripheral is successful, other manufacturers will follow suit. The cost of players will drop even further and the consumer market for videodisc education and games will finally get a chance. That's when the fun will really begin, because the demand for interactive entertainment, effective courseware, and generic videodiscs will be great.

Will the videodisc survive? Yes, in one form or another. It is not the end-all, be-all technology; it will eventually evolve into a recordable, very rapid access memory device. But that's all right, because the ideas will survive. Videodiscs will influence the way we learn, the way we play games, the way we interact with machines—even the way we think. It's a tremendously exciting time for our industry! Let's get to work.

Ken Christie is a videodisc designer for WED Enterprises in California. He had overall responsibility for design and development of over 80 disc projects for the EPCOT Center. Prior to his work with Disney, Ken was Unit Director with the Nebraska Videodisc Group. He worked as an on-air reporter for the CBS affiliate in Omaha. Ken holds a BS in mass communications from the University of Nebraska-Lincoln.

**Future Developments**

# ARCHIVAL PERSPECTIVES

### Henry A. Bertram

*"For now, the videodisc is a true wonder ... waiting for our imaginations to catch up with its capabilities."*[1]

Archives are often perceived as seldom-used, musty vaults containing the records of human history. Huge amounts of time and effort go into preparing, cataloging, and storing those records. That task grows larger every day. Information is generated so quickly that traditional storage methods are hard pressed to keep pace with the volume and the need to quickly access data.

Recognizing the shortcomings of traditional archival techniques, various governmental agencies, research groups, and commercial enterprises are now engaged in archival projects using one or more of the emerging videodisc technologies.

## ARCHIVAL VIDEODISC EXAMPLES

NASA/JPL already has one million plus planetary images on ten videodiscs collected from JPL spacecraft since the 1960s. The Smithsonian Institution's Air and Space Museum is producing a series of discs containing over a million photographs of airplanes, aviators, airports, etc. A service known as "Patsearch" has more than 700,000 recent patents on disc. The National Library of Medicine has a disc containing its pre-1865 History of Medicine collection of over 100,000 prints and photographs. An Optical Disk Pilot Program is under way at the Library of Congress, consisting of a print and nonprint program. The print program preserves and accesses text materials while the nonprint portion preserves and accesses still photographs, drawings, motion pictures, television programs, and sound recordings. The National Gallery of Art is marketing a disc offering a tour of its halls.

Universities have also been pursuing the archival uses of videodisc. Several departments within the University of Iowa joined in producing a disc that contains thousands of images from art, journalism, and pediatric medicine. Some 10,000 textile stills are stored on a disc used at the University of Wisconsin. Real estate photographs are kept on disc for use by the tax assessor's office in Massachusetts. The common denominator of all these examples is the need to store print and pictorial images in such a way that with minimal effort one can *rapidly* access them.

## ON THE HORIZON

In the immediate future we can expect more projects similar to those already mentioned. Of special note, though, is the announcement by Grolier, Inc. of an electronic encyclopedia, which will combine laser videodisc technology with computer-generated text to create an audiovisual encyclopedia. Technically an expansion of *Academic American*, a multivolume general encyclopedia currently available via videotex from BRS and others, the coming audiovisual version will be an important step towards a truly complete archive of information. Not only will a textual description of an historic event, such as President Roosevelt's appearance at the New York World's Fair on May 1, 1939, be available, but also a filmed recording can provide the sights and sounds of the actual event.

Business applications have been given a boost by the mid-1983 introduction of both digital and optical disc recorders from Panasonic. Future offices will store 10,000 letter-size documents that have been recorded *on-site* on an optical disc. The documents can then be sent to remote locations (intraoffice, transcontinental, etc.) to be printed by facsimile terminals. Thus, documents will be stored on disc for instant retrieval and subsequent transmission.

Discussions are under way regarding the use of videodisc for storage in the following areas:

- weather and land-sensing satellite data
- significant medical phototransparency and film/video collections (pathology, histology, neurology, anatomy, etc.)
- antique automobile photographs and motion footage
- major motion picture studio prop and costume inventories
- coin and stamp collector catalogs
- police "mug" shots
- television programs
- architectural history
- "living" genealogical charts

Undoubtedly, many more uses have either not yet been publicly announced, or will continue to be shrouded in secrecy as in the case of Defense Department work.

---

[1] T. Onosko, "Vision of the Future," *Creative Computing*, January 1982.

## IN THE CRYSTAL BALL

Many good ideas for archival use of videodiscs are yet to be developed, such as accessing libraries of videodiscs from our home telecenters. You would use microcomputer terminals with communications ports, and fifth generation "intelligent" computers could help you access videodisc archives of classic movies, debiting your bank account for the "admission" cost! You might also use your terminal to search videodisc archives of collectibles and locate items such as that rare painting needed to round out your collection. Or perhaps you'd like to browse through an archive of the complete creations of a notable political cartoonist indexed by subject, character, theme, and major political figures depicted.

The increased use of two-way video communication may someday enable you to store an audio/video data base of all your valuable household items on a remote disc-for-hire system owned by your insurance company.

Most towns and cities have title insurance companies whose job it is to research the history of a parcel of land, and who maintain photographic records of their communities. Through a network of the records of such companies stored on videodiscs, an architect/researcher could easily determine the look and feel of an entire block during a given decade. This capability would save an enormous amount of research time for redevelopment projects.

The possibilities are endless. Imagination truly is our final frontier and only boundary.

Henry A. Bertram has worked in the educational media field for over thirteen years. For the past five years he has been Division Head of the Learning Resources Center, School of Medicine, University of California at San Diego. Mr. Bertram studied architecture and communications at the University of Hawaii and earned his undergraduate degree in Telecommunications and Film from San Diego State University. His graduate work in Educational Technology was also done at SDSU.

## *TRENDS IN THE INTEGRATION OF COMPUTER AND VIDEO TECHNOLOGY*

*Mark Heyer*

### INTRODUCTION

In the 1890s a balloon-tired car won the Paris to Madrid powered-vehicle race against the expert opinions of all knowledgeable observers. "Clearly a 12-passenger, steam-powered omnibus, with kitchen, bedrooms, and observation deck will be superior for such journeys," they said. "These automobiles are so flimsy, and you mean everyone has to buy their own? Never happen," they said. Cars still use balloon tires today.

History will dutifully record the future uses of videodisc technology in large document storage systems, schools, training and consumer programming. Like many inventions of the past, however, the real benefits will come from uses that were not foreseen by the original inventors. The phonograph, originally invented as a dictating machine (because it wasn't clear why anyone would want to record a voice anyway), languished for years before the consumer possibilities were realized.

The two great information steam engines of today are the computer and the television. On one hand, computers give us virtually unlimited control of text and numbers, but not pictures and sound. Broadcast (including cable) television, on the other hand, is a very good "mass transit" information system, but is beyond our direct personal control.

We will find that the future, which now seems like a chaotic rush of new technology, is actually a process of exploiting fundamental capabilities and techniques, which, like Leonardo's wood and cloth helicopters, have always been available to us in conceptual form if not in practice. For the first time in history, our visual information environment can be put directly under the control of our individual minds and wills.

The combination of computers and videodiscs can be used to increase personal processing capabilities, perhaps based on whole new languages using spatial graphic symbology rather than text. In effect, we may now be inventing electronic hieroglyphs.

We should be looking today for those elements of new information technology that will be standard features of these devices for the indefinite future.

### COMPUTER CONTROL

Watching television is like taking a train. To do either you first have to go to the station and be there on schedule. At the station you are likely to witness murders and muggings. Once on board you had better like the destination, because trains and television do not deviate from their assigned courses. Both are "regulated" by governments and operated by unwieldly bureaucracies.

Private ownership is the historic power of both publishing and automobiles. But with video/computer publishing, our personal information environment is now defined by what we choose to own (rent, borrow, etc.,) and like a car we can travel to any intel-

# Future Developments

lectual destination at all, as often as we like and by whatever route we decide to take, with our right of ownership and freedom of use protected by the Constitution.

The engine for our new information and entertainment automobiles is the microprocessor. The videodisc provides a worthy artistic and theatrical medium with which to define our excursions and flights of fancy. What are the basic features of our personal information automobiles? Where are the balloon tires? "Hello, Mr. Firestone?"

The exact future of computer control is difficult to predict, since most of the applications have yet to be invented. But we can see that the present trends will continue to reinforce each other. Computers will get smarter, faster, and cheaper, while videodiscs get smaller, denser, and recordable.

Certainly the desktop assembly of various boxes and wires that we now think of as a computer/disc system is not the future. Imagine suggesting to the public that they assemble their own cars from parts made by different manufacturers. We need integrated devices that are specific to applications: portable and rugged units for industry and the military; cheap and reliable for consumers, and so on. The crude horseless carriage computers of today will become the information automobiles of tomorrow.

In the course of these developments, the videodisc and computer will lose their individual identities. The ultimate interface between computers, video, and humans is the screen. When both computers and video start using the same screen, both are changed completely and forever. As in the Wizard of Oz, the Scarecrow has been given a brain and the Tin Man a heart.

## VIDEO PLUS COMPUTER GRAPHICS

For as long as people are using computers and video together, we will be combining images from computer and video sources. This technique is commonly used in television production, especially for sports and news. Now, for the first time, the combination can take place in a low-cost presentation system in which the display is interactively determined by the computer, video, and viewer.

In the arcades, "superimpose" means computer-generated adversaries coming at you out of Star Wars level video special effects. In automated diagnostic systems, for example, pictorial images of circuit boards can be presented with computer-generated circles, arrows, and prompts, such as "Put your probe here and read the voltage."

The use of superimposed graphics is not in itself startling. What is surprising, though, is the range of new options that this simple change produces.

Superimpose is the technique that blurs distinctions between computers and television. Computers can obtain their images from cameras or videodisc visual memories, manipulate the image, add or subtract information, and present it to the viewer. Every personal computer will have it built in automatically as standard equipment.

Along with this new freedom to manipulate video images and computer graphics, we are also seeing a revolution in our ways of thinking about the audio portion of our intelligent information systems.

## COMPRESSED AUDIO

Like superimpose, compressed audio, sometimes called still-frame audio (SFA), is one of those techniques that will be used in some form forever. The original premise was that if you could attach audio to each of the video still frames on a disc, then the playing time, and hence the usefulness, would be improved. Presumably the audio would somehow be "buried" in the video picture.

This assumption is unrealistic, given the physics and mechanics of video, audio, and discs. The method that does work, however, is another example of a small change in our way of thinking, which creates vast possibilities for the future. In essence, some video frames can be used to store audio information in compressed digitized form instead of in video. This "compressed" audio can be dumped from the videodisc to a buffer memory, where it resides until commanded to play back. The buffer may hold 40 seconds or more of audio at one time. Since 50-to-1 compression is practical with good playback fidelity today, a single disc side could hold more than 20 hours of audio, but with no video frames. But we are not even close to the limits of data density today. Using data storage technology available even now, 1000-to-1 compression is entirely possible.

In fact, some combination of video and audio is generally used. But since the audio is now living in computer memory, it can be played back with one video frame, a sequence of frames, or a motion sequence. It can also be internally random accessed so that the SFA can be used for computer-directed responses quite independent of the video.

The disassociation of video and audio is another of the conceptual breakthroughs in understanding the technology of the future. Up until now, films and television have proceeded at fixed speeds for the convenience of the audio track, which in both cases runs along the side of the picture area and must be kept in constant uniform motion to produce output. (A secondary, but important reason is that mechanical film projectors were hard to control dynamically; and broadcast and cable systems are required by physics to send out a new frame every 1/30 sec.)

Now, with separate audio and video storage and playback, we can foresee an entire new branch in the taxonomy of video/computers based on dynamic and independent control of each.

For example, a small dictionary of words could be contained in one audio segment. Since the computer can assemble these brief segments in real time, a form of computer-generated speech would be produced using real voices and sounds, not some arcane algorithm. Thus, small computers can be given the power of realistically assembled speech while the process of creation requires nothing more complex than a microphone and a tape recorder.

For example, sound effects can be created by recording an entire event, like a jet engine going from idle to full throttle. For playback, the effect is produced by dumping the entire event into the audio buffer, and then continuously repeating a small "window" of sound over and over. Viewer control of the throttle (perhaps on a touch screen) moves the window position and thus changes the effect.

With the advent of faster searching disc players and larger audio memories, it is clear that the way is open for computers to say anything we want them to in any voice or person we wish to give them. Here we see the power of the medium arising from the computer's ability to manipulate basic information taken directly from the real world. The videodisc SFA speech synthesizer, requiring only simple audio recordings, can produce any language, sound, or music we can imagine. Compare this to the difficulty of providing a computer-only synthesizer with anything even resembling inflection or dialect, much less multiple languages.

There are so many ramifications to the separation of audio and visual information that we will spend the next twenty years just trying to sort them all out. There is, however, yet another disc technology that will be basic in reshaping our relationships with video and computers. That is the technology of recordable and/or erasable videodiscs.

## DRAW AND EDRAW

Like "conventional" videodisc technology, the promise of record-once Direct Read After Write (DRAW) and Erasable (EDRAW) videodiscs lies most provocatively in those areas that are completely unavailable when using traditional methods.

First, let's think a little about office automation. The biggest holdup in office automation is that it works perfectly in a paperless world, but doesn't work at all if the outside world sends any paper whatever. At least half of all the information in the business world does not go peacefully into computers. By allowing the actual images of paper documents to be stored and integrated into the computerized office information flow, the problem is solved. Workers can then store and send visual documents via the office automation system. I believe that all successful future office automation systems will incorporate this element. As a matter of fact, the future of the office itself is in doubt, but that's another subject altogether.

Present technology in recordable videodiscs indicates that typical systems will be able to store 20,000 or so documents on a single disc, with resolution comparable to a photocopy. Clearly we will see very large systems put to use in companies like American Express and the insurance industry. The major computer companies are already staking out the territory, but this is not where the excitement is.

## THE REAL USES FOR DRAW

Still photography cameras are designed to take a picture every now and then. Video cameras record at 30 frames per second no matter what. With DRAW machines, we can record at any frame rate that complements the action: one frame per second, or one frame every 10 seconds. With a portable recorder and palm-sized solid state camera, we could literally document our entire lives in exact detail. We could record every transaction we made, every person met, job done, or place visited; the video version of a yellow pad.

Mounted in a car, the DRAW recorder would allow us to record and replay trips. Connected to a small computer that monitors our mileage, it becomes a visual prompting system that will also allow us to rehearse our trips in visual detail before making them. With the addition of graphics superimpose, we could make notes on the visual images.

Erasable EDRAW discs add even more fuel to the fire, since computers could now record everything all the time if desired, keeping only those images that are useful. This "reality data base" is the practical definition of short term and long term memory. Combined with even simple speech recognition and the ability to remember, recall, and play back realistic sounds from the environment and a few other simple

# Future Developments

techniques, and with no more computer power than is available today, we have gone a long way toward producing machines that might be considered quite perceptive and intelligent.

## CONCLUSION

When we look around at our business world we see it dominated by apparently invincible giants: IBM, AT&T, GM, RCA, etc. We tend to think that these mature and powerful companies will rule forever. What chance is there if you are not one of the giants? Will no one ever build another great company?

History tells us that things can change very rapidly indeed, given the right circumstances. One hundred and fifty years ago the Watt Steam Engine Company was a giant. Less than 100 years ago, railroad companies dominated the American landscape. Each was displaced by new companies that exploited new technology in a way that was inaccessible to the giants. By the time these corporations caught on to what was happening, it was too late.

Certainly the giants will try to bend the new techniques to their ends, but they will bend them into the comfortable shapes of the past. Such has been the fate of videotex, cable, and perhaps the most bizarre contortion of all, the laser videodisc used as a medium for publishing old movies.

The phonograph (early models were recordable) was originally invented as a dictating machine and failed. The "authorities" of the time were absolutely certain that it had no redeeming entertainment value whatsoever. Imagine their reaction to hi-fi, multiple track recording, the Beatles, and a multibillion dollar record industry.

I believe that we are at a critical point today. In the seeds of the enterprises we see sprouting around us are the makings of the information industry giants of tomorrow. The door is open for a modern Edison to develop a whole set of new tools for cultural growth, expression, and exchange. Thousands of companies and millions of careers will be made in the course of exploiting the combination of computer and video technology.

The message is that we shouldn't be too impatient, or too hard on the trail blazers. The power is clearly in the technology, even if we can't reach for all of its subtlety and sophistication today. Watching or, better yet, participating in the unfolding will be one of the great thrills of all time. Stay tuned.

Mark Heyer presently works out of his management company, The Braden Group, to develop systems for efficient creation and delivery of video/computer publishing products. Mr. Heyer became interested in videodiscs when they were first shown in the early 1970s. His background spans from physics and computers to multimedia publishing. Mr. Heyer was employed by the Sony Corporation for development of the industrial videodisc market and was one of three national market development managers assigned specifically to the task of developing videodisc applications. He has been an active speaker and writer on the future of videodiscs and computers, and his credentials include the Nebraska Symposium, the Department of Defense, and many invitations by national educational organizations.

## *INDUSTRIAL PERSPECTIVES*

### *Perry Reeves*

David Hon, developer of the now famous CPR videodisc training course, asked in a recent article, "What are the 'space invaders' telling us?" The "us" he refers to is the industrial/educational community interested in bringing the vast potential of interactive video into the home and the work place. My principal interests, as a representative of a major manufacturer of videodisc players and related systems, are both home consumer and nonhome applications of videodisc technology; thus, my "industrial" perspective.

From my point of view, I would answer Mr. Hon by suggesting that "space invaders" has told us a great deal. Videodisc-based arcade games have presented new challenges to the manufacturers of videodisc-related hardware, and also created new opportunities for the developers of videodisc programming. For example, in the manufacturing community the quality of disc mastering and replication has improved. This is due to increased demand for visual clarity equalling that of arcade games as well as the need for a faster turnaround time; that is, the time it takes to make videodisc copies from an edited premaster tape.

Many of the major manufacturers can now offer a twenty-four-hour turnaround service, and even this may improve as demand increases, which I anticipate it will.

With regard to the creative side of videodisc technology, videodisc-based games in the arcades have been so successful that we are now seeing "home versions" coming into the marketplace. The narrow-minded among us will consider this as yet another, more insidious corruption of young people. But I say that the manufacturers who are backing the games with huge investments in special computer/videodisc hardware made for the home are presenting a golden opportunity to educators to produce interactive educational discs for home use.

Simply put, all of us in the manufacturing community need software to help sell our systems in this new "level 3" market for the home. And when I say level 3 I am referring to totally integrated computer/videodisc systems with superimpose (the capability to mix computer text and graphics with video), autodial modems, and touch screens—and all selling for a price of $1500 or less. It is very possible that videodiscs will help sell computers and modems in the home rather than vice versa. Moreover, I am excited to learn that some major book publishers are becoming increasingly involved with videodisc technology. For example, one can expect on-line videodisc encyclopedias to reach the home market in the near future.

Of course, videodiscs are thriving in environments other than the home. Thanks to the notoriety that videodiscs have gained in the arcades, educational institutions are finally beginning to use games and simulations as instructional strategies. Educators, especially those who have the money to fund projects, are more excited about discs right now than they ever have been. Clearly the tools are now in place for supplying these groups, and at a much lower cost.

Not surprisingly, business and professional people are also excited about the videodisc. With level 3 systems penetrating the home market, interactive catalogs will experience explosive growth. Related to this, the so-called P-O-P (point of purchase) network will attract even the most conservative businesses. Banks and insurance companies will offer various financial services using computer-controlled videodisc networks located in airports, hotel lobbies, etc. In fact, the hotels themselves will see discs as an opportunity to introduce guests to in-house services and to the community at large. The earlier, less successful attempts at using interactive videodiscs to sell merchandise have undergone a massive rethink and we are beginning to see well-conceived, more clearly targeted systems in and around the major department stores nationwide, with the home market not far behind.

Trade show displays and exhibits are exciting new markets for interactive videodisc modules. The compact size, attention-getting format, and easily updated material make this a natural market area. Touch screens allow easy interaction with the trade show attendees. Large products, either costly or difficult to demonstrate in an exhibit hall, are convincingly displayed by means of the disc. Multimedia-type displays using several disc players and rear screen projection systems are being introduced.

Thus, I am enthusiastic about the future of videodisc technology. Sales are good, but nothing compared to what they will be in the coming years. And although they are merely entertainment, the videodisc arcade games are pointing the way to an exciting future for videodiscs in our homes, schools, and communities.

Perry Reeves is Western Region Market Development Manager for Sony Video Communications. He moved to California in September 1983 from Sony's Federal Government Marketing Group in Chicago, where he had been located for three years. Perry has pioneered in the development of creative new applications of videodisc technology.

## GAME TECHNOLOGY: AN INSTRUCTIONAL DESIGNER'S WISH

*James J. L'Allier*

As an instructional designer, some of my most valuable insights into design theory have come from observing learners who were using various instructional products. The following four scenes represent some of these observations and their related insights.

After you have read these four scenes, analyze their similarities and differences for yourself before going on with the rest of this article. In examining the underlying assumptions of each scenario, you may make some connections that either go beyond, support, or contradict my own observations.

SCENE I: A fifth grade student is sitting in front of a microcomputer. On the screen are the words: "John hit the ball." Just below this is the prompt: "Is this a sentence or a fragment?" The student types in "fragment" and the program informs her that her response was wrong. Next she is given some remedial information for that item. Then she moves on to another question on sentences and fragments.

SCENE II: An auto mechanic is using an interactive video system to learn about the repair procedures of a new carburetor. A still frame shows the carburetor and a caption prompts: "With your light pen touch the fuel line." The mechanic touches the air intake and the system indicates his error by using a graphic overlay to outline the fuel line. Next, a rolling video gives the learner some remedial information and moves on to the next item.

SCENE III: A young friend of mine is using his home computer to play an interactive detective game. His job is to find out who murdered the victim. In this game our would-be Sherlock Holmes can examine fingerprints, analyze suspicious substances, question suspects, and when enough evidence has been gathered, make an arrest. All of this is accomplished by the typing in of whole sentences, which are read and reacted to by the program. After much deliberation an arrest is made, a trial is held, and the case is tossed out of court on the grounds of insufficient evidence.

SCENE IV: A user of a new computer-based business telephone system is learning its various features by playing an interactive videodisc game. In this game the user takes on the role of the communications officer of a large star ship that is under attack by an alien force. The communications officer's job is to coordinate the ship's defenses through the effective use of the communication system. As various features are used and misused, the ship's condition and the story itself changes—with each new communication situation presenting new challenges and consequences that affect the storyline of this interactive drama.

To me these four scenes represent two design philosophies as seen across two interactive media. Scenes I and II contain static Skinnerian structures that view the learner as someone that you "do learning to"—preferably in small sequential steps. Scenes III and IV represent more dynamic algorithmic structures that see the learner/game player as someone who gains knowledge by manipulating an environment that contains consequences. Across these two design philosophies are different media. Scenes I and III use the medium of the computer while Scenes II and IV use interactive videodisc as the delivery system. So what do these contrasts mean? Simply that without a basic assessment of our design biases and related views of the learner we may be transferring an inappropriate structure from the media of computer-assisted instruction to that of interactive video. Electronic page turning is electronic page turning and it really doesn't matter if the page has text, still frames, or rolling video.

Now I will admit that this is an extreme position, and I will further admit that there are times when a Skinnerian model is appropriate. But I take this position in order to underscore one basic and disturbing fact: static page turning structures are more the rule than the exception in much of what we see today in computer-aided instruction and interactive video.

So where do all of us who are involved with interactive video turn to for new design models that are worthy of the attributes of this new media? Read the literature on artificial intelligence and expert systems. Better still, walk into any game arcade or computer software store and look around for the answer. You will see no page turning and no static structures. Look beyond the graphics, joysticks, and the sounds, and what you will see are dynamic environments that are in constant change; environments in which actions are followed by immediate consequences that bring about powerful engagements of the emotions. True, the skill being learned may not have the highest social value, but that is not the point. The point is that learning *is* taking place. That very fact alone demands that we start to incorporate game structures in the design of a new generation of interactive video learning products.

One final point. I am firmly convinced that learning breakthroughs resulting from the dynamic use of interactive video will not come about because we have faster search times, better graphics, and field kicking. The real breakthroughs will take place in the minds of designers as they discover and incorporate into their programs new structures that put the learner in control of the learning environment and also use all of the attributes of the technology to bring about "real" consequences within that environment. This is the challenge our industry faces; game technology has only raised the issue.

Jim L'Allier is the director of research and development for Wilson Learning Corporation's interactive technologies group. He has been involved in the design and development of numerous videodisc projects that employ game technology. As well as having experience with interactive video, he has also designed a number of computer-based instructional packages for both micro and mainframe environments and has also done research in the field of learning algorithms. Mr. L'Allier holds a Ph.D. in instructional systems design and is a senior fellow of the College of Education of the University of Minnesota. Recently he edited a special issue of *Performance and Instruction* devoted to the topic of interactive video.

## *PUBLISHING PERSPECTIVES*

*Gary Carlson*

In this brief article an attempt will be made to outline the opportunities created by videodiscs for publishers, the potential role of publishers within the videodisc industry, the obstacles facing publishers who might wish to participate in this small but growing industry, and how these obstacles might be overcome. At that point I will offer a few guesses as to future developments.

## VIDEODISC OPPORTUNITIES FOR PUBLISHERS

One need only browse in a library or a bookstore to gain a sense of the opportunity videodisc technology offers publishers. Much of what exists in print can be augmented or presented more effectively via interactive videodisc, such as identification of things alive or inanimate, lessons on skiing, fly-tying, tennis, macrame, and dozens of other skills or crafts. Moreover, it is not difficult to imagine children's expanded use of videodiscs as sources of adventure (Mystery Disc), games, knowledge of physical or biological principles, or even their forming videodisc clubs. People involved in training, education and other fields have already begun to use videodiscs (alone or with other media) and it can be expected that their needs will expand in the near future.

## IS THERE A ROLE FOR BOOK PUBLISHERS IN THE VIDEODISC INDUSTRY?

Several factors suggest that book publishers will be ideal conduits for the development, production, and marketing of videodiscs. Supporting the general notion that book publishers might become successful videodisc publishers are the following facts.

- Publishers have experience in managing and directing relatively large scale multimedia productions.
- Depending upon specialization, publishers have created or have access to marketing channels that lead to most or all of the customers videodisc producers would like to reach.
- Many of the large publishers have international divisions that promote products overseas.
- Many publishers have earned reputations in their specific areas of strength as producers of high-quality materials (brand recognition).
- Publishers have demonstrated skills in molding products for specific markets (adding value to the products).

Underlying all of the above is book publishing's traditional sense of confidence in dealing with products that rest on ideas.

## FROM BOOKS TO VIDEODISCS: WHAT ARE THE OBSTACLES?

At this time there are less than a handful of videodiscs published by the larger publishing houses. That there are no more indicates major obstacles to be overcome before book publishers embrace videodisc technology. In order to better understand the problem, let's take a look at the following inhibitors.

- Publishers generally lack a clear understanding of what videodiscs are and the power of the medium.
- Very small (several hundred thousand) installed base of videodisc players of all types are in all markets at present.
- Failure of a "standard" in hardware to evolve that will allow economies of scale in manufacturing (reminiscent of the early days of microcomputers).
- Failure of hardware companies (videodisc and computer) to see the potential synergy between videodisc and microcomputer hardware and software; also failure to create simple and inexpensive interfaces.
- Low level of publisher support by videodisc hardware companies.
- Videodisc proposals are in competition with books, software, and other ventures for publishers' capital.
- Most publishers lack in-house technical expertise required to deal with the complexities of videodisc development and production.

Perhaps the most encouraging sign is that in spite of these obstacles, some publishers have been moving ahead with videodisc projects on an experimental basis, investing in the learning experience and the promise of future sales. Nonetheless, there are pockets of activity that represent genuine commitment: for example, Wilson Learning of Eden Prairie, Minnesota (a subsidiary of John Wiley & Sons).

## HOW CAN OBSTACLES AND INERTIA BE OVERCOME SO BOOK PUBLISHERS CAN ENTER THE VIDEODISC MARKET?

It is obvious that publishing houses have been reluctant to loosen their purse strings, either to develop or encourage development of videodiscs. To spur development, government agencies and other funding sources have provided capital for experimental projects, but few of these discs ever make it to market.

The crux of the developmental dilemma is this: videodisc development and production houses seem anxious to work with publishers, who they correctly view as prime candidates to help shape and market their products. Publishers are split between groups that know little or nothing about videodiscs and those who recognize their potential, but either don't know enough about the medium or simply wish to postpone commitment until many more disc players are sold. If microcomputer software offers any precedent, most publishers will wait until there are enough success stories (small groups of entrepreneurs) and a large number of machines in the marketplace before they make substantial commitments. On the plus side, the recent commitments made by publishers to microcomputer software publishing means that a core of editorial talent is slowly emerging that is capable of easily dealing with one component (software) of level 3 videodiscs.

If an early move by publishers to develop and publish videodiscs is to occur, much greater communication, cooperation and support (technical support, the loan of hardware, manufacturing incentives, bulk purchases of discs, etc.) must be forthcoming from major videodisc player manufacturers. An example of this kind of relationship is Apple, Inc.'s support program for educators and educational publishers. An example of the effect of that kind of support is Apple's clear dominance of hardware sales in educational markets. In addition, publishers, at all levels, from the editorial to the CEO, will need to more fully comprehend the videodisc's potential power and impact on those markets which are critical to the long-term success of their firms such as education, consumer, trade, training, entertainment, and so on. Of course, the surest way to ensure that publishers will open their purse strings for videodisc development is to demonstrate that their customers will buy competing and superior products enhanced and delivered via videodiscs.

It seems likely that eventually publishers, developers, and hardware companies will achieve that degree of cooperation, (perhaps in the form of co-ventures) required to launch some unique and provocative videodisc packages for a variety of markets. The danger for publishers is that developers will step around them and find or create their own channels to the publisher's customers.

## FUTURE DEVELOPMENTS

Let's speculate a bit and presume that book publishers are a major force in videodisc publishing a few years from now. Within the larger houses, art, photos, and text will be digitized and available for easy retrieval on optical discs and driven by software held on optical memory cards or discs. In some cases entire books may be digitized and held on optical discs for electronic reproduction and binding at the bookstore. Data bases on optical discs will be commonplace. Interactive videodiscs will no longer need to be driven by limited programs on an optical disc or a floppy disc. Sophisticated programs will occupy a significant functional area integral to the disc itself. "Advanced" software packages of today will be available on or written to optical discs to be used with motion sequences, slides, computer graphics, and audio in several languages.

Standard videodisc fare for publishers will be "how to" packages of all types, lessons and demonstrations in English and other languages for courses within a variety of curricula, and special purpose videodiscs of many types: those that utilize touch panel screens, mice and other peripherals, and holographic recreations of anything from biomolecules to computer-generated art. And of course, to accompany these videodiscs we will have books, which one will still cradle in one's lap on a snowy night as printed words fade to dreams.

Gary Carlson is a graduate of the University of Michigan with a BA in Far Eastern Languages (Mandarin Chinese). His current title is Publisher, Wiley Educational Software, John Wiley & Sons, New York, New York. Gary was one of the motivating forces behind Wiley's entrance into new technologies. He sponsored activities which led to the publication by Wiley of "The Puzzle of the Tacoma Narrows Bridge Collapse." He is currently involved in several educational software projects.

# Selected Bibliography of Videodisc Articles

Ahl, David. Aurora Systems Videodisc Controller. *Creative Computing*, January 1982, Vol. 8, No. 1, pp.56-57.

Ahl, David. Magazines of the Future. *Creative Computing*, January 1982, Vol. 8, No. 1, p.14.

Ahl, David. Rollercoaster Game Dissected. *Creative Computing*, January 1982, Vol. 8, No. 1, pp.80-83.

Allard, Kim E. *The Videodisc and Implications for Interactivity.* Paper presented at the Annual Meeting of the American Educational Research Association, New York, March 1982, 11pp. (ERIC Document Reproduction Service No. ED 217 884)

Anderson, Paul & Carr, Everett. Computer/Video Disk Combo That Really Works! *Kilobaud Microcomputing*, January 1982, Vol. 6, No. 1, pp.102-108.

Andriessen, J. J. & Kroon, D. J. Individualized Learning by Videodisc. *Educational Technology*, March 1980, Vol. 20, No. 3, pp.21-25.

Bejar, Isaac. Videodiscs in Education: Integrating the Computer and Communications Technologies. *BYTE*, June 1982, Vol. 7, No. 6, pp.78-104.

Bennion, Junius L. *Possible Applications of Optical Video Discs to Individualized Instruction Systems.* Provo, Utah: Brigham Young University, Division of Instructional Research, Development, and Evaluation, 14 February 1974, 10pp. Sponsoring Agency: Mitre Corporation, McLean, Va. National Science Foundation, Washington, DC. (ERIC Document Reproduction Service No. ED 158 721)

Bennion, Junius L. & Schneider, Edward W. *Interactive Video Disc Systems for Education.* Provo, Utah: Brigham Young University, Division of Instructional Research, Development, and Evaluation, 30 October 1975, 19pp. (ERIC Document Reproduction Service No. ED 158 719)

*Biological & Medical Sciences Libraries Section. Special Libraries Division. Papers.* The Hague (Netherlands): International Federation of Library Associations, August 1982, 29pp. Papers presented at the Annual Meeting of the International Federation of Library Associations (48th, Montreal, Canada, August 22-28, 1982). (ERIC Document Reproduction Service No. ED 229 037)

Blizek, John. First National Kidisc - TV Becomes a Plaything. *Creative Computing*, January 1982, Vol. 8, No. 1, pp.106-110.

Bork, Alfred & Others. *Conference on Intelligent Videodisc Systems.* Brief Informal Summary. Irvine: California University, December 1977, 14pp. Sponsoring Agency: National Science Foundation, Washington, DC. (ERIC Document Reproduction Service No. ED 163 978)

Brandt, Richard C. & Knapp, Barbara H. *Extension of TVCAI Project to Include Demonstration of Intelligent Videodisc System. Hardware, Software, and Courseware Implementation Component. Final Report.* Salt Lake City: Utah University, Department of Computer Science, 1982, 34pp. Sponsoring Agency: National Science Foundation, Washington, DC. (ERIC Document Reproduction Service No. ED 229 236)

Branson, Robert K. & Foster, Richard W. Educational Applications Research and Videodisc Technology. *Journal of Educational Technology Systems*, 1979, Vol. 8, No. 3, pp.241-62.

Braun, Ludwig. Microcomputers and Video Disc Systems: Magic Lamps for Educators? *People's Computers*, January-February 1978, Vol. 6, No. 4, pp.14-5, 47; *Calculators/Computers Magazine*, January 1978, Vol. 2, No. 1, pp.21-25.

Bunderson, C. Victor. Instructional Strategies for Videodisc Courseware: The McGraw Hill Disc. *Journal of Educational Technology Systems*, 1979, Vol. 8, No. 3, pp.207-10.

Bunderson, C. Victor. Will Technology Improve Instructional Productivity? *Instructional Innovator*, February 1982, Vol. 27, No. 2, pp.29, 44.

Bunderson, C. Victor & Others. *Instructional Systems Development Model for Interactive Videodisc Training Delivery Systems. Volume I: Hardware, Software and Procedures.* Orem, Utah: WICAT, Inc., June 1980, 108pp. Sponsoring Agency: Army Research Institute for the Behavioral and Social Sciences, Alexandria, Va. (ERIC Document Reproduction Service No. ED 220 071)

Bunderson, C. Victor & Others. *Proof-of-Concept Demonstration and Comparative Evaluation of a Prototype Intelligent Videodisc System. Final Report.* Orem, Utah: WICAT, Inc., 2 January 1981, 74pp. Sponsoring Agency: National Science Foundation, Washington, DC. (ERIC Document Reproduction Service No. ED 228 989)

Bunderson, C. Victor & Others. *A Study of Authoring Alternatives for Training-Oriented Videodiscs.* Orem, Utah: WICAT, Inc., September 1979, 80pp. Sponsoring Agency: Navy Personnel Research and Development Center, San Diego, Calif. (ERIC Document Reproduction Service No. ED 181 887)

Buterbaugh, James G. *Alternative Media Storage/Retrieval Systems—A Futeristic Forecast for Educational Technologists.* Pullman, Wash.: Information Futures, November 1980, 36pp. (ERIC Document Reproduction Service No. ED 217 885)

Butler, David W. 5 Caveats for Videodiscs in Training. *Instructional Innovator*, February 1981, Vol. 26, No. 2, pp.16-18.

Carr, Claire & Carr, Everett. VIC Intelligent Video Disc System. *Compute!*, June 1982, Vol. 4, No. 6, pp.46-49.

Ciarcia, Steve. Build an Interactive-Videodisc Controller (Pioneer VP-1000). *BYTE*, June 1982, Vol. 7, No. 6, pp.60-74.

Dann, Michael H. *The Videocassette and Video Disc in the Development of the Communications Media.* New York: Children's Television Workshop, 3 October 1973, 25pp. Paper presented at the International Market for Videocassette and Videodisc Programmes and Equipment (3rd, Cannes, France, September 28 through October 3, 1973). (ERIC Document Reproduction Service No. ED 084 790)

Daynes, Rod. Experimenting with Videodisc. *Instructional Innovator*, February 1982, Vol. 27, No. 2, pp.24-25, 44.

Daynes, Rod. Videodisc Interfacing Primer. *BYTE*, June 1982, Vol. 7, No. 6, pp.48-59.

Daynes, Rodney R. *Videodisc Technology Use Through 1986: A Delphi Study.* December 1976, 48pp. Sponsoring Agency: Navy Personnel Research and Development Center, San Diego, CA. (ERIC Document Reproduction Service No. ED 145 823)

DeChenne, James & Evans, Robert. Simulating Medical Emergencies. *Instructional Innovator*, January 1982, Vol. 27, No. 1, p.23.

Eastwood, Lester F., Jr. Motivations and Deterrents to Educational Use of "Intelligent Videodisc" Systems. *Journal of Educational Technology Systems*, 1978, Vol. 7, No. 4, pp.303-35.

Emmens, Carol A. Videodisc Software: Current Developments. *School Library Journal*, January 1982, Vol. 28, No. 5, p.39.

Evans, Art. Videodisc on the Horizon. *Audiovisual Instruction*, May 1975, Vol. 20, No. 5, pp.31-33.

Farr, Roger & Wolf, Robert L. ABC/NEA SCHOOLDISC: A More Careful Examination. *Phi Delta Kappan*, March 1981, Vol. 26, No. 7, pp.516-18.

Fedale, Scott. Interactive Video in the Pacific Northwest. *T.H.E. Journal*, September 1982, Vol. 10, No. 1, pp.124-126.

Frenzel, Louis. Microcomputer and the Video Disk. *Interface Age*, April 1981, Vol. 6, No. 4, pp.40-41.

Galbraith, Gary & Others. Interfacing an Inexpensive Home Computer to the Videodisc: Educational

Applications for the Hearing Impaired. *American Annals of the Deaf, (Educational Technology for the '80's)*, September 1979, Vol. 124, No. 5, pp.536-41.

Gale, Larrie E. Montevidisco: An Anecdotal History of an Interactive Videodisc. *CALICO Journal*, June 1983, Vol. 1, No. 1, pp.42-46.

Glenn, Allen D. Videodiscs and the Social Studies Classroom. *Social Education*, May 1983, Vol. 47, No. 5, pp.328-30.

Glenn, Allen & Kehrberg, Kent. Intelligent Videodisc: An Instructional Tool for the Classroom. *Educational Technology*, October 1981, Vol. 2, No. 10, pp.60-63.

Grabowski, Barbara. *A Symposium: Relevant Cue Research, a Program of Systematic Evaluation: Considerations for Sustaining Instructional Design Research Using an Integrated Learning System.* May 1982, 9pp. Paper presented at the Annual Meeting of the Association for Educational Communications and Technology, Research and Theory Division (Dallas, TX, May 1982). (ERIC Document Reproduction Service No. ED 223 198)

Gray, Robert A. Watch Out for Videodiscs in Your Future. *Audiovisual Instruction*, December 1978, Vol. 23, No. 9, p.19.

Gray, Robert A. Videodisc Technology: Its Potential for Libraries. *School Library Journal*, January 1978, Vol. 24, No. 5, pp.22-24.

Hiraki, Joan & Garcia, Oscar N. Setting the Stage for the Interactive Classroom of the 1980s. *Educational Computer Magazine*, May-June 1981, Vol. 1, No. 1, pp.20-22.

Hiscox, Michael D. *Integrating Testing and Instruction Using the Videodisc.* 12 August 1981, 12pp. Paper presented at the Annual Conference of the Society for Applied Learning Technology (Los Angeles, CA, August 12, 1981). (ERIC Document Reproduction Service No. ED 208 877)

Hiscox, Michael D. A Summary of the Practicality and Potential of Videodiscs in Education. *Videodisc/Videotex*, Spring 1982, Vol. 2, No. 2, pp.99-109.

Hofmeister, Alan M. & Thorkildsen, Ron J. Videodisc Technology and the Preparation of Special Education Teachers. *Teacher Education and Special Education*, Summer 1981, Vol. 4, No. 3, pp.34-39.

Holloway, Robert E. The Videodisc in Education: A Case for Technological Gradualism. *Videodisc/Videotex*, Spring 1982, Vol. 2, No. 2, pp.132-37.

Holmgren, J. E. & Others. The Effectiveness of Army Training Extension Course Lessons on Videodisc. *Journal of Educational Technology Systems*, 1979, Vol. 8, No. 3, pp.263-74.

Hon, David. Interactive Training in Cardiopulmonary Resuscitation. *BYTE*, June 1982, Vol. 7, No. 6, pp.108-138.

Hon, David C. Space Invaders, Videodiscs and the "Bench Connection." *Training and Development Journal*, December 1981, Vol. 35, No. 12, pp.10-17.

Ingalls, Richard E. *Intelligent Video Disc as a Major Component of Individualized Instruction.* 1977, 11pp. (ERIC Document Reproduction Service No. ED 154 837)

Ingraham, Leonard W. Is There a Videodisc in Your Future? *Social Education*, November-December 1980, Vol. 44, No. 7, pp.641-43.

*Interactive Videodisc for Special Education Technology. Final Report.* Utah State University, Logan, 1982, 256pp. Sponsoring Agency: Office of Special Education and Rehabilitative Services (ED), Washington, DC. (ERIC Document Reproduction Service No. ED 230 187)

Keener, John R. & Bright, Larry K. Micros and Interactive Videodiscs for Improving Access to Health Education. *Health Education*, October 1983, Vol. 4, No. 6, pp.47-50.

Kehrberg, Kent T. & Pollack, Richard A. Videodiscs in the Classroom: An Interactive Economics Course. *Creative Computing*, January 1982, Vol. 8, No. 1, pp.98-102.

Kellner, Charlie. V is for Videodisc. *Creative Computing*, January 1982, Vol. 8, No. 1, pp.104-105.

Kemph, Jeff. Videodisc Comes to School. *Educational Leadership*, May 1981, Vol. 38, No. 8, pp.646-49.

Kirchner, Glenn. Simon Fraser University Videodisc Project: Part One: Design and Production of an Interactive Videodisc for Elementary School Children. *Videodisc/Videotex*, Fall 1982, Vol. 2, No. 4, pp.275-87.

Kribs, H. Dewey. Authoring Techniques for Interactive Videodisc Systems. *Journal of Educational Technology Systems*, 1979, Vol. 8, No. 3, pp.211-19.

LaGow, Robert L. Instructional Development Related to an Intelligent Videodisc. *Journal of Educational Technology Systems*, 1979, Vol. 8, No. 3, pp.231-39.

Laumer, Mike. Interpretive Language Used to Program the CPR System. *BYTE*, June 1982, Vol. 7, No. 6, pp.126-130.

Levenson, Phyllis M. Interactive Video: A New Dimension in Health Education. *Health Education*, October 1983, Vol. 14, No. 6, pp.36-38.

Leveridge, Leo L. Experience in Educational Design for Interactive Videodisc and QuadraSync Presentations. *Journal of Educational Technology Systems*, 1979, Vol. 8, No. 3, pp.221-30.

Leveridge, Leo L. The Interactive Videodisc. *MOBIUS, (Technology and the Future of Continuing Education)*, April 1983, Vol. 3, No. 2, pp.68-72.

Lipson, Joseph. Design and Development of Programs for the Videodisc. *Journal of Educational Technology Systems*, 1980, Vol. 9, No. 3, pp.277-85.

Lipson, Joseph I. & Fisher, Kathleen M. *Public Education and Electronic Technologies.* Madison: Wisconsin Center for Education Research, June 1982, 43pp. Sponsoring Agency: National Inst. of Education (ED), Washington, DC. (ERIC Document Reproduction Service No. ED 226 725)

Long, Harvey S. The Videodisc: A Picture Book in the Round. *Technological Horizons in Education*, May 1981, Vol. 8, No. 4, pp.39-42.

Love, John. The Videodisc: Television's New Horn of Plenty. *Media and Methods*, October 1979, Vol. 16, No. 2, pp.16-18, 22-23, 89.

Lubar, David. Adventures in Videoland. *Creative Computing*, January 1982, Vol. 8, No. 1, pp.60-78.

Lubar, David. Videodiscs in Education (Dawn of the Disc). *Creative Computing*, January 1981, Vol. 7, No. 1, pp.50-54.

Markoff, John. Interactive Videodiscs Act as New Learning Tool for IBM PC Users. *InfoWorld*, January 3, 1983, Vol. 5, No. 1 & 2, p.18.

Marsh, Fred E., Jr. Videodisc Technology. *Journal of the American Society for Information Science*, July 1982, Vol. 33, No. 4, pp.237-44.

Marsh, Patrick. Videodisc/Microcomputer Courseware Design. *Educational Technology*, March 1983, Vol. 23, No. 3, pp.51-52.

Marx, Raymond J. Videodisc-based Training: Does it Make Economic Sense? *Training*, March 1982, Vol. 19, No. 3, pp.56-58, 60-61, 65.

McClain, Larry. A Friendly Introduction to Videodiscs. *Popular Computing*, April 1983, Vol. 2, No. 6, pp.79-82.

Merrill, Paul F. & Bennion, Junius L. Videodisc Technology in Education: The Current Scene. *NSPI Journal*, November 1979, Vol. 18, No. 9, pp.18-19, 22-26.

Merrill, Paul F. & Bunderson, C. Victor. *Guidelines for Employing Graphics in a Videodisc Training Delivery System. ISD for Videodisc Training Systems. First Annual Report. Vol. III.* Orem, Utah: WICAT, Inc., July 1979, 61pp. Sponsoring Agency: Army Research Inst. for the Behavioral and Social Sciences, Alexandria, Va. (ERIC Document Reproduction Service No. ED 196 413)

Micro-Video Interlace Generator Improves Apple Graphics. *InfoWorld*, November 8, 1982, Vol. 4, No. 44, p.13.

Mole, Dennis. The Videodisc as a Pilot Project of the Public Archives of Canada. *Videodisc/Videotex*, Summer 1981, Vol. 1, No. 3, pp.154-61.

Molnar, Andrew R. Intelligent Videodisc and the Learning Society. *Journal of Educational Technology Systems*, 1979, Vol. 8, No. 1, pp.31-40.

Molnar, Andrew R. Microcomputers and Videodisc: Innovations of the Second Kind. *Technological Horizons in Education*, November 1980, Vol. 7, No. 6, pp.58-62.

Negroponte, Nicholas. *The Impact of Optical Videodiscs on Filmmaking.* June 1979, 32pp. (ERIC Document Reproduction Service No. ED 216 674)

Nugent, Gwen C. & Stone, Casey G. Videodisc Instructional Design. *Educational Technology*, May 1980, Vol. 20, No. 5, pp.29-32.

Nugent, Gwen C. & Stone, Casey G. The Videodisc Meets the Microcomputer. *American Annals of the Deaf*, September 1982, Vol. 127, No. 5, pp.569-72.

Onosko, Tim. Vision of the Future (Videodiscs). *Creative Computing*, January 1982, Vol. 8, No. 1, pp.84-94.

Otto, Sue E. K. Videodisc Material Retrieval for Language Teaching. *System*, 1983, Vol. 11, No. 1, pp.47-52.

Paris, Judith. The Interfaces (Videodiscs). *Popular Computing*, April 1983, Vol. 2, No. 6, pp.83-86.

Post, Dan. Epcot Information System/Carroll Touch Technology Input System. *Interface Age*, January 1983, Vol. 8, No. 1, p.27.

Propp, George & Others. Satellite Demonstration: The Videodisc Technology. *American Annals of the Deaf*, (Educational Technology for the '80's), September 1979, Vol. 124, No. 5, pp.652-55.

# Selected Bibliography of Videodisc Articles

Propp, George & Others. *Videodisc: An Instructional Tool for the Hearing Impaired.* Lincoln: Nebraska University, 1980, 43pp. Sponsoring Agency: Bureau of Education for the Handicapped (DWEW/OE), Washington, DC. Media Services and Captioned Films Branch. (ERIC Document Reproduction Service No. ED 200 227) Technology for the '80's, September 1979, Vol. 124, No. 5, pp. 652-55.

Rice, James, Jr. There's a Videodisc in Your Future. *Library Journal,* January 15, 1978, Vol. 103, No. 2, pp.143-4.

Robson, Walt. Is It True What They Say About Videodiscs? *Training,* December 1975, Vol. 12, No. 12, pp.80-2.

Schipma, Peter B. Videodisc for Storage of Text. *Videodisc/Videotex,* Summer 1981, Vol. 1, No. 3, pp.168-72.

Schneider, E. W. Videodiscs, or the Individualization of Instructional Television. *Educational Technology,* May 1976, Vol. 16, No. 5, pp.53-58.

Schneider, Edward W. *Applications of Videodisc Technology to Individualized Instruction.* Provo, Utah: Brigham Young University, Division of Instructional Research, Development, and Evaluation, September 1975, 22pp. Paper presented at the National Science Foundation Conference for Ten Year Forecast for Computers and Communications (Warrenton, Va., September 16- 18, 1975). (ERIC Document Reproduction Service No. ED 158 722)

Schneider, Edward W. & Bennion, Junius L. Veni, Vidi, Vici Via Videodisc: A Simulator for Instructional Conversations. *System,* 1983, Vol. 11, No. 1, pp.41-46.

Schulman, Jacque-Lynne. Video PATSEARCH: A Mixed-Media System. *Information Technology and Libraries,* June 1982, Vol. 1, No. 2, pp.150-56.

Shea, Tom. Interactive Videodisc Debuts as Teaching Tool on Iowa Campus. *InfoWorld,* September 6, 1982, Vol. 4, No. 35, p.22.

Sonnemann, Sabine S. The Videodisc as a Library Tool. *Special Libraries,* January 1983, Vol. 74, No. 1, pp.7-13.

Stone, David E. & Others. *State of the Art Developments in Established CAI Efforts—The TICCIT System.* March 1982, 18pp. Paper presented at the Annual Meeting of the American Educational Research Association (New York, NY, March 19-23, 1982). (ERIC Document Reproduction Service No. ED 217 882)

Survey Shows Disc Training Works Well for American Bell. *Educational-Industrial Television,* June 1983, Vol. 15, No. 6, p.57.

Sustik, Joan M. An Interactive Videodisc Project in Art History. *Performance and Instruction,* November 1981, Vol. 20, No. 9, pp.10-13.

Thorkildsen, Ron. *Microcomputer/Videodisc Authoring System for Instructional Programming.* March 1982, 24pp. Paper presented at the Annual Meeting of the American Educational Research Association (New York, NY, March 19-23, 1982). Sponsoring Agency: Department of Education, Washington, DC. (ERIC Document Reproduction Service No. ED 227 841)

Thorkildsen, Ron. A Microcomputer/Videodisc System for Delivering Computer Assisted Instruction to Mentally Handicapped Students. July 1982, 2pp. In: *The Computer: Extension of the Human Mind. Proceedings, Annual Summer Conference,* College of Education, University of Oregon (3rd, Eugene, OR, July 21-23, 1982). (ERIC Document Reproduction Service No. ED 219 867)

Throkildsen, Ron; Allard, Kim; & Reid, Bob. The Interactive Videodisc for Special Education Project: Providing CAI for the Mentally Retarded. *Computing Teacher,* April 1983, Vol. 10, No. 8, pp.73-76.

Tross, Glenn & Di Stefano, Mary F. Interactive Video at Miami-Dade *Community College.* 1983, 10pp. (ERIC Document Reproduction Service No. ED 230 256)

Troutner, Joanne. How to Produce an Interactive Video Program. *Electronic Learning,* January 1983, Vol. 2, No. 4, pp.70-75.

The Video PATSEARCH System: An Interview with Peter Urbach. *Videodisc/Videotex,* Winter 1982, Vol. 2, No. 1, pp.30-37.

Videodisc Markets Make an Amazing About-Face. *Business Week,* September 20, 1982, No. 2757, pp.119-121.

Videodiscs: A Revolution That Isn't. *Canadian Library Journal,* December 1982, Vol. 39, No. 6, pp.357-64.

Videodiscs and Computers: A Dynamic Duo. *Business Week,* February 7, 1983, No. 2776, pp.109-111.

Videodiscs for Customized Newspapers. *Popular Computing,* January 1983, Vol. 2, No. 3, p.20.

*Videodiscs in Special Education.* Falls Church, Va.: Education Turnkey Systems, Inc., April 1983, 20pp.

Sponsoring Agency: Special Education Programs (ED/OSERS), Washington, DC. (ERIC Document Reproduction Service No. ED 231 177)

Walker, C. L. Computer Controlled Videodisc/Videoplayer System. *Journal of Educational Technology Systems*, 1979, Vol. 8, No. 3, pp.201-06.

Winslow, Ken. A Videodisc in Your Future. *Educational and Industrial Television*, May 1975, Vol. 7, No. 5, pp.21-22.

Winslow, Ken. Videodiscs; Reality Begins (Almost). *Educational and Industrial Television*, November 1974, Vol. 6, No. 11, pp.36-40.

Wise, Deborah. How to Turn Video Recorder into an Interactive Tool. *InfoWorld*, May 31, 1982, Vol. 4, No. 21, p.22.

Wise, Deborah. Interactive Video: Video + Computers = New Medium. *InfoWorld*, May 31, 1982, Vol. 4, No. 21, p.20.

Wollman, Jane. Videodisc: A New Educational Technology Takes Off. *Electronic Learning*, November/December 1981, Vol. 1, No. 2, pp.38-40.

Wollman, Jane. Videodisc Applications. *Popular Computing*, April 1983, Vol. 2, No. 6, pp.81-82.

Wood, R. Kent & Stephens, Kent G. An Educator's Guide to Videodisc Technology. *People's Computers*, January-February 1978, Vol. 6, No. 4, pp.12-3.

Wood, R. Kent & Woolley, Robert D. *An Overview of Videodisc Technology and Some Potential Applications in the Library, Information, and Instructional Sciences.* Syracuse, NY: ERIC Clearinghouse on Information Resources, December 1980, 37pp. Sponsoring Agency: National Inst. of Education (DHEW), Washington, DC. (ERIC Document Reproduction Service No. ED 206 328)

Zollman, Dean & Fuller, Robert. The Puzzle of the Tacoma Narrows Bridge Collapse: An Interactive Videodisc Program for Physics Instruction. *Creative Computing*, October 1982, Vol. 8, No. 10, pp.100-09.

Zullo, Allan. Disney Goes Interactive. *Popular Computing*, April 1983, Vol. 2, No. 6, pp.86-88.

# PART TWO

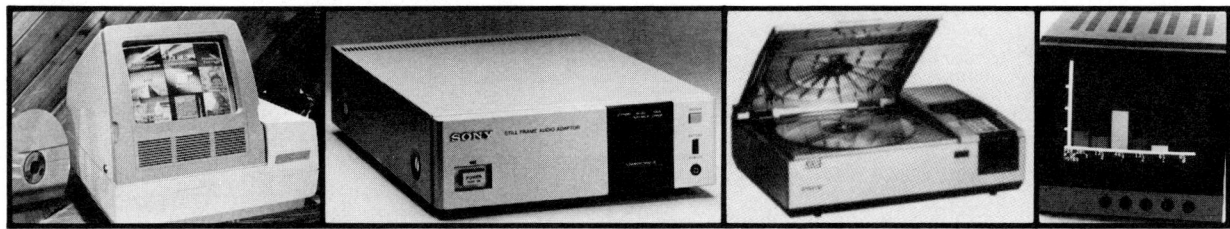

# THE DIRECTORY

# Production and Development Resources

| | |
|---|---|
| ORGANIZATION | **AARON MARCUS AND ASSOCIATES** |
| TYPE | Independent Consultant, Company, Research and Development |
| ADDRESS | 1196 Euclid Ave.<br>Berkeley, CA 94708 |
| CONTACT | Mr. Aaron Marcus,<br>Principal |
| PHONE | (415) 527-6224/6225 |
| PREPRODUCTION SERVICES | Analysis, Design, Seminars |
| DESCRIPTION | Aaron Marcus is a firm of information-oriented, graphic designers for computer-based telecommunications. They design presentational graphics, human-computer interfaces, and program visualization schema. They have designed low and high resolution video/crt displays and interfaces. They work with system programmers and with clients. They also do storyboarding for complete information/entertainment narratives for video display. |

| | |
|---|---|
| ORGANIZATION | **ACTRONICS, INC.** |
| TYPE | Company |
| ADDRESS | 810 River Ave.<br>Pittsburgh, PA 15212 |
| CONTACT | Ms. Joyce Hallas |
| PHONE | (412) 231-6200 |
| PREPRODUCTION SERVICES | Analysis, Design, Scripting |
| PRODUCTION SERVICES | Producing, Computer Programming |

| | |
|---|---|
| POSTPRODUCTION SERVICES | System Design and Configuration |
| DESCRIPTION | Actronics, Inc. has capabilities for complete analysis, design, and implementation of computer-based, interactive videodisc training programs. Their particular area of expertise is in medical training products. |

| | |
|---|---|
| ORGANIZATION | **ADVANCED IMAGE TECHNOLOGY, INC.** |
| TYPE | Company |
| ADDRESS | 122 East 42nd St., Suite 623<br>New York, NY 10017 |
| CONTACT | Dr. Jean-Pierre Isbouts |
| PHONE | (212) 986-2861 |
| PREPRODUCTION SERVICES | Analysis, Design, Scripting, Seminars |
| PRODUCTION SERVICES | Producing, Directing, Special Effects, Computer Programming, Seminars, Workshops |
| POSTPRODUCTION SERVICES | Premastering, System Design and Configuration, Seminars |
| DESCRIPTION | Advanced Image Technology, Inc. has a research and development program for the development of "GPS Imagebase" software for management of very high volumes of imagery and data combined. This is integrated with development of AIT'S proprietary backend processor. ATI's production services include high-volume image capture and animation. They have postproduction facilities in the U.S. and Europe for disc production in both PAL and NTSC. |

| | |
|---|---|
| ORGANIZATION | **ADVANCED VIDEO, INC.** |
| TYPE | Company |
| ADDRESS | 2131 Las Palmas Dr., Suite D<br>Carlsbad, CA 92008 |
| CONTACT | Mr. Rick Fisher or Mr. Jim Burley |
| PHONE | (619) 438-8985 |
| PRODUCTION SERVICES | Computer Programming |
| POSTPRODUCTION SERVICES | Premastering, System Design and Configuration |
| DESCRIPTION | Advanced Video, Inc. is involved in the development of a videodisc NFL football arcade game, as well as additional disc production. |

| | |
|---|---|
| ORGANIZATION | **ALBERTA EDUCATIONAL COMMUNICATIONS CORP. (ACCESS)** |
| TYPE | Provincially funded corporation |
| ADDRESS | 16930—114 Ave., Edmonton<br>T5M 3S2 Alberta, Canada |
| CONTACT | Mr. Russell Sawchuk |
| PHONE | (403) 451-3160 |
| PREPRODUCTION SERVICES | Analysis, Design, Scripting, Funding, Seminars, Workshops |
| PRODUCTION SERVICES | Producing, Directing, Computer Programming, Seminars, Workshops |
| POSTPRODUCTION SERVICES | Premastering, System Design and Configuration |
| DESCRIPTION | Although ACCESS Alberta's main responsibililty is to provide educational communications support to the province's educational institutions, the Corporation, subject to availability of resources, is interested in making its facilities and expertise available to other agencies and organizations. |

# Production and Development

| | |
|---|---|
| ORGANIZATION | **ALEC GROUP** |
| TYPE | Independent Consultant, Research and Development |
| ADDRESS | Suite 300, 95 So. Market St. <br> San Jose, CA 95113 |
| CONTACT | Mr. Lawrence Dietz, <br> President |
| PHONE | (408) 297-3789 |
| PREPRODUCTION SERVICES | Seminars, Workshops |
| PRODUCTION SERVICES | Seminars, Workshops |
| POSTPRODUCTION SERVICES | Seminars, Workshops |
| DESCRIPTION | The Alec Group conducts reseach on specific uses of videodiscs in particular industries or markets, and provides security and threat awareness analysis for electronic information. |

| | |
|---|---|
| ORGANIZATION | **ALPHATEL SYSTEMS LTD.** |
| TYPE | Company |
| ADDRESS | #102, 11430—168 St., Edmonton <br> T5M 3T9, Alberta, Canada |
| CONTACT | Dr. R.A. (Bob) Abell |
| PHONE | (403) 451-5761 |
| PREPRODUCTION SERVICES | Analysis, Design, Seminars, Workshops |
| PRODUCTION SERVICES | Computer Programming, Seminars, Workshops |
| POSTPRODUCTION SERVICES | System Design and Configuration, Seminars, Workshops |
| DESCRIPTION | ALPHATEL Systems Ltd. specializes in the custom design of interactive videotex information, training and directory systems. They also provide consulting services, product development, applications software, courseware, and equipment for the general field of technology-based education and training, with an emphasis on computer, videotex, and laser videodisc applications. |

| | |
|---|---|
| ORGANIZATION | **AMERICAN VIDEO INSTITUTE/ VIDEODISC & ELECTRONIC PUBLISHING LAB.** |
| TYPE | Not-for-Profit, Research and Development |
| ADDRESS | Rochester Institute of Technology <br> P.O. Box 9887, 1 Lomb Memorial Dr. <br> Rochester, NY 14623 |
| CONTACT | Mr. John Ciampa |
| PHONE | (716) 475-2779/6625 |
| PREPRODUCTION SERVICES | Analysis, Design, Scripting, Seminars |
| PRODUCTION SERVICES | Producing, Directing, Computer Programming |
| POSTPRODUCTION SERVICES | System Design and Configuration, Seminars |

| | |
|---|---|
| ORGANIZATION | **APPLIED VIDEO TECHNOLOGY, INC.** |
| TYPE | Company |
| ADDRESS | 5118 Westminster Pl. <br> St. Louis, MO 63108 |
| CONTACT | Mr. Charles Kindleberger <br> President |
| PREPRODUCTION SERVICES | Analysis, Design, Scripting, Funding, Seminars |
| PRODUCTION SERVICES | Producing, Computer Programming |
| POSTPRODUCTION SERVICES | System Design and Configuration |
| DESCRIPTION | Applied Video Technology, Inc. is a firm specializing in consulting |

assistance, market analysis, and systems design with special emphasis on microcomputer and videodisc-based applications. Mr. Kindleberger is a contributing editor to *Interactive Video Technology*, a monthly industry newsletter produced by Heartland Communications, as well as an occasional contributor to other video-related publications.

Applied Video Technology offers a number of publications, the best known of which is *Interactive Video: A Guide to Users, Vendors, Producers, Researchers and Observers concerned with Interactive Videodiscs and Tape*.

| | |
|---|---|
| ORGANIZATION | **BALL COMMUNICATIONS, INC.** |
| TYPE | Company |
| ADDRESS | 1101 N. Fulton Ave.<br>Evansville, IN 47710 |
| CONTACT | Ms. Donna Leader,<br>Vice President |
| PHONE | (812) 428-2300 |
| PREPRODUCTION SERVICES | Analysis, Design, Scripting, Seminars |
| PRODUCTION SERVICES | Producing, Directing, Special Effects, Computer Programming, Seminars |
| POSTPRODUCTION SERVICES | Seminars |
| DESCRIPTION | Ball Communications, Inc. specializes in the conceptualization, development, and production of education and motivation programs for business and industry. Advanced technology is incorporated into the design, development, and production of these programs. Interactive videodisc and videotape educational programs have been produced for major pharmaceutical companies in the U.S. |

| | |
|---|---|
| ORGANIZATION | **BALL TELEVISION GROUP** |
| TYPE | Company |
| ADDRESS | 2501 Hillsboro Rd.<br>Nashville, TN 37212 |
| CONTACT | Mr. Todd Staff |
| PHONE | (615) 292-2800 |
| PREPRODUCTION SERVICES | Analysis, Design, Scripting, Seminars |
| PRODUCTION SERVICES | Producing, Directing, Special Effects, Computer Programming, Seminars |
| POSTPRODUCTION SERVICES | Premastering, Program Simulation, System Design and Configuration |
| DESCRIPTION | Ball Television Group provides creative and technical expertise in television, film, audio, and interactive video production and distribution. Types of production include high-quality corporate and educational programming. Ball Television Group also specializes in the design and production of interactive videotape and videodisc courses. |

| | |
|---|---|
| ORGANIZATION | **BASHISTA, MICHAEL** |
| TYPE | Independent Consultant |
| ADDRESS | 320 Soquel Ave. C-1<br>Santa Cruz, CA 95062 |
| CONTACT | Mr. Michael Bashista |
| PHONE | (408) 426-7316 |
| PREPRODUCTION SERVICES | Analysis, Scripting |

## Production and Development

| | |
|---|---|
| ORGANIZATION | **BBC OPEN UNIVERSITY PRODUCTION CENTRE** |
| TYPE | Educational Institution |
| ADDRESS | Walton Hall, Bletchley |
| | Milton Keynes MK7 6BH |
| | United Kingdom |
| CONTACT | Mr. David Nelson |
| PHONE | 0908 655468 |
| PREPRODUCTION SERVICES | Analysis, Design, Scripting, Funding, Seminars, Workshops |
| PRODUCTION SERVICES | Producing, Directing, Special Effects, Computer Programming, Seminars, Workshops |
| POSTPRODUCTION SERVICES | Premastering, System Design and Configuration, Seminars, Workshops |
| DESCRIPTION | This BBC department produced 300 television and video programs and 400 audiovisual and radio packs per year across the widest spectrum of both undergraduate courses and continuing education packs. It has been actively involved in videodisc production for two years; offers a comprehensive service with in-house creative and technical expertise; and is totally outfitted with studios, locations, equipment, and postproduction facilities. |

| | |
|---|---|
| ORGANIZATION | **BNA COMMUNICATIONS INC.** |
| TYPE | Independent Consultant, Company |
| ADDRESS | 9439 Key West Ave. |
| | Rockville, MD 20850 |
| CONTACT | Mr. Cliff Witt |
| PHONE | (301) 948-0540 |
| PREPRODUCTION SERVICES | Analysis, Design, Scripting |
| PRODUCTION SERVICES | Producing, Directing, Special Effects |
| POSTPRODUCTION SERVICES | Premastering, System Design and Configuration |

| | |
|---|---|
| ORGANIZATION | **BOSTON MEDIA CONSULTANTS** |
| TYPE | Independent Consultant, Research and Development |
| ADDRESS | 19 Damon Rd. |
| | Scituate, MA 02066 |
| CONTACT | Mr. David P. Allen |
| PHONE | (617) 545-2696 |
| PREPRODUCTION SERVICES | Analysis, Design, Scripting, Funding, Seminars, Workshops |
| PRODUCTION SERVICES | Producing, Directing, Special Effects, Computer Programming, Seminars, Workshops |
| POSTPRODUCTION SERVICES | Premastering, Program Simulation, System Design and Configuration, Seminars, Workshops |
| DESCRIPTION | Boston Media Consultants are involved in ongoing research assistance and production for RCA Laboratories. |

| | |
|---|---|
| ORGANIZATION | **BRIEN LEE & COMPANY** |
| TYPE | Company |
| ADDRESS | 2025 N. Summit Ave. |
| | Milwaukee, WI 53202 |
| CONTACT | Mr. Brien Lee |
| PHONE | (414) 277-7600 |
| PREPRODUCTION SERVICES | Design, Scripting |
| PRODUCTION SERVICES | Producing, Directing |

| | |
|---:|:---|
| ORGANIZATION | **BUTLER, MADELEINE L.** |
| TYPE | Independent Consultant |
| ADDRESS | 402 Highland Ave. #30 |
| | Somerville, MA 02144 |
| CONTACT | Ms. Madeleine Butler |
| PHONE | (617) 625-6588 or 491-8865 |
| PREPRODUCTION SERVICES | Analysis, Design, Scripting |
| PRODUCTION SERVICES | Producing, Directing, Special Effects, Computer Programming |
| POSTPRODUCTION SERVICES | Premastering, System Design and Configuration |
| DESCRIPTION | Ms. Butler provides a full range of design, production, and consulting services. |

| | |
|---:|:---|
| ORGANIZATION | **CABLESHARE INC.** |
| TYPE | Company |
| ADDRESS | P.O. Box 5880 |
| | 20 Enterprise Dr. |
| | London, Ontario |
| | Canada N6A 4L6 |
| CONTACT | Mr. George S. McCabe |
| PHONE | (519) 686-2900 |
| PREPRODUCTION SERVICES | Design, Scripting |
| PRODUCTION SERVICES | Computer Programming |
| POSTPRODUCTION SERVICES | System Design and Configuration |

| | |
|---:|:---|
| ORGANIZATION | **CARLOS RAMIREZ & ALBERT H. WOODS INC.** |
| TYPE | Company |
| ADDRESS | 18 East 53rd St. |
| | New York, NY 10022 |
| CONTACT | Mr. Thomas J. Nicholson |
| PHONE | (212) 751-2650 |
| PREPRODUCTION SERVICES | Analysis, Design, Scripting |
| PRODUCTION SERVICES | Producing, Directing, Special Effects, Computer Programming |
| POSTPRODUCTION SERVICES | Premastering, Program Simulation, System Design and Configuration |
| DESCRIPTION | The firm of Carlos Ramirez & Albert H. Woods Inc. has for nearly twenty years specialized in the planning and design of exhibitions—environments where visitors can explore subjects actively through a variety of media. The final design solution for each project is shaped both by the subject matter and the media that will communicate most effectively with the anticipated audience. Taking a flexible approach to developing design solutions, Ramirez & Woods has produced exhibitions incorporating graphics, three-dimensional forms, still images, film, computer programs and interactive videodiscs. |

| | |
|---:|:---|
| ORGANIZATION | **CAT ZURICH-COMPUTER ASSISTED TELEVIDEO AG** |
| TYPE | Company |
| ADDRESS | Freudenbergstrasse 30 |
| | Zurich, Switzerland 8044 |
| PHONE | (01) 251.28.22 |
| | U.S. Offices: |
| | CAT Corporation Detroit |
| | 13530 Michigan Ave., Suite 226 |
| | Dearborn, MI 48126 |
| | (313) 584-3271 |

**Production and Development**

                        CAT Corporation Los Angeles
                        1875 Century Park East, Suite 930
                        Los Angeles, CA 90067
                        (213) 551-1052

                        CAT Corporation New York
                        131 Perry St., Loft 6-B
                        New York, NY 10014
                        (212) 924-6981

| | |
|---|---|
| PREPRODUCTION SERVICES | Analysis, Design, Scripting, System Architecture, System Engineering |
| PRODUCTION SERVICES | Producing, Directing, Special Effects, Computer Programming |
| POSTPRODUCTION SERVICES | Premastering, Program Simulation, System Design and Configuration |
| DESCRIPTION | CAT Computer Assisted Televideo is an international system house for complete interactive audiovisual information systems (hardware, software, installation, and service). CAT has developed seven different systems covering most industrial marketing needs. Everything from simple level 1 up to multiplayer systems, combinations with telecommunications, touch screen monitors, etc., is available. |

| | |
|---|---|
| ORGANIZATION | **CEDI INFORMATIQUE** |
| TYPE | Company |
| ADDRESS | 539 Cherrier St., Montreal<br>H2L 1H2 Quebec, Canada |
| CONTACT | Ms. Isabelle Mahy |
| PHONE | (514) 598-8217 |
| PREPRODUCTION SERVICES | Analysis, Design, Scripting |
| POSTPRODUCTION SERVICES | System Design and Configuration |
| DESCRIPTION | CEDI provides level 3 programs that control the disc as well as the scenario behind it. They have also developed experimental applications using videodisc, the Apple II Plus (level 3 software), and touch panel. |

| | |
|---|---|
| ORGANIZATION | **CENTRO TELEVISIVO UNIVERSITARIO (CTU)** |
| TYPE | Not-for-Profit, Research and Development, Educational Institution |
| ADDRESS | Universita degli Studi<br>Via Celoria<br>20—20133 Milano, Italy |
| CONTACT | Ms. Patrizia Ghislandi |
| PHONE | 02 / 23.64.504 |
| PREPRODUCTION SERVICES | Analysis, Design, Scripting, Seminars |
| PRODUCTION SERVICES | Producing, Directing, Computer Programming, Seminars |
| DESCRIPTION | CTU is working on several level 3 Laser CAV videodisc projects including a file of art images (frescoes, architecture of Italy, Switzerland, France) managed by an external computer, and an electronics encyclopedia on videodisc. |

| | |
|---|---|
| ORGANIZATION | **CENTURY III TELEPRODUCTIONS** |
| TYPE | Company |
| ADDRESS | 651 Beacon St.<br>Boston, MA 02215 |
| CONTACT | Ms. Judith B. Downes, Executive Producer of Corporate Communications Division |
| PHONE | (617) 267-6400 |

| | |
|---|---|
| PREPRODUCTION SERVICES | Scripting, Seminars |
| PRODUCTION SERVICES | Producing, Directing, Special Effects, Seminars |
| POSTPRODUCTION SERVICES | Premastering, Seminars |

| | |
|---|---|
| ORGANIZATION | **CITIDISC PRODUCTIONS** |
| TYPE | Company |
| ADDRESS | Suite 209, 1437 Rhode Island Ave., N.W. |
| | Washington, DC 20005 |
| CONTACT | Mr. Garri Garripoli |
| PHONE | (202) 483-9369 |
| PREPRODUCTION SERVICES | Analysis, Design, Scripting |
| PRODUCTION SERVICES | Producing, Directing, Special Effects, Computer Programming |
| POSTPRODUCTION SERVICES | Premastering, System Design and Configuration |
| DESCRIPTION | CITIDISC is researching new game applications for interactive entertainment. They specialize in developing unique solutions to difficult and challenging videodisc problems. They have many years of film and video production experience to ensure high quality images and graphics along with advanced, high quality hardware. |

| | |
|---|---|
| ORGANIZATION | **CLARK, DAVID R.** |
| TYPE | Independent Consultant |
| ADDRESS | Hartlands, 100 Crouch Hill |
| | London, England N8 9EA |
| CONTACT | Dr. David R. Clark |
| PREPRODUCTION SERVICES | Analysis, Design, Scripting, Seminars |
| PRODUCTION SERVICES | Producing, Directing, Special Effects, Computer Programming |
| POSTPRODUCTION SERVICES | System Design and Configuration |
| DESCRIPTION | Dr. Clark is a specialist in the design of interactive videodisc programs. He is a patent holder for digital data in analogue video format. He is conducting research, with government funding, on the laser disc as an archive of both digital and analogue data. |

| | |
|---|---|
| ORGANIZATION | **CLINICAL INTERACTIONS, INC.** |
| TYPE | Company |
| ADDRESS | 40 N. Van Brunt St. |
| | Englewood, NJ 07631 |
| CONTACT | Mr. Donald Stark or Mr. Don Monaco |
| PHONE | (201) 568-0446 |
| PREPRODUCTION SERVICES | Analysis, Design, Scripting, Consulting |
| PRODUCTION SERVICES | Producing, Directing, Special Effects, Computer Programming |
| POSTPRODUCTION SERVICES | Premastering, Program Simulation, System Design and Configuration, Technical Support at Exhibits |
| DESCRIPTION | Clinical Interactions specializes in medical and technical design and writing, on-site technical support for meetings and exhibits, and programming such applications as multiuser systems, data collection, etc. |

| | |
|---|---|
| ORGANIZATION | **COAST COMMUNICATIONS** |
| TYPE | Company |
| ADDRESS | P.O. Box 1591 |
| | Palo Alto, CA 94025 |

# Production and Development

| | |
|---|---|
| CONTACT | Dr. Ronald Trugman |
| PHONE | (415) 366-9694 |
| PREPRODUCTION SERVICES | Analysis, Design, Scripting, Funding, Seminars, Workshops |
| PRODUCTION SERVICES | Producing, Directing, Special Effects, Computer Programming, Seminars, Workshops |
| POSTPRODUCTION SERVICES | Premastering, Program Simulation, System Design and Configuration, Seminars, Workshops |
| DESCRIPTION | Coast Communications draws on the resources of Stanford University and SRI International for the production and development of videodiscs. |

| | |
|---|---|
| ORGANIZATION | **COASTAL VIDEO COMMUNICATIONS** |
| TYPE | Company |
| ADDRESS | 4663 Haygood Rd., Suite 202 |
| | Virginia Beach, VA 23455 |
| CONTACT | Mr. Paul Michels |
| PHONE | (804) 499-9228 |
| PREPRODUCTION SERVICES | Analysis, Design, Scripting, Workshops |
| PRODUCTION SERVICES | Producing, Directing, Special Effects |
| POSTPRODUCTION SERVICES | Program Simulation |
| DESCRIPTION | Coastal Video Communications specializes in the interactive design and production of skilled trades training materials. They have worked with both military and corporate clients on trades training subjects. |

| | |
|---|---|
| ORGANIZATION | **COMBUSTION ENGINEERING, INC.** |
| TYPE | Company |
| ADDRESS | 1000 Prospect Hill Rd. |
| | Windsor, CT 06095 |
| CONTACT | Mr. James F. Church or Mr. W.B. Renfro |
| PHONE | (203) 688-1911 |
| PREPRODUCTION SERVICES | Analysis, Design, Scripting |
| PRODUCTION SERVICES | Producing, Directing, Special Effects, Computer Programming |
| POSTPRODUCTION SERVICES | Premastering, System Design and Configuration |
| DESCRIPTION | Combustion Engineering is actively involved in the development of interactive videodisc training for nuclear power and other high technology applications. |

| | |
|---|---|
| ORGANIZATION | **COMMUNICATION STUDIO** |
| TYPE | Company |
| ADDRESS | 154 West 27th St.#7E |
| | New York, NY 10001 |
| CONTACT | Mr. John Vaughan |
| PHONE | (212) 924-3729 |
| PREPRODUCTION SERVICES | Design, Scripting |
| PRODUCTION SERVICES | Producing, Directing, Special Effects, Computer Programming, Videotex and Graphics |
| POSTPRODUCTION SERVICES | Premastering, System Design and Configuration |

| | |
|---|---|
| ORGANIZATION | **COMPUTER SCIENCES CORPORATION** |
| TYPE | Company, Research and Development |
| ADDRESS | P.O. Box M |
| | Ft. Eustis, VA 23604 |

|  |  |
|---|---|
| CONTACT | Dr. William A. Stembler or J.G. Palmer |
| PHONE | (804) 887-1677 |
| PREPRODUCTION SERVICES | Analysis, Design, Scripting, Seminars, Workshops, Automated Storyboarding, Videodisc Authoring |
| PRODUCTION SERVICES | Producing, Directing, Special Effects, Computer Programming, Seminars, Workshops, Computerized Production Management |
| POSTPRODUCTION SERVICES | Premastering, Program Simulation, System Design and Configuration, Seminars, Workshops, Computerized Postproduction Managment |
| DESCRIPTION | Computer Sciences is involved in the Patriot Project. They are developing a prototype maintenance videodisc with emphasis on automation of the development process, as well as production of an interactive videodisc to be tested in the field. Software is being designed to automate all phases of design, storyboarding, programming, and to assist in planning production and postproduction. |

|  |  |
|---|---|
| ORGANIZATION | **COMPUTER VIDEO PRODUCTIONS, INC.** |
| TYPE | Company |
| ADDRESS | 1317 Clover Dr. South<br>Minneapolis, MN 55420 |
| CONTACT | Mr. Dean N. Sutliff |
| PHONE | (612) 888-2388 |
| PREPRODUCTION SERVICES | Analysis, Design, Scripting, Seminars, Workshops |
| PRODUCTION SERVICES | Producing, Directing, Special Effects, Computer Programming, Seminars, Workshops, Full 1- and 3/4-inch studio and location facilities |
| POSTPRODUCTION SERVICES | Premastering, System Design and Configuration, Seminars, Workshops, Full 1- and 3/4-inch U-matic Computer-Controlled Editing, OPTISCAN Evaluation |
| DESCRIPTION | Computer Video Productions, Inc. is the only postproduction facility offering the OPTISCAN service, which can potentially save the cost of making an incorrect videodisc master. OPTISCAN is a premastering service that scans a 1-inch premaster using computer technology, and identifies dropouts, color frame errors, changes in field dominance, control track errors, and errors in SMPTE time code. OPTISCAN identifies the problem and then provides the error location in its computer software print-out. Call Computer Video for price and availability of service. |

|  |  |
|---|---|
| ORGANIZATION | **CONSULTANT'S CHOICE, INC.** |
| TYPE | Company, Research and Development |
| ADDRESS | 5002 Shadow Glen Court<br>Atlanta, GA 30338 |
| CONTACT | Mr. Billy B. Wise |
| PHONE | (404) 396-6202 or 394-0629 |
| PREPRODUCTION SERVICES | Analysis, Design, Scripting |
| PRODUCTION SERVICES | Producing, Directing, Computer Programming |
| POSTPRODUCTION SERVICES | Program Simulation, System Design and Configuration |
| DESCRIPTION | Consultant's Choice is a small business involved with the development of command and control systems and devices. Generic functions include system integration, software development, and some unique hardware development. |

**Production and Development**

| | |
|---:|:---|
| ORGANIZATION | **CONTROL DATA CORPORATION** |
| TYPE | Company |
| ADDRESS | 8100 34th Ave. So. |
| | Mailing Address/Box 0 |
| | Minneapolis, MN 55440 |
| CONTACT | Mr. Charles W. Loney |
| PHONE | (612) 853-5022 |
| PREPRODUCTION SERVICES | Analysis, Design, Scripting, Other |
| PRODUCTION SERVICES | Producing, Directing, Special Effects, Computer Programming |
| DESCRIPTION | Control Data has been involved in videodisc research, development, and delivery since 1975. In joint ventures with Phillips, Thompson, CSF, and others, they did much of the original work that has led to the present state of the art. They have developed proprietary training courseware involving interactive videodisc in Germany, France, Canada, and the U.S. Several divisions offer videodisc and computer-based instructional development. These include the Professional Services Division, the PLATO Training and Education Division, and the Agribusiness Division. |

| | |
|---:|:---|
| ORGANIZATION | **CORNELL UNIVERSITY/DEPARTMENT OF EDUCATION** |
| TYPE | Educational Institution |
| ADDRESS | Stone Hall |
| | Ithaca, NY 14853 |
| CONTACT | Mr. Robert Bruce or Ms. Geri Gay |
| PHONE | (607) 256-2015 |
| PREPRODUCTION SERVICES | Analysis, Design, Scripting, Workshops |
| PRODUCTION SERVICES | Computer Programming, Seminars, Workshops |
| POSTPRODUCTION SERVICES | System Design and Configuration, Seminars, Workshops |
| DESCRIPTION | The Department of Education is primarily a graduate teaching program concerned with applications of videodisc and other interactive technology to formal and nonformal education. In addition to preproduction planning applications, they are interested in development of criteria and method of applications decisions; i.e., when, where, for what audiences, and for what materials, etc., that the technology should be used. |

| | |
|---:|:---|
| ORGANIZATION | **CORPORATION FOR PUBLIC BROADCASTING** |
| TYPE | Not-for-Profit |
| ADDRESS | 1111 16th St. NW |
| | Washington, DC 20036 |
| CONTACT | Mr. Richard Grefe, Director, |
| | Policy Development and Planning |
| PHONE | (202) 293-6160 |
| DESCRIPTION | Conducts research into the educational use and effectiveness of videodisc technology, including experiments and demonstrations. |

| | |
|---:|:---|
| ORGANIZATION | **COURSEWARE, INC.** |
| TYPE | Company |
| ADDRESS | 10075 Carroll Canyon Rd. |
| | San Diego, CA 92131 |

| | |
|---:|:---|
| CONTACT | Ms. Pomela Flanigan |
| PHONE | (619) 578-1700 |
| PREPRODUCTION SERVICES | Analysis, Design, Scripting, Workshops |
| PRODUCTION SERVICES | Computer Programming |
| POSTPRODUCTION SERVICES | System Design and Configuration |

| | |
|---:|:---|
| ORGANIZATION | **CREATIVE UNIVERSAL, INC.** |
| TYPE | Company |
| ADDRESS | Suite 1200, Tower 14 |
| | 21700 Northwestern Hwy. |
| | Southfield, MI 48075 |
| CONTACT | Mr. Larry Short, |
| | President |
| PHONE | (313) 557-4100 |
| PREPRODUCTION SERVICES | Analysis, Design, Scripting, Seminars, Workshops |
| PRODUCTION SERVICES | Producing, Directing, Computer Programming |
| POSTPRODUCTION SERVICES | Program Simulation, System Design and Configuration |

| | |
|---:|:---|
| ORGANIZATION | **CREATIVISION, INC.** |
| TYPE | Company |
| ADDRESS | P.O. Drawer 1070 |
| | 456A 11th St. |
| | Holly Hill, FL 32017 |
| CONTACT | Mr. Rick Glasby |
| | Manager, Videodisc Projects |
| PHONE | (904) 255-0911 |
| PREPRODUCTION SERVICES | Analysis, Design, Scripting |
| PRODUCTION SERVICES | Producing, Directing, Special Effects, Computer Programming |
| POSTPRODUCTION SERVICES | Premastering, Program Simulation, System Design and Configuration |

| | |
|---:|:---|
| ORGANIZATION | **CUTTING EDGE PRODUCTIONS, INC.** |
| TYPE | Company |
| ADDRESS | 115 West 88th St. |
| | New York, NY 10024 |
| CONTACT | Ms. Judy Bergsma-Owen |
| PHONE | (212) 799-2174 |
| PREPRODUCTION SERVICES | Analysis, Design, Scripting |
| PRODUCTION SERVICES | Producing, Directing |
| DESCRIPTION | Cutting Edge is a production company that generally produces only customized programs for its corporate clients. |

| | |
|---:|:---|
| ORGANIZATION | **DAVID O. MCKAY INSTITUTE** |
| TYPE | Independent Consultant, Not-for-Profit, Research and Development |
| ADDRESS | 215 MCKB |
| | Brigham Young University |
| | Provo, UT 84602 |
| CONTACT | Dr. Larrie E. Gale, |
| | Coordinator of Research |
| PHONE | (801) 378-7082 |
| PREPRODUCTION SERVICES | Analysis, Design, Scripting, Seminars, Workshops |
| PRODUCTION SERVICES | Producing, Directing, Computer Programming, Seminars, Workshops |

**Production and Development**

| | |
|---:|:---|
| POSTPRODUCTION SERVICES | Premastering, System Design and Configuration |
| DESCRIPTION | The David O. McKay Institute offers a unique, interactive scripting service that obviates the need for flowcharts when producing interactive video or random access audio. They develop computer-assisted production software. Their designs and computer authoring systems are based on learning research. They also research results in the design of equipment configurations and delivery systems. |

| | |
|---:|:---|
| ORGANIZATION | **DAVIDSON, STUART P.** |
| TYPE | Independent Consultant |
| ADDRESS | 2 Soldiers Field Park—415 |
| | Boston, MA 02163 |
| CONTACT | Mr. Stuart Davidson |
| PHONE | (617) 498-8675 |
| PREPRODUCTION SERVICES | Analysis, Design, Scripting |
| PRODUCTION SERVICES | Producing, Directing, Special Effects, Computer Programming |
| POSTPRODUCTION SERVICES | Premastering |
| DESCRIPTION | As an employee of Warner Amex and independent producers, Mr. Davidson has produced level 2 programs with program dumps on videodisc. He has been through the mastering process at DiscoVision, prior to its closing, and has premastered a number of discs using Sony equipment. He feels his greatest value as an independent consultant is in the planning and production phases of interactive entertainment and games, and sales/promotional videodiscs. |

| | |
|---:|:---|
| ORGANIZATION | **DAVIS AUDIO-VISUAL, INC.** |
| TYPE | Independent Consultant |
| ADDRESS | 1801 Federal Blvd. |
| | Denver, CO 80204 |
| CONTACT | Mr. Robert A. Newman, III, |
| | Director of Interactive Systems Division |
| PHONE | (303) 455-1122 |
| PREPRODUCTION SERVICES | Analysis, Design, Scripting, Funding, Seminars, Workshops |
| PRODUCTION SERVICES | Producing, Directing, Computer Programming, Seminars, Workshops |
| POSTPRODUCTION SERVICES | Program Simulation, Seminars, Workshops |
| DESCRIPTION | Bob Newman is an independent consultant on a full-time basis to Davis Audio-Visual and is Director of Interactive Systems Division. Bob lectures on interactive systems design and utilization across the country. A partial list of past clients includes United Airlines, Mansville Corporation, Bell Laboratories, Pepsi-Cola Bottling Co. and many others. |

| | |
|---:|:---|
| ORGANIZATION | **DAVIS, BARBARA GROSS** |
| TYPE | Independent Consultant |
| ADDRESS | 932 Hilldale Ave. |
| | Berkeley, CA 94708 |
| CONTACT | Ms. Barbara Davis |
| PREPRODUCTION SERVICES | Instructional Design |
| POSTPRODUCTION SERVICES | Evaluation |

| | |
|---:|:---|
| ORGANIZATION | **DBS FILMS, INC.** |
| TYPE | Independent Consultant, Company |
| ADDRESS | 3 Great Valley Parkway East<br>Malvern, PA 19355 |
| CONTACT | Mr. Allen Dugan |
| PHONE | (215) 296-5850 |
| ADDRESS | 30 East 40th St.<br>New York, NY 10016 |
| CONTACT | Mr. Jay Silber |
| PHONE | (212) 684-3270 |
| PREPRODUCTION SERVICES | Analysis, Design, Scripting |
| PRODUCTION SERVICES | Producing, Directing, Special Effects, Computer Programming |
| POSTPRODUCTION SERVICES | Premastering, Program Simulation, System Design and Configuration |

| | |
|---:|:---|
| ORGANIZATION | **DEPARTMENT OF NATIONAL DEFENCE** |
| TYPE | Research and Development |
| ADDRESS | P.O. Box 2000, Downsview<br>M3M 3B9 Ontario, Canada |
| CONTACT | Mr. Bruce Ferguson |
| PREPRODUCTION SERVICES | Analysis, Design, Scripting, Funding |
| PRODUCTION SERVICES | Special Effects, Computer Programming |
| DESCRIPTION | This is a governmental military research lab involved in simulation and training for combat systems, maintenance, and operation. |

| | |
|---:|:---|
| ORGANIZATION | **DEVLIN PRODUCTIONS, INC.** |
| TYPE | Independent Consultant, Company |
| ADDRESS | 150 West 55th St.<br>New York, NY 10019 |
| CONTACT | Ms. Sandra Devlin,<br>President |
| PHONE | (212) 582-5572 |
| PREPRODUCTION SERVICES | Analysis, Design, Scripting, Seminars |
| PRODUCTION SERVICES | Producing, Directing, Computer Programming |
| POSTPRODUCTION SERVICES | Premastering, Program Simulation, System Design and Configuration, Seminars |
| DESCRIPTION | Devlin's facilities include a multiframe storage management system with real time i/o, an intelligent videodisc scratch pad, and a Sony interactive videodisc emulator. |

| | |
|---:|:---|
| ORGANIZATION | **DIGITAL EQUIPMENT CORPORATION** |
| TYPE | Company, Research and Development |
| ADDRESS | Educational Services,<br>12 Crosby Drive, Mail Stop BUO<br>Bedford, MA 01730 |
| CONTACT | Mr. Colin Allan |
| PHONE | (617) 276-1408 |
| PREPRODUCTION SERVICES | Analysis, Design, Scripting, Seminars, Funding, Customer Training for Courseware Development |
| PRODUCTION SERVICES | Producing, Directing, Special Effects, Computer Programming, Customer Training for Courseware Development |
| POSTPRODUCTION SERVICES | Premastering, Program Simulation, System Design and Configuration, Customer Training for Courseware Development |

# Production and Development

| | |
|---:|:---|
| ORGANIZATION | **DIGITAL VIDEO CORPORATION** |
| TYPE | Company |
| ADDRESS | 369 North Orange Ave. |
| | Orlando, FL 32801 |
| CONTACT | Mr. Stephen Matheny |
| PHONE | (305) 425-1999 |
| PREPRODUCTION SERVICES | Analysis, Instructional Design, Scripting, Seminars, Workshops |
| PRODUCTION SERVICES | Producing, Directing, Special Effects, Computer Programming, Digital Graphics, Film-to-tape Transfer, Seminars, Workshops |
| POSTPRODUCTION SERVICES | Premastering, Program Simulation, System Design and Configuration, Seminars, Workshops |

| | |
|---:|:---|
| ORGANIZATION | **D'VIDEO AND ASSOCIATES** |
| TYPE | Independent Consultant |
| ADDRESS | 21381 Stans Lane |
| | Laguna Beach, CA 92651 |
| CONTACT | Mr. David McElhatten |
| PHONE | (714) 497-3019 |
| PREPRODUCTION SERVICES | Analysis, Design, Seminars, Workshops |
| POSTPRODUCTION SERVICES | Premastering, System Design and Configuration |
| DESCRIPTION | D'Video and Associates specialize in product information delivery systems. |

| | |
|---:|:---|
| ORGANIZATION | **EASYWAY COMPUTER SOLUTIONS** |
| TYPE | Company |
| ADDRESS | 9373 Activity Rd., #H |
| | San Diego, CA 92126 |
| CONTACT | Mr. Robert Pannoni |
| PHONE | (619) 693-0080 |
| PREPRODUCTION SERVICES | Analysis, Design, Scripting, Seminars, Workshops |
| PRODUCTION SERVICES | Producing, Special Effects, Computer Programming, Seminars, Workshops |
| POSTPRODUCTION SERVICES | Program Simulation, System Design and Configuration |
| DESCRIPTION | Easyway has a learning center providing instruction in basic computer skills, use of authoring languages, and videodisc production. |

| | |
|---:|:---|
| ORGANIZATION | **EDITEL-CHICAGO** |
| TYPE | Company |
| ADDRESS | 301 East Erie St. |
| | Chicago, IL 60611 |
| CONTACT | Mr. Lenard Pearlman |
| PHONE | (312) 440-2360 |
| PRODUCTION SERVICES | Seminars |
| POSTPRODUCTION SERVICES | Premastering |
| DESCRIPTION | Editel-Chicago is part of the Editel Group with additional facilities in New York and Los Angeles. |

| | |
|---:|:---|
| ORGANIZATION | **EDITEL-NEW YORK** |
| TYPE | Company |
| ADDRESS | 222 East 44th St. |
| | New York, NY 10017 |

| | |
|---:|:---|
| CONTACT | Mr. Frank Dobyns |
| PHONE | (212) 867-4600 |
| PREPRODUCTION SERVICES | Analysis, Design, Scripting |
| PRODUCTION SERVICES | Producing, Directing, Special Effects, Computer Programming |
| POSTPRODUCTION SERVICES | Premastering, Program Simulation, System Design and Configuration |

| | |
|---:|:---|
| ORGANIZATION | **EDUCATIONAL BROADCASTING CORPORATION (WNET/13)** |
| TYPE | Company, Not-for-Profit |
| ADDRESS | 356 West 58 St.<br>New York, NY 10019 |
| CONTACT | For full-scale instructional project development,<br>Mr. Stephen L. Salyer,<br>Vice President and Director,<br>Education Division |
| PHONE | (212) 664-7172 |
| CONTACT | For line production and videodisc design,<br>Mr. Jeffrey Honeyman,<br>Producer/Programmer |
| PHONE | (212) 664-7194 |
| CONTACT | For hardware/technical advice,<br>Mr. David Sit,<br>Chief Engineer |
| PHONE | (212) 560-3523 |
| PREPRODUCTION SERVICES | Analysis, Design, Scripting |
| PRODUCTION SERVICES | Producing, Directing, Special Effects |
| POSTPRODUCTION SERVICES | Premastering |
| DESCRIPTION | WNET/13 is a public television station with facilities and personnel available for both profit-making and not-for-profit projects. Full production services are available as well as videodisc design services. |

| | |
|---:|:---|
| ORGANIZATION | **EDUCATIONAL COMPUTER CORPORATION** |
| TYPE | Company |
| ADDRESS | 175 Strafford Ave.<br>Strafford, PA 19087 |
| CONTACT | Mr. Nick Siecko |
| PHONE | (215) 687-2600 |
| PREPRODUCTION SERVICES | Analysis, Design, Scripting |
| PRODUCTION SERVICES | Directing, Computer Programming |
| POSTPRODUCTION SERVICES | System Design and Configuration |

| | |
|---:|:---|
| ORGANIZATION | **EUGENE PUBLIC SCHOOLS** |
| TYPE | Educational Institution |
| ADDRESS | 200 North Monroe St.<br>ATTN: Cal Young<br>Eugene, OR 97401 |
| CONTACT | Mr. Phil Kessinger |
| PHONE | (503) 687-3234 |
| PREPRODUCTION SERVICES | Analysis, Design, Scripting, Workshops |
| PRODUCTION SERVICES | Producing, Directing |
| DESCRIPTION | The Eugene Public Schools produce in-service activities for educators (K-12, Adult Education) on system applications of videodisc technology. They are also working on instructional design for public school |

curricula, using level 3 players with touch screen interactions. A *Teacher Guide* is available, designed to introduce educators to the theory and practice of videodisc technology with the emphasis on practice. They are also currently working on a Travel-Area Study disc on the Pacific Northwest, which is part of a series.

| | |
|---|---|
| ORGANIZATION | **EXHIBIT TECHNOLOGY, INC.** |
| TYPE | Company |
| ADDRESS | 200 Park Ave., Suite 2555<br>New York, NY 10166 |
| CONTACT | Mr. Don Cochrane,<br>President, Marketing |
| PHONE | (212) 692-9211 |
| PREPRODUCTION SERVICES | Analysis, Design, Scripting, Workshops |
| PRODUCTION SERVICES | Producing, Directing, Special Effects, Computer Programming, Workshops |
| POSTPRODUCTION SERVICES | Premastering, Program Simulation, System Design and Configuration, Workshops |
| DESCRIPTION | Exhibit Technology offers a wide range of videodisc design and production services. Some of their more recent projects include interactive videodisc presentations for AT&T, EXXON, and Sperry at EPCOT, a point-of-sales presentation for CitiBank, a series of presentations for IBM Canada, and an interactive "interview" presentation for the Bowling Hall of Fame. |

| | |
|---|---|
| ORGANIZATION | **FIRST CAPITAL CORPORATION OF CHICAGO** |
| TYPE | Venture capital for videodisc-related companies |
| ADDRESS | 200 Clarendon St.<br>Boston, MA 02116 |
| CONTACT | Mr. Darius G. Nevin |
| PHONE | (617) 247-4856 |

| | |
|---|---|
| ORGANIZATION | **FLORIDA STATE UNIVERSITY/<br>CENTER FOR EDUCATIONAL TECHNOLOGY** |
| TYPE | Research and Development |
| ADDRESS | 1 A TULLY, Florida State University<br>Tallahassee, FL 32306 |
| CONTACT | Dr. Robert Branson or Dr. Brent Hewlett |
| PHONE | (904) 644-4720 |
| PREPRODUCTION SERVICES | Analysis, Design, Scripting, Seminars, Workshops |
| PRODUCTION SERVICES | Producing, Directing, Special Effects, Computer Programming, Seminars, Workshops |
| POSTPRODUCTION SERVICES | Program Simulation, Seminars, Workshops |
| DESCRIPTION | The center is developing a large computer-based training program on basic skills for the Army, which will include a videodisc. The program should be completed in May 1984. |

| | |
|---|---|
| ORGANIZATION | **FORRESTER PRODUCTIONS** |
| TYPE | Company |
| ADDRESS | 3900 Greystone Ave.<br>Riverdale, NY 10463 |

| | |
|---|---|
| CONTACT | Mr. David Kappes |
| PHONE | (212) 884-1919 |
| PREPRODUCTION SERVICES | Analysis, Design, Scripting |
| PRODUCTION SERVICES | Producing, Directing, Special Effects, Computer Programming |
| POSTPRODUCTION SERVICES | Premastering, Program Simulation, System Design and Configuration |

| | |
|---|---|
| ORGANIZATION | **FRANKLIN RESEARCH CENTER** |
| TYPE | Not-for-Profit, Research and Development |
| ADDRESS | 20th & Race Sts.<br>Philadelphia, PA 19103 |
| CONTACT | Mr. Ken Fertner |
| PHONE | (215) 448-1340 |
| PREPRODUCTION SERVICES | Analysis, Design |
| PRODUCTION SERVICES | Computer Programming |
| POSTPRODUCTION SERVICES | Premastering, System Design and Configuration |

| | |
|---|---|
| ORGANIZATION | **FROST & SULLIVAN, INC.** |
| TYPE | Company |
| ADDRESS | 106 Fulton St.<br>New York, NY 10038 |
| ADDRESS | 104-112 Marylebone Lane<br>London, England W1M 5FU |
| CONTACT | Mr. Robert A. Sanzo,<br>Vice President, Marketing |
| PHONE | (212) 233-1080 |
| DESCRIPTION | Frost & Sullivan is a market research supplier, which publishes industry studies and market research reports on U.S. and overseas markets. They produce the following titles for the videodisc industry: "Interactive Videodisc Software Market in the U.S." (A1171); "Interactive Videodisc Hardware Market in the U.S." (A1187); "Videodisc Hardware and Software Market in the U.S." (A1219); "Videodisc Hardware and Software Market in Western Europe" (E593). The reports include forecasts of market size by product, market share by company and end use, and forecasts of technological and financial development. |

| | |
|---|---|
| ORGANIZATION | **FULL CIRCLE COMMUNICATIONS, INC.** |
| TYPE | Company |
| ADDRESS | 13530 Michigan Ave., Suite 226<br>Dearborn, MI 48126 |
| CONTACT | Mr. Stan Williams |
| PHONE | (313) 584-3200 |
| PREPRODUCTION SERVICES | Analysis, Design, Scripting, Seminars |
| PRODUCTION SERVICES | Producing, Directing, Special Effects, Computer Programming, Seminars |
| POSTPRODUCTION SERVICES | Premastering, Program Simulation, System Design and Configuration, Seminars |
| DESCRIPTION | Full Circle Communications, Inc. is a company of designers and producers of custom interactive video systems and software. |

| | |
|---|---|
| ORGANIZATION | **FUSION MEDIA, INC.** |
| TYPE | Company, Research and Development |
| ADDRESS | 8 West 95th St.<br>New York, NY 10025 |

**Production and Development**

|  |  |
|--|--|
| CONTACT | Mr. Jeffrey Silverstein, Mr. Craig A. Mengel, Mr. Leslie Barry, or Mr. J. Taylor Klotz |
| PHONE | (212) 864-4100 |
| PREPRODUCTION SERVICES | Analysis, Design, Scripting, Seminars, Workshops |
| PRODUCTION SERVICES | Producing, Directing, Special Effects, Computer Programming, Seminars, Workshops |
| POSTPRODUCTION SERVICES | Premastering, Program Simulation, System Design and Configuration, Seminars, Workshops |

|  |  |
|--|--|
| ORGANIZATION | **FUTURE SYSTEMS, INC.** |
| TYPE | Company |
| ADDRESS | Office located at:<br>2209 North Quintana<br>Arlington, VA 22205<br>Mailing Address:<br>P.O. Box 26<br>Falls Church, VA 22046 |
| CONTACT | Mr. Rockley Miller, President,<br>or John Sayers |
| PHONE | (703) 241-1799 |
| PREPRODUCTION SERVICES | Analysis, Seminars |
| DESCRIPTION | Future Systems publishes *The Videodisc Monitor*, and provides, on a consulting basis, front end concept evaluation and also seminars presenting in-house technical and market overviews. |

|  |  |
|--|--|
| ORGANIZATION | **FUTUREDAY** |
| TYPE | Independent Consultant |
| ADDRESS | 902 Willow Lane<br>S. Milwaukee, WI 53172 |
| CONTACT | Mr. James A. Lisowski |
| PHONE | (414) 762-5253 |
| PREPRODUCTION SERVICES | Analysis, Design, Workshops |
| PRODUCTION SERVICES | Special Effects, Computer Programming, Seminars |
| POSTPRODUCTION SERVICES | System Design and Configuration, Seminars, Workshops |

|  |  |
|--|--|
| ORGANIZATION | **GERSTENMAIER, WILLIAM** |
| TYPE | Independent Consultant |
| ADDRESS | 690 Washington St. (2-B)<br>New York, NY 10014 |
| CONTACT | Mr. William Gerstenmaier |
| PHONE | (212) 243-1343 |
| PREPRODUCTION SERVICES | Analysis, Design |
| PRODUCTION SERVICES | Producing, Directing |
| DESCRIPTION | Mr. Gerstenmaier specializes in the writing, planning and production of complete video, film, slide-tape, and multimedia training packages, interdepartmental presentations, and promotional/public relations materials. He has extensive "hands-on" production skills, sub-contractor supervision, and client contact experience. |

| | |
|---|---|
| ORGANIZATION | **GLYN GROUP, INC.** |
| TYPE | Company |
| ADDRESS | 258 West Fourth St.<br>New York, NY 10014 |
| CONTACT | Mr. Joseph McCarthy,<br>Vice President |
| PHONE | (212) 255-5156 |
| PREPRODUCTION SERVICES | Analysis, Design, Scripting |
| PRODUCTION SERVICES | Producing, Directing, Special Effects, Computer Programming |
| POSTPRODUCTION SERVICES | Premastering, Program Simulation, System Design and Configuration |
| DESCRIPTION | The Glyn Group offers consulting services in a variety of areas related to videodisc. Their work has ranged from course design to merchandising consulting for point-of-purchase applications and large library archiving applications. |

| | |
|---|---|
| ORGANIZATION | **GRUMMAN AEROSPACE CORPORATION** |
| TYPE | Company |
| ADDRESS | Mail Stop C01-47<br>Bethpage, NY 11714 |
| CONTACT | Mr. Al Cella |
| PHONE | (516) 435-6805 |
| PREPRODUCTION SERVICES | Analysis, Design, Scripting, Seminars, Workshops |
| PRODUCTION SERVICES | Producing, Directing, Special Effects, Computer Programming, Seminars, Workshops |

| | |
|---|---|
| ORGANIZATION | **HAYES PRODUCTIONS, INC.** |
| TYPE | Company |
| ADDRESS | 710 South Bowie<br>San Antonio, TX 78205 |
| CONTACT | G. Nelson, D. White, or B. Hayes |
| PHONE | (512) 224-9565 |
| PRODUCTION SERVICES | Producing, Directing, Special Effects |
| POSTPRODUCTION SERVICES | Premastering |

| | |
|---|---|
| ORGANIZATION | **HEARTLAND COMMUNICATIONS** |
| TYPE | Independent Consultant, Newsletter Publisher |
| ADDRESS | 223 Sunrise Dr.<br>Shreve, OH 44676 |
| CONTACT | Mr. Michael Butler |
| PHONE | (216) 567-3732 |
| PREPRODUCTION SERVICES | Analysis, Design, Scripting |
| PRODUCTION SERVICES | Producing, Directing |
| POSTPRODUCTION SERVICES | System Design and Configuration |
| DESCRIPTION | Heartland Communications has published *Interactive Video Technology* since October 1982. This monthly newsletter covers the interactive video and videodisc industry, and costs $45 a year. It is also available through the NEWSNET electronic newspaper database. Heartland Communications has also published a report entitled "The Interactive Videodisc: Past, Present, and Future," which is available for $150. |

# Production and Development

| | |
|---:|:---|
| ORGANIZATION | **HELGERSON ASSOCIATES** |
| TYPE | Company |
| ADDRESS | 6609 Rosecroft Pl. |
| | Falls Church, VA 22043 |
| CONTACT | Ms. Linda W. Helgerson |
| PHONE | (703) 237-0682 |
| PREPRODUCTION SERVICES | Analysis, Design, Scripting, Seminars, Workshops, Market Research |
| DESCRIPTION | Feasibility studies for associations, government agencies, publishers, and venture capital firms to determine appropriate use, content, delivery systems, and distribution mechanisms for video and optical disc systems. |

| | |
|---:|:---|
| ORGANIZATION | **HEWLETT, BRENT** |
| TYPE | Independent Consultant |
| ADDRESS | 205 East Sinclair Rd. |
| | Tallahassee, FL 32312 |
| CONTACT | Dr. Brent Hewlett |
| PHONE | (904) 385-5263 |
| PREPRODUCTION SERVICES | Analysis, Design, Scripting, Seminars, Workshops |
| PRODUCTION SERVICES | Producing, Seminars, Workshops |
| DESCRIPTION | Dr. Hewlett is currently developing level 3 materials on basic skills utilizing the IBM PC, Apple, PLATO, and TICCIT as delivery systems. He has been associated with the development of over twelve level 3 projects in the last two years. |

| | |
|---:|:---|
| ORGANIZATION | **HORIZONTAL EDITING STUDIOS** |
| TYPE | Company |
| ADDRESS | 2625 West Olive Ave. |
| | Burbank, CA 91505-4598 |
| CONTACT | Mr. William Carlquist, |
| | Owner |
| PHONE | (818) 841-6750 |
| PREPRODUCTION SERVICES | Analysis, Design, Scripting |
| PRODUCTION SERVICES | Producing, Directing, Special Effects, Computer Programming |
| POSTPRODUCTION SERVICES | Premastering, Program Simulation, System Design and Configuration, Seminars |
| DESCRIPTION | Horizontal Editing Studios deals extensively in electronic publishing with major companies. They have their own facilities for production, editing, etc. |

| | |
|---:|:---|
| ORGANIZATION | **HUMAN RESOURCES RESEARCH ORGANIZATION (HUMRRO)** |
| TYPE | Company, Not-for-Profit, Research and Development |
| ADDRESS | 1100 S. Washington St. |
| | Alexandria, VA 22314 |
| CONTACT | Ms. Carol Hargan |
| PHONE | (703) 549-3611 |
| PREPRODUCTION SERVICES | Analysis, Design, Scripting |
| PRODUCTION SERVICES | Producing, Directing, Special Effects, Computer Programming |
| POSTPRODUCTION SERVICES | Premastering, Program Simulation, System Design and Configuration |

| | |
|---:|:---|
| ORGANIZATION | **I.S.D. #279/PUBLIC SCHOOL SYSTEM** |
| TYPE | Other |
| ADDRESS | 11200 93 Ave. N. |
| | Maple Grove, MN 55369 |
| CONTACT | Mr. Brian Doyle |
| PHONE | (612) 425-4131 |
| PREPRODUCTION SERVICES | Analysis, Design, Scripting, Seminars, Workshops |
| PRODUCTION SERVICES | Producing, Directing, Computer Programming, Seminars, Workshops |
| DESCRIPTION | They are in the process of producing a fully interactive videodisc/microcomputer course in trigonometry. They have produced three other videodiscs, one titled "Apostrophe" and two for the Minnesota Educational Computing Consortium on Economics. All of the discs they have done are level 3. |

| | |
|---:|:---|
| ORGANIZATION | **ICS-INTEXT, A DIVISION OF NATIONAL EDUCATION CORPORATION** |
| TYPE | Company |
| ADDRESS | 315 Post Rd., West |
| | Westport, CT 06880 |
| CONTACT | Ms. Nina Selz, |
| | Director of Videodisc Development |
| PHONE | (203) 227-0891 |
| PREPRODUCTION SERVICES | Analysis, Design, Scripting |
| PRODUCTION SERVICES | Producing, Directing |
| POSTPRODUCTION SERVICES | Premastering, Program Simulation, System Design and Configuration |
| DESCRIPTION | ICS-Intext provides consultation on the unique applications of ActionCode, their highly interactive videodisc-based training system. ICS-Intext provides a full range of services that will allow a company to use the ActionCode system on their own. ActionCode is being used in a continuing education program for the American Medical Association. Numerous other applications of the ActionCode system are being researched and/or developed, including high-tech training, electronic cataloging and point-of-purchase sales support. ActionCode represents a joint venture between ICS-Intext and Perceptronics, Inc. |

| | |
|---:|:---|
| ORGANIZATION | **ILLINOIS STATE UNIVERSITY/ COLLEGE OF FINE ARTS** |
| TYPE | Independent Consultant, Research and Development |
| ADDRESS | Art Department |
| | College of Fine Arts |
| | Illinois State University |
| | Normal, IL 61761 |
| CONTACT | Dr. Frances Anderson, Professor of Art, |
| | Project Director |
| PHONE | (309) 438-5621 |
| PREPRODUCTION SERVICES | Analysis, Design, Workshops |
| POSTPRODUCTION SERVICES | System Design and Configuration, Workshops |
| DESCRIPTION | Dr. Anderson has received a small grant from Illinois State University to research the "development and evaluation of interactive videodisc programming for visual arts instruction." The project will entail developing interactive programming for use with a commercially |

# Production and Development

available disc, such as "Vincent Van Gogh." A series of experimental treatments will be developed and supplemented by material from art history and criticism. After a small pilot study and further testing, recommendations will be made about optimal formatting of interactive videodiscs for visual arts instructional use.

---

| | |
|---|---|
| ORGANIZATION | **IMAGE PREMASTERING SERVICES** |
| TYPE | Company, Research and Development |
| ADDRESS | Suite 707, 245 E. 6th St. |
| | St. Paul, MN 55101 |
| CONTACT | Mr. Gene Ramsay |
| PHONE | (612) 292-0192 |
| PREPRODUCTION SERVICES | Analysis, Design, Scripting |
| PRODUCTION SERVICES | Producing, Directing, Special Effects, Computer Programming |
| POSTPRODUCTION SERVICES | Premastering, Program Simulation, System Design and Configuration |
| DESCRIPTION | Image premastering specializes in the design of systems for high speed on-location premastering of large volumes of single-frame still images. Simultaneous image inventory, titling, and frame recording are included. |

---

| | |
|---|---|
| ORGANIZATION | **INDIANA UNIVERSITY RADIO AND TELEVISION SERVICE** |
| TYPE | Educational Institution |
| ADDRESS | Radio and Television Building |
| | Indiana University |
| | Bloomington, IN 47405 |
| CONTACT | Claire Gregory, |
| | Director of Special Projects |
| PHONE | (812) 335-6953 |
| PREPRODUCTION SERVICES | Analysis, Design, Scripting |
| PRODUCTION SERVICES | Producing, Directing, Computer Programming |
| POSTPRODUCTION SERVICES | Premastering |

---

| | |
|---|---|
| ORGANIZATION | **INDIS INTERNATIONAL** |
| TYPE | Company |
| ADDRESS | 250 W. 57 St., Suite 1527 |
| | New York, NY 10019 |
| CONTACT | Mr. Eric Donne |
| PHONE | (212) 582-0734 |
| PREPRODUCTION SERVICES | Analysis, Design, Scripting |
| PRODUCTION SERVICES | Producing, Directing, Computer Programming |
| POSTPRODUCTION SERVICES | System Design and Configuration |
| DESCRIPTION | INDIS INTERNATIONAL was formed in June 1983 after ten years of developing interactive programming, producing film and video, and controlling video by computer. They have created the technique of "artificial personality™" for sophisticated game design, interactive drama, and tutorial simulations. They helped design "Laser Shuffle," and are currently developing a proprietary line of interactive video products. They are also working on game design and training projects as consultants. |

| | |
|---:|:---|
| ORGANIZATION | **INFOTECH CONSULTANTS** |
| TYPE | Partnership |
| ADDRESS | #1840, 10123—99 St., |
| | Sun Life Pl. |
| | Edmonton, Alberta, |
| | Canada |
| CONTACT | Ms. Jan Belzile or Ms. Cindy Gordon |
| PHONE | (403) 425-9253 |
| PREPRODUCTION SERVICES | Analysis, Design, Scripting, Seminars, Workshops |
| DESCRIPTION | InfoTech Consultants conducted the literature review for the Access Alberta Laserdisc Project which resulted in the completion of a laserdisc, which was released October 31, 1983. |

| | |
|---:|:---|
| ORGANIZATION | **INFORMATION DELIVERY SYSTEMS, INC.** |
| TYPE | Company |
| ADDRESS | 10 West 35th St. |
| | Chicago, IL 60616 |
| CONTACT | Mr. Don Fenhagen |
| PHONE | (312) 567-4904 |
| PREPRODUCTION SERVICES | Analysis, Design |
| PRODUCTION SERVICES | Producing, Special Effects, Computer Programming, Computer reformatting of textual and/or graphic charts for inclusion on the analog videodisc. Can work from either hard copy or machine readable copy. |
| POSTPRODUCTION SERVICES | Premastering, System Design and Configuration |
| DESCRIPTION | Information Delivery Systems specializes in hybrid digital/analog disc development and interface to on-line retrieval systems. |

| | |
|---:|:---|
| ORGANIZATION | **INNOTECH** |
| TYPE | Independent Consultant, Research and Development |
| ADDRESS | 18552 MacArthur Blvd., Suite 475 |
| | Irvine, CA 92715 |
| CONTACT | Mr. Stuart Krasney |
| PHONE | (714) 476-2051 |
| PREPRODUCTION SERVICES | Analysis, Design |
| PRODUCTION SERVICES | Computer Programming |
| DESCRIPTION | Innotech specializes in stress management training and point-of-sale displays. |

| | |
|---:|:---|
| ORGANIZATION | **INNOVATIVE TECHNOLOGY ASSOCIATES (ITA)** |
| TYPE | Independent Consultant, Research and Development |
| ADDRESS | P.O. Box 637 |
| | Sierra Madre, CA 91024 |
| CONTACT | Dr. Joseph Gaynor, |
| | President |
| PHONE | (818) 355-6868 |
| DESCRIPTION | Innovative Technology is working on optical disc recording media and processes. They also consult with companies working in the same areas. |

# Production and Development

| | |
|---|---|
| ORGANIZATION | **INNOVISION** |
| TYPE | Company |
| ADDRESS | Weidenallee 37 |
| | 2000 Hamburg 6, Germany |
| CONTACT | Mr. Jurgen Becker |
| PHONE | (040 / 43 11 26 |
| PREPRODUCTION SERVICES | Analysis, Design, Scripting |
| PRODUCTION SERVICES | Producing, Directing, Computer Programming |
| POSTPRODUCTION SERVICES | Premastering, Program Simulation, System Design and Configuration |

| | |
|---|---|
| ORGANIZATION | **INSTRUCTIONAL SCIENCE AND DEVELOPMENT, INC.** |
| TYPE | Company, Research and Development |
| ADDRESS | 5059 Newport Ave. |
| | San Diego, CA 92107 |
| CONTACT | Dr. H. Dewey Kribs |
| PHONE | (619) 226-1882 |
| PREPRODUCTION SERVICES | Analysis, Design, Scripting |
| DESCRIPTION | Instructional Science and Development performs feasibility studies, economic analyses, and evaluations of alternative training delivery and production systems. They also develop custom job and training analyses with follow-up of training system design. |

| | |
|---|---|
| ORGANIZATION | **INTERACT, INC.** |
| TYPE | Company |
| ADDRESS | 350 Townsend St. |
| | San Francisco, CA 94107 |
| CONTACT | Mr. Art Porter |
| PHONE | (415) 495-0910 |
| PREPRODUCTION SERVICES | Analysis, Design, Scripting |
| PRODUCTION SERVICES | Producing, Directing, Special Effects, Computer Programming |
| POSTPRODUCTION SERVICES | Premastering, Program Simulation, System Design and Configuration |

| | |
|---|---|
| ORGANIZATION | **INTERACTIVE ARTS & SCIENCE/IAS, INC.** |
| TYPE | Company |
| ADDRESS | 5199 E. Pacific Coast Hwy., Suite #510 |
| | Long Beach, CA 90804 |
| CONTACT | Mr. Jack D. Cooper, |
| | President |
| PHONE | (213) 494-3327 |
| PREPRODUCTION SERVICES | Design, Scripting, Seminars, Workshops, Game Design (rules, game play, scoring, sound effects, music, etc.) |
| PRODUCTION SERVICES | Producing, Directing, Special Effects, Computer Programming, Seminars, Workshops |
| POSTPRODUCTION SERVICES | System Design and Configuration |

| | |
|---|---|
| ORGANIZATION | **INTERACTIVE ARTS INTERNATIONAL** |
| TYPE | Independent Consultant, Company, Not-for-Profit, Research and Development |
| ADDRESS | 25785 Bassett Lane |
| | Los Altos Hills, CA 94022 |

| | |
|---|---|
| CONTACT | Mr. Allen Lee Adkins |
| PHONE | (415) 941-9686 |
| PREPRODUCTION SERVICES | Analysis, Design, Scripting, Funding, Seminars, Workshops |
| PRODUCTION SERVICES | Producing, Directing, Special Effects, Computer Programming, Seminars, Workshops |
| POSTPRODUCTION SERVICES | System Design and Configuration, Seminars, Workshops |

| | |
|---|---|
| ORGANIZATION | **INTERACTIVE AUTHORING, INC.** |
| TYPE | Company |
| ADDRESS | 277 West End Ave.<br>New York, NY 10023 |
| CONTACT | Mr. Raymond J. Bouley |
| PHONE | (212) 877-8169 |
| PREPRODUCTION SERVICES | Analysis, Design, Scripting, Workshops |
| PRODUCTION SERVICES | Producing, Directing, Special Effects |
| DESCRIPTION | Interactive Authoring is dedicated to the imaginative design of interactive videodisc systems for a wide range of applications. They specialize in cross-cultural training, foreign language tutorials, surrogate travel via disc, conversion of existing media materials to videodisc, and education in the humanities. |

| | |
|---|---|
| ORGANIZATION | **INTERACTIVE IMAGE TECHNOLOGIES INC.** |
| TYPE | Company |
| ADDRESS | Suite 401, 49 Bathurst St.<br>Toronto, Ontario<br>M5V 2P2 Canada |
| CONTACT | Mr. Patrick Lee |
| PHONE | (416) 361-0333 |
| PREPRODUCTION SERVICES | Analysis, Design, Scripting |
| PRODUCTION SERVICES | Producing, Directing, Special Effects, Computer Programming |

| | |
|---|---|
| ORGANIZATION | **INTERACTIVE RESEARCH CORPORATION** |
| TYPE | Company |
| ADDRESS | 3080 Olcott St., Suite 200B<br>Santa Clara, CA 95051 |
| CONTACT | Ms. Sandie Paddock |
| PHONE | (408) 986-1420 |
| DESCRIPTION | Interactive Research has developed and is in the process of licensing an advanced authoring system, which includes control code generation, pert chart, and full videodisc project management software. This program software is more than just an authoring system. It encompasses every aspect of an interactive video project. It is written in the "C" programming language. It is extendable so it can be modified if a new feature needs to be added. |

| | |
|---|---|
| ORGANIZATION | **INTERACTIVE TECHNOLOGIES CORPORATION** |
| TYPE | Company |
| ADDRESS | 1131 G St.<br>San Diego, CA 92101 |
| CONTACT | Mr. Rod Daynes |
| PHONE | (619) 231-7774 |

# Production and Development

| | |
|---|---|
| PREPRODUCTION SERVICES | Analysis, Design, Scripting |
| PRODUCTION SERVICES | Producing, Directing, Special Effects, Computer Programming |
| POSTPRODUCTION SERVICES | Premastering, Program Simulation, System Design and Configuration |
| DESCRIPTION | Interactive Technologies Corporation serves the needs of the electronic publishing, microcomputer, and videodisc industries. They combine into a single entity instructional design, videodisc design/production, and advanced computer programming skills, all of which are necessary for the development of the highest quality interactive microcomputer and videodisc products. |

| | |
|---|---|
| ORGANIZATION | **INTERACTIVE TELEVISION COMPANY** |
| TYPE | Company, Research and Development |
| ADDRESS | 1901 N. Moore St., #1100<br>Arlington, VA 22209 |
| CONTACT | Dr. Steven L. Levin |
| PHONE | (703) 525-5625 |
| PREPRODUCTION SERVICES | Analysis, Design, Scripting |
| PRODUCTION SERVICES | Producing, Directing, Special Effects, Computer Programming |
| POSTPRODUCTION SERVICES | Premastering, Program Simulation, System Design and Configuration |
| DESCRIPTION | The Interactive Television Company is involved in several research and development projects: developing selected part task videodisc-based trainers for ashore and on-board use, and exploring the application of surrogate travel to military operations in urban terrain. |

| | |
|---|---|
| ORGANIZATION | **INTERACTIVE TRAINING SYSTEMS, INC.** |
| TYPE | Company |
| ADDRESS | 4 Cambridge Center<br>Cambridge, MA 02142 |
| CONTACT | Mr. David Snow,<br>Sales Manager |
| PHONE | (617) 497-6100 |
| PREPRODUCTION SERVICES | Analysis, Design, Scripting |
| PRODUCTION SERVICES | Producing, Directing, Special Effects, Computer Programming |
| POSTPRODUCTION SERVICES | Premastering, System Design and Configuration |
| DESCRIPTION | Interactive Training Systems was founded in 1981 by Dr. Harry M. Lasker and Dr. David A. Lubin. They bring to the company a wealth of expertise in the fields of cognitive development, adult learning and video-based education. Interactive Training Systems was founded on the premise that this rich background could be combined with the revolution in affordable microcomputer and video technologies to provide significant advances in applied learning. |

| | |
|---|---|
| ORGANIZATION | **INTERACTIVE VIDEO CONCEPTS, INC.** |
| TYPE | Company |
| ADDRESS | Suite 105, Wilford Building<br>101 North 33rd St.<br>Philadelphia, PA 19104 |
| CONTACT | Ms. Peg Callahan,<br>Vice President |
| PHONE | (215) 387-0707 |
| PREPRODUCTION SERVICES | Analysis, Design, Scripting, Funding, Seminars, Workshops |

| | |
|---|---|
| PRODUCTION SERVICES | Producing, Directing, Special Effects, Computer Programming, Seminars, Workshops |
| POSTPRODUCTION SERVICES | Premastering, Program Simulation, System Design and Configuration, Seminars, Workshops |

| | |
|---|---|
| ORGANIZATION | **INTERACTIVE VIDEO CORPORATION** |
| TYPE | Independent Consultant, Company, Research andDevelopment |
| ADDRESS | 7500 San Felipe, Suite 100<br>Houston, TX 77063 |
| CONTACT | Ms. Sharon Brown |
| PHONE | (713) 781-6984 |
| PREPRODUCTION SERVICES | Analysis, Design, Scripting, Workshops, University of Houston Interactive Video class |
| PRODUCTION SERVICES | Producing, Directing, Special Effects, Computer Programming |
| POSTPRODUCTION SERVICES | Premastering, Program Simulation, System Design and Configuration |
| DESCRIPTION | Interactive Video Corporation is working with special effects on the Apple Computer for transfer via VB-3 to tape. |

| | |
|---|---|
| ORGANIZATION | **INTERACTIVE VIDEO PROGRAMMES (DRAGONFLAIR LTD)** |
| TYPE | Independent Consultant |
| ADDRESS | 24 Millway<br>London NW7 3RB<br>England |
| CONTACT | Mr. Roger Wilson |
| PHONE | (01) 959-0724 |
| PREPRODUCTION SERVICES | Analysis, Design, Scripting |
| PRODUCTION SERVICES | Producing |
| POSTPRODUCTION SERVICES | Premastering, System Design and Configuration |
| DESCRIPTION | Interactive Video is involved in the continuing development of the VHD system in Europe as a sub- contractor to Thorn EMI Videodisc. They are the only European producer to have produced level 1 and level 3 discs on two different systems. |

| | |
|---|---|
| ORGANIZATION | **INTERACTOR COMPUTER-MEDIA SYSTEMS, INC./APAL FOUNDATION** |
| TYPE | Independent Consultant, Company, Not-for-Profit |
| ADDRESS | 7510 Brava St.<br>Carlsbad, CA 92008 |
| CONTACT | Dr. L. Lotecka |
| PHONE | (619) 436-9248 |
| PREPRODUCTION SERVICES | Analysis, Design, Scripting |
| DESCRIPTION | Interactor Computer-Media specializes in applied behavioral science and communication content: decision-making attitude change, enjoyable learning, problem-solving processes. |

| | |
|---|---|
| ORGANIZATION | **INTERMEDIA** |
| TYPE | Company |
| ADDRESS | 1600 Dexter Ave. N.<br>Seattle, WA 98109 |
| CONTACT | Ms. Susan Hoffman or Mr. Stan Barnes |
| PHONE | (206) 282-7262 |
| PREPRODUCTION SERVICES | Analysis, Design, Scripting, Funding, Workshops |

**Production and Development**

| | |
|---|---|
| PRODUCTION SERVICES | Producing, Directing, Special Effects, Computer Programming |
| POSTPRODUCTION SERVICES | Premastering, Program Simulation, System Design and Configuration |
| DESCRIPTION | Intermedia produces a level 3 interactive product demonstration system featuring a product feature comparison mode. The initial system includes five of the most popular word processing computer software packages, and is intended for point-of-purchase use in retail computer software stores. |

| | |
|---|---|
| ORGANIZATION | **INTERNATIONAL RESOURCE DEVELOPMENT INC. (IRD)** |
| TYPE | Company, Research and Development |
| ADDRESS | 30 High St.<br>Norwalk, CT 06851 |
| CONTACT | Ms. Joann K. Niedzwiecki,<br>Research Associate |
| PHONE | (203) 866-6914 |
| TOLL FREE | (800) 243-5008 |
| PREPRODUCTION SERVICES | Analysis, Design, Scripting |
| POSTPRODUCTION SERVICES | System Design and Configuration |
| DESCRIPTION | IRD is an independent consulting firm specializing in market research and product/service planning. IRD's primary activity is the generation of market research reports, covering a wide range of generally high tech-related topics. |

| | |
|---|---|
| ORGANIZATION | **IT, INC.** |
| TYPE | Company |
| ADDRESS | 2009 Pacheco St.<br>Santa Fe, NM 87501 |
| CONTACT | Mr. Gary Quinlin, President or<br>Ms. Constance Buffalo, Vice President for Production |
| TOLL FREE | (800) 328-7937 |
| PREPRODUCTION SERVICES | Analysis, Design, Scripting, Funding |
| PRODUCTION SERVICES | Producing, Directing, Special Effects, Computer Programming |
| POSTPRODUCTION SERVICES | Premastering, Program Simulation, System Design and Configuration |

| | |
|---|---|
| ORGANIZATION | **ITM SYSTEMS** |
| TYPE | Company |
| ADDRESS | P.O. Box 774<br>Sunset Beach, CA 90742 |
| CONTACT | Mr. Kurt Zimmerman,<br>President |
| PHONE | (213) 592-3044 |
| PREPRODUCTION SERVICES | Analysis, Funding |
| PRODUCTION SERVICES | Directing, Computer Programming |
| DESCRIPTION | ITM Systems is involved in the development and acquisition of videodisc-based interactive teaching systems, including project research, program license, and consulting. |

| | |
|---|---|
| ORGANIZATION | **ITTELSON, JOHN C.** |
| TYPE | Independent Consultant |
| ADDRESS | 811 Grass Ct.<br>Chico, CA 95926 |

| | |
|---|---|
| CONTACT | Dr. John C. Ittelson |
| PHONE | (916) 893-9023 or 895-5751 |
| PREPRODUCTION SERVICES | Analysis, Design, Scripting, Seminars, Workshops |
| PRODUCTION SERVICES | Producing, Directing, Seminars |
| DESCRIPTION | Dr. Ittelson is an associate professor for Information and Communication Studies at California State University at Chico, and is a consultant for DesignWare, Inc. in San Francisco. He was a member of the authoring team for "Computer Discovery," a course in computer literacy published by Science Research Associates. He has worked on many instructional design projects for both education and industry. |

| | |
|---|---|
| ORGANIZATION | **JACK LIEB PRODUCTIONS, INC.** |
| TYPE | Company |
| ADDRESS | 200 E. Ontario St.<br>Chicago, IL 60611 |
| CONTACT | Mr. Warren Lieb, Charles R. Kite, and Sharon Spence |
| PHONE | (312) 943-1440 |
| PREPRODUCTION SERVICES | Analysis, Design, Scripting |
| PRODUCTION SERVICES | Producing, Directing, Special Effects, Computer Programming |
| POSTPRODUCTION SERVICES | Premastering, Program Simulation, System Design and Configuration |
| DESCRIPTION | Jack Lieb Productions custom designs interactive audiovisual programs to fit the specialized needs of its clients. |

| | |
|---|---|
| ORGANIZATION | **JAM, INC.** |
| TYPE | Company |
| ADDRESS | 42 East Ave.<br>Rochester, NY 14604 |
| CONTACT | Mr. Michael Yampolsky<br>Vice President, Research and Development |
| PHONE | (716) 232-5500 |
| PREPRODUCTION SERVICES | Analysis, Design, Scripting, Funding, Seminars, Workshops |
| PRODUCTION SERVICES | Producing, Directing, Special Effects, Computer Programming |
| POSTPRODUCTION SERVICES | Premastering, Program Simulation, System Design and Configuration, Seminars, Workshops |
| DESCRIPTION | Jam, Inc. is developing a library of programs for microcomputer training, to be released in 1984. |

| | |
|---|---|
| ORGANIZATION | **JOSEPH KEYERLEBER PRODUCTIONS** |
| TYPE | Independent Consultant |
| ADDRESS | 318 W. North Ave.<br>Pittsburgh, PA 15212 |
| CONTACT | Mr. Joe Keyerleber |
| PHONE | (412) 322-2367 |
| PREPRODUCTION SERVICES | Analysis, Design, Scripting |
| PRODUCTION SERVICES | Producing, Directing |

| | |
|---|---|
| ORGANIZATION | **KELBAUGH, LARRY E.** |
| TYPE | Research and Development |
| ADDRESS | 5707 Seminary Rd.<br>Bailey's Crossroads, VA 22041 |

# Production and Development

| | |
|---|---|
| CONTACT | Mr. Larry E. Kelbaugh |
| PHONE | (703) 820-1000 |
| PREPRODUCTION SERVICES | Analysis, Design, Scripting |
| PRODUCTION SERVICES | Computer Programming |
| POSTPRODUCTION SERVICES | System Design and Configuration |
| DESCRIPTION | Larry E. Kelbaugh has been involved in such research and development projects as PLATO programming for the Falcon Jet and basic math skills, and ACCP subcourses for the U.S. Army. |

| | |
|---|---|
| ORGANIZATION | **KEN S. CHRISTIE AND ASSOCIATES** |
| TYPE | Independent Consultant |
| ADDRESS | 565 W. Stocker St., #306 |
| | Glendale, CA 91202 |
| CONTACT | Mr. Ken Christie |
| PHONE | (818) 242-4672 |
| PREPRODUCTION SERVICES | Analysis, Design, Scripting, Flowcharting |
| DESCRIPTION | Ken S. Christie and Associates have experience in the design and production of over 100 interactive and linear videodisc programs. They translate the client's ideas into working flowcharts and production/postproduction scripts. They also assist in all phases of the project when necessary. |

| | |
|---|---|
| ORGANIZATION | **LAKE, PETER A.** |
| TYPE | Independent Consultant |
| ADDRESS | 27 Outrigger St. |
| | Marina Del Rey, CA 90292 |
| CONTACT | Mr. Peter A. Lake |
| PHONE | (213) 306-7367 |
| PREPRODUCTION SERVICES | Design, Scripting |
| PRODUCTION SERVICES | Producing, Directing |
| DESCRIPTION | Peter A. Lake has been involved in development with major suppliers. |

| | |
|---|---|
| ORGANIZATION | **LASER 7 GROUP** |
| TYPE | Company |
| ADDRESS | 8950 Via La Jolla Dr., Suite 1200 |
| | La Jolla, CA 92037 |
| CONTACT | Mr. Philip H. Rapp |
| PHONE | (619) 453-0622 |
| PREPRODUCTION SERVICES | Design, Scripting |
| PRODUCTION SERVICES | Producing, Directing, Special Effects, Computer Programming |
| POSTPRODUCTION SERVICES | Premastering, System Design and Configuration |
| DESCRIPTION | The Laser 7 Group is a unique blend of professional and business talents brought together in a common goal of developing outstanding interactive laser disc products. The initial drive is directed at producing videodiscs that put the emphasis on the subject in a manner that is both entertaining and educational. They are dedicated to supplying both the consumer and educational markets with creative products. |

| | |
|---|---|
| ORGANIZATION | **LEARNINGWARE CORP.** |
| TYPE | Company |
| ADDRESS | 135 Charles St. |
| | New York, NY 10014 |

|  |  |
|---|---|
| CONTACT | Mr. Bob Rosensweet |
| PREPRODUCTION SERVICES | Analysis, Design, Scripting, Funding, Seminars, Workshops |
| PRODUCTION SERVICES | Producing, Directing, Special Effects, Computer Programming, Seminars, Workshops |
| POSTPRODUCTION SERVICES | Premastering, Program Simulation, System Design and Configuration |
| DESCRIPTION | Seminars and workshops are available through:<br>Interactive Author's Guild (Not-for-Profit)<br>80 Wall St.<br>New York, NY 10005<br>(212) 903-4300 |

|  |  |
|---|---|
| ORGANIZATION | **LENFEST & ASSOCIATES** |
| TYPE | Independent Consultant, Company |
| ADDRESS | Box 8, El Rancho<br>Golden, CO 80401 |
| CONTACT | Dr. David Lenfest |
| PHONE | (303) 239-7867 |
| PREPRODUCTION SERVICES | Analysis, Design, Scripting |
| DESCRIPTION | Lenfest & Associates is interested in analysis, design, and scripting for interactive video projects that have to do with human motivation. They also produce "high-touch" documentation and training for emerging software products. They view interactive video as the perfect training medium for the computer industry. |

|  |  |
|---|---|
| ORGANIZATION | **LEO BEISER INC.** |
| TYPE | Independent Consultant, Research and Development |
| ADDRESS | 151-77 28 Ave.<br>Flushing, NY 11354 |
| CONTACT | Mr. Leo Beiser |
| PHONE | (212) 353-7298 |
| PREPRODUCTION SERVICES | Analysis, Design |
| DESCRIPTION | Leo Beiser is involved in the formatting of digital data on video and compact discs. |

|  |  |
|---|---|
| ORGANIZATION | **LIGHT TRACKS** |
| TYPE | Company |
| ADDRESS | 1242 10th St., Suite #10<br>Santa Monica, CA 90401 |
| CONTACT | Joey Silvian |
| PHONE | (213) 395-6962 |
| PREPRODUCTION SERVICES | Analysis, Design, Scripting, |
| PRODUCTION SERVICES | Producing, Directing, Computer Programming |
| POSTPRODUCTION SERVICES | Premastering |

|  |  |
|---|---|
| ORGANIZATION | **LOGIVISION** |
| TYPE | Company |
| ADDRESS | 46, rue du Docteur Charcot<br>92000 Nanterre, France |
| CONTACT | Mr. Alain Dupuy or Mr. Christian Durix |
| PHONE | (33) (1) 725 14 14 |
| PREPRODUCTION SERVICES | Analysis, Design |

# Production and Development

| | |
|---|---|
| DESCRIPTION | LOGIVISION, a videodisc systems engineering corporation, is strongly involved with analysis and design in the preproduction process. They also provide access to production and postproduction facilities as a total service to customers. |

| | |
|---|---|
| ORGANIZATION | **LUCASFILM, LTD.** |
| TYPE | Research and Development |
| ADDRESS | P.O. Box 2009<br>San Rafael, CA 94912 |
| CONTACT | Mr. Ralph Guggenheim |
| PHONE | (415) 499-0239 |
| POSTPRODUCTION SERVICES | Lucasfilm provides videodisc editing systems for film and television production. |

| | |
|---|---|
| ORGANIZATION | **LUMIERE PRODUCTIONS** |
| TYPE | Company |
| ADDRESS | 860 Second St.<br>San Francisco, CA 94107 |
| CONTACT | Mr. Charles Halpern |
| PHONE | (415) 974-1242 |
| PREPRODUCTION SERVICES | Design, Scripting |
| PRODUCTION SERVICES | Producing, Directing, Special Effects |
| POSTPRODUCTION SERVICES | Premastering |
| DESCRIPTION | Lumiere Productions is a film and video production company which is scheduled to design and produce an interactive videodisc game for the new RCA interactive player. They designed an interactive videodisc program for VHD programs, but VHD decided to close their doors before the program was produced. |

| | |
|---|---|
| ORGANIZATION | **M. J. ZINK PRODUCTIONS, INC.** |
| TYPE | Company |
| ADDRESS | 132 West 31st St.<br>New York, NY 10001 |
| CONTACT | Ms. Jan Morgan Zink,<br>Vice President |
| PHONE | (212) 563-5777 |
| PREPRODUCTION SERVICES | Analysis, Design, Scripting |
| PRODUCTION SERVICES | Producing, Directing, Special Effects, Computer Programming |
| POSTPRODUCTION SERVICES | System Design and Configuration |
| DESCRIPTION | M. J. Zink Productions is currently working on the development of an archival application pilot as well as game shows/graphics computer applications. |

| | |
|---|---|
| ORGANIZATION | **MARITZ COMMUNICATIONS COMPANY** |
| TYPE | Company |
| ADDRESS | Main Office:<br>Maritz Inc.<br>1375 North Highway Dr.<br>Fenton, MO 63026 |
| CONTACT | Mr. Ed Smith |
| PHONE | (314) 225-4000 |

| | |
|---:|:---|
| ADDRESS | Maritz Video Center |
| | 4925 Cadieux Rd. |
| | Detroit, MI 48224 |
| CONTACT | Mr. Jerry Sundt |
| PHONE | (313) 882-9100 |
| PREPRODUCTION SERVICES | Analysis, Design, Scripting, Seminars, Workshops |
| PRODUCTION SERVICES | Producing, Directing, Special Effects, Computer Programming, Seminars, Workshops |
| POSTPRODUCTION SERVICES | Premastering, Program Simulation, System Design and Configuration, Seminars, Workshops |
| DESCRIPTION | Martiz Communications Company plans to complete 100 level 2 and level 3 discs per year. From 1981 to 1983 they have produced over 200 discs. They specialize in training and point-of-purchase discs. All resources, people, and facilities needed from preproduction through premastering are available in-house. |

| | |
|---:|:---|
| ORGANIZATION | **MARK IV DESIGNS** |
| TYPE | Independent Consultant |
| ADDRESS | 2315 So. Canterbury Lane |
| | Lincoln, NE 68512 |
| CONTACT | Mr. Mark Hansen |
| PHONE | (402) 423-0363 |
| PRODUCTION SERVICES | Computer Programming |
| DESCRIPTION | Mark IV Designs specializes in custom hardware and software design and programming. Mark Hansen is an electrical engineer with ten years experience in microprocessor, digital design, and software development. |

| | |
|---:|:---|
| ORGANIZATION | **MARTIN MARIETTA DATA SYSTEMS/ TECHNOLOGY SYSTEMS GROUP** |
| TYPE | Company, Research and Development |
| ADDRESS | 106 Inverness Circle East |
| | Englewood, CO 80112 |
| CONTACT | Dr. Darrell F. Humphrey |
| PHONE | (303) 790-3339 |
| PREPRODUCTION SERVICES | Analysis, Design |
| PRODUCTION SERVICES | Computer Programming |
| POSTPRODUCTION SERVICES | System Design and Configuration |
| DESCRIPTION | The Technology Systems Group at Martin Marietta is involved in experiments with machine independent videodisc-driven software. They are also experimenting with applications for level 3 videodisc systems including: surrogate travel, parts catalog information retrieval, digital infrared imaging, helicopter missile gunnery simulations, electron beam microscopic image storage, and retrieval with full text keyword retrieval. |

| | |
|---:|:---|
| ORGANIZATION | **MARYLAND CENTER FOR PUBLIC BROADCASTING** |
| TYPE | Not-for-Profit |
| ADDRESS | 11767 Bonita Ave. |
| | Owings Mills, MD 21117 |
| CONTACT | Mr. Michael Styer |
| PHONE | (301) 356-5600 |

# Production and Development

| | |
|---|---|
| PREPRODUCTION SERVICES | Analysis, Design, Scripting |
| PRODUCTION SERVICES | Producing, Directing, Special Effects, Computer Programming |
| POSTPRODUCTION SERVICES | Premastering |
| DESCRIPTION | The Maryland Center for Public Broadcasting has used videodisc for broadcast origination. |

| | |
|---|---|
| ORGANIZATION | **MARYLAND INSTRUCTIONAL TELEVISION** |
| TYPE | Not-for-Profit |
| ADDRESS | 11767 Bonita Ave.<br>Owings Mills, MD 21117 |
| CONTACT | Dr. Martha Cammarata |
| PHONE | (301) 337-4221 |
| PREPRODUCTION SERVICES | Analysis, Design, Scripting |
| PRODUCTION SERVICES | Producing, Directing, Special Effects, Computer Programming |
| DESCRIPTION | Maryland Instructional Television is developing a level 3 interactive videodisc for teachers and secondary school students that should be completed by Fall 1986. |

| | |
|---|---|
| ORGANIZATION | **MATRIX INFORMATION SYSTEMS** |
| TYPE | Company, Research and Development |
| ADDRESS | 11728 Avon Way<br>Los Angeles, CA 90066 |
| CONTACT | Mr. Michael North,<br>Director |
| PHONE | (213) 391-0243 |
| PREPRODUCTION SERVICES | Analysis, Design, Scripting, Workshops |
| PRODUCTION SERVICES | Producing, Directing, Computer Programming, Workshops |
| POSTPRODUCTION SERVICES | Premastering, System Design and Configuration, Workshops |

| | |
|---|---|
| ORGANIZATION | **MATROX ELECTRONIC SYSTEMS LTD.** |
| TYPE | Company |
| ADDRESS | 5800 Andover Ave.<br>Town of Mount Royal, Quebec<br>Canada H4T 1H4 |
| CONTACT | Mr. Chris Morin |
| PHONE | (514) 735-1182 |
| POSTPRODUCTION SERVICES | Premastering |
| DESCRIPTION | Matrox is developing a videodisc premastering station. The station is used to encode digital data with proper error correcting code and to digitize audio signal. The formatted digital data is then coded into the active portion of the video signal, which is delivered to a videotape recorder. The mastering station is thus used to edit a master videotape, with digital data to be read by a Matrox videodisc interface system. Total disk capacity is 1.2 G bytes of raw data. |

| | |
|---|---|
| ORGANIZATION | **MECKLER PUBLISHING** |
| TYPE | Company |
| ADDRESS | 520 Riverside Ave.<br>Westport, CT 06880 |
| CONTACT | Mr. Alan Meckler, Mr. Anthony Abbott, or Ms. Judith Paris |
| PHONE | (203) 226-6967 |

| | |
|---|---|
| PREPRODUCTION SERVICES | Seminars |
| PRODUCTION SERVICES | Seminars |
| POSTPRODUCTION SERVICES | Seminars |
| DESCRIPTION | Meckler Publishing, under its division Meckler Communications, sponsors an annual three-day conference on new developments in videodisc and optical disk technologies, including applications, production techniques (not workshops, however), preproduction funding, scripting, design, production programming, testing, and other aspects of the process of producing a disc, and using video discs and optical disks in business.

Meckler Publishing also publishes a bimonthly journal (formerly quarterly), *Videodisc/Videotex*, with full-length articles on disc and diskette applications, including illustrations, flowcharts, a substantial news column, book reviews, publication announcements, and a calendar. In addition, MP also publishes *Videodisc Update*, a monthly companion newsletter to *V/V*, which covers new developments in the technology, staff changes, product announcements, and a substantial calendar. |

---

| | |
|---|---|
| ORGANIZATION | **MEDIA CONCEPTS, INC.** |
| TYPE | Company |
| ADDRESS | 331 N. Broad St.<br>Philadelphia, PA 19107 |
| CONTACT | Ms. Judy Haupt |
| PHONE | (215) 923-2545 |
| PREPRODUCTION SERVICES | Analysis, Design, Scripting |
| PRODUCTION SERVICES | Producing, Directing, Special Effects, Computer Programming |
| POSTPRODUCTION SERVICES | Premastering, Program Simulation, System Design and Configuration |
| DESCRIPTION | MEDIA CONCEPTS has completed six interactive videodisc programs in the fields of medicine, automotives, and manufacturing. |

---

| | |
|---|---|
| ORGANIZATION | **MEMOREX/BURROUGHS—OPTICAL MEMORY TECHNOLOGY** |
| TYPE | Research and Development |
| ADDRESS | 5411 No. Lindero Canyon Rd.<br>Westlake Village, CA 91362 |
| CONTACT | Mr. Neville Lee |
| PHONE | (213) 706-5143 |
| DESCRIPTION | Memorex/Burroughs has an ongoing research and development project to produce optical memory disks for use with either helium-neon lasers or laser diode light sources. |

---

| | |
|---|---|
| ORGANIZATION | **MICHAEL J. PETRO LTD.** |
| TYPE | Company |
| ADDRESS | 181 Shepherd St. East<br>Windsor, Ontario<br>Canada N8X 2K4 |
| CONTACT | Mr. Michael J. Petro |
| PHONE | (519) 258-9008 |
| PREPRODUCTION SERVICES | Analysis, Design, Scripting, Funding, Seminars, Workshops |
| PRODUCTION SERVICES | Producing, Directing, Special Effects, Computer Programming, Seminars, Workshops |

**Production and Development**

|  |  |
|---|---|
| POSTPRODUCTION SERVICES | Program Simulation, System Design and Configuration, Seminars, Workshops |
| DESCRIPTION | Michael J. Petro Ltd. has been involved in over 40 videodiscs for government and industry, and in continuing training and promotional programs for both General Motors of Canada and American Motors of Canada. They have produced pilot videodiscs for the National Film Board of Canada and the National Library of Canada. They are also involved in the technical research and writing of a major study, "Technological Assessment of Present and Near Term Videodisc Systems," for the Secretary of Canada. |

|  |  |
|---|---|
| ORGANIZATION | NATIONAL EVALUATION SYSTEMS, INC. |
| TYPE | Company |
| ADDRESS | Interactive Technology Division<br>30 Gatehouse Rd., Box 226<br>Amherst, MA 01004 |
| CONTACT | Ms. Elizabeth Wright,<br>Project Director |
| PHONE | (413) 256-0444 |
| PREPRODUCTION SERVICES | Analysis, Design, Scripting |
| PRODUCTION SERVICES | Producing, Directing, Special Effects, Computer Programming |
| DESCRIPTION | The Interactive Technology Division of National Evaluation Systems, Inc., is a videodisc development house specializing in custom training and educational programs. The experienced videodisc development team is internationally recognized for its contribution to the field of instructional design in videodisc technology. Programming and graphics are done in-house with extensive mainframe to multilanguage, mini-computer capabilities. Video production (one-inch) is directed in-house, utilizing experienced videodisc cinematographers and the most up-to-date techniques. Editing is done at the University of Nebraska Educational Network facilities or at other like-quality editing houses. |

|  |  |
|---|---|
| ORGANIZATION | NAVAL PERSONNEL RESEARCH & DEVELOPMENT CENTER (NPRDC) |
| TYPE | Research and Development<br>San Diego, CA 92152 |
| CONTACT | Mr. George Lahey |
| PHONE | (619) 225-7122 |
| DESCRIPTION | NPRDC is involved in training research. They are using videodiscs for single-based, computer-controlled, interactive trainer simulators. They do not market or provide products to nongovernment organizations. |

|  |  |
|---|---|
| ORGANIZATION | NIEMEYER, JOHN J. |
| TYPE | Independent Consultant |
| ADDRESS | 709 Marshall Ave.<br>Webster Groves, MO 63119 |
| CONTACT | Mr. John J. Niemeyer |
| PHONE | (314) 968-7486 |
| PREPRODUCTION SERVICES | Analysis, Design |
| DESCRIPTION | Mr. Niemeyer assists colleges and universities in the analysis and design of materials and programs for "Developmental Education." |

| | |
|---|---|
| ORGANIZATION | **OMNICOM ASSOCIATES** |
| TYPE | Company |
| ADDRESS | 908 Steam Mill Rd.<br>Ithaca, NY 14850 |
| CONTACT | Mr. David Williams or Ms. Diane Gayeski |
| PHONE | (607) 277-0405 |
| PREPRODUCTION SERVICES | Analysis, Design, Scripting, Seminars, Workshops |
| PRODUCTION SERVICES | Producing, Directing, Computer Programming, Seminars, Workshops, |
| POSTPRODUCTION SERVICES | Program Simulation, System Design and Configuration, Seminars, Workshops |
| DESCRIPTION | OmniCom Associates are fully qualified and experienced professionals in corporate/industrial training, media production, behavioral research, and human resource development. They specialize in designing and producing informational and instructional programs in a variety of media formats, including videotape, interactive video, personal/mainframe computer-assisted instruction; and in training their clients to do so through workshops and "instructed productions." They conduct research and provide consultation on the impact of the new information technologies on individuals and organizations, especially in the area of interactive media. |

| | |
|---|---|
| ORGANIZATION | **ONE PASS FILM AND VIDEO** |
| TYPE | Company |
| ADDRESS | One China Basin Building<br>San Francisco, CA 94107 |
| CONTACT | Client Services |
| PHONE | (415) 777-5777 |
| PREPRODUCTION SERVICES | Analysis, Design, Scripting |
| PRODUCTION SERVICES | Producing, Directing, Special Effects |
| POSTPRODUCTION SERVICES | Premastering |
| DESCRIPTION | One Pass Film and Video provides full-service production, postproduction and premastering of levels 1, 2, and 3 videodiscs. Among others, they have completed two-sided level 3 discs for Interactive Research Corporation and Apple Computer, Inc. |

| | |
|---|---|
| ORGANIZATION | **ONTARIO INSTITUTE FOR STUDIES IN EDUCATION** |
| TYPE | Research and Development |
| ADDRESS | 252 Bloor St. W.<br>Toronto, Ontario, Canada |
| CONTACT | Ms. Mary Pat Carroll |
| PHONE | (416) 923-6641 |
| PREPRODUCTION SERVICES | Analysis, Design |
| PRODUCTION SERVICES | Producing, Special Effects, Computer Programming |
| POSTPRODUCTION SERVICES | System Design and Configuration |
| DESCRIPTION | The Institute is working on interactive videodiscs in high school level economic geography and geometry. |

| | |
|---|---|
| ORGANIZATION | **OPTICAL SOCIETY OF AMERICA** |
| TYPE | Association |
| ADDRESS | 1816 Jefferson Pl., NW<br>Washington, DC 20036 |

# Production and Development

| | |
|---|---|
| CONTACT | Ms. Joan Carlisle |
| PHONE | (202) 223-8130 |
| PREPRODUCTION SERVICES | Seminars, Workshops |
| PRODUCTION SERVICES | Seminars, Workshops |
| POSTPRODUCTION SERVICES | Seminars, Workshops |
| DESCRIPTION | The Optical Society of America jointly sponsored the Optical Data Storage Meeting in Monterey in April 18-20, 1984. Optical recording media, optical recording and readout systems, and components and techniques were the major topic areas discussed. |

| | |
|---|---|
| ORGANIZATION | **ORENDA, INCORPORATED** |
| TYPE | Company |
| ADDRESS | 3754 Overland Ave.<br>Los Angeles, CA 90034 |
| CONTACT | Mr. Andrew H. Mishkin,<br>Vice President |
| PHONE | (213) 202-0770 |
| PREPRODUCTION SERVICES | Analysis, Design, Scripting |
| PRODUCTION SERVICES | Producing, Directing, Special Effects, Computer Programming |
| POSTPRODUCTION SERVICES | Program Simulation, System Design and Configuration |
| DESCRIPTION | Orenda, Inc. specializes in videodisc-based game development. They provide comprehensive consulting in a wide range of applications areas from games to data distribution. They are currently working on an employee training disc for a large nonprofit organization to be completed in Spring 1984. |

| | |
|---|---|
| ORGANIZATION | **PARALLEL COMMUNICATIONS** |
| TYPE | Company |
| ADDRESS | 115 East 57th St.<br>New York, NY 10022 |
| CONTACT | Mr. Rudy Dsabbarzadeh |
| PHONE | (212) 308-5200 |
| PREPRODUCTION SERVICES | Funding |
| PRODUCTION SERVICES | Producing, Directing |

| | |
|---|---|
| ORGANIZATION | **PERCEPTRONICS, INC.** |
| TYPE | Company, Research and Development |
| ADDRESS | Washington D.C. Operations<br>1911 N. Fort Meyer Dr., Suite 308<br>Arlington, VA 22209 |
| CONTACT | Mr. Steven Johnston |
| PHONE | (703) 525-0184 |
| PREPRODUCTION SERVICES | Analysis, Design, Scripting |
| PRODUCTION SERVICES | Producing, Directing, Special Effects, Computer Programming |
| POSTPRODUCTION SERVICES | Premastering, System Design and Configuration |
| DESCRIPTION | Perceptronics has been involved in the development of interactive videodisc mapping systems for military and civil agencies. They have also developed three generic interactive training systems using videodisc, as follows: ACTIONCODE™; DELTA™ (Disc Editor, Lesson Trainer, and Authoring); and VIMAD™ (Voice Interpretive Maintenance Aiding Device). |

| | |
|---:|:---|
| ORGANIZATION | **PHILIPS INTERNATIONAL/**<br>**AUDIO DIVISION/LASERVISION DEPARTMENT** |
| TYPE | Company |
| ADDRESS | Building SFF-1<br>Eindhoven, Holland |
| PREPRODUCTION SERVICES | Analysis |
| PRODUCTION SERVICES | Computer Programming |
| POSTPRODUCTION SERVICES | Premastering, Program Simulation, System Design and Configuration, Seminars, Workshops |

| | |
|---:|:---|
| ORGANIZATION | **PIONEER VIDEO, INC.** |
| TYPE | Company |
| ADDRESS | 5150 E. Pacific Coast Hwy #300<br>Long Beach, CA 90804 |
| CONTACT | L. A. Yeazel,<br>Director, Marketing Support |
| PHONE | (213) 498-0300 |
| PREPRODUCTION SERVICES | Seminars, Workshops |
| PRODUCTION SERVICES | Computer Programming, Seminars, Workshops, |
| POSTPRODUCTION SERVICES | Premastering, Program Simulation, System Design and Configuration, Seminars, Workshops |

| | |
|---:|:---|
| ORGANIZATION | **PREF, THE PLACEMENT REFERENCE NETWORK** |
| TYPE | Company |
| ADDRESS | P.O. Box 3416<br>Durham, NC 27702 |
| CONTACT | Ms. Aileen McKenna |
| PHONE | (919) 493-3479 |
| TOLL FREE | (800) 334-1638 |
| PREPRODUCTION SERVICES | Design, Scripting |

| | |
|---:|:---|
| ORGANIZATION | **PRESIDIO VIDEO** |
| TYPE | Company |
| ADDRESS | 121 S. Cherry Ave.<br>Tucson, AZ 85719 |
| CONTACT | Mr. Sam Behrend |
| PHONE | (602) 792-2266 |
| PREPRODUCTION SERVICES | Analysis, Design, Scripting |
| PRODUCTION SERVICES | Producing, Directing, Computer Programming |
| POSTPRODUCTION SERVICES | Premastering |

| | |
|---:|:---|
| ORGANIZATION | **PRODUCTION ASSOCIATES, INC.** |
| TYPE | Company |
| ADDRESS | 5456 West Crenshaw<br>Tampa, FL 33614 |
| CONTACT | Mr. Robert Gordon |
| PHONE | (813) 884-3000 or 447-3074 |
| PREPRODUCTION SERVICES | Analysis, Design, Scripting |
| PRODUCTION SERVICES | Producing, Directing, Special Effects |
| POSTPRODUCTION SERVICES | Premastering, Program Simulation, System Design and Configuration |

# Production and Development

| | |
|---|---|
| ORGANIZATION | **PRODUCTION CONSULTANTS** |
| TYPE | Company |
| ADDRESS | 11 Sherman St.<br>Fairfield, CT 06430 |
| CONTACT | Ms. Barbara Wright |
| PREPRODUCTION SERVICES | Design, Scripting |
| PRODUCTION SERVICES | Producing, Directing |
| POSTPRODUCTION SERVICES | Premastering |

| | |
|---|---|
| ORGANIZATION | **PRODUCTIVE LEARNING, INC.** |
| TYPE | Company |
| ADDRESS | 9207 5th Ave. So.<br>Bloomington, MN 55420 |
| CONTACT | Dr. Kent O. Stever<br>General Manager |
| PHONE | (612) 888-7281 |
| PRODUCTION SERVICES | Computer Programming |
| DESCRIPTION | Productive Learning is involved in the research and development of computer programming for computer/videodisc-based market research systems. |

| | |
|---|---|
| ORGANIZATION | **PROPERTY TECHNOLOGY, INC.** |
| TYPE | Company, Research and Development |
| ADDRESS | 283 Dartmouth St.<br>Boston, MA 02116 |
| CONTACT | Mr. Walter R. Crosby,<br>Vice President of Research and Development |
| PHONE | (617) 266-2115 |
| PREPRODUCTION SERVICES | Analysis, Design |
| PRODUCTION SERVICES | Producing, Directing, Computer Programming, Seminars, Workshops |
| POSTPRODUCTION SERVICES | Premastering, System Design and Configuration, Seminars, Workshops |

| | |
|---|---|
| ORGANIZATION | **RAYTHEON SERVICE COMPANY** |
| TYPE | Company, Research and Development |
| ADDRESS | 2 Wayside Rd.<br>Burlington, MA 01803 |
| CONTACT | Mr. Ed Dextraze or Mr. Howard Lavin |
| PHONE | (617) 272-9300 |
| TWX | 710-322-0157 |
| TELEX | 094-9490 |
| PREPRODUCTION SERVICES | Analysis, Design, Scripting, Seminars, Workshops, Full Support for Validation & Evaluation Services Related to Videodisc-based Training |
| PRODUCTION SERVICES | Producing, Directing, Special Effects, Computer Programming, Seminars, Workshops |
| POSTPRODUCTION SERVICES | Premastering, Program Simulation, System Design and Configuration, Seminars, Workshops, Full On-site Validation and Testing of Videodisc-based Courseware |
| DESCRIPTION | Raytheon is currently developing a comprehensive model for in-service training for interactive videodisc-based production. This study will cover all phases of design and production as well as critical organizational concerns and job-performance identification. They are |

also experimenting with better cost-effective methods of disc premastering and postproduction.

|  |  |
|---|---|
| ORGANIZATION | **RCA VIDEODISCS** |
| TYPE | Company |
| ADDRESS | 1133 Ave. of the Americas<br>New York, NY 10036 |
| CONTACT | Mr. Frank McCann |
| PREPRODUCTION SERVICES | Analysis |
| PRODUCTION SERVICES | Producing |
| POSTPRODUCTION SERVICES | Premastering |

|  |  |
|---|---|
| ORGANIZATION | **RECENT PRODUCTIONS LIMITED** |
| TYPE | Company, Research and Development |
| ADDRESS | 10 Chapel St.<br>Belgrave Square<br>London SW1X 7BY UK |
| CONTACT | Mr. Chris Marsh |
| PHONE | (01) 235-6744 |
| PREPRODUCTION SERVICES | Analysis, Design, Scripting, Funding, Seminars, Delivery Hardware Systems Analysis |
| PRODUCTION SERVICES | Producing, Directing, Special Effects, Computer Programming, Specialize in Still Sequencing |
| POSTPRODUCTION SERVICES | Program Simulation, System Design and Configuration |
| DESCRIPTION | Recent Productions Limited just completed "Videodiscs—An Investigation and Recommendations" for the Ford Motor Company, UK/Europe Research and Engineering Department. They are now implementating their recommendations resulting in a project with touch screen, personal computer, and laservision player intended for 80 sites. They are also beginning a similar military project, a children's disc, and a games oriented sailing training disc. |

|  |  |
|---|---|
| ORGANIZATION | **REEVES, THOMAS C.** |
| TYPE | Independent Consultant |
| ADDRESS | Route 2, Box 310A<br>Belle Spring Woods<br>Athens, GA 30607 |
| CONTACT | Dr. Thomas C. Reeves |
| PHONE | (404) 542-4244 or 354-4834 |
| PREPRODUCTION SERVICES | Analysis, Design, Scripting, Funding, Seminars, Workshops, Evaluation |
| DESCRIPTION | Thomas C. Reeves has experience in the design and development of videodiscs for military training, university faculty development, and public school education. He specializes in evaluation and provides comprehensive evaluation design, implementation, and reporting; including project documentation, formative evaluation, instructional effectiveness, impact assessment, and cost effectiveness. His clients include the U.S. Army, the University of Maryland, Emory University Hospital, and the University of Georgia. |

# Production and Development

| | |
|---:|:---|
| ORGANIZATION | **RENAN PRODUCTIONS** |
| TYPE | Company |
| ADDRESS | 2253 Pontius Ave. |
| | Los Angeles, CA 90064 |
| CONTACT | Mr. Peter Bloch |
| PHONE | (213) 478-0393 |
| PREPRODUCTION SERVICES | Analysis, Design, Scripting, Funding |
| PRODUCTION SERVICES | Producing, Directing, Special Effects |
| DESCRIPTION | Renan Productions has considerable experience in medical education films, special effects production, and planning/producing discs for the consumer market. Two shows, "Blackjack" and "Emergency First Aid," were completed for VHD programs before they liquidated and at that time four other discs were in development. The company is currently involved in developing three interactive games for use with personal computers, and also arcade games with two different sponsors. |

| | |
|---:|:---|
| ORGANIZATION | **ROBERT DRUCKER & COMPANY** |
| TYPE | Company |
| ADDRESS | 8901 Zelzah Ave. |
| | Northridge, CA 91325 |
| CONTACT | Mr. Robert Drucker |
| PHONE | (818) 349-1451 |
| PREPRODUCTION SERVICES | Analysis, Design, Scripting, Funding |
| PRODUCTION SERVICES | Producing, Directing, Special Effects, Computer Programming |
| POSTPRODUCTION SERVICES | Premastering, Program Simulation, System Design and Configuration |
| DESCRIPTION | Robert Drucker & Company is presently engaged in the production of six hours of material, designed for the laser videodisc and dealing with energy utilization and energy management. Portions of this program dealing with the maintenance and servicing of energy systems and components will be available for sale to building operators and owners in 1984. They have also completed the production of 40 1/2-hour laser videodisc sides for the sheet metal industry to use in their apprentice and journeyman training programs. The latter are not available for sale or rental to any organizations or individuals. |

| | |
|---:|:---|
| ORGANIZATION | **SANDERS ASSOCIATES, INC.** |
| TYPE | Company, Research and Development |
| ADDRESS | 95 Canal St. |
| | Nashua, NH 03061 |
| CONTACT | Mr. Ralph H. Baer |
| PHONE | (603) 885-4112 |
| DESCRIPTION | Sanders Associates, Inc. developed and patented a basic videodisc of computer graphics generation sync-lock methods (U.S.P. 4,346,407). They also hold U.S. Patent 4,359,223 covering real-time data residing on videodisc. They have developed most of the basic interactive videodisc/VTR technology over the past ten years, as well as video game technology, and have had about 150 patents issued to them worldwide in these general areas. All video game manufacturers and all interactive videodisc manufacturers either currently operate under license to them (directly or via exclusive licenses) or soon will. |

| | |
|---:|:---|
| ORGANIZATION | **SAPAN** |
| TYPE | Independent Consultant |
| ADDRESS | 425A Bornt Hill Rd. |
| | Endicott, NY 13760 |
| CONTACT | Mr. Mike Szabo |
| PREPRODUCTION SERVICES | Analysis, Design |
| PRODUCTION SERVICES | Computer Programming |
| POSTPRODUCTION SERVICES | Program Simulation, System Design and Configuration, Seminars |

| | |
|---:|:---|
| ORGANIZATION | **SCHECTER, ELLEN** |
| TYPE | Independent Consultant |
| ADDRESS | 260 West End Ave. |
| | New York, NY 10023 |
| CONTACT | Ms. Ellen Schecter |
| PHONE | (212) 666-1800 or 580-7744 |
| PREPRODUCTION SERVICES | Scripting, Research |
| DESCRIPTION | Ms. Schecter served as researcher and script writer for "Fun & Games," an interactive home videodisc for children produced by ScholasticProductions. The disc won a 1981 Grammy nomination. |

| | |
|---:|:---|
| ORGANIZATION | **SCOTT DOBBINS PRODUCTIONS** |
| TYPE | Independent Consultant, Company, Research and Development |
| ADDRESS | 162 West 13th St., Suite 54 |
| | New York, NY 10011 |
| CONTACT | Mr. Scott Dobbins, Producer/Director |
| PHONE | (212) 741-1383 |
| PREPRODUCTION SERVICES | Analysis, Design, Scripting |
| PRODUCTION SERVICES | Producing, Directing, Special Effects |
| POSTPRODUCTION SERVICES | Premastering |
| DESCRIPTION | Scott Dobbins is currently working on production of sales/marketing presentations for an international microcomputer corporation. He is also consulting on the creation of an interactive training environment for a major medical group. |

| | |
|---:|:---|
| ORGANIZATION | **SCRIPTECH** |
| TYPE | Research and Development |
| ADDRESS | 33 Winwood Place |
| | Mill Valley, CA 94941 |
| CONTACT | Mr. William Claxton |
| PHONE | (415) 459-8896 |
| PREPRODUCTION SERVICES | Design, Proposal Writing |
| PRODUCTION SERVICES | Producing, Editing |
| POSTPRODUCTION SERVICES | System Design and Configuration |
| DESCRIPTION | Scriptech is currently developing a script editing system, which will enable the developer of interactive materials to see interaction at the script and storyboard level. |

# Production and Development

| | |
|---:|:---|
| ORGANIZATION | **SIGNAL COMMUNICATIONS** |
| TYPE | Company, Research and Development |
| ADDRESS | 600 East Court Ave. |
| | Des Moines, IA 50306 |
| CONTACT | Mr. Jim Kreiser or Wes Ritchie |
| PHONE | (515) 246-3500 |
| PREPRODUCTION SERVICES | Analysis, Design, Scripting |
| PRODUCTION SERVICES | Producing, Directing |

| | |
|---:|:---|
| ORGANIZATION | **SIGNATURE PRODUCTIONS** |
| TYPE | Association |
| ADDRESS | 10108 W. 69th Terrace |
| | Merriam, KS 66203 |
| CONTACT | Ms. Dorothy Yaworski |
| PHONE | (913) 831-4803 |
| PREPRODUCTION SERVICES | Analysis, Design, Scripting |
| PRODUCTION SERVICES | Producing, Directing, Seminars |
| POSTPRODUCTION SERVICES | Premastering, System Design and Configuration |
| DESCRIPTION | Signature is currently involved in a production for the Hallmark Cards, Inc. Visitor Center in Kansas City, MO. Visitors will view five individual interactive programs, which will highlight major aspects in the creating, manufacturing and distribution of Hallmark products. |

| | |
|---:|:---|
| ORGANIZATION | **SIMON FRASER UNIVERSITY/INSTRUCTIONAL MEDIA CENTRE** |
| TYPE | Educational Institution |
| ADDRESS | Simon Fraser University |
| | Burnaby, B.C., Canada |
| | V5A 1S6 |
| | c/o Instructional Media Centre |
| CONTACT | Mr. Christopher Hildred |
| | Manager, Photographic Services |
| PHONE | (604) 291-4752, or |
| CONTACT | Dr. Glenn Kirchner |
| PHONE | (604) 291-3395 |
| PREPRODUCTION SERVICES | Analysis, Design, Scripting, Workshops |
| PRODUCTION SERVICES | Producing, Directing, Special Effects, Computer Programming |
| POSTPRODUCTION SERVICES | System Design and Configuration, Seminars, Workshops |

| | |
|---:|:---|
| ORGANIZATION | **SMELOFF TELEPRODUCTIONS (SMELOFF, INC.)** |
| TYPE | Company |
| ADDRESS | 8201 E. Pacific Place, Suite 502 |
| | Denver, CO 80231 |
| CONTACT | Mr. Nick Smeloff |
| PHONE | (303) 750-5000 |
| PREPRODUCTION SERVICES | Scripting |
| PRODUCTION SERVICES | Producing, Directing, Special Effects, Seminars, Workshops |
| POSTPRODUCTION SERVICES | Premastering, Program Simulation, System Design and Configuration, Seminars, Workshops |

| | |
|---|---|
| ORGANIZATION | **SOCIETY FOR VISUAL EDUCATION, INC.** |
| ADDRESS | 1345 Diversey Parkway<br>Chicago, IL 60614 |
| CONTACT | Mr. Harry A. Reilly,<br>Director, Product Development |
| PHONE | (312) 525-1500 |
| DESCRIPTION | SVE has worked with Micro-Ed, Inc. to develop videodisc/micro software programs marketed under the trade name LaserSoft. |

| | |
|---|---|
| ORGANIZATION | **SOFI INC. (INTERACTIVE TRAINING & INFO. CORP.)** |
| TYPE | Company |
| ADDRESS | 906 Cherrier St., Montreal<br>Quebec, Canada H2L 1H7 |
| CONTACT | Mr. William Lacoursiere, Nicole Laurier,<br>or B. Michaud |
| PHONE | (514) 524-6781/9444 |
| PREPRODUCTION SERVICES | Analysis, Design, Scripting, |
| PRODUCTION SERVICES | Producing, Directing, Special Effects, Computer Programming |
| POSTPRODUCTION SERVICES | Premastering, Program Simulation, System Design and Configuration |
| DESCRIPTION | SOFI is a company specializing in the development and application of interactive communication concepts and techniques in the fields of training and information. SOFI makes use of microcomputer technology, which it combines by means of its own authoring system to its audiovisual productions, thus definitively improving the communication effectiveness and impact. SOFI's staff is multilingual (French, English, and Spanish). |

| | |
|---|---|
| ORGANIZATION | **SONY COMMUNICATIONS PRODUCTS COMPANY** |
| TYPE | Company |
| ADDRESS | Sony Dr.<br>Park Ridge, NJ 07656 |
| CONTACT | Mr. John Hartigan, National Marketing Manager,<br>Interactive Products |
| PHONE | (201) 930-1000 |
| TOLL FREE | (800) 222-SONY |
| PREPRODUCTION SERVICES | Workshops |
| PRODUCTION SERVICES | Workshops |
| POSTPRODUCTION SERVICES | Premastering, Program Simulation, Workshops |
| DESCRIPTION | Workshops are conducted throughout the country by Sony Video Utilization Services. Premastering is done in Japan. Simulation is done at three emulator sites: Maritz Communications Company; Devlin Productions; and Merlin Video Publishing, Inc. |

| | |
|---|---|
| ORGANIZATION | **SONY CORPORATION OF AMERICA** |
| TYPE | Company |
| ADDRESS | Sony Dr.<br>Parkridge, NJ 07656 |
| CONTACT | Ms. Mary Ann Axt or Mr. Jason Farrow |
| PHONE | (201) 930-1000 |
| PREPRODUCTION SERVICES | Analysis, Seminars, Workshops |
| PRODUCTION SERVICES | Computer Programming, Seminars, Workshops |

**Production and Development**

| | |
|---|---|
| POSTPRODUCTION SERVICES | Premastering, Program Simulation, System Design and Configuration, Seminars, Workshops |
| DESCRIPTION | Materials are available concerning videodisc premastering and formatting and videodisc program emulation. Computer programming and consulting services are available on a fee basis. |

| | |
|---|---|
| ORGANIZATION | **SONY VIDEO UTILIZATION SERVICES** |
| TYPE | Company |
| ADDRESS | P.O. Box 29906 |
| | Los Angeles, CA 90029 |
| CONTACT | Dr. Tom Dargan |
| PREPRODUCTION SERVICES | Analysis, Design, Scripting, Funding, Seminars, Workshops, Walk-through Consultations |
| PRODUCTION SERVICES | Computer Programming, Seminars, Workshops |
| POSTPRODUCTION SERVICES | Program Simulation, System Design and Configuration, Seminars, Workshops |
| DESCRIPTION | Sony Video Utilization Services has experience consulting and producing on many projects including military, Library Of Congress, and point-of-purchase. They will send brochures describing seminars and consultations upon request. |

| | |
|---|---|
| ORGANIZATION | **SOUTHERN ALBERTA INSTITUTE OF TECHNOLOGY LEARNING RESOURCES CENTRE** |
| TYPE | Educational Institution |
| ADDRESS | 1801—16TH Ave., NW, Calgary |
| | Alberta, Canada T2M 0L4 |
| CONTACT | Ms. Christine Sammon or Mr. Jim Armstrong |
| PHONE | (403) 284-8665/8617 |
| PREPRODUCTION SERVICES | Analysis |
| DESCRIPTION | They are currently investigating videodisc technology for possible library applications. |

| | |
|---|---|
| ORGANIZATION | **SPERRY MEDIA ENGINEERING CLEARWATER** |
| TYPE | Company |
| ADDRESS | P.O. Box 4648 M.S. 138 |
| | Clearwater, FL 33518 |
| CONTACT | Mr. Glenn L. Wescott |
| PHONE | (813) 577-1900 ext. 2249 |
| PREPRODUCTION SERVICES | Analysis, Design, Scripting |
| PRODUCTION SERVICES | Producing, Directing, Special Effects |
| POSTPRODUCTION SERVICES | Premastering (3/4-inch only) |
| DESCRIPTION | Sperry Media Engineering has produced seven disc sides in support of the Sperry Interactive Training System. The material on these discs ranges from basic skills to complex technical training programs. |

| | |
|---|---|
| ORGANIZATION | **SPIE-THE INTERNATIONAL OPTICAL ENGINEERING SOCIETY** |
| TYPE | Association |
| ADDRESS | P.O. Box 10 |
| | Bellingham, WA 98227-0010 |
| CONTACT | Dr. Chelcie Liu, Technical Director |

| | |
|---|---|
| PHONE | (206) 676-3290 |
| DESCRIPTION | While SPIE does not currently produce videodiscs, it does sponsor conferences and short courses on optical storage technology and applications, as well as approximately 70 other conferences and 100 short courses annually. There is a possibility that these courses and conferences will be recorded on videodisc in the future, for distribution. |

| | |
|---|---|
| ORGANIZATION | **STOREY, MEG** |
| TYPE | Independent Consultant |
| ADDRESS | 375A Harvard St. #22A<br>Cambridge, MA 02138 |
| CONTACT | Ms. Meg Storey |
| PHONE | (617) 492-8135 |
| PREPRODUCTION SERVICES | Design, Scripting |
| DESCRIPTION | Meg Storey is an educational game specialist, and is particularly interested in discs for children. |

| | |
|---|---|
| ORGANIZATION | **SUN INTERACTIVE VIDEO** |
| TYPE | Company |
| ADDRESS | 1488 Sierra Creek Way<br>San Jose, CA 95132 |
| CONTACT | Mr. Chen Sun |
| PHONE | (408) 272-4385 |
| PREPRODUCTION SERVICES | Analysis, Design, Scripting |
| PRODUCTION SERVICES | Producing, Computer Programming |
| POSTPRODUCTION SERVICES | System Design and Configuration |

| | |
|---|---|
| ORGANIZATION | **SYSCON CORPORATION, INC.** |
| TYPE | Company |
| ADDRESS | 133 Gaither Dr., Suite G<br>Mt. Laurel, NJ 08054 |
| CONTACT | Mr. Thomas Koehl |
| PHONE | (609) 234-5510 |
| PREPRODUCTION SERVICES | Design, Scripting |
| PRODUCTION SERVICES | Producing, Directing |
| POSTPRODUCTION SERVICES | Premastering |
| DESCRIPTION | SYSCON is currently producing a level 3 CAV disc for the U.S. Navy Aegis Shipbuilding project. The disc mixes graphics with live action footage to introduce fleet users to the many elements, or subsystems, of the Aegis Combat System. This program is a joint development of SYSCON and WICAT, using a PR 7820 player and a 3M-mastered disc. As a full video production facility, SYSCON's Mt. Laurel, NJ office is editing the premaster tape. ISD work is the responsibility of SYSCON's Dahlgren, VA office. WICAT is providing software development. |

| | |
|---|---|
| ORGANIZATION | **TCT TECHNICAL TRAINING, INC. ("T-CUBED")** |
| TYPE | Company |
| ADDRESS | 599 North Mathilda Ave.<br>Sunnyvale, CA 94086 |
| CONTACT | Mr. Saroj K. Kar |
| PHONE | (408) 735-9990 |

# Production and Development

| | |
|---|---|
| PREPRODUCTION SERVICES | Analysis, Design, Scripting, Workshops |
| PRODUCTION SERVICES | Producing, Directing, Computer Programming |
| POSTPRODUCTION SERVICES | Premastering, Program Simulation, System Design and Configuration |
| DESCRIPTION | TCT Technical Training, Inc. ("T-CUBED") specializes in providing data communications (and related) training solutions, using computer-based interactive video (both videotape and videodisc) as the principal medium. T-CUBED products include off-the-shelf courseware in data communications, custom course design service, and customized authoring systems. |

| | |
|---|---|
| ORGANIZATION | **TECNETRA-TECHNOLOGY TRANSFER COMPANY** |
| TYPE | Company |
| ADDRESS | Piazza Morandi 2<br>Milano, Italy |
| CONTACT | Mr. Umberto Pellegrini |
| PREPRODUCTION SERVICES | Analysis, Design, Scripting, Seminars |
| PRODUCTION SERVICES | Computer Programming, Seminars |
| POSTPRODUCTION SERVICES | Program Simulation, System Design and Configuration, Seminars |

| | |
|---|---|
| ORGANIZATION | **TELEMATIC SYSTEMS** |
| TYPE | Company |
| ADDRESS | 55 Wheeler St.<br>Cambridge, MA 02138 |
| CONTACT | Mr. Fredric J. Raab |
| PHONE | (617) 492-2881 |
| PREPRODUCTION SERVICES | Analysis, Design, Scripting |
| PRODUCTION SERVICES | Producing, Directing, Special Effects, Computer Programming, Creation of Computer Graphic Overlays |
| POSTPRODUCTION SERVICES | Program Simulation, System Design and Configuration |

| | |
|---|---|
| ORGANIZATION | **TELEMEDIA CORPORATION** |
| TYPE | Independent Consultant, Company, Research and Development |
| ADDRESS | P.O. Box 1190<br>Princeton, NJ 08542 |
| CONTACT | Mr. William Ring |
| PHONE | (609) 799-6222 |
| PREPRODUCTION SERVICES | Analysis, Design, Scripting, Seminars, Workshops |
| PRODUCTION SERVICES | Producing, Directing, Special Effects, Computer Programming |
| POSTPRODUCTION SERVICES | Premastering, System Design and Configuration |
| DESCRIPTION | William Ring was Executive Vice President of ISO Communications, Inc. until September 1983. ISO produced the first interactive CED prototype courseware for CBS Publishing as well as research and development videodisc projects for ABC laserdisc (Schooldisc), RCA (first interactive CED engineering demo disc), and InterDigital, Inc. They also were contracted by Digital Equipment Corporation to assist in the market planning and product planning of IVIS, DEC's interactive video system, and to produce the first videodiscs for the system. This project was discontinued by ISO in September 1983. Telemedia Corporation designs and produces interactive videodiscs and videotex systems for the information industry. Feasibility studies and project consulting services are also available. |

| | |
|---:|:---|
| ORGANIZATION | **TELEMEDIA GMBH** |
| TYPE | Company |
| ADDRESS | Carl Bertlesmann Strabe 161 |
| | D-4830 Gutersloh 1 |
| | West Germany |
| CONTACT | Dr. Franz Netta |
| PHONE | Gutersloh (05241) / 80-3324 |
| TELEX | 933 822 |
| POSTPRODUCTION SERVICES | Premastering, System Design and Configuration, Seminars |

| | |
|---:|:---|
| ORGANIZATION | **TRANS-LINGUAL COMMUNICATIONS, INC.** |
| TYPE | Company, Research and Development |
| ADDRESS | Suite 1200, Eight South Michigan Ave. |
| | Chicago, IL 60603 |
| CONTACT | Mr. Lyric Hughes, |
| | President |
| PHONE | (312) 332-6555 |
| TOLL FREE | (800) 621-2167 |
| PREPRODUCTION SERVICES | Analysis, Design, Scripting |
| PRODUCTION SERVICES | Producing, Directing, Special Effects, Computer Programming, Foreign-language Translation |
| POSTPRODUCTION SERVICES | Program Simulation, System Design and Configuration |
| DESCRIPTION | TRANS-LINGUAL COMMUNICATIONS, INC. is headquartered in Chicago with offices in Los Angeles and Hong Kong. The company is comprised of three major divisions. The Communications Division offers translation services, international advertising and marketing, and foreign language film and slide presentations. The Education Division includes educational software, as well as videodisc and voice synthesis systems designed to meet a variety of educational needs. The Consulting Division, which coordinates major projects in the other two divisions, is particularly active in the computer and health-care industries, both in the United States and abroad. |

| | |
|---:|:---|
| ORGANIZATION | **TRILOGIC, INC.** |
| TYPE | Company |
| ADDRESS | 9430 Washington Blvd. |
| | Culver City, CA 90230 |
| CONTACT | Ms. Virginia Barreto |
| PHONE | (213) 838-1140 |
| PREPRODUCTION SERVICES | Analysis, Design, Scripting |
| PRODUCTION SERVICES | Producing, Directing, Special Effects, Computer Programming |
| POSTPRODUCTION SERVICES | Premastering, Program Simulation, System Design and Configuration |

| | |
|---:|:---|
| ORGANIZATION | **U.S. ARMY RESEARCH INSTITUTE** |
| TYPE | Not-for-Profit, Research and Development, Government |
| ADDRESS | Fort Benning Field Unit |
| CONTACT | Dr. James E. Schroeder |
| PHONE | (404) 545-1278 |
| PREPRODUCTION SERVICES | Analysis, Design, Scripting |
| PRODUCTION SERVICES | Computer Programming |
| POSTPRODUCTION SERVICES | System Design and Configuration |

**Production and Development**                                                                                  169

|  |  |
|---|---|
| ORGANIZATION | **U.S. PUBLIC HEALTH SERVICE**<br>**NATIONAL HANSEN'S DISEASE CENTER** |
| TYPE | Not-for-Profit, Research and Development |
| ADDRESS | Carville, LA 70721 |
| CONTACT | Dr. Richard O'Connor |
| PHONE | (504) 642-7771 ext. 281 |
| DESCRIPTION | This office will collaborate with other medical and educational institutions in development of software and evaluation studies, especially with authoring in PILOT. |

|  |  |
|---|---|
| ORGANIZATION | **UNIVERSITY COLLEGE, CARDIFF/**<br>**CENTRE FOR EDUCATIONAL TECHNOLOGY** |
| TYPE | Educational Institution |
| ADDRESS | 8 North Rd.<br>Cardiff CF1 3DY<br>United Kingdom |
| CONTACT | Dr. D.K. Roach |
| PHONE | 0222-44211 ext. 2428 |
| PREPRODUCTION SERVICES | Analysis, Design, Scripting |
| PRODUCTION SERVICES | Producing, Directing, Special Effects, Computer Programming |
| POSTPRODUCTION SERVICES | System Design and Configuration |
| DECRIPTION | The Centre combines the college's media services and educational technology service, providing production, evaluation and related services to the teaching departments. The Centre has been involved in a pilot disc project that has four segments: an instructional program for library staff and users, a 600-frame still picture "resource", an Interaction Analysis program based on a school French lesson, and a program on Electronic Music synthesizers. Negotiations are about to be completed on several contracts for interactive training materials for industrial and commercial companies, and a college Interactive Learning organization is being set up to offer training consultancies to industry and business. |

|  |  |
|---|---|
| ORGANIZATION | **UNIVERSITY OF ARIZONA/**<br>**HEALTH SCIENCES CENTER**<br>**BIOMEDICAL COMMUNICATIONS** |
| TYPE | Not-for-Profit, Educational Institution |
| ADDRESS | Tucson, AZ 85724 |
| CONTACT | Sam Behrend |
| PHONE | (602) 626-7343 |
| PREPRODUCTION SERVICES | Analysis, Design, Scripting |
| PRODUCTION SERVICES | Producing, Directing, Special Effects, Computer Programming |

|  |  |
|---|---|
| ORGANIZATION | **UNIVERSITY OF ILLINOIS AT CHICAGO/**<br>**HEALTH SERVICES CENTER**<br>**CENTER FOR EDUCATIONAL DEVELOPMENT** |
| TYPE | Independent Consultant, Not-for-Profit, Research and Development |
| ADDRESS | 808 South Wood<br>Chicago, IL 60612 |
| CONTACT | Dr. A.B. Ockerse |
| PHONE | (312) 996-3590 |
| PREPRODUCTION SERVICES | Analysis, Design, Scripting, Seminars |

| | |
|---|---|
| PRODUCTION SERVICES | Producing |
| POSTPRODUCTION SERVICES | Premastering |
| DESCRIPTION | The Center for Educational Development is an internationally recognized medical education department. It is also a collaborative Health Manpower Training Center for the World Health Organization. The Center has a large, well-equipped audiovisual production facility. The Center is involved in a project called "Electronic Curriculum for Medical Education and Health Care." The principal goal of this project is development of a suitable content structure for instructional material used in medical education and health care delivery, for use with interactive electronic media such as videodiscs and videotex. |

| | |
|---|---|
| ORGANIZATION | **UNIVERSITY OF IOWA/ COMPUTER ASSISTED INSTRUCTION LABORATORY** |
| TYPE | Research and Development |
| ADDRESS | WEEG Computing Center<br>Iowa City, IA 52242 |
| CONTACT | Dr. Joan Sustik Huntley |
| PHONE | (319) 353-7253 |
| PREPRODUCTION SERVICES | Analysis, Design, Scripting |
| PRODUCTION SERVICES | Computer Programming |
| DESCRIPTION | The Computer Assisted Instruction Laboratory conducts research and development on uses of computers in the instructional process. It has supported computer-controlled videodisc development since 1978, originally with transmissive systems, and now with reflective laser systems. Applications include: tutorial (pediatrics); drill (ballet); psychological research (psychology); visual retrieval (art, english, language lab, etc.); simulation (film and broadcasting); and patient education (gynecology). Developed initially for on-campus use, many applications are of interest beyond the campus. |

| | |
|---|---|
| ORGANIZATION | **UNIVERSITY OF IOWA/ UNIVERSITY VIDEO CENTER** |
| TYPE | Not-for-Profit |
| ADDRESS | E205 Seashore Hall<br>University of Iowa<br>Iowa City, IA 52242 |
| CONTACT | Mr. Daniel G. Lind,<br>General Manager |
| PHONE | (319) 353-4333 |
| PREPRODUCTION SERVICES | Design, Scripting |
| PRODUCTION SERVICES | Producing, Directing, Special Effects |
| POSTPRODUCTION SERVICES | Premastering |
| DESCRIPTION | The University Video Center has developed a process called AFI (Alternate Field Interpolation), which eliminates the interfield jitter that occurs during a freeze frame of a motion sequence on videodisc or videotape. The Video Center is also working on high-density storage of slide collections for the university, and a child psychiatry training disc for the University of Iowa hospitals and clinics. |

**Production and Development**

| | |
|---|---|
| ORGANIZATION | **UNIVERSITY OF LONDON AUDIO-VISUAL CENTRE** |
| TYPE | Research and Development, Educational Institution |
| ADDRESS | North Wing Studios, Senate House<br>Malet St.<br>London WC1E 7JZ<br>United Kingdom |
| CONTACT | Dr. David R. Clark |
| PHONE | 01-636 8000 |
| PREPRODUCTION SERVICES | Analysis, Design, Scripting, Funding, Seminars, Workshops |
| PRODUCTION SERVICES | Producing, Directing, Special Effects, Computer Programming, Seminars, Workshops |
| DESCRIPTION | The University of London Audio-Visual Centre exists primarily to service the academic requirements of the university, but may undertake joint projects, of educational relevance, with appropriate bodies from time to time. |

| | |
|---|---|
| ORGANIZATION | **UNIVERSITY OF MINNESOTA, TWIN CITIES CAMPUS/<br>UNIVERSITY MEDIA RESOURCES TELEVISION** |
| TYPE | Not-for-Profit, Educational Institution |
| ADDRESS | 540 Rarig Center<br>Minneapolis, MN 55455 |
| CONTACT | Mr. Arnold Walker,<br>Associate Director for Television |
| PHONE | (612) 373-3863 |
| PREPRODUCTION SERVICES | Analysis, Design, Funding |
| PRODUCTION SERVICES | Producing, Directing, Special Effects |
| POSTPRODUCTION SERVICES | Premastering, System Design and Configuration |
| DESCRIPTION | University Media Resources Television is a large broadcast-level production facility specializing in training and instructional television programs. They work with other university departments and selected not-for-profit organizations in the design and production of programs. |

| | |
|---|---|
| ORGANIZATION | **UNIVERSITY OF NEBRASKA/<br>NEBRASKA VIDEODISC DESIGN/PRODUCTION GROUP** |
| TYPE | Not-for-Profit |
| ADDRESS | Box 83111<br>Lincoln, NE 68501 |
| CONTACT | Mr. Ron Nugent |
| PHONE | (402) 472-3611 |
| PREPRODUCTION SERVICES | Analysis, Design, Scripting, Seminars, Workshops |
| PRODUCTION SERVICES | Producing, Directing, Special Effects, Computer Programming, Seminars, Workshops |
| POSTPRODUCTION SERVICES | Premastering, Program Simulation, System Design and Configuration, Seminars, Workshops |
| DESCRIPTION | The Nebraska Group is well-known for its pioneering work with interactive videodiscs and its seminars and workshops. Gerry Broad, who is on sabbatical from Teesside Polytechnic, Middleborough, England, has joined the group on an intern status to research application of videodisc technology towards management and education. |

| | |
|---|---|
| ORGANIZATION | **UNIVERSITY OF WEST FLORIDA/ OFFICE FOR INTERACTIVE TECHNOLOGY AND TRAINING** |
| TYPE | Not-for-Profit, Research and Development, Educational Institution |
| ADDRESS | University of West Florida<br>Pensacola, FL 32514 |
| CONTACT | Ms. Patricia Lynett<br>Director, Social Work Contracts & Grants |
| PHONE | (904) 474-2251/2253 |
| PREPRODUCTION SERVICES | Analysis, Design, Scripting, Seminars, Workshops |
| PRODUCTION SERVICES | Computer Programming |
| DESCRIPTION | The Office for Interactive Technology and Training is a development house that specializes in custom-designed training programs. They have completed a project that involved the transfer of an entire new employee classroom training program into approximately 160 hours of computer-assisted interactive videodisc training. Objective and subjective data were collected during a pilot test of the program. |

| | |
|---|---|
| ORGANIZATION | **UNIVERSITY OF WISCONSIN/ EDUCATIONAL TECHNOLOGY PROGRAM** |
| TYPE | Educational Institution |
| ADDRESS | 528F Teacher Education Building<br>225 N. Mills St.<br>Madison, WI 53706 |
| CONTACT | Dr. Michael J. Streibel |
| PHONE | (608) 263-4674 |
| PREPRODUCTION SERVICES | Analysis, Design, Scripting, Seminars, Workshops, College Courses |
| PRODUCTION SERVICES | Producing, Directing, Computer Programming, Seminars, Workshops, College Courses |
| POSTPRODUCTION SERVICES | Seminars, Workshops, College Courses |

| | |
|---|---|
| ORGANIZATION | **VCM SYSTEMS, INC.** |
| TYPE | Company, Research and Development |
| ADDRESS | 427 Sixth Ave. S.E., P.O. Box 2789<br>Cedar Rapids, IA 52406 |
| CONTACT | Mr. Thomas Hedges,<br>Executive Producer |
| PHONE | (319) 364-6959 |
| TOLL FREE | (800) 553-8879 |
| PREPRODUCTION SERVICES | Analysis, Design, Scripting |
| PRODUCTION SERVICES | Producing, Directing, Computer Programming |
| POSTPRODUCTION SERVICES | Premastering, System Design and Configuration |
| DESCRIPTION | In July 1983, VCM Systems, Inc. completed a national field test of a prototype point-of-sale interactive information service. As part of this test, VCM Systems, Inc. conducted extensive qualitative and quantitative market research. In addition to their research and development activities, VCM Systems, Inc. has worked with a hardware manufacturer in designing cost-effective kiosk hardware to be used in sales of integrated point of sale interactive information services. VCM Systems, Inc. is currently researching and developing point-of-sale interactive information systems for other industries. |

# Production and Development

| | |
|---|---|
| ORGANIZATION | **VIDEO IMAGE CONSULTANTS** |
| TYPE | Independent Consultant |
| ADDRESS | 1912 South University Blvd.<br>Denver, CO 80210 |
| CONTACT | Mr. Truxton Simmons |
| PHONE | (303) 777-1940 |
| PREPRODUCTION SERVICES | Design, Scripting |
| PRODUCTION SERVICES | Producing, Directing, Special Effects, Computer Programming |
| POSTPRODUCTION SERVICES | Premastering |
| DESCRIPTION | Video Image Consultants has been creating custom video presentations in the Rockies since 1977. The firm specializes in the use of video to inform, motivate, instruct, and educate. Video Image Consultants has offered interactive video design/production services since 1980. |

| | |
|---|---|
| ORGANIZATION | **VIDEO MASTERS, INC.** |
| TYPE | Company, Research and Development |
| ADDRESS | 1616 Broadway<br>P.O. Box 1963<br>Kansas City, MO 64141 |
| CONTACT | Mr. Paul Dark,<br>Vice President |
| PHONE | (816) 474-8530 |
| POSTPRODUCTION SERVICES | System Design and Configuration |
| DESCRIPTION | Video Masters, Inc. is a sophisticated video systems supplier and contractor, and a custom product manufacturer. Interactive video is a market that VMI has addressed with national installations, including custom computer interfaces, systems hardware, and touch screens and their control. |

| | |
|---|---|
| ORGANIZATION | **VIDEO PARK, INC.** |
| TYPE | Company |
| ADDRESS | 8116 One Calais St. 1A<br>Baton Rouge, LA 70806 |
| CONTACT | Mr. Park Seward |
| PHONE | (504) 766-3163 |
| PREPRODUCTION SERVICES | Design, Scripting, Seminars, Workshops |
| PRODUCTION SERVICES | Producing, Directing, Special Effects, Seminars, Workshops |
| POSTPRODUCTION SERVICES | Premastering, Program Simulation |

| | |
|---|---|
| ORGANIZATION | **VIDEO SOFTWARE ASSOCIATES, INC.** |
| TYPE | Company |
| ADDRESS | 1215 S. 26th Rd.<br>Arlington, VA 22202 |
| CONTACT | Mr. David Hopwood or Ms. Susan Brooks Kelly |
| PHONE | (703) 892-6988 or (202) 833-9800 |
| PREPRODUCTION SERVICES | Analysis, Design, Scripting |
| PRODUCTION SERVICES | Producing, Directing, Special Effects |
| DESCRIPTION | Video Software Associates, Inc. is a television and videodisc producing company. Besides producing programming for national cable television and industrial training, Video Software has produced fourteen level 3 videodiscs under a subcontract with the Human Resources Research Organization for the military. VSA's staff served as instructional |

designers, executive producers, production manager, writer, producer, and director on these discs. The disc programming used a variety of formats—games, dramas, dramatic comedy, documentary, vicarious travel, and a variety of techniques that exploit the joystick-based spatial data management system developed for microcomputer by Interactive Television Company. Discs currently in development involve content in tactical decision-making and land navigation.

| | |
|---|---|
| ORGANIZATION | **VIDEO VISION ASSOCIATES, LTD.** |
| TYPE | Company |
| ADDRESS | 7 Waverly Pl.<br>Madison, NJ 07940 |
| CONTACT | Mr. Bill Clark or Mr. Ralph Heigl |
| PHONE | (201) 377-0302 |
| PREPRODUCTION SERVICES | Design |
| PRODUCTION SERVICES | Producing |
| DESCRIPTION | Video Vision Associates are the producers of the Space Disc Series of educational videodiscs, and the Space Archive Series of consumer discs. |

| | |
|---|---|
| ORGANIZATION | **VIDEOACTIVE, INC.** |
| TYPE | Company |
| ADDRESS | Box 95783, University Station<br>Seattle, WA 98145-2783 |
| CONTACT | Dr. James Laffey or Dr. Gerald Gillmore |
| PHONE | (415) 469-2693 or (206) 543-1170 |
| PREPRODUCTION SERVICES | Analysis, Design, Scripting, Seminars, Workshops |
| PRODUCTION SERVICES | Producing, Directing, Special Effects, Computer Programming, Seminars, Workshops |
| POSTPRODUCTION SERVICES | Premastering, Program Simulation, System Design and Configuration, Seminars, Workshops |
| DESCRIPTION | Videoactive, Inc. is a corporation dedicated to combining audiovisual and microcomputer technology to produce attractive and effective self-paced learning systems for schools, government, and industry. Such systems display a wide variety of television images on one video monitor and computer-generated text on another to provide interactive lessons or training sessions. They can provide a full range of videodisc/videotape production services and complete CAI programming support, using their own authoring language. They are currently offering technical support in the development of the "Math in Biology" project of the University of Washington's Educational Assessment Center. |

| | |
|---|---|
| ORGANIZATION | **VIDEODISC TECHNOLOGY LIMITED** |
| TYPE | Company |
| ADDRESS | Roquebrune House<br>South Park, South Godstone<br>Surrey, RH9 8LE<br>United Kingdom |
| CONTACT | Mr. Don Tombs,<br>Managing Director |
| PHONE | (0) 342.852250 |
| PREPRODUCTION SERVICES | Analysis, Design, Scripting, Seminars, Workshops |

# Production and Development

| | |
|---:|:---|
| ORGANIZATION | **VIDEODISCDESIGN** |
| TYPE | Company |
| ADDRESS | Box 319 B, Route 3 |
| | Escondido, CA 92025 |
| CONTACT | Mr. Larry Lowe |
| PHONE | (619) 489-8183 |
| PREPRODUCTION SERVICES | Analysis, Design, Scripting, Seminars, Workshops |
| PRODUCTION SERVICES | Directing, Computer Programming |
| POSTPRODUCTION SERVICES | Premastering, System Design and Configuration |
| DESCRIPTION | VIDEODISCDESIGN specializes in sophisticated level 3 entertainment and educational systems. They are currently designing and producing a 48-player educational game for America's Electric Energy Exhibit at the Louisiana World Exposition, New Orleans, LA. Projects under development are a second generation laser disc arcade game, and an interactive experience for consumer release. |

| | |
|---:|:---|
| ORGANIZATION | **VIDEODISCOVERY** |
| TYPE | Independent Consultant, Company, Research and Development |
| ADDRESS | P.O. Box 85878 |
| | Seattle, WA 98145-1878 |
| CONTACT | Dr. D. Joseph Clark |
| PHONE | (206) 524-4007 |
| PREPRODUCTION SERVICES | Analysis, Design, Scripting, Seminars, Workshops |
| PRODUCTION SERVICES | Producing, Directing |
| POSTPRODUCTION SERVICES | Premastering, System Design and Configuration |
| DESCRIPTION | VIDEODISCOVERY is involved with a study to use a generic disc as an element to encourage computer literacy by allowing students to make their own program under computer control, using the visual data base present on the Bio Sci videodisc. They are also involved in marketing research for videodisc materials and developing methods to involve educators in the utilization of videodisc technology. |

| | |
|---:|:---|
| ORGANIZATION | **WDC MARKETING ASSOCIATES, INC.** |
| TYPE | Independent Consultant |
| ADDRESS | 244 Goodwin Crest Dr. |
| | Suite 106 |
| | Birmingham, AL 35209 |
| CONTACT | Mr. David Carrington |
| PHONE | (205) 945-8951 |
| PREPRODUCTION SERVICES | Analysis, Design, Scripting |
| PRODUCTION SERVICES | Computer Programming |
| POSTPRODUCTION SERVICES | Premastering, Program Simulation, System Design and Configuration |
| DESCRIPTION | WDC Marketing Associates is a retail management consulting firm specializing in merchandising and marketing. |

| | |
|---:|:---|
| ORGANIZATION | **WECHSLER, JUDITH** |
| TYPE | Independent Consultant, Not-for-Profit |
| ADDRESS | 375A Harvard St. #22A |
| | Cambridge, MA 02138 |
| CONTACT | Ms. Meg Storey |
| PHONE | (617) 492-8135 |
| PREPRODUCTION SERVICES | Design, Scripting |

| | |
|---:|:---|
| ORGANIZATION | **WESTINGHOUSE ELECTRIC CORP.** <br> **INTEGRATED LOGISTICS SUPPORT DIVISIONS** <br> **(TECHNICAL TRAINING OPERATIONS)** |
| TYPE | Company |
| ADDRESS | MS2410—CCC#20 <br> 10400 Patuxent Pkwy. <br> Columbia, MD 21044 |
| CONTACT | Mr. Steve G. Kukla |
| PHONE | (301) 995-5482 |
| PREPRODUCTION SERVICES | Analysis, Design, Scripting |
| PRODUCTION SERVICES | Producing, Directing, Special Effects, Computer Programming |
| POSTPRODUCTION SERVICES | Premastering, System Design and Configuration |

| | |
|---:|:---|
| ORGANIZATION | **WGBH EDUCATIONAL FOUNDATION** |
| TYPE | Not-for-Profit |
| ADDRESS | 125 Western Ave. <br> Boston, MA 02134 |
| CONTACT | Mr. Rob Lippincott |
| PHONE | (617) 492-2777 |
| PREPRODUCTION SERVICES | Analysis, Design, Scripting, Funding, Seminars, Workshops |
| PRODUCTION SERVICES | Producing, Directing, Special Effects, Computer Programming, Seminars, Workshops |
| POSTPRODUCTION SERVICES | Premastering, System Design and Configuration |
| DESCRIPTION | WGBH has the following projects in development: 1) Medical Database Storage Project. They are seeking funding for a co-venture with Tufts New England Medical Birth Defects Information Service to store a multi-megabyte digital database, consisting of 35,000 images and the control software on a single laser disc; and 2) Time Independent Video Art. They are seeking marketing funds for an interactive narrative quest through the animated collaboration of 12 videodiscs. WGBH is also continuing a research study of still frame production/transfer/editing problems. |

| | |
|---:|:---|
| ORGANIZATION | **WICAT SYSTEMS, INC.** |
| TYPE | Company |
| ADDRESS | 1875 South State St. <br> P.O. Box 539 <br> Orem, UT 84057 |
| CONTACT | Mr. Richard M. Cavagnol |
| PHONE | (801) 224-6400 ext. 282 |
| PREPRODUCTION SERVICES | Analysis, Design, Scripting, Seminars |
| PRODUCTION SERVICES | Producing, Directing, Computer Programming |
| POSTPRODUCTION SERVICES | Premastering, Program Simulation, System Design and Configuration |

| | |
|---:|:---|
| ORGANIZATION | **WICKERWORKS VIDEO PRODUCTIONS, INC.** |
| TYPE | Company |
| ADDRESS | 7342 S. Alton Way <br> Denver, CO 80112 |
| CONTACT | Ms. Terry Wickre |
| PHONE | (303) 741-3400 |
| PRODUCTION SERVICES | Production Support (studio, camera, 1-inch VTR) |
| POSTPRODUCTION SERVICES | Premastering |

# Production and Development

| | |
|---|---|
| ORGANIZATION | **WILLIAMS, MARK B., INTERACTIVE VIDEO DESIGN** |
| TYPE | Independent Consultant |
| ADDRESS | 2 Genoa Pl., Suite 12 |
| | San Francisco, CA 94133 |
| CONTACT | Mr. Mark B. Williams |
| PHONE | (415) 362-2242 |
| PREPRODUCTION SERVICES | Analysis, Design, Scripting, Seminars, Workshops |
| PRODUCTION SERVICES | Computer Programming, Seminars, Workshops |
| POSTPRODUCTION SERVICES | Program Simulation, Seminars, Workshops |

| | |
|---|---|
| ORGANIZATION | **WINDMILL PRODUCTIONS, INC.** |
| TYPE | Company |
| ADDRESS | 1820 Briarwood Industrial Court |
| | Atlanta, GA 30329 |
| CONTACT | Mr. Larry Goddard |
| PHONE | (404) 634-9060 |
| PREPRODUCTION SERVICES | Analysis, Design, Scripting |
| PRODUCTION SERVICES | Producing, Directing, Special Effects, Computer Programming |
| POSTPRODUCTION SERVICES | Premastering, System Design and Configuration |
| DESCRIPTION | Windmill Productions developed the Slidescan transfer process for transferring multiprojector slide presentations to videotape and film. They are currently involved in developing a sophisticated, high-density slide transfer process for fast, high-quality transfers of slides to videotape for videodisc premasters. |

| | |
|---|---|
| ORGANIZATION | **WREN ASSOCIATES INC.** |
| TYPE | Company |
| ADDRESS | Communications Park |
| | Bunn Dr. |
| | Princeton, NJ 08540 |
| CONTACT | Mr. Karl Faller, President or |
| | Ms. Susan Firestone, Videodisc Manager |
| PHONE | (609) 924-8085 |
| PREPRODUCTION SERVICES | Analysis, Design, Scripting, Seminars |
| PRODUCTION SERVICES | Producing, Directing, Special Effects, Computer Programming, Seminars |
| POSTPRODUCTION SERVICES | Premastering, Program Simulation, Seminars |

| | |
|---|---|
| ORGANIZATION | **XENON** |
| TYPE | Company |
| ADDRESS | 3417 West Cahuenga Blvd. |
| | Hollywood, CA 90068 |
| CONTACT | Mr. Brent Oakley |
| PHONE | (213) 851-8650 |
| PRODUCTION SERVICES | Producing, Directing, Special Effects, Computer Programming, Special Production Support. |
| POSTPRODUCTION SERVICES | Premastering, System Design and Configuration |
| DESCRIPTION | XENON is currently producing special effects and production support for VIDEODISCDESIGN. In addition, XENON continues to produce training systems for Merle Norman, and is developing projects for Sega, Mylstar, Southgate, and Perceptronics. |

# Hardware/Software/Integrated Systems

| | |
|---|---|
| ORGANIZATION | **ACTRONICS, INC.** |
| ADDRESS | 810 River Ave. |
| | Pittsburgh, PA 15212 |
| CONTACT | Ms. Joyce Hallas or Jeffrey A. Howell |
| PHONE | (412) 231-6200 |
| SOFTWARE PRODUCED | Drivers/Video Utilities, Authoring Systems |
| INTEGRATED SYSTEMS OFFERED | Training Work Stations/Carrels |

| | |
|---|---|
| ORGANIZATION | **ADVANCED IMAGE TECHNOLOGY, INC.** |
| ADDRESS | 122 East 42nd St., Suite 623 |
| | New York, NY 10017 |
| CONTACT | Dr. Jean-Pierre Isbouts |
| PHONE | (212) 986-2861 |
| HARDWARE MANUFACTURED | Overlays/Superimpose Units, Still-frame Audio, Input Devices, Interfaces |
| SOFTWARE PRODUCED | Drivers/Video Utilities, Authoring Systems, Graphics Editors/Titlers |
| INTEGRATED SYSTEMS OFFERED | Training Work Stations/Carrels, Information Displays, Point-of-Purchase Systems |
| DESCRIPTION | Advanced Image Technology, Inc. also provides large archival systems for storage/retrieval of high-volume imagery within data base management environments, and network systems for distributed image and data base access through terminals with optical drives. |

| | |
|---|---|
| ORGANIZATION | **ALLEN COMMUNICATION** |
| ADDRESS | 140 Lakeside Plaza II |
| | 5225 Wiley Post Way |
| | Salt Lake City, UT 84116 |

|  |  |
|---|---|
| CONTACT | Mr. Steven Allen or Rex Allen |
| PHONE | (801) 537-7800 |
| HARDWARE MANUFACTURED | Overlays/Superimpose Units, Still-frame Audio, Interfaces |
| SOFTWARE PRODUCED | Drivers/Video Utilities, Authoring Systems |
| INTEGRATED SYSTEMS OFFERED | Training Work Stations/Carrels, Information Displays, Point-of-Purchase Systems |

|  |  |
|---|---|
| ORGANIZATION | **ALPHATEL SYSTEMS LTD.** |
| ADDRESS | 102, 11430-168 St., Edmonton<br>Alberta, Canada T5M 3T9 |
| CONTACT | Dr. R.A. (Bob) Abell |
| PHONE | (403) 451-5761 |
| SOFTWARE PRODUCED | Drivers/Video Utilities, Authoring Systems |
| INTEGRATED SYSTEMS OFFERED | Training Work Stations/Carrels |
| DESCRIPTION | ALPHATEL Systems Ltd. specializes in the custom design of interactive videotex information, training, and directory systems. They are currently working on BASIC and PASCAL versions of a BIOP videodisc controller for the Apple II Plus and IIe, using available authoring systems. An elementary Apple version is available. In addition, an enhanced version for the IBM PC is under development. |

|  |  |
|---|---|
| ORGANIZATION | **AMERICAN VIDEO INSTITUTE/**<br>**VIDEODISC & ELECTRONIC PUBLISHING LAB.** |
| ADDRESS | Rochester Institute of Technology<br>P.O. Box 9887, 1 Lomb Memorial Dr.<br>Rochester, NY 14623 |
| CONTACT | Mr. John Ciampa or Lamont Seckman |
| PHONE | (716) 475-6625/2779 |
| HARDWARE MANUFACTURED | Interfaces |
| INTEGRATED SYSTEMS OFFERED | Information Displays |

|  |  |
|---|---|
| ORGANIZATION | **ANTHRO-DIGITAL, INC.** |
| ADDRESS | P.O. Box 1385<br>Pittsfield, MA 01202 |
| PHONE | (413) 448-8278 |
| HARDWARE MANUFACTURED | Interfaces |

|  |  |
|---|---|
| ORGANIZATION | **BCD ASSOCIATES, INC.** |
| ADDRESS | 5809 S.W. 5th St., Suite 101<br>Oklahoma City, OK 73128 |
| CONTACT | Ms. Diane Howard,<br>President |
| PHONE | (405) 948-1293 |
| TELEX | 499-1435 |
| HARDWARE MANUFACTURED | Interfaces |
| SOFTWARE PRODUCED | Authoring Systems, Interactive Videotape |

|  |  |
|---|---|
| ORGANIZATION | **BELL TELEPHONE LABORATORIES/**<br>**INTERACTIVE VIDEO SYSTEMS RESEARCH DEPARTMENT** |
| ADDRESS | 600 Mountain Ave., Room 3D-474<br>Murray Hill, NJ 07974 |

# Hardware/Software/Integrated Systems

| | |
|---|---|
| CONTACT | Mr. Ronald Gordon |
| PHONE | (201) 582-4099 |
| HARDWARE MANUFACTURED | Overlays/Superimpose Units, Still-frame Audio, Digital Audio, Input Devices |
| SOFTWARE PRODUCED | Drivers/Video Utilities, Authoring Systems, Graphics Editor/Titlers, Disc Simulators |
| INTEGRATED SYSTEMS OFFERED | Training Work Stations/Carrels, Information Displays, Point-of-Purchase Systems, Videogames |
| DESCRIPTION | The Interactive Video Systems Research Department currently has research efforts under way in the areas of advanced videodisc-based information retrieval systems, videodisc authoring and production tools, and knowledge-based expert systems used to aid production tasks as well as information retrieval. Prototype systems are built to explore proposed services and evaluate the suitability of various human-machine interfaces. Application areas include training and education, advertising, entertainment, electronic catalogs, and medical imagery. The purpose of their research is to develop fundamental principles and insights, which may form the basis for new services that integrate information in the form of pictures, graphics, text, sound, and motion video. |

| | |
|---|---|
| ORGANIZATION | **BLUE LAKES COMPUTING** |
| ADDRESS | 3240 University Ave.<br>Madison, WI 53705 |
| PHONE | (608) 233-6502 |
| HARDWARE MANUFACTURED | Interfaces |

| | |
|---|---|
| ORGANIZATION | **BOSTON MEDIA CONSULTANTS** |
| ADDRESS | 19 Damon Rd.<br>Scituate, MA 02066 |
| CONTACT | Mr. David P. Allen |
| PHONE | (617) 545-2696 |
| SOFTWARE PRODUCED | Graphics Editors/Titlers |

Microtex by Cableshare, Inc., is a revolutionary new communication tool. It combines full color computer graphics, interactive videodiscs, and touch sensitive terminals into an easy-to-use automatic salesperson or public information terminal.

| | |
|---|---|
| ORGANIZATION | **CABLESHARE INC.** |
| ADDRESS | P.O. Box 5880, 20 Enterprise Dr.<br>London, Ontario<br>Canada N6A 4L6 |

| | |
|---|---|
| CONTACT | Mr. George S. McCabe |
| | Product Manager, Elec. Marketing Div. |
| PHONE | (519) 686-2900 |
| TELEX | 064-5693 |
| HARDWARE MANUFACTURED | Overlays/Superimpose Units |
| SOFTWARE PRODUCED | Drivers/Video Utilities, Authoring Systems, Graphics Editor/Titlers, Integrated videotex/videodisc data base runs on IBM PC, DEC Micro-II. |
| INTEGRATED SYSTEMS OFFERED | Information Displays, Point-of-Purchase Systems |
| DESCRIPTION | Cableshare produces an authoring system for graphics on their interactive videotex/videodisc system. This system uses touch-sensitive screens to access information—permanent information for videodisc, flexible from videotex graphics. It is currently used by banks and auto dealers. |

---

| | |
|---|---|
| ORGANIZATION | **CAT ZURICH-COMPUTER ASSISTED TELEVIDEO AG** |
| ADDRESS | Freudenbergstrasse 30 |
| | Zurich, Switzerland 8044 |
| PHONE | (01) 251.28.22 |
| SOFTWARE PRODUCED | Drivers/Video Utilities, Authoring Systems, Graphics Editor/Titlers, Disc Simulators |
| INTEGRATED SYSTEMS OFFERED | Training Work Stations/Carrels, Information Displays, Point-of-Purchase Systems |
| DESCRIPTION | CAT Computer Assisted Televideo is an international system house for complete interactive audiovisual information systems. CAT has developed seven different systems, covering most industrial needs. Everything from simple level 1 up to multiplayer systems, combinations with telecommunication, and touch screen monitors, is available. CAT also provides mastering and replication services, in cooperation with European mastering plants, as a general contractor. |

---

| | |
|---|---|
| ORGANIZATION | **CAVRI SYSTEMS, INC.** |
| ADDRESS | 26 Trumbull St. |
| | New Haven, CT 06511 |
| PHONE | (203) 562-4979 |
| HARDWARE MANUFACTURED | Interfaces |

---

| | |
|---|---|
| ORGANIZATION | **CENTRE FOR REMOTE SENSING** |
| ADDRESS | The Blackette Laboratory |
| | Imperial College of Science and Technology |
| | Prince Consort Rd. |
| | London SW7 2BZ |
| CONTACT | Dr. David R. Clark |
| PHONE | 01-589 5111 ext. 2573 |
| HARDWARE MANUFACTURED | Input Devices, Interfaces |
| DESCRIPTION | The Centre for Remote Sensing has a formatting interface to store satellite digital image data as CCIR video, allowing storage of 1 gigabyte, or approximately 130 full-resolution landsat images, per disc side. |

---

| | |
|---|---|
| ORGANIZATION | **CITIDISC PRODUCTIONS** |
| ADDRESS | Suite 209, 1437 Rhode Island Ave. N.W. Washington, DC 20005 |
| CONTACT | Mr. Garri Garripoli |
| PHONE | (202) 483-9369 |

# Hardware/Software/Integrated Systems

|  |  |
|---|---|
| SOFTWARE PRODUCED | Authoring Systems, Graphics Editor/Titlers |
| INTEGRATED SYSTEMS OFFERED | Information Displays, Video Games |
| DESCRIPTION | CITIDISC specializes in developing unique solutions to difficult and challenging videodisc applications. They have many years of film and video production experience that ensures high-quality images and graphics along with advanced, high-quality hardware. |

|  |  |
|---|---|
| ORGANIZATION | **COMMUNICATION STUDIO** |
| ADDRESS | 154 West 27th St. 7E<br>New York, NY 10001 |
| CONTACT | Mr. John Vaughan |
| PHONE | (212) 924-3729 |
| SOFTWARE PRODUCED | Drivers/Video Utilities, Authoring Systems, Graphics Editor/Titlers |
| DESCRIPTION | The Communication Studio is currently developing a universal level 3 authoring system to support several currently available drivers. They will also be offering a unique utility package to support flexible systems implementation. They are currently offering aunique music composition and production capability specifically for random access media. |

|  |  |
|---|---|
| ORGANIZATION | **COMPUTER CENTER/WARNER AMEX QUBE** |
| ADDRESS | 1201 Olentangy River Rd.<br>Columbus, OH 43212 |
| CONTACT | Mr. David Seibold |
| INTEGRATED SYSTEMS OFFERED | Information Displays |

|  |  |
|---|---|
| ORGANIZATION | **COMPUTER HARDWARE, INC.** |
| ADDRESS | P.O. Box 1824<br>Kearney, NE 68847 |
| CONTACT | Mr. Tom Myer |
| PHONE | (308) 234-9335 |
| INTEGRATED SYSTEMS OFFERED | Authoring System-Driven Videodisc Interface with Apple Computer |

|  |  |
|---|---|
| ORGANIZATION | **COMPUTER SCIENCES CORPORATION** |
| ADDRESS | P.O. Box M<br>Ft. Eustis, VA 23604 |
| CONTACT | Dr. William A. Stembler or Mr. J.G. Palmer |
| PHONE | (804) 887-1677 |
| SOFTWARE PRODUCED | Drivers/Video Utilities, Authoring Systems, Graphics Editor/Titlers, Disc Simulators |
| INTEGRATED SYSTEMS OFFERED | Training Work Stations/Carrels |
| DESCRIPTION | Computer Sciences Corporation is currently working on automating the process of converting from paper to videodisc. |

|  |  |
|---|---|
| ORGANIZATION | **COMPUTER VIDEO PRODUCTIONS, INC.** |
| ADDRESS | 1317 Clover Dr. South<br>Minneapolis, MN 55420 |
| CONTACT | Mr. Dean N. Sutliff |
| PHONE | (612) 888-2388 |
| INTEGRATED SYSTEMS OFFERED | Training Work Stations/Carrels, Information Displays, Point-of-Purchase Systems |

| | |
|---|---|
| ORGANIZATION | **CONTROL DATA CORPORATION** |
| ADDRESS | 8100 34th Ave. South |
| | Mailing Address/Box 0 Minneapolis, MN 55440 |
| CONTACT | Mr. Charles W. Loney |
| PHONE | (612) 853-5022 |
| SOFTWARE PRODUCED | Drivers/Video Utilities, Authoring Systems, Graphics Editor/Titlers |
| INTEGRATED SYSTEMS OFFERED | Training Work Stations/Carrels |
| DESCRIPTION | For several years, Control Data has offered an authoring service for videodisc-supported interactive computer-based instruction as part of their PLATO system. They plan to expand their authoring offerings to a broad line of authoring tools for micros, minis, and mainframes. |

| | |
|---|---|
| ORGANIZATION | **COOPERLASER SONICS** |
| ADDRESS | 3420 Central Expressway |
| | Santa Clara, CA 95051 |
| CONTACT | Mr. Joseph J. Batcho |
| PHONE | (408) 720-8844 |
| HARDWARE MANUFACTURED | Argon Ion Lasers for Mastering Discs |

| | |
|---|---|
| ORGANIZATION | **COURSEWARE, INC.** |
| ADDRESS | 10075 Carroll Canyon Rd. |
| | San Diego, CA 92131 |
| CONTACT | Ms. Pomela Flanigan |
| PHONE | (619) 578-1700 |
| SOFTWARE PRODUCED | Drivers/Video Utilities, Authoring Systems, Graphics Editor/Titlers |

| | |
|---|---|
| ORGANIZATION | **CRANBROOK INSTITUTE OF SCIENCE** |
| ADDRESS | 500 Lone Pine Rd. |
| | P.O. Box 801 |
| | Bloomfield Hills, MI 48013 |
| CONTACT | Dr. V. Elliott Smith |
| PHONE | (313) 645-3260 |
| DESCRIPTION | The Cranbrook Institute specializes in interactive microcomputer programs, based on existing videodiscs, for science museum exhibits. They will advise other museums about setting up similar exhibits. They are preparing a new program for the Apple IIe and Pioneer 8210 player called "Introducing Whales." It will be an interactive exhibit based on the "Whales" videodisc of the National Geographic Society. |

| | |
|---|---|
| ORGANIZATION | **CREATIVISION, INC.** |
| ADDRESS | P.O. Drawer 1070 |
| | 456A 11th St. |
| | Holly Hill, FL 32017 |
| CONTACT | Mr. Rick Glasby, Manager Videodisc Projects or Mr. James Williams, Marketing Manager |
| PHONE | (904) 255-0911 |
| INTEGRATED SYSTEMS OFFERED | Training Work Stations/Carrels, Information Displays, Point-of-Purchase Systems |

# Hardware/Software/Integrated Systems

| | |
|---:|:---|
| ORGANIZATION | **CUBIC CORPORATION DEFENSE SYSTEMS DIVISION** |
| ADDRESS | P.O. Box 80787<br>San Diego, CA 92138 |
| CONTACT | Mr. J. R. Wilson |
| PHONE | (619) 277-6780 |
| HARDWARE MANUFACTURED | Overlays/Superimpose Units, Input Devices, Interfaces, Hardware Packaging |
| SOFTWARE PRODUCED | Drivers/Video Utilities, Authoring Systems, Graphics Editor/Titlers |
| INTEGRATED SYSTEMS OFFERED | Training Work Stations/Carrels, Information Displays |

| | |
|---:|:---|
| ORGANIZATION | **DAVID O. MCKAY INSTITUTE** |
| ADDRESS | 215 MCKB<br>Brigham Young University<br>Provo, UT 84602 |
| CONTACT | Mr. Larrie Gale,<br>Coordinator of Research |
| PHONE | (801) 378-7082 |
| HARDWARE MANUFACTURED | Interfaces |
| SOFTWARE PRODUCED | Authoring Systems |
| DESCRIPTION | The David O. McKay Institute is currently involved in developing computer-assisted production software for producing video to be used for interactive video training programs. |

| | |
|---:|:---|
| ORGANIZATION | **DAVIS AUDIO-VISUAL, INC.** |
| ADDRESS | 1801 Federal Blvd.<br>Denver, CO 80204 |
| CONTACT | Mr. Robert Newman III<br>Independent Consultant |
| PHONE | (303) 455-1122 |
| SOFTWARE PRODUCED | Drivers/Video Utilities, Disc Simulators |
| INTEGRATED SYSTEMS OFFERED | Training Work Stations/Carrels, Information Displays, Point-of-Purchase Systems, Videogames |
| DESCRIPTION | Davis Audio-Visual specializes in integration of videotape and videodisc interactive systems including one-on-one, classroom, point-of-purchase, etc. They have been involved with sales and training systems for BDC Associates, Panasonic Interactive Systems, and others. |

| | |
|---:|:---|
| ORGANIZATION | **DEPT. OF NATIONAL DEFENCE** |
| ADDRESS | P.O. Box 2000, Downsview<br>M3M 3B9 Ontario, Canada |
| CONTACT | Mr. Bruce Ferguson |
| HARDWARE MANUFACTURED | Overlays/Superimpose Units, Input Devices, Interfaces |
| SOFTWARE PRODUCED | Simulation software |
| DESCRIPTION | This is a government research lab involved in simulation and military training. They are working on one-off devices for research and development, possibly to be produced by a third party. |

| | |
|---:|:---|
| ORGANIZATION | **DIGITAL CONTROLS INC.** |
| ADDRESS | 5555 Oakbrook Parkway, Suite 200<br>Norcross, GA 30093 |
| PHONE | (800) 441-3332 |
| HARDWARE MANUFACTURED | Interfaces |

At the IVIS (Interactive Video Information System) learning station, multi-media instruction greatly increases student learning speed and retention.

| | |
|---|---|
| ORGANIZATION | **DIGITAL EQUIPMENT CO.** |
| ADDRESS | Main St. Mail Stop ML03-6/E94 |
| | Maynard, MA 01754 |
| CONTACT | Mr. Charles T. Smith |
| | Manager of Optical Memory Products |
| PHONE | (617) 493-4861 |
| HARDWARE MANUFACTURED | Overlays/Superimpose Units, Still-frame Audio, Input Devices, Interfaces |
| SOFTWARE PRODUCED | Authoring Systems, Graphics Editor/Titlers, Disc Simulators |
| INTEGRATED SYSTEMS OFFERED | Training Work Stations/Carrels, Information Displays, Point-of-Purchase Systems |

---

DTI's TOUCHCOM™ systems are used internationally in Customer Service Automation, Structured Training, Database Access, Executive workstation, and other end-user applications.

| | |
|---|---|
| ORGANIZATION | **DIGITAL TECHNIQUES INC. (DTI)** |
| ADDRESS | 209 Middlesex Turnpike |
| | Burlington, MA 01803 |
| CONTACT | Mr. Jonathan Fleischmann |
| | Director of Marketing |
| PHONE | (617) 273-3495 |

# Hardware/Software/Integrated Systems

| | |
|---|---|
| HARDWARE MANUFACTURED | Overlays/Superimpose Units, Interfaces |
| SOFTWARE PRODUCED | Drivers/Video Utilities, Authoring Systems, Graphics Editor/Titlers |
| INTEGRATED SYSTEMS OFFERED | Training Work Stations/Carrels, Information Displays, Point-of-Purchase Systems |
| DESCRIPTION | Digital Techniques is a total systems company, including hardware and software development, and the manufacture of interactive user systems and authoring systems. They offer a total interactive video system that includes a high resolution color monitor, 16-bit microprocessor, touch screen, authoring station, and such peripherals as a hard disk, card reader, and printer. Their complete authoring station has machine-drawn and digitized graphics, and an interactive language and font generator. |

| | |
|---|---|
| ORGANIZATION | **DIGITAL VIDEO CORPORATION** |
| ADDRESS | 369 North Orange Ave.<br>Orlando, FL 32801 |
| CONTACT | Mr. Stephen Matheny |
| PHONE | (305) 425-1999 |
| HARDWARE MANUFACTURED | Input Devices, Interfaces |
| SOFTWARE PRODUCED | Drivers/Video Utilities, Authoring Systems |
| INTEGRATED SYSTEMS OFFERED | Training Work Stations/Carrels, Information Displays, Point-of-Purchase Systems |

| | |
|---|---|
| ORGANIZATION | **DISCOVISION ASSOCIATES** |
| ADDRESS | 3300 Hyland Ave., P.O. Box 6600<br>Costa Mesa, CA 92626 |
| CONTACT | Ms. Bonnie D. Sala,<br>Licensing Representative |
| PHONE | (714) 957-3000 |
| DESCRIPTION | Discovision Associates (DVA) is no longer manufacturing videodiscs, but currently manages a worldwide patent portfolio of over 1,000 patents relating to laser optical disc technology. DVA is interested in encouraging the production efforts of individual companies in the field through the open licensing of their state-of-the-art patents and technology.<br><br>DVA's technology available for license includes laser videodisc substrate manufacturing and mastering process know-how. Additionally, DVA technology is transferable to the manufacture of digital audio discs and DRAW (Direct Read After Write) substrates. Digital audio disc players, digital audio discs, videodisc players, along with consumer and industrial videodiscs, are all included in DVA'S patent coverage.<br><br>Discovision Associates was formed in 1979 as a joint venture partnership between MCA and IBM. Known as MCA Laboratories and then later MCA Disco Vision, the company has been involved in research and development of laser optical technology since the late 1960s, developing not only disc manufacturing but also videodisc players such as the Disco Vision PR-7820 (now known as the Pioneer 7820), the mainstay of industrial videodisc players. |

| | |
|---:|:---|
| ORGANIZATION | **EDITEL-NEW YORK** |
| ADDRESS | 222 East 44th St. |
| | New York, NY 10017 |
| CONTACT | Mr. Frank Dobyns |
| PHONE | (212) 867-4600 |
| SOFTWARE PRODUCED | Authoring Systems, Graphics Editors/Titlers, Disc Simulators |
| INTEGRATED SYSTEMS OFFERED | Point-of-Purchase Systems, Video Games |

| | |
|---:|:---|
| ORGANIZATION | **EDUCATIONAL COMPUTER CORPORATION** |
| ADDRESS | 175 Strafford Ave. |
| | Strafford, PA 19087 |
| CONTACT | Mr. Nick Siecko |
| PHONE | (215) 687-2600 |
| HARDWARE MANUFACTURED | Interfaces |
| SOFTWARE PRODUCED | Operating System Internal-to-Maintenance Simulator to Operate Disc Player |
| INTEGRATED SYSTEMS OFFERED | Training Work Stations/Carrels |
| DESCRIPTION | Educational Computer Corporation is working on maintenance training simulators, using videodisc for the USMC LVT-7A1. Once delivered, the simulators will be USMC property. They are also developing programs using the MicroTICCIT system. |

| | |
|---:|:---|
| ORGANIZATION | **EECO INCORPORATED** |
| ADDRESS | 1601 East Chestnut Ave. |
| | Santa Ana, CA 92702 |
| CONTACT | Mr. George Treneer, |
| | Marketing Manager |
| PHONE | (714) 835-6000 |
| TOLL FREE | (800) 854-3808 |
| HARDWARE MANUFACTURED | Still-frame Audio, Premaster Videotape Editing System, including Still-frame Audio Capability |

| | |
|---:|:---|
| ORGANIZATION | **ENERGY, MINES & RESOURCES** |
| ADDRESS | 580 Booth St., Altona |
| | Ontario, Canada K1AO34 |
| CONTACT | P. Cheffins |
| PHONE | (613) 995-3065 |
| HARDWARE MANUFACTURED | Videodisc Players |
| SOFTWARE PRODUCED | Drivers/Video Utilities, Graphics Editor/Titlers |
| INTEGRATED SYSTEMS OFFERED | Information Displays, Video Games |
| DESCRIPTION | This agency is responsible for informing and educating the public about energy issues. An interactive computerized system is being used as an informational/educational tool in a large exhibit program, which reaches over one million viewers per year. |

| | |
|---:|:---|
| ORGANIZATION | **EXHIBIT TECHNOLOGY, INC.** |
| ADDRESS | 200 Park Ave., Suite 2555 |
| | New York, NY 10166 |
| CONTACT | Mr. Don Cochrane |
| | President, Marketing |

# Hardware/Software/Integrated Systems

|     |     |
| --- | --- |
| PHONE | (212) 692-9211 |
| HARDWARE MANUFACTURED | Interfaces |
| SOFTWARE PRODUCED | Drivers/Video Utilities, Authoring Systems |
| INTEGRATED SYSTEMS OFFERED | Training Work Stations/Carrels, Information Displays, Point-of-Purchase Systems, Videogames |

|     |     |
| --- | --- |
| ORGANIZATION | **FRANKLIN RESEARCH CENTER** |
| ADDRESS | 20th & Race Sts. <br> Philadelphia, PA 19103 |
| CONTACT | Mr. Ken Fertner |
| PHONE | (215) 448-1340 |
| SOFTWARE PRODUCED | Drivers/Video Utilities, Authoring Systems |
| INTEGRATED SYSTEMS OFFERED | Training Work Stations/Carrels |

|     |     |
| --- | --- |
| ORGANIZATION | **FULL CIRCLE COMMUNICATIONS, INC.** |
| ADDRESS | 13530 Michigan Ave., St. 226 <br> Dearborn, MI 48126 |
| CONTACT | Mr. Stan Williams |
| PHONE | (313) 584-3200 |
| SOFTWARE PRODUCED | Drivers/Video Utilities, Authoring Systems |
| INTEGRATED SYSTEMS OFFERED | Training Work Stations/Carrels, Information Displays, Point-of-Purchase Systems |
| DESCRIPTION | All software or integrated system work done by Full Circle Communications is customized. They offer no "off-the-shelf" products. They are currently providing the systems and software for projects with the Ford Motor Co. and the Cadillac Motor Division of General Motors. |

GenTech combines the benefits of computerized interactive television (CIT) with all the advantages of videodisc. GenTech's Laser Disc Interface, LDI-2100, extends the features of interactive television to include instant access, instant response and high quality still-action video. The LDI-2100 complements GenTech's full line of computerized hardware and software for interactive television and live-instructor presentations.

|     |     |
| --- | --- |
| ORGANIZATION | **GENERAL TECHNICAL CORPORATION (GENTECH)** |
| ADDRESS | 4101 N. St. Joseph Ave. <br> Evansville, IN 47712 |
| CONTACT | Mr. Bill Noelting |
| PHONE | (812) 423-4200 |
| HARDWARE MANUFACTURED | Interfaces |
| SOFTWARE PRODUCED | Authoring Systems |
| INTEGRATED SYSTEMS OFFERED | Training Work Stations/Carrels |
| DESCRIPTION | GenTech produces electronic design, manufacturing, and computer programming for computer-video products. A current research project involves the development of an RS 232 universal interface for computers to videomixers/switchers for videotape or videodisc. GenTech also manufactures a computerized interactive group response system for use with tape or disc. |

The GenTech Interactive Learning Center is a custom-designed instructional delivery system for computerized interactive television (CIT). Each student receives personalized, individual instruction while progressing at his/her own pace. The free-standing unit is a durable cabinet with wood-grain finish and features a large storage area for student materials, door locks, master on/off switch, 4-outlet power strip, two 5 1/2-inch speakers with 1.5 watts per channel amplifier, and rolling casters. The system components include an Apple II Plus (tm) microcomputer and disk drive, Panasonic 12-inch color monitor, videotape or videodisc player, and GenTech CIT interface package.

| | |
|---|---|
| ORGANIZATION | **GREGG, D. P.** |
| ADDRESS | 3650 Helms Ave.<br>Culver City, CA 90230 |
| CONTACT | D.P. Gregg, Engineer |
| PHONE | (213) 202-7390 |
| DESCRIPTION | Mr. Gregg specializes in the design, construction, operation, and improvement of disc-replicating plants (Laser, CED). |

| | |
|---|---|
| ORGANIZATION | **GRUMMAN AEROSPACE CORPORATION** |
| ADDRESS | Mail Stop C01-47<br>Bethpage, NY 11714 |
| CONTACT | Mr. Al Cella |
| PHONE | (516) 435-6805 |
| SOFTWARE PRODUCED | Authoring Systems |
| INTEGRATED SYSTEMS OFFERED | Training Work Stations/Carrels |

| | |
|---|---|
| ORGANIZATION | **HAZELTINE CORPORATION/<br>TRAINING SYSTEMS CENTER** |
| ADDRESS | 7680 Old Springhouse Rd.<br>McLean, VA 22102 |
| CONTACT | Ms. L. Zingg |
| PHONE | (703) 827-2300 |
| HARDWARE MANUFACTURED | Overlays/Superimpose Units, Interfaces |
| SOFTWARE PRODUCED | Authoring Systems |
| INTEGRATED SYSTEMS OFFERED | Training Work Stations/Carrels |
| DESCRIPTION | Hazeltine's next generation of the TICCIT training system, MicroTICCIT, includes all of the comprehensive CBT features of its field-proven predecessor and adds a variety of new important features. |

Sophisticated capabilities have been included to integrate videodisc-based materials with computer-based training.

The MicroTICCIT workstation, based on the IBM PC, provides combined full color displays of videodisc overlaid by computer-generated text and graphics.

The introduction of MicroTICCIT includes the introduction of ADAPT, a new standard for user-friendly yet comprehensive interactive authoring systems. ADAPT includes capabilities for both beginning and experienced authors to develop effective training materials that combine simultaneous displays of video and computer-generated screens, as well as comprehensive capabilities to control video and computer branching for truly interactive presentations.

| | |
|---|---|
| ORGANIZATION | **HITACHI SALES CORPORATION OF AMERICA** |
| ADDRESS | 1290 Wall St. |
| | West Lyndhurst, N.J. 07071 |
| CONTACT | Dave M. Fukuda |
| | Manager OEM Business Division |
| PHONE | (201) 935-8980 |
| HARDWARE MANUFACTURED | Videodisc Players |

| | |
|---|---|
| ORGANIZATION | **HUMAN RESOURCES RESEARCH ORGANIZATION (HUMRRO)** |
| ADDRESS | 1100 S. Washington St. |
| | Alexandria, VA 22314 |
| CONTACT | Mr. William Underhill |
| PHONE | (703) 549-3611 |
| SOFTWARE PRODUCED | Authoring Systems |

| | |
|---|---|
| ORGANIZATION | **IBM** |
| ADDRESS | 1801 S. Federal Hwy. |
| | Del Rey Beach, FL 33444 |
| CONTACT | Mr. Paul Frauenhoffer |
| PHONE | (305) 241-7632 |
| HARDWARE MANUFACTURED | Overlays/Superimpose Units, Interfaces |
| SOFTWARE PRODUCED | Drivers/Video Utilities, Authoring Systems |
| INTEGRATED SYSTEMS OFFERED | Training Work Stations/Carrels, Point-of-Purchase Systems |

| | |
|---|---|
| ORGANIZATION | **ICS-INTEXT, A DIVISION OF NATIONAL EDUCATION CORPORATION** |
| ADDRESS | 315 Post Rd., West |
| | Westport, CT 06880 |
| CONTACT | Mr. Colin Harris, National Marketing Manager |
| PHONE | (203) 227-0891 |
| TOLLFREE | (800) 233-0259 |
| HARDWARE MANUFACTURED | ActionCode System |
| SOFTWARE PRODUCED | Authoring Systems, Special Microprocessor Firmware and Barcode Technology |
| INTEGRATED SYSTEMS OFFERED | Training Work Stations/Carrels, Information Displays, Point-of-Purchase Systems |

| | |
|---|---|
| DESCRIPTION | ActionCode is a highly interactive transfer-of-knowlege/information system, well-suited to a variety of applications. It consists of some off-the shelf components, and some unique components, provided for ICS-Intext by Perceptronics, Inc. The ActionCode system comes complete with existing courseware and/or authoring specifications for companies wanting to develop their own proprietary discs. |

| | |
|---|---|
| ORGANIZATION | **INNOTECH** |
| ADDRESS | 18552 MacArthur Blvd., Suite 475 |
| | Irvine, CA 92715 |
| CONTACT | Mr. Stuart Krasney |
| PHONE | (714) 476-2051 |
| SOFTWARE PRODUCED | Disc Simulators |
| INTEGRATED SYSTEMS OFFERED | Point-of-Purchase Systems |

| | |
|---|---|
| ORGANIZATION | **INTEGRATED AUTOMATION** |
| ADDRESS | 2121 Allston Way |
| | Berkeley, CA 94704 |
| CONTACT | Mr. Bert King |
| PHONE | (415) 843-8227 |
| HARDWARE MANUFACTURED | Input Devices, Interfaces Resolution Scanners |
| INTEGRATED SYSTEMS OFFERED | Information Displays |
| DESCRIPTION | Integrated Automation is also working on the following products: digital video; high resolution (>2000 lines) scanners; CRT displays; laser printers; read/write digital optical disk drives for document storage applications; optical disc "jukebox" storing 100 discs for random access storage. |

| | |
|---|---|
| ORGANIZATION | **INTERACTIVE ARTS INTERNATIONAL** |
| ADDRESS | 25785 Bassett Lane |
| | Los Altos Hills, CA 94022 |
| CONTACT | Mr. Allen Lee Adkins |
| PHONE | (415) 941-9686 |
| INTEGRATED SYSTEMS OFFERED | Video Games |
| DESCRIPTION | This company was founded by Allen Lee Adkins to produce advanced videodisc based games, entertainment and educational products, using the interactive video/computer medium. Emphasis is placed on supporting creative artists in the interactive video field. Special projects may be funded by a nonprofit division, or associated government and other nonprofit agencies who promote program development in the areas of education, science, industry, and the arts. |

| | |
|---|---|
| ORGANIZATION | **INTERACTIVE RESEARCH CORPORATION** |
| ADDRESS | 3080 Olcott St., Suite 200B |
| | Santa Clara, CA 95051 |
| CONTACT | Ms. Sandie Paddock |
| PHONE | (415) 986-1420 |
| HARDWARE MANUFACTURED | Digital Audio, Interfaces, Control Unit, Video/Graphics Switch |
| SOFTWARE PRODUCED | Drivers/Video Utilities, Authoring Systems, Graphics Editor/Titlers, Disc Simulators |

**Hardware/Software/Integrated Systems**

|   |   |
|---|---|
| DESCRIPTION | IRC produces a control unit that includes digital voice synthesis for still-frame audio, a video/graphics switch which allows switching from video out to computer-graphics under external computer control for level 3 systems. This unit also contains up to 128k RAM for voice data storage. |
|   | IRC has also developed an authoring system, which is proprietary at present. Plans are under way to make this available on a licensing basis for companies who need a complete authoring system along with complete videodisc project management. |

|   |   |
|---|---|
| ORGANIZATION | **INTERACTIVE TELEVISION COMPANY** |
| ADDRESS | 1901 N. Moore St. Arlington, VA 22209 |
| CONTACT | Dr. Steven L. Levin |
| PHONE | (703) 525-5625 |
| HARDWARE MANUFACTURED | Input Devices, Interfaces, Joysticks |
| SOFTWARE PRODUCED | Drivers/Video Utilities, Authoring Systems |
| INTEGRATED SYSTEMS OFFERED | Training Work Stations/Carrels, Information Displays |
| DESCRIPTION | ITC manufactures a special eight-button joystick (3 degrees of freedom) compatible with the Apple II and several other computer systems. For the Sony SMC-70, ITC manufactures a compatible joystick interface expansion unit. ITC also designs and manufactures a series of secure component workstations for environments requiring EMI/RFI shielding. These workstations are available with several options including microcomputers, videodisc systems, graphic processors, and a variety of hard copy units. |

|   |   |
|---|---|
| ORGANIZATION | **INTERACTIVE TRAINING SYSTEMS, INC.** |
| ADDRESS | 4 Cambridge Center<br>Cambridge, MA 02142 |
| CONTACT | Mr. David Snow<br>Sales Manager |
| PHONE | (617) 497-6100 |
| HARDWARE MANUFACTURED | Overlays/Superimpose Units, Interfaces |
| SOFTWARE PRODUCED | Authoring Systems |
| INTEGRATED SYSTEMS OFFERED | Training Work Stations/Carrels, Information Displays, Point-of-Purchase Systems |

|   |   |
|---|---|
| ORGANIZATION | **INTERACTIVE VIDEO CONCEPTS, INC.** |
| ADDRESS | Suite 105, Wilford Building<br>101 North 33rd St.<br>Philadelphia, PA 19104 |
| CONTACT | Ms. Peg Callahan<br>Vice President |
| PHONE | (215) 387-0707 |
| SOFTWARE PRODUCED | Authoring Systems, Software for Level 2 Programming (VELOS) |

|   |   |
|---|---|
| ORGANIZATION | **INTERMEDIA** |
| ADDRESS | 1600 Dexter Ave. North<br>Seattle, WA 98109 |
| CONTACT | Ms. Susan Hoffman or Mr. Stan Barnes |
| PHONE | (206) 282-7262 |

|     |     |
| --- | --- |
| HARDWARE MANUFACTURED | Input Devices, Interfaces |
| INTEGRATED SYSTEMS OFFERED | Information Displays, Point-of-Purchase Systems |

|     |     |
| --- | --- |
| ORGANIZATION | **IT, INC.** |
| ADDRESS | 2009 Pacheco St.<br>Sante Fe, NM 87501 |
| CONTACT | Mr. Gary Quinlin, President or<br>Ms. Constance Buffalo, V.P. Production |
| TOLL FREE | (800) 328-7937 |
| INTEGRATED SYSTEMS OFFERED | Training Work Stations/Carrels, Point-of-Purchase Systems |

|     |     |
| --- | --- |
| ORGANIZATION | **ITM SYSTEMS** |
| ADDRESS | P.O. Box 774<br>Sunset Beach, CA 90742 |
| CONTACT | Mr. Kurt Zimmerman, President |
| PHONE | (213) 592-3044 |
| INTEGRATED SYSTEMS OFFERED | Training Work Stations/Carrels, Video Games |
| DESCRIPTION | ITM Systems has developed videotape and videodisc interactive instructional systems such as medical (emergency room) procedures, computer-assisted instructional programs, and general system/applications software. |

|     |     |
| --- | --- |
| ORGANIZATION | **JACK LIEB PRODUCTIONS, INC.** |
| ADDRESS | 200 E. Ontario St.<br>Chicago, IL 60611 |
| CONTACT | Mr. Warren Lieb, Charles R. Kite, or Sharon Spence |
| PHONE | (312) 943-1440 |
| INTEGRATED SYSTEMS OFFERED | Information Displays, Point-of-Purchase Systems |
| DESCRIPTION | Jack Lieb Productions custom designs interactive audiovisual programs to fit the specialized needs of its clients. |

|     |     |
| --- | --- |
| ORGANIZATION | **JAM, INC.** |
| ADDRESS | 42 East Ave.<br>Rochester, NY 14604 |
| CONTACT | Mr. Michael Yampolsky<br>Vice President, Research & Development |
| PHONE | (716) 232-5500 |
| HARDWARE MANUFACTURED | Interfaces |
| SOFTWARE PRODUCED | Authoring Systems |
| INTEGRATED SYSTEMS OFFERED | Training Work Stations/Carrels |
| DESCRIPTION | The JAM videodisc interface card is a single board, which interfaces a microcomputer to an industrial laser disc player. A single card will support every laser disc player on the market currently (including all Pioneer industrial models, Sony, Hitachi, and Phillips). The card also contains a soft video switch for switching between computer and video output on a single screen. The JAM card is available for both the Apple II or IIe and the IBM PC or XT.<br><br>DiscWriter is a course authoring system for level 3 videodisc program generation. It provides a user interface for control and authoring of interactive videodisc programs, and allows educators and course |

planners to customize existing programs to meet their specific requirements. DiscWriter is a menu-driven program, designed for use by nonprogrammers. When attached to a videodisc player, DiscWriter will enter frame numbers automatically without being keyed in, and will allow instantaneous confirmation of video/audio sequences.

| | |
|---|---|
| ORGANIZATION | **LIGHT TRACKS** |
| ADDRESS | 1242 10th St., Suite 10<br>Santa Monica, CA 90401 |
| CONTACT | Joey Silvian |
| PHONE | (213) 395-6962 |
| DESCRIPTION | Light Tracks produces computer graphic overlay videodisc games for the arcade and home environments, software for point-of-purchase displays, and training programs for corporate and school use. Light Tracks also offers consulting services in all of these areas. |

| | |
|---|---|
| ORGANIZATION | **LOGIVISION** |
| ADDRESS | 46, Rue du Docteur Charcot<br>92000 Nanterre, France |
| CONTACT | Mr. Alain Dupuy or Mr. Christian Durix |
| HARDWARE MANUFACTURED | Interfaces |
| INTEGRATED SYSTEMS OFFERED | Training Work Stations/Carrels, Information Displays, Point-of-Purchase Systems, Videogames |
| DESCRIPTION | LOGIVISION's field of activity is twofold: videodisc system engineering, including design of computer software and hardware, interfaces, electronics; and related services, including videodisc production and updating, maintenance, software or hardware extensions. |

| | |
|---|---|
| ORGANIZATION | **MARK IV DESIGNS** |
| ADDRESS | 2315 So. Canterbury Lane<br>Lincoln, NE 68512 |
| CONTACT | Mr. Mark Hansen |
| PHONE | (402) 423-0363 |
| HARDWARE MANUFACTURED | Interfaces, Custom Design of Digital Hardware for Interfacing to Microcomputers |
| SOFTWARE PRODUCED | Drivers/Video Utilities, Authoring Systems |
| DESCRIPTION | Mark IV Designs specializes in custom hardware and/or software for videodisc development projects. Mark Hansen is an electrical engineer with ten years experience in microprocessor hardware design and software development. He is experienced in total system development from concept to production release. |

| | |
|---|---|
| ORGANIZATION | **MARTIN MARIETTA DATA SYSTEMS/<br>TECHNOLOGY SYSTEMS GROUP** |
| ADDRESS | 106 Inverness Circle East<br>Englewood, CO 80112 |
| CONTACT | Dr. Darrell F. Humphrey |
| PHONE | (303) 790-3339 |
| SOFTWARE PRODUCED | Drivers/Video Utilities, Videodisc Image and Document Retrieval System Software (IBM PC/XT with full text keyword retrieval) |

| | |
|---:|:---|
| ORGANIZATION | **MATRIX INFORMATION SYSTEMS** |
| ADDRESS | 11728 Avon Way |
| | Los Angeles, CA 90066 |
| CONTACT | Mr. Michael North, Director |
| PHONE | (213) 391-0243 |
| SOFTWARE PRODUCED | Drivers/Video Utilities, Authoring Systems |
| INTEGRATED SYSTEMS OFFERED | Training Work Stations/Carrels |

| | |
|---:|:---|
| ORGANIZATION | **MATROX ELECTRONIC SYSTEMS LTD.** |
| ADDRESS | 5800 Andover Way |
| | Town of Mount Royal, Quebec |
| | Canada H4T 1H4 |
| CONTACT | Mr. Chris Morin |
| PHONE | (514) 735-1182 |
| HARDWARE MANUFACTURED | Overlays/Superimpose Units, Still-frame Audio, Digital Audio, Interfaces |
| INTEGRATED SYSTEMS OFFERED | Training Work Stations/Carrels |
| DESCRIPTION | Matrox Electronic Systems Ltd. announces the development of a Multibus board set incorporating revolutionary new features for advanced applications, using standard NTSC laser videodiscs. These new features greatly enhance the capabilities and increase the range of application for videodisc technology. The board set provides three unique modes of system operation 1) retrieval of a very large amount of digital information, providing different error correction capacities; 2) "sound-over-still" operation, allowing audio to be played while displaying a single television frame, with audio output right on the board; and 3) color computer graphics and color alphanumerics overlay on top of a videopicture. |

| | |
|---:|:---|
| ORGANIZATION | **MATSUSHITA TECHNOLOGY CENTER** |
| ADDRESS | One Panasonic Way |
| | Secaucus, NJ 07094 |
| CONTACT | Mr. Anthony Jasionowski |
| PHONE | (201) 348-7777 |
| HARDWARE MANUFACTURED | Videodisc Players and Recorders, Still-frame Audio, Digital Audio |
| DESCRIPTION | Products, including the optical memory disc recorder, are marketed under the Panasonic brand name. |

| | |
|---:|:---|
| ORGANIZATION | **MAZ & MOVIE** |
| ADDRESS | Betriebsgesellschaft mbH |
| | Weidenallee 372000 Hamburg 6 |
| | Germany |
| CONTACT | Mr. Jurgen Becker |
| PHONE | 040 /43 80 43 /44 |
| SOFTWARE PRODUCED | Disc Simulators |
| INTEGRATED SYSTEMS OFFERED | Training Work Stations/Carrels, Point-of-Purchase Systems |

| | |
|---:|:---|
| ORGANIZATION | **MEDICAL COLLEGE OF GEORGIA/ RESEARCH & EDUCATIONAL SUPPORT** |
| ADDRESS | Division of Systems & Computer Services |
| | Augusta, GA 30912 |
| CONTACT | Dr. Richard E. Pogue |

**Hardware/Software/Integrated Systems**  197

|  |  |
|---|---|
| PHONE | (404) 828-3786 |
| SOFTWARE PRODUCED | Authoring Systems, Graphics Editor/Titlers |

|  |  |
|---|---|
| ORGANIZATION | **MICHAEL J. PETRO LTD.** |
| ADDRESS | 181 Shepherd St. East<br>Windsor, Ontario<br>Canada N8X 2K4 |
| CONTACT | Mr. Michael J. Petro |
| PHONE | (519) 258-9008 |
| SOFTWARE PRODUCED | Drivers/Video Utilities, Authoring Systems, Disc Simulators |
| INTEGRATED SYSTEMS OFFERED | Training Work Stations/Carrels, Information Displays, Point-of-Purchase Systems |
| DESCRIPTION | Michael J. Petro's interface system for the Apple II includes an authoring program, a disc simulator, an interactor program, a disc memory reader, a disc programmer, and a demonstration program. The single screen output is selectable from disc or computer output via soft switching. |

Digital Optical Memory
Disc Recorder

Panasonic Optical
Memory Disc Recorder

|  |  |
|---|---|
| ORGANIZATION | **MULTI IMAGE INTERNATIONAL/<br>A DIVISION OF CHENAULT/** |
| ADDRESS | 605 Third Ave.<br>New York, NY 10158 |
| CONTACT | Dr. David Buckley, President or Jack Berry, Vice President |
| PHONE | (212) 557-0598 |
| TOLLFREE | (800) 824-7888 ext. A3062 |
| HARDWARE MANUFACTURED | Analog and Digital Videodisc Recorders |
| SOFTWARE PRODUCED | Peripheral Integration Software for Specific Applications |
| INTEGRATED SYSTEMS OFFERED | Visual Information Management Systems |
| DESCRIPTION | Multi Image International delivers solutions for the management of visual information: optical memory disc recorders, video and microcomputer equipment, as well as consulting services; including customized system planning, operational techniques, and network strategies. |

| | |
|---:|:---|
| ORGANIZATION | **NAVAL PERSONNEL RESEARCH & DEVELOPMENT CENTER (NPRDC)** |
| ADDRESS | San Diego, CA 92152 |
| CONTACT | Mr. George Lahey |
| PHONE | (619) 225-7122 |
| SOFTWARE PRODUCED | Confidential |
| INTEGRATED SYSTEMS OFFERED | Confidential |
| DESCRIPTION | Work done at NPRDC is for government use only. |

| | |
|---:|:---|
| ORGANIZATION | **NEW MEDIA GRAPHICS CORPORATION** |
| ADDRESS | 279 Cambridge St. #5<br>Burlington, MA 01803 |
| CONTACT | Mr. Arthur Franke |
| PHONE | (617) 272-8844 |
| HARDWARE MANUFACTURED | Overlays/Superimpose Units, Input Devices, Interfaces |
| SOFTWARE PRODUCED | Drivers/Video Utilities, Authoring Systems, Graphics Editor/Titlers, Disc Simulators |
| INTEGRATED SYSTEMS OFFERED | Training Work Stations/Carrels, Information Displays, Point-of-Purchase Systems, Videogames |

| | |
|---:|:---|
| ORGANIZATION | **NORTH AMERICAN PHILIPS CORPORATION** |
| ADDRESS | 100 E. 42nd St.<br>New York, NY 10017 |
| CONTACT | Mr. Robert Moes, Director of Marketing or Mr. John Messerschmitt, Vice President |
| PHONE | (212) 850-5011 |
| HARDWARE MANUFACTURED | Videodisc Players, Still-frame Audio, Digital Audio, Input Devices |
| SOFTWARE PRODUCED | Sound-over-Still Videodisc developed by Joshua Levine, Philips Laboratories, Briarcliff, NY |

| | |
|---:|:---|
| ORGANIZATION | **NUVATEC, INC.** |
| ADDRESS | 261 Eisenhower Lane, South<br>Lambard, IL 60148 |
| PHONE | (312) 620-4830 |
| HARDWARE MANUFACTURED | Interfaces |

| | |
|---:|:---|
| ORGANIZATION | **OMNICOM ASSOCIATES** |
| ADDRESS | 908 Steam Mill Rd.<br>Ithaca, NY 14850 |
| CONTACT | Mr. David Williams or Ms. Diane Gayeski |
| PHONE | (607) 277-0405 |
| SOFTWARE PRODUCED | Authoring Systems, Disc Simulators, Interactive Instructional Software/Courseware |
| INTEGRATED SYSTEMS OFFERED | Training Work Stations/Carrels, Information Displays, Point-of-Purchase Systems |

| | |
|---:|:---|
| ORGANIZATION | **ONLINE COMPUTER SERVICES** |
| ADDRESS | 20010 Century Blvd.<br>Germantown, MD 20767 |
| PHONE | (301) 429-3700 |
| HARDWARE MANUFACTURED | Interfaces |

Hardware/Software/Integrated Systems

| | |
|---|---|
| ORGANIZATION | **OPTICAL RECORDING PROJECT/3M** |
| ADDRESS | 223-5S-01 3M Center |
| | St. Paul, MN 55144 |
| CONTACT | Mr. Frank Price, Sales and Marketing Manager |
| PHONE | (612) 733-3906 or |
| | Ms. Evie Homme (612) 733-4435 |
| TOLL FREE | (800) 328-1300 ext. 3-2142 |
| | East Coast Office |
| ADDRESS | 136 Fairview Ave. |
| | Fairfield, CT 06430 |
| CONTACT | Mr. Mike King |
| PHONE | (203) 255-7978 |
| HARDWARE MANUFACTURED | 3M Cue Inserter |

| | |
|---|---|
| ORGANIZATION | **ORENDA, INCORPORATED** |
| ADDRESS | 3754 Overland Ave. |
| | Los Angeles, CA 90034 |
| CONTACT | Mr. Andrew H. Mishkin |
| | Vice President |
| PHONE | (213) 202-0770 |
| HARDWARE MANUFACTURED | Overlays/Superimpose Units, Input Devices, Interfaces |
| SOFTWARE PRODUCED | Drivers/Video Utilities |
| INTEGRATED SYSTEMS OFFERED | Training Work Stations/Carrels, Information Displays, Point-of-Purchase Systems, Videogames |
| DESCRIPTION | Orenda, Inc. products are produced on a custom basis and exist within the products of other companies. They specialize in videodisc-based game development. |

| | |
|---|---|
| ORGANIZATION | **OWL MICRO-COMMUNICATIONS** |
| ADDRESS | The Maltings, Station Rd. |
| | Sabridgeworth, Herts CM21 9CY |
| | United Kingdom |
| HARDWARE MANUFACTURED | Interfaces |

| | |
|---|---|
| ORGANIZATION | **PERCEPTRONICS, INC.** |
| ADDRESS | 6271 Variel Ave. |
| | Woodland Hills, CA 91367 |
| CONTACT | Dr. Gershow Weltman |
| PHONE | (818) 884-7470 |
| SOFTWARE PRODUCED | Authoring Systems |
| INTEGRATED SYSTEMS OFFERED | Training Work Stations/Carrels, Information Displays, Video games |

| | |
|---|---|
| ORGANIZATION | **PERCEPTRONICS, INC.** |
| ADDRESS | 1911 N. Fort Meyer Dr., Suite 308 |
| | Arlington, VA 22209 |
| CONTACT | Mr. Steven Johnston |
| PHONE | (703) 525-0184 |
| HARDWARE MANUFACTURED | Overlays/Superimpose Units, Input Devices |
| SOFTWARE PRODUCED | Drivers/Video Utilities, Authoring Systems, Graphics Editor/Titlers |
| INTEGRATED SYSTEMS OFFERED | Training Work Stations/Carrels, Information Displays, Point-of-Purchase Systems, Videogames |

| | |
|---|---|
| ORGANIZATION | **PERFORMANCE TECHNOLOGY, INC.** |
| ADDRESS | Suite 21, 11250 Roger Bacon Dr. Reston, VA 22090 |
| CONTACT | Ms. Vivian L. Hegg, Courseware Developer |
| PHONE | (703) 437-1122 |
| HARDWARE MANUFACTURED | Input Devices |
| DESCRIPTION | As agents for FingerTouch, Performance Technology provides systems that allow the user to interact directly with screen images. With these systems they offer applications software that permits nonprogrammers to prepare and access displays quickly and easily. Their staff will custom design materials, or provide training and support for user-developed materials. |

| | |
|---|---|
| ORGANIZATION | **PERGAMON INTERNATIONAL INFORMATION CORPORATION** |
| ADDRESS | 1340 Old Chain Bridge Rd. McLean, VA 22101 |
| PHONE | (703) 442-0900 |
| HARDWARE MANUFACTURED | Interfaces |

| | |
|---|---|
| ORGANIZATION | **PHILIPS INTERNATIONAL/AUDIO DIVISION/LASERVISION DEPARTMENT** |
| ADDRESS | Building SFF-1 Eindhoven, Holland |
| HARDWARE MANUFACTURED | Videodisc Players, Overlays/Superimpose Units, Interfaces |
| SOFTWARE PRODUCED | Drivers/Video Utilities, Authoring Systems, Disc Simulators |
| INTEGRATED SYSTEMS OFFERED | Training Work Stations/Carrels, Information Displays, Point-of-Purchase Systems, Videogames |

| | |
|---|---|
| ORGANIZATION | **PIONEER VIDEO, INC.** |
| ADDRESS | Corporate Headquarters 200 West Grand Ave. Montvale, NJ 07645 |
| PHONE | (201) 573-1122 |
| ADDRESS | Industrial Sales Division 5150 East Pacific Coast Hwy., #300 Long Beach, CA 90804 |
| CONTACT | Mr. Ron Butler, Vice President, Industrial Sales |
| PHONE | (213) 498-0300 |
| HARDWARE MANUFACTURED | Videodisc Players, Still-frame Audio, Input Devices, Interfaces |
| SOFTWARE PRODUCED | Authoring Systems, Disc Simulators |

| | |
|---|---|
| ORGANIZATION | **POSITRON** |
| ADDRESS | 30 Lincoln Plaza, Suite 35 New York, NY 10023 |
| PHONE | (212) 586-1666 |
| HARDWARE MANUFACTURED | Interfaces |

### Hardware/Software/Integrated Systems

|  |  |
|--|--|
| ORGANIZATION | **PRIMARIUS** |
| ADDRESS | 4186-C Sorrento Valley Blvd. |
|  | San Diego, CA 92121 |
| CONTACT | Mr. Dan Brancaccio |
| PHONE | (619) 455-6841 |
| HARDWARE MANUFACTURED | Still-frame Audio, Interfaces |
| SOFTWARE PRODUCED | Drivers/Video Utilities, Authoring Systems, Graphics Editor/Titlers |
| INTEGRATED SYSTEMS OFFERED | Training Work Stations/Carrels, Information Displays, Medical Education, Patient Education |

---

|  |  |
|--|--|
| ORGANIZATION | **PROPERTY TECHNOLOGY, INC.** |
| ADDRESS | 283 Dartmouth St. |
|  | Boston, MA 02116 |
| CONTACT | Mr. Walter R. Crosby, |
|  | Vice President of Research & Development |
| PHONE | (617) 266-2115 |
| SOFTWARE PRODUCED | Control Software for Specific Applications |
| INTEGRATED SYSTEMS OFFERED | Training Work Stations/Carrels, Information Displays |

---

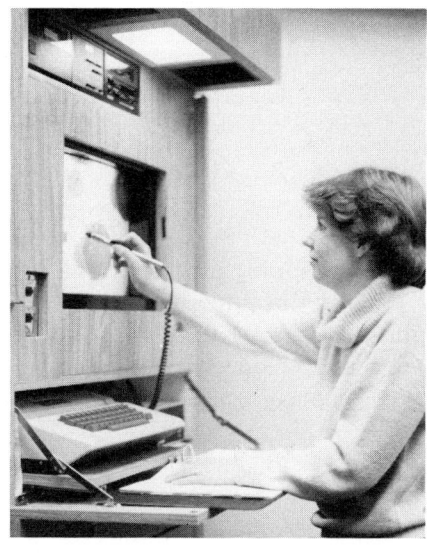

Raytheon's RAYTRAIN™ interactive videodisc training system.

|  |  |
|--|--|
| ORGANIZATION | **RAYTHEON SERVICE COMPANY** |
| ADDRESS | 2 Wayside Rd. |
|  | Burlington, MA 01803 |
| CONTACT | Mr. Ed Dextraze or Mr. J.H. Zickel |
| PHONE | (617) 272-9300 |
| HARDWARE MANUFACTURED | Overlays/Superimpose Units, Other |
| SOFTWARE PRODUCED | Authoring Systems |
| INTEGRATED SYSTEMS OFFERED | Training Work Stations/Carrels |

| | |
|---:|:---|
| DESCRIPTION | The RAYTRAIN Interactive Videodisc System is a computer-aided videodisc training system. The system will allow technical training, or human resources development personnel to develop interactive training programs that can be used in business, education, or industrial applications. It provides sound-on-still frame capability, off-disc audio, as well as computer-generated graphics and text overlays. |

---

| | |
|---:|:---|
| ORGANIZATION | **RCA VIDEODISCS** |
| ADDRESS | 1133 Ave. of the Americas<br>New York, NY 10036 |
| CONTACT | Mr. Frank McCann |
| HARDWARE MANUFACTURED | Videodisc Players |

---

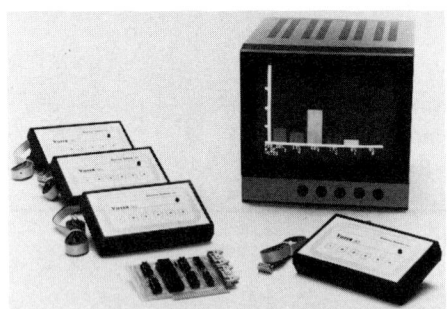

| | |
|---:|:---|
| ORGANIZATION | **REACTIVE SYSTEMS, INC.** |
| ADDRESS | 40 N. Van Brunt St.<br>Englewood, NJ 07631 |
| CONTACT | Mr. Don Monaco |
| PHONE | (201) 568-0446 |
| HARDWARE MANUFACTURED | Input Devices |
| DESCRIPTION | Reactive Systems, Inc. also provides technical support for meetings and conventions, and designs software that enables the videodisc to be controlled by a single- or multiple-response system. With this system, up to 1000 users in a seminar setting can respond simultaneously to questions, a bar graph of responses can be displayed, and appropriate scenes on the videodisc can be accessed. |

---

| | |
|---:|:---|
| ORGANIZATION | **RECENT PRODUCTIONS LIMITED** |
| ADDRESS | 10 Chapel St.<br>London SW1X 7B7 UK |
| CONTACT | Mr. Chris Marsh |
| PHONE | (01) 235-6744 |
| SOFTWARE PRODUCED | Disc Simulators, Other |
| DESCRIPTION | Recent Productions Limited produces a digital origination and retrieval of images system (DORIS), a disc mimic. |

# Hardware/Software/Integrated Systems

Quarterhorse consists of a laser videodisc, a PR-8210 laser player, and dual sound tracks harnessed to a 19-inch race monitor and a 13-inch tote board.

| | |
|---|---|
| ORGANIZATION | **RODESCH DEVELOPMENT CORP.** |
| ADDRESS | 157-2B E. Valley Parkway |
| | Escondido, CA 92025 |
| CONTACT | Mr. Dale Rodesch |
| PHONE | (619) 743-7681 |
| INTEGRATED SYSTEMS OFFERED | Video Games |
| DESCRIPTION | Rodesch Development Corp. is involved in the production of the world's first commercial laser videodisc-based game called "Quarterhorse," which was initially shipped in January 1982. Currently there are over 2000 machines in the field in the United States, the United Kingdom, France, Australia, Japan, and Central America. The machines are manufactured by various licensees. |

| | |
|---|---|
| ORGANIZATION | **SCRIPTECH** |
| ADDRESS | 33 Winwood Pl. |
| | Mill Valley, CA 94941 |
| CONTACT | Mr. William Claxton, |
| | Producer |
| PHONE | (415) 459-8896 |
| HARDWARE MANUFACTURED | Interfaces, Time Code Reader for Microcomputers |
| SOFTWARE PRODUCED | Authoring Systems, Disc Simulators |
| INTEGRATED SYSTEMS OFFERED | Script Editing System |
| DESCRIPTION | Script editing is a method of linking text with scenes on videotape for development of a rough cut. This has applications in the creation of both linear and nonlinear programming, and it can generate edit lists for film or video. The equipment used in the prototype system includes an IBM PC interfaced with a Sony Beta video editing system. |

| | |
|---|---|
| ORGANIZATION | **SOCIETY FOR VISUAL EDUCATION, INC.** |
| ADDRESS | 1345 Diversey Parkway<br>Chicago, IL 60614 |
| CONTACT | Mr. Harry Reilly<br>Director, Product Development |
| PHONE | (312) 525-1500 |
| SOFTWARE PRODUCED | Discs for interactive video/micro learning |

| | |
|---|---|
| ORGANIZATION | **SOFI INC. (INTERACTIVE TRAINING & INFO. CORP.)** |
| ADDRESS | 906 Cherrier St., Montreal<br>Quebec, Canada H2L 1H7 |
| CONTACT | Mr. William Lacoursiere, Nicole Laurier, or B. Michaud |
| PHONE | (514) 524-6781/9444 |
| SOFTWARE PRODUCED | Drivers/Video Utilities, Authoring Systems |
| INTEGRATED SYSTEMS OFFERED | Information Displays, Point-of-Purchase Systems |

Sony Model LDP-1000

Sony SFA-1000 Still-frame Audio Adaptor

| | |
|---|---|
| ORGANIZATION | **SONY COMMUNICATIONS PRODUCTS COMPANY** |
| ADDRESS | Sony Dr.<br>Park Ridge, NJ 07656 |
| CONTACT | Mr. John Hartigan,<br>National Marketing Manager, Interactive Products |
| PHONE | (201) 930-1000 |
| TOLL FREE | (800) 222-SONY |
| HARDWARE MANUFACTURED | Videodisc Players, Overlays/Superimpose Units, Still-frame Audio, Digital Audio, Interfaces, Videotex Terminals |
| SOFTWARE PRODUCED | Drivers/Video Utilities, Graphics Editor/Titlers, Disc Simulators |

| | |
|---|---|
| ORGANIZATION | **SONY CORPORATION OF AMERICA** |
| ADDRESS | Sony Dr.<br>Park Ridge, NJ 07656 |
| CONTACT | Mr. Jason Farrow |
| PHONE | (201) 930-4000 |
| HARDWARE MANUFACTURED | Videodisc Players, Overlays/Superimpose Units, Still-frame Audio, Digital Audio, Input Devices, Interfaces |
| SOFTWARE PRODUCED | Drivers/Video Utilities, Graphics Editor/Titlers, Disc Simulators |
| INTEGRATED SYSTEMS OFFERED | Training Work Stations/Carrels, Information Displays, Point-of-Purchase Systems, Videogames |

# Hardware/Software/Integrated Systems

| | |
|---|---|
| ORGANIZATION | **SONY OF CANADA LTD.** |
| ADDRESS | 411 Gordon Baker Rd., Willowdale<br>Ontario, Canada 2H 2S6 |
| CONTACT | Mr. Tom Windsor<br>Manager, Market Development Department<br>Communications Products Division |
| PHONE | (416) 499-1414 |
| HARDWARE MANUFACTURED | Videodisc Players, Overlays/Superimpose Units, Still-frame Audio, Digital Audio, Input Devices, Interfaces |
| SOFTWARE PRODUCED | Drivers/Video Utilities, Authoring Systems, Graphic Editor/Titlers, Disc Simulators |

| | |
|---|---|
| ORGANIZATION | **SOUND CONCEPTS** |
| ADDRESS | 27 Newell Rd.<br>Brookline, MA 02146 |
| CONTACT | Mr. Joel Cohen |
| PHONE | (617) 566-0110 |
| HARDWARE MANUFACTURED | CX Decoder for use with non-CX players (Laser and CED), Stereo synthesizer for use with non-stereo discs, Image Expander to widen stereo effect, Time Delay for rear channel sound. |

| | |
|---|---|
| ORGANIZATION | **SOUTHWEST TEXAS STATE UNIVERSITY/<br>BIOLOGY DEPARTMENT** |
| ADDRESS | San Marcos, TX 78666 |
| CONTACT | Mr. David Huffman |
| PHONE | (512) 245-2284 |
| DESCRIPTION | The Biology Department at Southwest Texas State is writing Biology lessons in Apple Super Pilot that are interfaced to VIDEODISCOVERY's *Bio Sci* videodisc through the VMI interface and VB-3 keyer. The projected completion date for the first module is August 1984. They will also be adapting drill-and-practice lessons to the disc. |

| | |
|---|---|
| ORGANIZATION | **SUN INTERACTIVE VIDEO** |
| ADDRESS | 1488 Sierra Creek Way<br>San Jose, CA 95132 |
| CONTACT | Mr. Chen Sun |
| PHONE | (408) 272-4385 |
| SOFTWARE PRODUCED | Authoring Systems |
| INTEGRATED SYSTEMS OFFERED | Training Work Stations/Carrels, Information Displays, Point-of-Purchase Systems |

| | |
|---|---|
| ORGANIZATION | **SUN RESEARCH, INC.** |
| ADDRESS | Box 210<br>New Durham, NH 03855 |
| CONTACT | Mr. Michael Mathieson,<br>Sales Manager |
| PHONE | (603) 859-7110 |
| DESCRIPTION | Sun Research, Inc. manufactures and distributes the Mayday and Sun line of uninterruptible power systems and line conditioners/voltage regulators. These devices protect hardware and discs from blackouts, brownouts, surges, and spikes. Units range in power output of 50 watts to 1500 watts and cost $90 and up. |

| | |
|---:|:---|
| ORGANIZATION | **SYMTEC, INC.** |
| ADDRESS | 15933 West 8 Mile |
| | Detroit, MI 48235 |
| HARDWARE MANUFACTURED | Interfaces |

| | |
|---:|:---|
| ORGANIZATION | **SYNSOR CORPORATION** |
| ADDRESS | 14241 N.E. 200th St. |
| | Woodinville, WA 98072 |
| PHONE | (206) 481-6600 |
| TOLL FREE | (800) 426-0193 |
| INTEGRATED SYSTEMS OFFERED | Training Work Stations/Carrels |

| | |
|---:|:---|
| ORGANIZATION | **TCT TECHNICAL TRAINING, INC. ("T-CUBED")** |
| ADDRESS | 599 North Mathilda Ave. |
| | Sunnyvale, CA 94086 |
| CONTACT | Mr. Saroj K. Kar |
| PHONE | (408) 735-9990 |
| SOFTWARE PRODUCED | Drivers/Video Utilities, Authoring Systems |
| INTEGRATED SYSTEMS OFFERED | Training Work Stations/Carrels |
| DESCRIPTION | TCT Technical Training, Inc. ("T-CUBED") specializes in providing data communications (and related) training solutions, using computer-based interactive video (both videotape and videodisc) as the principal medium. T-CUBED products include: off-the-shelf courseware in data communications, custom course design service, and customized authoring systems. |

| | |
|---:|:---|
| ORGANIZATION | **TECNETRA-TECHNOLOGY TRANSFER COMPANY** |
| ADDRESS | Piazza Morandi 2 |
| | Milano, Italy |
| CONTACT | Mr. Umberto Pellegrini |
| PHONE | 702015 |
| HARDWARE MANUFACTURED | Interfaces |
| SOFTWARE PRODUCED | Disc Simulators |
| INTEGRATED SYSTEMS OFFERED | Training Work Stations/Carrels |

| | |
|---:|:---|
| ORGANIZATION | **TELEMATIC SYSTEMS** |
| ADDRESS | 55 Wheeler St. |
| | Cambridge, MA 02138 |
| CONTACT | Mr. Fredric Raab |
| PHONE | (617) 492-2881 |
| SOFTWARE PRODUCED | Drivers/Video Utilities, Authoring Systems, Graphics Editor/Titlers, Disc Simulators |
| INTEGRATED SYSTEMS OFFERED | Training Work Stations/Carrels, Point-of-Purchase Systems |

| | |
|---:|:---|
| ORGANIZATION | **TELEMEDIA GmbH** |
| ADDRESS | Carl Bertlesmann Strabe 161 |
| | D-4830 Gutersloh 1 |
| | West Germany |
| CONTACT | Dr. Franz Netta |
| PHONE | Gutersloh (05241) /80-3324 |
| INTEGRATED SYSTEMS OFFERED | Information Displays, Point-of-Purchase Systems |

# Hardware/Software/Integrated Systems

| | |
|---|---|
| ORGANIZATION | **TELEVISION AUDIO SUPPORT ACTIVITY** |
| ADDRESS | Sacramento Army Depot SELTV-C |
| | Sacramento, CA 95813 |
| CONTACT | Mr. Robert H. Vincelette, Jr. |
| PHONE | (916) 388-3216 |
| DESCRIPTION | The Television Audio Support Activity provides engineering services for videodisc and all other video equipment to all Department of Defense agencies. They are the procurement office for all commercial off-the-shelf or modified off-the-shelf video equipment, including videodiscs, for all Department of Defense agencies. |

| | |
|---|---|
| ORGANIZATION | **TERAK CORPORATION** |
| ADDRESS | 14151 N. 76th St. |
| | Scottsdale, AZ 85260 |
| CONTACT | Mr. George Garant, VGM Product Manager or |
| | Mr. Nick Temple, Marketing Communications |
| PHONE | (602) 998-4800 |
| TOLL FREE | (800) 647-1655 |
| HARDWARE MANUFACTURED | Overlays/Superimpose Units |
| SOFTWARE PRODUCED | Authoring Systems, Graphics Editor/Titlers |
| INTEGRATED SYSTEMS OFFERED | Training Work Stations/Carrels, Information Displays, Business Graphics |
| DESCRIPTION | The Videographics System from Terak is a complement to their color graphics computer. Standard NTSC signals from either a videodisc, tape, or camera are decoded into RGB signals and are displayed on the high-resolution color monitor. Once the external video image is displayed on the monitor, it can be overlaid with text and graphics input from an integral graphics software package. Included with the system is a standard authoring system, which allows the programmer/user to select either still frames, frame sequences, or random sequences of frames for display while also calling up the computer images that overlay the images. This authoring system can also be used to adjust the tint and hue of the external image, designate touch targets (for systems equipped with the Touchpanel Option), and selectively turn on and off the two audio channels. |

| | |
|---|---|
| ORGANIZATION | **TEXAS INSTRUMENTS** |
| ADDRESS | Consumer Relations/Personal Computers |
| | P.O. Box 53 |
| | Lubbock, TX 79408 |
| HARDWARE MANUFACTURED | Interfaces |

| | |
|---|---|
| ORGANIZATION | **TRANS-LINGUAL COMMUNICATIONS, INC.** |
| ADDRESS | Suite 1200, Eight South Michigan Ave. |
| | Chicago, IL 60603 |
| CONTACT | Mr. Lyric Hughes, |
| | President |
| PHONE | (312) 332-6555 |
| TOLL FREE | (800) 621-2167 |
| SOFTWARE PRODUCED | Authoring Systems, Disc Simulators |

TriLogic's TouchMe Center℠ is the first and only completely portable, fully integrated interactive videodisc sales terminal available today. This sleek stand-alone unit combines 256K of memory, high resolution graphics, motion picture and still photographs, with an easy to use touch sensitive monitor. Customers simply touch the surface of the screen to retrieve product information-no cumbersome peripherals. Just plug into any A.C. power and the TouchMe℠ is ready to "interact" with your customers.

| | |
|---|---|
| ORGANIZATION | **TRILOGIC, INC.** |
| ADDRESS | 9430 Washington Blvd.<br>Culver City, CA 90230 |
| CONTACT | Mr. Jay Payne or Ms. Virginia Barreto |
| PHONE | (213) 838-1140 |
| INTEGRATED SYSTEMS OFFERED | Training Work Stations/Carrels, Information Displays, Point-of-Purchase Systems |
| HARDWARE MANUFACTURED | Interfaces |
| DESCRIPTION | TriLogic specializes in the design and production of interactive videodisc programs and computer videodisc systems for sales applications. TriLogic's veteran team of videodisc producers can provide you with powerful scripts, vivid pictures, special effects, and imaginitive programming that will make your videodisc point-of-purchase display a show stopper at any convention or on any sales floor. |

| | |
|---|---|
| ORGANIZATION | **VCM SYSTEMS, INC.** |
| ADDRESS | 427 Sixth Ave. S.E., P.O. Box 2789<br>Cedar Rapids, IA 52406 |
| CONTACT | Mr. Thomas Hedges,<br>Executive Producer |
| PHONE | (319) 364-6959 |
| TOLL FREE | (800) 553-8879 |
| INTEGRATED SYSTEMS OFFERED | Information Displays, Point-of-Purchase Systems |

| | |
|---|---|
| ORGANIZATION | **VHD CORPORATION OF AMERICA** |
| ADDRESS | 2602 McGaw Ave.<br>Irvine, CA 92714 |
| CONTACT | Mr. Joe McHugh |
| PHONE | (714) 857-0414 or 606-9294 |
| HARDWARE MANUFACTURED | Videodisc Players, Overlays/Superimpose Units, Digital Audio, Digital Audio Processor, RGB Distributor |
| SOFTWARE PRODUCED | Graphics Editor/Titlers |
| INTEGRATED SYSTEMS OFFERED | Information Displays |
| DESCRIPTION | Using a Digital Audio Processor, a certain amount of audio signal can be repeated for a specified amount of time which is synchronized to the vertical time interval of a tv signal. The RGB Distributor has one analog RGB input to seven analog outputs. |

### Hardware/Software/Integrated Systems

|  |  |
|---:|:---|
| ORGANIZATION | **VIDEO ASSOCIATES LABS** |
| ADDRESS | 3933 Steck Ave., Suite B109 |
|  | Austin, TX 78759 |
| CONTACT | Mr. Mike Dyer |
| PHONE | (512) 346-3379 |
| HARDWARE MANUFACTURED | Overlays/Superimpose Units |
| SOFTWARE PRODUCED | Drivers/Video Utilities |

|  |  |
|---:|:---|
| ORGANIZATION | **VIDEO MASTERS, INC.** |
| ADDRESS | 1616 Broadway |
|  | P.O. Box 1963 |
|  | Kansas City, MO 64141 |
| CONTACT | Mr. Paul Dark, |
|  | Vice President |
| PHONE | (816) 474-8530 |
| HARDWARE MANUFACTURED | Input Devices, Interfaces |
| INTEGRATED SYSTEMS OFFERED | Training Work Stations/Carrels, Information Displays, Point-of-Purchase Systems |

|  |  |
|---:|:---|
| ORGANIZATION | **VIDEO VISION ASSOCIATES, LTD.** |
| ADDRESS | 7 Waverly Pl. |
|  | Madison, NJ 07940 |
| CONTACT | Mr. Ralph Heigl |
| PHONE | (201) 377-0302 |
| HARDWARE MANUFACTURED | Interfaces |
| SOFTWARE PRODUCED | Authoring Systems |
| DESCRIPTION | Video Vision has an Apple/videodisc player interface and an audiotape/videodisc interface. |

|  |  |
|---:|:---|
| ORGANIZATION | **WANG LABORATORIES, INC.** |
| ADDRESS | One Industrial Ave. |
|  | Lowell, MA 01851 |
| CONTACT | Dr. Dolores E. Lesnick |
|  | Product Manager ILS/CAI |
| PHONE | (617) 459-5000 |
| INTEGRATED SYSTEMS OFFERED | Training Work Stations/Carrels |
| DESCRIPTION | WILS, the Wang Interactive Learning System, offers videodisc-based computer-assisted instruction in Wang Word Processing on OIS and VS/IIS systems. |

|  |  |
|---:|:---|
| ORGANIZATION | **WDC MARKETING ASSOCIATES, INC.** |
| ADDRESS | 244 Goodwin Crest Drive, Suite 106 |
|  | Birmingham, AL 35209 |
| CONTACT | Mr. David Carrington |
| PHONE | (205) 945-8951 |
| HARDWARE MANUFACTURED | Interfaces |
| SOFTWARE PRODUCED | Authoring Systems |
| INTEGRATED SYSTEMS OFFERED | Information Displays, Point-of-Purchase Systems |

| | |
|---:|:---|
| ORGANIZATION | **WESTINGHOUSE ELECTRIC CORP.** |
| | **INTEGRATED LOGISTICS SUPPORT DIVISIONS** |
| | **(TECHNICAL TRAINING OPERATIONS)** |
| ADDRESS | MS2410-CCC #20 |
| | 10400 Patuxent Parkway |
| | Columbia, MD 21044 |
| CONTACT | Mr. Steve G. Kukla |
| PHONE | (301) 995-5482 |
| SOFTWARE PRODUCED | Custom Courseware |
| INTEGRATED SYSTEMS OFFERED | Training Work Stations/Carrels, "Maintenance-Aiding" Consoles |

| | |
|---:|:---|
| ORGANIZATION | **WHITNEY EDUCATIONAL SYSTEMS** |
| ADDRESS | 1777 Borel Place, Suite 416 |
| | San Mateo, CA 94402 |
| PHONE | (414) 341-5818 |
| HARDWARE MANUFACTURED | Interfaces |

| | |
|---:|:---|
| ORGANIZATION | **WICAT SYSTEMS, INC.** |
| ADDRESS | 1875 South State St. |
| | P.O. Box 539 |
| | Orem, UT 84057 |
| CONTACT | Mr. Richard M. Cavagnol |
| PHONE | (801) 224-6400 ext. 282 |
| HARDWARE MANUFACTURED | Overlays/Superimpose Units, Input Devices, Interfaces |
| SOFTWARE PRODUCED | Authoring Systems |
| INTEGRATED SYSTEMS OFFERED | Training Work Stations/Carrels, Information Displays |

| | |
|---:|:---|
| ORGANIZATION | **WINDELOTT CO.** |
| ADDRESS | Drawer P |
| | West Winfield, NY 13491 |
| PHONE | (315) 894-3958 |
| HARDWARE MANUFACTURED | Interfaces |

# Mastering and Replication Resources

| | |
|---|---|
| ORGANIZATION | **CBS VIDEO ENTERPRISES** |
| ADDRESS | 51 West 52nd St.<br>New York, NY 10019 |
| SERVICES/PRICES | CBS Video Enterprises is a custom manufacturer of CBS Entertainment Division Videodiscs. They will send rate card upon request. |

| | |
|---|---|
| ORGANIZATION | **OPTICAL RECORDING PROJECT/3M** |
| ADDRESS | 223-5S 3M Center<br>St. Paul, MN 55144 |
| CONTACT | Ms. Evie Homme |
| PHONE | (612) 733-4435 |
| TOLL FREE | (800) 328-1300 ext. 3-2142 |
| SERVICES/PRICES | See following information. |

SCOTCH LASER VIDEODISC    Prices effective October 1, 1983.

*Levels one and three:* (No digital dumps—may have chapter codes and automatic frame stops.)

| Setup | One Side | Two Sides |
|---|---|---|
| CAV | $1,800 | $3,600 |
| CLV | 2,500 | 5,000 |
| Quantity | | |
| 1– 99 | $18 | $23 |
| 100– 499 | 14 | 20 |
| 500– 999 | 12 | 17 |
| 1000–1499 | 10 | 15 |

| No Setup | One Side | Two Sides | Third Side of Album |
|---|---|---|---|
| 1500–1999 | $10.00 | $15.00 | $8.50 |
| 2000–2499 | 8.50 | 12.50 | 6.75 |
| 2500–4999 | 6.75 | 10.50 | 5.75 |
| 5000–up | Prices on request | | |

*Level two:* (Prices include two digital dumps in either Pioneer or Sony format.)

| Setup | One Side | Two Sides |
|---|---|---|
| $2,500 per side, regardless of quantity | | |
| Quantity | | |
| 1– 999 | $18.00 | $23.00 |
| 1000–1999 | 15.00 | 21.00 |
| 2000–2999 | 12.00 | 17.00 |
| 3000–4999 | 6.75 | 10.50 |
| 5000–up | Prices on request | |

*Over 2 dumps (additional charges):*
Quantity 1–100—$200 per side per dump flat charge;
101 and more—$2 per dump per side (over 2 dumps per side).

*Tandem dumps (Sony and Pioneer dumps on same disc):*
Quantity 1–100—$200 flat charge for each tandem dump per side;
101 and more—$2 per each tandem dump per side per disc.

*Level two interactive disc development package\**: Frame-accurate check disc is first mastered without digital encoding. This allows a digital program to be written and frame accurate. Disc is then remastered once with approved program in Audio 2. Customer pays one setup charge for two masters.  $750 per side

*Proof disc\**: Frame-accurate disc for checking level one, per side two or three programs before quantity replication. Level two disc contains digital program in Audio 2, and can be remastered once at no charge if changes are required in the digital dump.  $750 per side

*Three-fourths-inch check tape\**: Frame-accurate videotape. Used for writing level two and three computer programs before setup charges are incurred.  $250 per side

*Master disc and premaster tape storage for 12 months\**: setup charge for disc reruns during storage period. Masters are not stored after initial replication.  $1000 per side

\* Requests for these services must accompany order for normal mastering and replication services.

- 3M cue inserter: $6,500, including installation.
- "Producing Interactive Videodiscs": An interactive instructional unit consisting of interactive disc and manual illustrating videodisc program production; $495.

- Special quote for charges for certification of each disc per order.
- Price based on delivery to 3M of a customer-approved 3/4-inch professional or 1-inch Type C-NTSC videotape.
- Prices do not include any production, editing, or postproduction work (i.e., film-to-tape transfer).
- Names of recommended postproduction studios or master tape specifications supplied upon request by 3M optical recording sales representative.
- Standard packaging included in price.
- All orders FOB manufacturing plant.
- Custom labels and jackets on special quote upon receipt of artwork and/or four-color negatives.
- Customer may supply jacket and label based on 3M specifications. No charge for packaging.

PRICES SUBJECT TO CHANGE WITHOUT NOTICE.

PRICES FOR SPECIAL SERVICES   Effective July 1, 1983

*Fast turnaround:*
CAV only One working day-maximum 25 discs. 3M checks in premaster tape any time one day, order shipped by 6:00 p.m. following work day.   CAV only
$3,600 set up per side\*\*

Three working days-maximum 100 discs. 3M checks in premaster tape any time one day, order shipped by 6:00 p.m. on third working day after tape check-in.   $2,500 set up per side\*\*

\*\* Plus customary charges per disc order, which are provided on the price sheet M-OR-PS.

*Level two program entry:*
Level two programs keypunched from written mnemonics   $225 per dump

*Technical service:*
Special technical service field support upon day, plus air fare customer request.   $500 a day, plus air fare

*Terms and conditions:*
Orders for fast turnaround must be scheduled and 3M St. Paul must have received a signed purchase order at least two weeks in advance.
Customer must have established 3M credit. (Normal 3M credit clearance takes about six weeks.)
If for some reason discs are not shipped after one working day, 3M will charge the three-day prices. If discs are not shipped after three working days, you will be charged the normal fee.

# Mastering and Replication Resources

|  |  |
|---|---|
| ORGANIZATION | **PIONEER VIDEO, INC.** |
|  | Corporate Headquarters |
| ADDRESS | 200 West Grand Ave. |
|  | Montvale, NJ 07645 |
| PHONE | (201) 573-1122 |
|  | Industrial Sales Division |
| ADDRESS | 5150 East Pacific Coast Hwy., #300 |
|  | Long Beach, CA 90804 |
| CONTACT | Mr. Ron Butler, |
|  | Vice President, Industrial Sales |
| PHONE | (213) 498-0300 |
| SERVICES/PRICES | See following information. |

PROGRAM DEVELOPMENT SERVICES   Prices effective November 15, 1983

| *Digital programming* | *Per dump* |
|---|---|
| • Digital coding from customer-supplied flowcharts conforming to PVI flowcharting specifications. PVI will encode the digital program and verify that it conforms to the flowchart provided by the customer. Flowcharting specifications are available at no charge from PVI account executives. | $600 |
| • Data entry from customer-supplied data entry sheets conforming to PVI specifications; PVI will enter the program and notify the customer of logic errors identified during verification. PVI data entry sheets and specifications are available at no charge from PVI account executives. | $200 |
| • Data transmission of properly formatted object code via diskette or telephone modem. | $70, beginning with the third dump |
| • Program design and implementation consulting. | RFQ-subject to program design evaluation |

| *Digital programming tools* | *Each* |
|---|---|
| • DiscAssemble™: Translates source programs into object code for Pioneer PR7820 LaserDisc™ players. (Includes library of macros, subroutines, and diagnostic programs on diskette, plus user's manual package.) | $2,250 |
| • DiscSimulate™: Tests program logic, using object code from DiscAssemble™ and a check disc. DiscSimulate™ includes a diskette and user's manual package. | $1,250 |
| • DiscMail™: Provides modem transmission at 300 or 1,200 Baud rates. (Diskette and user's manual package are included.) | $750 |
| • Check disc*: One disc replica of customer's edited master for use in testing program logic. (Audio and video quality not representative of final discs.) | $500 per side* |
| • Proof disc*: One disc replica of customer's edited master including digital control programs. (Audio and video quality not representative of final discs.) | $500 per side* |

*If a disc production run purchase order does not accompany the check/proof disc order, the fee for the first disc will be $2,500; additional discs will cost $500 each.

| *Transfers-coding-editing* | *Per hour* |
|---|---|
| • Film-to-tape transfer via Rank Cintel flying spot scanner, 35 mm or 16 mm, positive or negative film |  |
| —without mag track | $320 |
| —with mag track | $360 |
| • On-line editing (two machines) | $400 |
| —additional machines, each | $ 95 |
| • Dubbing (two machines) | $190 |
| • Off-line edit decision list preparation | $125 |
| • 3/2 field identification for film on tape; correction of mixed field editing | $125 |
| • Time code restriping | $125 |
| • Audio retracking/sweetening | $125 |
| • Character generator (including operator) | $125 |
| • Electronic still storage (ESS) | $400 |
| • Encoding of chapter intervals/picture stops in excess of 100 per side (maximum—79 chapter intervals per side) | $10 each |

Additional charges may be added for use of blank tape stock.

Note: All prices are subject to Pioneer Video, Inc. standard terms and conditions of sale. Prices are subject to change without prior notification. Delivery is subject to availability.

LINEAR CAV INDUSTRIAL DISC REPLICATION

| | Mastering | Disc Replication | | | | | |
|---|---|---|---|---|---|---|---|
| | | Two-Week Turnaround* | | Three-Week Turnaround* | | Four-Week Turnaround* | |
| Quantity | Setup Per Side | 1 Side | 2 Sides | 1 Side | 2 Sides | 1 Side | 2 Sides |
| 5– 99 | $3200 | $24 | $29 | $21 | $26 | $18 | $23 |
| 100– 199 | $2600 | $22 | $27 | $19 | $24 | $16 | $21 |
| 200– 499 | $2000 | $18 | $23 | $16 | $21 | $14 | $19 |
| 500– 999 | $1000 | $16 | $21 | $14 | $19 | $12 | $17 |
| 1000–1999 | No charge | $12 | $17 | $11 | $16 | $10 | $15 |
| 2000–4999 | No charge | $11 | $16 | $10 | $15 | $ 9 | $14 |
| 5000–up | No charge | Special quote | | Special quote | | Special quote | |

*From release to manufacturing.

For PVI to provide a flexible two-, three-, or four-week disc replication turnaround, customers must request these services in advance, must meet their scheduled media in-dates, and must meet agreed-upon review and approval dates. The following advance notice periods are required:

| Requested Turnaround | Advance Notice Required |
|---|---|
| 2 weeks | 4 weeks |
| 3 weeks | 3 weeks |
| 4 weeks | 2 weeks |

Advance notice must be in the form of a valid purchase order or other acceptable form of payment.

*Linear CAV Discs* have no data dumps recorded on disc. They may include encoding for chapter intervals and picture stops. Any combination of chapter intervals, picture stops, or significant frames up to a combined total of 100 per side is included in the prices listed. (Note, however, that current players will not accommodate more than 79 chapters.) Chapter intervals or picture stops in excess of 100 are subject to a video encoding charge of $10 each. Special quality control requirements or additional significant frames over 100 per side quoted on request.

These prices are based on the condition that PVI will receive an edited videotape master on 1-inch type C, 2-inch quad, or 2-inch helical. Other input formats (film, 1-inch A and B, 3/4-inch U-Matic) may require additional transfer charges. Special formatting information is available from your account executive.

Shipment charges are FOB Hawthorne, California, freight collect. Allow one to two days from completion date to Hawthorne shipment date.

*Standard PVI disc labels* are provided at no additional cost. Customer label copy must accompany the production order.

*Custom labels and jackets:* If PVI is to arrange for the printing of custom labels and/or jackets, board art and color separations must accompany or precede receipt of the customer's video input media. PVI will provide a quotation for such services upon request. Contact your PVI account executive.

Notes:
- Up to 30 minutes of recorded material per side on CAV.
- All prices subject to Pioneer Video, Inc. standard terms and conditions of sale. Prices subject to change without prior notice.
- Disc warranty limited to credit for or direct replacement of defective discs returned to PVI within 90 days from date of delivery.

INTERACTIVE CAV AND ALL CLV INDUSTRIAL DISC REPLICATION

| | Mastering | | Disc Replication | | | | |
|---|---|---|---|---|---|---|---|
| | | | Two-Week Turnaround* | | Three-Week Turnaround* | | Four-Week Turnaround* |
| Quantity | Setup Per Side | | 1 Side | 2 Sides | 1 Side | 2 Sides | 1 Side | 2 Sides |
| 5– 99 | $3200 | | $33 | $40 | $30 | $37 | $27 | $34 |
| 100– 199 | $2600 | | $28 | $34 | $25 | $31 | $22 | $28 |
| 200– 499 | $2000 | | $21 | $27 | $19 | $25 | $17 | $23 |
| 500– 999 | $1000 | | $19 | $24 | $17 | $22 | $15 | $20 |
| 1000–1999 | No charge | | $15 | $20 | $14 | $19 | $13 | $18 |
| 2000–4999 | No charge | | $13 | $18 | $12 | $17 | $11 | $16 |
| 5000–up | No charge | | Special quote | | Special quote | | Special quote | |

*From release to manufacturing.

For PVI to provide a flexible two-, three-, or four-week disc replication turnaround, customers must request these services in advance, must meet their scheduled media in-dates, and must meet agreed-upon review and approval dates. The following advance notice periods are required:

| Requested Turnaround | Advance Notice Required |
|---|---|
| 2 weeks | 4 weeks |
| 3 weeks | 3 weeks |
| 4 weeks | 2 weeks |

Advance notice must be in the form of a valid purchase order or other acceptable form of payment.

*Check disc:* one-disc replica of customer's edited master for use in testing program logic (audio and video quality not representative of final discs)     $500 per side**

*Proof disc:* one-disc replica of customer's edited master, including digital control programs (audio and video quality not representative of final discs)     $500 per side**

**If a disc production run purchase order does not accompany check/proof disc order, the fee for the first disc will be $2500, and additional discs will cost $500 each.

These prices are based on PVI receiving an edited videotape master on 1-inch type C, 2-inch quad, or 2-inch helical. Other input formats (film, 1-inch A and B, 3/4-inch U-Matic) may require additional transfer charges. Special formatting information is available from your account executive.

Shipment charges are FOB Hawthorne, California, freight collect. Allow one to two days from completion date to Hawthorne shipment date.

*Interactive CAV discs* have one or more digital data dumps stored on the disc. Prices include up to two dumps and 100 significant frames per side, when supplied by customers on diskette or by telephone transfer. The dump format must comply with PVI specifications. PVI can also encode dumps for Sony LDP-1000 players and tandem dumps. Two dumps are defined as two Pioneer, two Sony, or one tandem (Pioneer/Sony) dump. Special quality control requirements for additional significant frames over 100 per side quoted on request.

PVI also offers digital programming from customer supplied flowcharts conforming to PVI specifications, and data entry from customer supplied data entry sheets. More detailed information on these services is provided on the "PVI Program Development Services Price List."

*CLV discs,* either linear or for host computer control, may include chapter intervals. Prices include up to seventy-nine chapter intervals per side.

*One-check cassette* in 3/4-inch U-Matic, VHS, or BETAMAX format of interactive CAV material will be provided at customer request.

*Standard PVI disc labels* are provided at no additional cost. Customer label copy must accompany the production order.

*Custom labels and jackets:* If PVI is to arrange for the printing of custom labels and/or jackets, board art and color separations must accompany or precede receipt of the customer's video input media. PVI will provide a quotation for such services upon request. Contact your PVI account executive.

# Mastering and Replication Resources

Notes:

- Up to 30 minutes of recorded material per side on CAV. Up to 60 minutes of recorded material per side on CLV.
- All prices subject to Pioneer Video, Inc. standard terms and conditions of sale. Prices subject to change without prior notice.
- Disc warranty limited to credit for or direct replacement of defective discs returned to PVI within 90 days from date of delivery.

---

| | |
|---|---|
| ORGANIZATION | **RCA VIDEODISC OPERATIONS** |
| ADDRESS | P.O. Box 91079 |
| | 7900 Rockville Rd. |
| | Indianapolis, IN 46291 |
| CONTACT | Manager of Marketing Administration |
| PHONE | (317) 273-3431 |
| SERVICES/PRICES | Contact RCA for information. |

---

| | |
|---|---|
| ORGANIZATION | **SONY COMMUNICATIONS PRODUCTS COMPANY** |
| ADDRESS | Sony Dr. |
| | Park Ridge, NJ 07656 |
| CONTACT | Mr. John Hartigan, |
| | National Marketing Manager |
| | Interactive Products |
| PHONE | (201) 930-1000 |
| TOLL FREE | (800) 222-SONY |
| SERVICES/PRICES | See following information. |

DISC REPLICATION PRICING

Prices effective April 1, 1983.

Reflective Laser Optical Disc (Per Side)

0-1 DUMP

| Disc Quantity | Single Active Unit Price | Double Active Unit Price | Mastering Charge |
|---|---|---|---|
| 50– 249 | $19 | $30 | $2000 |
| 250– 499 | $18 | $26 | $2000 |
| 500– 999 | $16 | $25 | $2000 |
| 1000–1999 | $15 | $24 | $2000 |
| 2000–2999 | $14 | $23 | No charge |
| 3000–up | | By special quotation only | |

For 2 dumps, add $2 per disc; 3 dumps or more and/or critical frame check by special quotation only.

- Discs are shipped to customer in bulk in individual paper sleeves.
- Master tape and master disc storage for three months is available at no charge.
- After storage period, if customer requests that tape be returned, there will be a $40 shipping charge.
- Emulator verification of programs is available in New York, Detroit, and Los Angeles at an additional charge.
- All master tapes submitted for replication must be 1-inch Type C format with proper format, as set down in master tape cue sheet and Sony pamphlets "Pre-mastering and Formatting Laser Optical Disc."
- Payment terms: net 30 days.

No returns accepted without prior written authorization from Sony Corporation of America. All prices are subject to change without notice. All orders are subject to our final acceptance.

*Introductory package price list (preliminary)*

In order to assist corporations who want to try videodisc replication, Sony has prepared an introductory package wherein we offer to master and replicate five copies of a laser optical disc at a package price of $2500 per side (0-1 dump). This introductory package includes the following items and services.

- Disc master (master tape is supplied by customer on 1-inch Type C format, NTSC). Format as prescribed by Sony.
- Five copies of program replicated on laser optical disc.
- Information verification of tape and documents check.
- Three months storage of master disc and tape.
- Option to reorder within three months a minimum of 50 disc copies of the same program at Sony's published price with no mastering charge.
- FOB dealer or customer's location as requested.

Payment terms: net 30 days.

Complete information on videodisc replication is available in the "Videodisc Production Guide," available through the Sony Distribution Center.

|             |                          |
|-------------|--------------------------|
| ORGANIZATION | **TECHNIDISC** |
| ADDRESS | 2250 Meijer Dr. |
|         | Troy, MI 48084 |
| CONTACT | Mr. Leon Chick, |
|         | National Sales Manager |
| PHONE | (313) 435-7430 |
| TOLL FREE | (800) 321-9610, (800) 321-9652 |
| SERVICES/PRICES | See following information. |

Prices effective May 1, 1983.

*Formatting requirements:* Input media for videodisc mastering can include videotape, film, slides, or graphics. Videotape programming can be on either 16 mm or 35 mm formats (either single or double system) and can include filmstrips and slides (either plastic, cardboard, or glass mounted). All program material must be transferred to a 1-inch Type C composite NTSC or a 2-inch for mastering on TECHNIDISC'S reflective videodisc format. Postproduction transfer services are available at Producers Color Services' (PCS) Video Communications Division in Southfield, Michigan (313) 352-5353. Please call the sales department for further information.

DISC MASTERING AND REPLICATION

*Linear CAV and linear CLV:* No digital dumps. Prices include up to 10 chapter/picture stops with up to 30 minutes of recorded material per side.

| Setup | One Side | Two Sides |
|---|---|---|
| Linear CAV | $1,500 | $3,000 |
| Linear CLV | 2,000 | 4,000 |
| Quantity | | |
| 5– 49 | $18 | $23 |
| 50– 199 | $16 | $21 |
| 200– 499 | $14 | $19 |
| 500– 999 | $12 | $17 |
| 1000–1499 | $11 | $16 |
| 1500–1999 | $10 | $15 |
| 2000–up | Prices on request | Prices on request |

*Interactive CAV:* Prices include two digital dumps and up to 100 significant frames.

| Setup | One Side | Two Sides |
|---|---|---|
| INTERACTIVE CAV | $2,000 | $4,000 |
| Quantity | | |
| 5– 49 | $18.00 | $23.00 |
| 50– 199 | 17.50 | 22.50 |
| 200– 499 | 17.00 | 22.00 |
| 500– 999 | 16.50 | 21.50 |
| 1000–1499 | 16.00 | 21.00 |
| 1500–1999 | 15.00 | 20.00 |
| 2000–up | Prices on request | Prices on request |

*Three-quarter-inch check cassettes* on material to be replicated can be provided to customer. Setup charges are applicable after customer's approval of cassette. $250

*Proof disc:* If requested, a frame accurate disc is supplied to customer before quantity replication. $550

- The above pricing is based on receiving an edited master on 1-inch Type C or 2-inch and does not include any production, editing or postproduction work. If these services are required, please call PCS' Video Communications Division, Sales Department, (313) 352-5353.
- All prices subject to TECHNIDISC's terms and conditions.
- All orders are F.O.B. Troy, MI.
- Standard packaging includes sleeves and one-color labels. Custom labels and jackets are available on special quotes.

# Videodiscography

| | |
|---:|:---|
| TITLE | **A.C. SWITCHBOARD OPERATION** |
| DESCRIPTION | Complete setup of an A.C. switchboard including: energizing a dead bus, paralleling, securing, and correcting for abnormal conditions such as kilowatt or frequency imbalance. |
| TYPE | Technical, Instructional |
| FORMAT | Laser CAV |
| LEVEL | 3 |
| SYSTEM DESCRIPTION | Pioneer 7820 Model 2, Apple IIe with 2 drives, touch panel (infrared) monitor, all in a cabinet. |
| DEVELOPER | Creativision, Inc. |
| CONTACT FOR MORE INFO. | Mr. Rick Glasby, Creativision |
| FEATURES | Stereo, Dual Audio, Touch Panel Monitor |
| PRODUCER | Creativision, Inc. |
| DISTRIBUTOR | Creativision, Inc. |
| AVAILABILITY | Sale |
| SPECIAL CONDITIONS | Produced for the Navy Electrician's Mate "A" School with contract funds from the U.S. Navy. |

| | |
|---:|:---|
| TITLE | **ABBA** |
| TYPE | Music |
| FORMAT | Laser CAV |
| LEVEL | 1 |
| FEATURES | Stereo |
| PRODUCER | Warner Home Video |
| DISTRIBUTOR | Pioneer Video, Inc., 74-006, $24.95 |

| | |
|---:|:---|
| TITLE | **ABBA IN CONCERT** |
| TYPE | Music |
| FORMAT | Laser CLV |
| FEATURES | Stereo, CX Encoded |
| PRODUCER | Pioneer Video Imports |
| DISTRIBUTOR | Pioneer Video, Inc., MP-066US, $27.95 |

| | |
|---:|:---|
| TITLE | **ABC MANTRAP** |
| FORMAT | CED |
| FEATURES | Stereo |
| DISTRIBUTOR | RCA Videodiscs, 12217, $19.98 |

| | |
|---:|:---|
| TITLE | **ABC-NEA SCHOOLDISC** |
| DESCRIPTION | Schooldisc was a joint effort of the American Broadcasting Co. and the National Education Association. Its purpose was to model a possible elementary school curriculum for videodisc. |
| TYPE | Research and Development, Technical, Instructional |
| FORMAT | Laser CAV |
| LEVEL | 3 |
| SYSTEM DESCRIPTION | Proprietary |
| FEATURES | Chapters, Picture Stops, Digital Dump-Pioneer, Stereo, Dual Audio |
| RUNNING TIME | 60 minutes |
| PRODUCER | ABC |

| | |
|---:|:---|
| TITLE | **ABOUT COMPUTERS** |
| DESCRIPTION | A disc that explains how computers work, designed for children ages 8 through 15. |
| TYPE | Children, Adolescents, Instructional |
| RELEASE DATE | 1982 |
| FORMAT | Laser CAV |
| LEVEL | 2 |
| DEVELOPER | Exhibit Technology, Inc. |
| CONTACT FOR MORE INFO. | Don Cochrane, President—Marketing |
| FEATURES | Chapters, Picture Stops, Digital Dump-Pioneer, Dual Audio, Sound Over Still |
| PRODUCER | Exhibit Technology, Inc. Patrick Dillon and Lauri Halderman, Producers |
| AVAILABILITY | This disc is available from Exhibit Technology only as part of their product line sampling, but it is installed at EPCOT. |

| | |
|---:|:---|
| TITLE | **ABSENCE OF MALICE** |
| TYPE | Motion Picture (PG) |
| FORMAT | Laser CLV, CED |
| PRODUCER | RCA/Columbia Pictures Home Video |
| DISTRIBUTOR | Pioneer Video, Inc., VLD1015, $29.95 RCA Videodiscs, 03030, $29.98 |

| | |
|---:|:---|
| TITLE | **ABSENT-MINDED PROFESSOR, THE** |
| TYPE | Motion Picture (G) |
| FORMAT | Laser CLV, CED |
| PRODUCER | Walt Disney |
| DISTRIBUTOR | RCA Videodiscs, 00702, $24.98 |
| RETAIL OUTLET | (CLV) UltraVideo, 028AS, $34.95 |

| | |
|---|---|
| TITLE | **ACTION CODE COURSEWARE - INDUSTRIAL TECHNICIAN PROGRAM** |
| DESCRIPTION | 25 disc program covering electronics technology (basic and advanced), mechanical technology (hydraulics, pneumatics, and servomechanisms), and product specific applications in electronics (Gould-Modicon 584 Programmable Controller). |
| TYPE | Technical, Instructional, In-house Training |
| RELEASE DATE | The first discs are available now. All 25 discs are to be completed by July 1984. |
| FORMAT | Solid-state laser videodisc |
| LEVEL | 3 |
| SYSTEM DESCRIPTION | Special bar-coded workbooks and scanning wand (wand passed over bar-codes accesses videodisc); microprocessor to control interactivity (developed by Perceptronics, Inc.); infrared touch screen on the video monitor; and 3 control buttons for reversing, replaying, or stopping a video segment. |
| DEVELOPER | ICS-Intext, a Division of National Education Corporation |
| CONTACT FOR MORE INFO. | Steve Cline, ICS, (800) 233-0259 |
| FEATURES | Dual Audio |
| PRODUCER | ICS-Intext |
| DISTRIBUTOR | ICS-Intext |
| AVAILABILITY | Rent |
| SPECIAL CONDITIONS | Courseware will be provided at a per-student cost. The necessary Action Code system hardware will be provided at no additional charge. There is a minimum courseware lease required. |

| | |
|---|---|
| TITLE | **ACTRONICS, INC. - (NOT GIVEN)** |
| TYPE | Medical, Instructional |
| FORMAT | Laser CAV |
| LEVEL | 3 |
| SYSTEM DESCRIPTION | Turnkey computer-based CPR training system. Instrumental training mannequin provides feedback of skills. |
| DEVELOPER | American Heart Association/Actronics, Inc. |
| CONTACT FOR MORE INFO. | Joyce Hallas, Actronics, Inc., (412) 231-6200 |
| FEATURES | Sound Over Still |
| RUNNING TIME: | Varies from student to student. Range is four to six hours. |
| DISTRIBUTOR | Actronics, Inc. |
| AVAILABILITY | Sale, Lease |
| SPECIAL CONDITIONS | This disc only performs on an Actronics hardware/courseware integrated system. |
| SUGGESTED PRICE | $25000 |

| | |
|---|---|
| TITLE | **ADVANCED MAP INTERPRETATION AND TERRAIN ANALYSIS COURSE (ADVANCED MITAC)** |
| DESCRIPTION | Training course that follows Basic MITAC. It consists of 13 lessons for Army aviators, and provides additional instruction in Map-of-the- Earth (MOE) navigation, using topographic maps. |
| TYPE | Instructional |
| RELEASE DATE | January 1984 |
| FORMAT | Laser CAV |
| LEVEL | 0 |
| DEVELOPER | Anacapa Sciences Inc. |
| CONTACT FOR MORE INFO. | Anacapa Sciences Inc. |

| | |
|---|---|
| FEATURES | Chapters, Picture Stops |
| AVAILABILITY | Made under contract for the U.S. Army Research Institute, Ft. Rucker, AL. Military use only. |

| | |
|---|---|
| TITLE | **ADVANCED VIDEO MAINTENANCE INFORMATION** |
| DESCRIPTION | An explanation of videodisc capabilities, and a demonstration of the use of videodisc as a storage and presentation medium for maintenance information. |
| TYPE | Research and Development, Technical, Instructional, Archival, Catalog |
| FORMAT | Laser CAV |
| LEVEL | 2, 3 |
| DEVELOPER | Ken Pearce for McDonnell Douglas Corporation |
| CONTACT FOR MORE INFO. | Ken Pearce (213) 593-0593 |
| FEATURES | Chapters, Picture Stops, Digital Dump-Sony, Digital Dump-Pioneer, Dual Audio |
| PRODUCER | Ken Pearce for McDonnell Douglas Corporation |
| AVAILABILITY | Not for sale at present time. |

| | |
|---|---|
| TITLE | **ADVANCED VIDEO, INC. - (NOT GIVEN)** |
| DESCRIPTION | NFL football arcade game |
| TYPE | Game |
| RELEASE DATE | 12/83 |
| FORMAT | CED, Laser CAV |
| LEVEL | 3 |
| FEATURES | Dual Audio |

| | |
|---|---|
| TITLE | **ADVENTURE WITH DUSTY** |
| DESCRIPTION | Career education for hearing-impaired students. |
| TYPE | Adolescents, Instructional |
| FORMAT | Laser CAV |
| LEVEL | 3 |
| CONTACT FOR MORE INFO. | California School for the Deaf |
| PRODUCER | University of Nebraska/ Nebraska Videodisc Design/Production Group |
| AVAILABILITY | Not available for release or distribution. |

| | |
|---|---|
| TITLE | **ADVENTURES OF ROBIN HOOD, THE** |
| TYPE | Motion Picture |
| FORMAT | Laser CLV, CED |
| RUNNING TIME | 102 minutes |
| PRODUCER | Warner Home Video |
| DISTRIBUTOR | Pioneer Video, Inc., 4540-80, $34.98<br>RCA Videodiscs, 01446, $24.98 |

| | |
|---|---|
| TITLE | **ADVENTURES OF TOM SAWYER, THE** |
| TYPE | Motion Picture |
| FORMAT | CED |
| FEATURES | Stereo |
| RETAIL OUTLET | Movies Unlimited Inc., A1-4662, $29.98 |

| | |
|---|---|
| TITLE | **AEROBICISE** |
| TYPE | Instructional, Dance |
| FORMAT | Laser CLV, CED |
| FEATURES | Stereo, CX Encoded |

# Videodiscography

|              |                                           |
|---:|:---|
| RUNNING TIME | 113 minutes |
| PRODUCER | Paramount Home Video |
| DISTRIBUTOR | Pioneer Video, Inc., LV2308, $29.95 |
| | RCA Videodiscs, 13608, $24.98 |

|              |                                           |
|---:|:---|
| TITLE | **AEROBICISE: THE BEGINNING WORKOUT** |
| TYPE | Instructional, Dance |
| FORMAT | Laser CLV, CED |
| FEATURES | Stereo, CX Encoded |
| RUNNING TIME | 96 minutes |
| PRODUCER | Paramount Home Video |
| DISTRIBUTOR | Pioneer Video, Inc., LV2312, $29.95 |
| | RCA Videodiscs, 13617, $24.98 |

|              |                                           |
|---:|:---|
| TITLE | **AEROBICISE: THE ULTIMATE WORKOUT** |
| TYPE | Instructional, Dance |
| FORMAT | Laser CLV |
| FEATURES | Stereo, CX Encoded |
| RUNNING TIME | 105 minutes |
| PRODUCER | Paramount Home Video |
| DISTRIBUTOR | Pioneer Video, Inc., LV2313, $29.95 |

|              |                                           |
|---:|:---|
| TITLE | **AEROBICS - NEW VIDEO** |
| TYPE | Instructional, Dance |
| FORMAT | CED |
| RETAIL OUTLET | Movies Unlimited Inc., A3-3009, $29.95 |

|              |                                           |
|---:|:---|
| TITLE | **AFRICA SCREAMS** |
| TYPE | Motion Picture |
| FORMAT | CED |
| RETAIL OUTLET | Movies Unlimited Inc., A3-2014, $29.98 |

|              |                                           |
|---:|:---|
| TITLE | **AFRICAN QUEEN, THE** |
| TYPE | Motion Picture |
| FORMAT | Laser CLV, CED |
| RUNNING TIME | 105 minutes |
| PRODUCER | Warner Home Video |
| DISTRIBUTOR | Pioneer Video, Inc., 2025-80, $34.98 |
| | RCA Videodiscs, 01501, $24.98 |

|              |                                           |
|---:|:---|
| TITLE | **AGENCY** |
| TYPE | Motion Picture |
| FORMAT | CED |
| RETAIL OUTLET | Movies Unlimited Inc., A3-5008, $29.95 |

|              |                                           |
|---:|:---|
| TITLE | **AIDA** |
| TYPE | Music |
| FORMAT | Laser CLV |
| FEATURES | Stereo, CX Encoded, |
| RUNNING TIME | 154 minutes |
| PRODUCER | Pioneer Artists |
| DISTRIBUTOR | Pioneer Video, Inc., PA-82-017, $59.95 |

| | |
|---:|:---|
| TITLE | **AIR FORCE** |
| TYPE | Motion Picture |
| FORMAT | CED |
| PRODUCER | United Artists |
| DISTRIBUTOR | RCA Videodiscs, 03401, $24.98 |

| | |
|---:|:---|
| TITLE | **AIRPLANE** |
| TYPE | Motion Picture, Comedy (PG) |
| FORMAT | Laser CLV, CED |
| RUNNING TIME | 88 minutes |
| PRODUCER | Warner Home Video |
| DISTRIBUTOR | Pioneer Video, Inc., LV1305, $34.98 |
| | RCA Videodiscs, 00638, $29.98 |

| | |
|---:|:---|
| TITLE | **AIRPLANE II** |
| TYPE | Motion Picture, Comedy (R) |
| FORMAT | Laser CLV, CED |
| RUNNING TIME | 84 minutes |
| PRODUCER | Paramount Home Video |
| DISTRIBUTOR | Pioneer Video, Inc., LV1489, $29.95 |
| | RCA Videodiscs, 03613, $29.98 |

| | |
|---:|:---|
| TITLE | **AIRPORT** |
| TYPE | Motion Picture |
| FORMAT | CED (2 discs) |
| DISTRIBUTOR | RCA Videodiscs, 03307, $29.98 |

| | |
|---:|:---|
| TITLE | **ALAMO, THE** |
| TYPE | Motion Picture |
| FORMAT | CED (2 discs) |
| RETAIL OUTLET | Movies Unlimited Inc., A1-4561, $39.98 |

| | |
|---:|:---|
| TITLE | **ALICE DOESN'T LIVE HERE ANYMORE** |
| TYPE | Motion Picture |
| FORMAT | CED |
| DISTRIBUTOR | RCA Videodiscs, 03114, $24.98 |

| | |
|---:|:---|
| TITLE | **ALICE IN WONDERLAND** |
| TYPE | Motion Picture (G) |
| FORMAT | Laser CLV, CED |
| RUNNING TIME | 75 minutes |
| PRODUCER | Walt Disney |
| DISTRIBUTOR | Pioneer Video, Inc., 42036AS, $34.95 |
| | RCA Videodiscs, 00733, $24.98 |

| | |
|---:|:---|
| TITLE | **ALIEN** |
| TYPE | Motion Picture (R) |
| FORMAT | Laser CLV, CED |
| FEATURES | Stereo, CX Encoded |
| RUNNING TIME | 116 minutes |

|                |                                           |
|---------------:|:------------------------------------------|
| PRODUCER       | Warner Home Video                         |
| DISTRIBUTOR    | Pioneer Video, Inc., 1090-80, $34.98      |
| RETAIL OUTLET  | (CED) Movies Unlimited Inc., A1-1090, $29.98 |

|                |                                           |
|---------------:|:------------------------------------------|
| TITLE          | **ALL THAT JAZZ**                         |
| TYPE           | Motion Picture, Music, Dance (R)          |
| FORMAT         | Laser CLV, CED                            |
| FEATURES       | Stereo                                    |
| RUNNING TIME   | 123 minutes                               |
| PRODUCER       | Warner Home Video                         |
| DISTRIBUTOR    | Pioneer Video, Inc., 1095-80, $34.98      |
| RETAIL OUTLET  | (CED) Movies Unlimited Inc., A1-1095, $29.98 |

|                |                                           |
|---------------:|:------------------------------------------|
| TITLE          | **ALL THE MARBLES**                       |
| TYPE           | Motion Picture                            |
| FORMAT         | CED                                       |
| RETAIL OUTLET  | Movies Unlimited Inc., A2-1069, $29.95    |

|                |                                           |
|---------------:|:------------------------------------------|
| TITLE          | **ALL THE PRESIDENT'S MEN**               |
| TYPE           | Motion Picture                            |
| FORMAT         | CED (2 discs)                             |
| DISTRIBUTOR    | RCA Videodiscs, 03115, $39.98             |

|                         |                                           |
|------------------------:|:------------------------------------------|
| TITLE                   | **ALLEGHENY INTERNATIONAL INFOSOURCE I**  |
| DESCRIPTION             | Touch screen activated point-of-purchase system, featuring 29 houseware items from Sunbeam, Oster, Wilkinson Sword. Helps user select gift items, demonstrates products, answers questions, and suggests recipes. |
| TYPE                    | Point-of-Purchase                         |
| RELEASE DATE            | 9/83                                      |
| FORMAT                  | Laser CAV                                 |
| LEVEL                   | 3                                         |
| SYSTEM DESCRIPTION      | Sony LDP 1000, IBM PC, touch screen       |
| DEVELOPER               | MetaMedia and Allegheny International     |
| CONTACT FOR MORE INFO.  | Mary Wiemann, Allegheny International, (412) 562-5857 |
| PRODUCER                | MetaMedia                                 |
| SPECIAL CONDITIONS      | Will exhibit to selected groups.          |

|                |                                           |
|---------------:|:------------------------------------------|
| TITLE          | **ALLMAN BROTHERS BAND, THE**             |
| TYPE           | Music                                     |
| FORMAT         | CED                                       |
| FEATURES       | Stereo                                    |
| DISTRIBUTOR    | RCA Videodiscs, 12140, $19.98             |

|                |                                           |
|---------------:|:------------------------------------------|
| TITLE          | **ALONE IN THE DARK**                     |
| TYPE           | Motion Picture (R)                        |
| FORMAT         | CED                                       |
| PRODUCER       | New Line Cinema                           |
| DISTRIBUTOR    | RCA Videodiscs, 02171, $19.98             |

| | |
|---:|:---|
| TITLE | **ALTERED STATES** |
| TYPE | Motion Picture |
| FORMAT | CED |
| FEATURES | Stereo |
| DISTRIBUTOR | RCA Videodiscs, 13142, $29.98 |

| | |
|---:|:---|
| TITLE | **AMARCORD** |
| TYPE | Motion Picture |
| FORMAT | CED |
| DISTRIBUTOR | RCA Videodiscs, 00904, $24.98 |

| | |
|---:|:---|
| TITLE | **AMATEUR, THE** |
| TYPE | Motion Picture |
| FORMAT | CED |
| RETAIL OUTLET | Movies Unlimited Inc., A1-1147, $29.98 |

| | |
|---:|:---|
| TITLE | **AMAZING SPIDERMAN, THE** |
| TYPE | Motion Picture |
| FORMAT | CED |
| DISTRIBUTOR | Movies Unlimited Inc., A1-6114, $29.98 |

| | |
|---:|:---|
| TITLE | **AMERICA LIVE IN CENTRAL PARK** |
| TYPE | Music |
| FORMAT | Laser CAV |
| LEVEL | 1 |
| FEATURES | Stereo, CX Encoded |
| PRODUCER | Pioneer Artists |
| DISTRIBUTOR | Pioneer Video, Inc., PA-82-013, $24.95 |

| | |
|---:|:---|
| TITLE | **AMERICAN ALCOHOLIC/READING, WRITING AND REEFER** |
| TYPE | Motion Picture |
| FORMAT | CED |
| DISTRIBUTOR | RCA Videodiscs, 01208, $14.98 |

| | |
|---:|:---|
| TITLE | **AMERICAN GIGOLO** |
| TYPE | Motion Picture (R) |
| FORMAT | Laser CLV, CED |
| RUNNING TIME | 117 minutes |
| PRODUCER | Warner Home Video |
| DISTRIBUTOR | Pioneer Video, Inc., LV8989, $29.95 |
| | RCA Videodiscs, 00637, $29.98 |

| | |
|---:|:---|
| TITLE | **AMERICAN GRAFFITI** |
| TYPE | Motion Picture (PG) |
| FORMAT | Laser CLV, CED |
| FEATURES | Stereo, (CED) Banded |
| RUNNING TIME | 110 minutes |
| PRODUCER | Warner Home Video |
| DISTRIBUTOR | Pioneer Video, Inc., 16-001, $29.95 |
| | RCA Videodiscs, 13304, $29.98 |

# Videodiscography

| | |
|---:|:---|
| TITLE | **AMERICAN HOT WAX** |
| TYPE | Motion Picture, Music (R) |
| FORMAT | CED |
| RUNNING TIME | 91 minutes |
| PRODUCER | Paramount Pictures |
| DISTRIBUTOR | RCA Videodiscs, 00681, $24.98 |

| | |
|---:|:---|
| TITLE | **AMERICAN IN PARIS, AN** |
| TYPE | Motion Picture, Music, Dance |
| FORMAT | Laser CLV, CED |
| RUNNING TIME | 102 minutes |
| PRODUCER | MGM/UA Home Video |
| DISTRIBUTOR | Pioneer Video, Inc., ML100006, $34.95 |
| RETAIL OUTLET | (CED) Movies Unlimited Inc., A2-1001, $29.95 |

| | |
|---:|:---|
| TITLE | **AMERICAN WEREWOLF IN LONDON, AN** |
| TYPE | Motion Picture (R) |
| FORMAT | Laser CLV, CED |
| DISTRIBUTOR | RCA Videodiscs, 02113, $24.98 |
| RETAIL OUTLET | (CLV) UltraVideo, 40-017, $32.95 |

| | |
|---:|:---|
| TITLE | **AMITYVILLE HORROR, THE** |
| TYPE | Motion Picture (R) |
| FORMAT | Laser CLV, CED |
| RUNNING TIME | 118 minutes |
| PRODUCER | Vestron Video |
| DISTRIBUTOR | Pioneer Video, Inc., VL4022, $34.95 |
| | RCA Videodiscs, 01902, $24.98 |

| | |
|---:|:---|
| TITLE | **AMITYVILLE 2: THE POSSESSION** |
| TYPE | Motion Picture (R) |
| FORMAT | Laser CLV, CED |
| RUNNING TIME | 110 minutes |
| PRODUCER | Embassy Home Entertainment |
| DISTRIBUTOR | Pioneer Video, Inc., 17095, $29.95 |
| RETAIL OUTLET | (CED) Movies Unlimited Inc., A4-7096, $29.95 |

| | |
|---:|:---|
| TITLE | **AMTESS: AUTOMOTIVE** |
| DESCRIPTION | Troubleshooting the engine electrical system of the M110 self-propelled Howitzer. |
| TYPE | Technical, Instructional |
| RELEASE DATE | 6/83 |
| FORMAT | Laser CAV |
| LEVEL | 3 |
| SYSTEM DESCRIPTION | AMTESS, an integrated training system. Price quoted in response to official request for quote. |
| DEVELOPER | Grumman Aersopace Corporation |
| CONTACT FOR MORE INFO. | Chuck Myers |
| FEATURES | Dual Audio |
| PRODUCER | Grumman Aerospace Corporation |
| DISTRIBUTOR | U.S. Army |
| AVAILABILITY | Limited to distribution by U.S. Army. |

| | |
|---:|:---|
| TITLE | **AMTESS: MISSILE** |
| DESCRIPTION | Troubleshooting the transmitter of the improved high-power illuminator radar of the HAWK missile system. |
| TYPE | Technical, Instructional |
| RELEASE DATE | 8/83 |
| FORMAT | Laser CAV |
| LEVEL | 3 |
| SYSTEM DESCRIPTION | AMTESS, an integrated training system. Price quoted in response to official request for quote. |
| DEVELOPER | Grumman Aerospace Corporation |
| CONTACT FOR MORE INFO. | Jim Stonge |
| FEATURES | Dual Audio |
| PRODUCER | Grumman Aerospace Corporation |
| DISTRIBUTOR | U.S. Army |
| AVAILABILITY | Limited to distribution by U.S. Army. |

| | |
|---:|:---|
| TITLE | **AN/WLR-1 SYSTEM REPAIRABLE IDENTIFICATION** |
| DESCRIPTION | Disassembly from the top down, starting with the major assembly for the repairable units associated with AN/WLR-IG and its peripheral system. |
| TYPE | Technical, Instructional, In-house Training, Catalog |
| FORMAT | Laser CAV |
| LEVEL | 3 |
| SYSTEM DESCRIPTION | Disc operates with a Texas Instruments Professional Computer. The operator identifies a part using a reference symbol number, part number, or stock number. A visual presentation of the part and the associated supply data is presented to the operator. |
| DEVELOPER | Jonathan Corporation |
| CONTACT FOR MORE INFO. | William A. Goodrich, (703) 556-0100 |
| AVAILABILITY | Sale. Price on request. |

| | |
|---:|:---|
| TITLE | **AND GOD CREATED WOMAN** |
| TYPE | Motion Picture, Adult (R) |
| FORMAT | Laser CLV, CED |
| RUNNING TIME | 90 minutes |
| PRODUCER | Vestron Video |
| DISTRIBUTOR | Pioneer Video, Inc., VL3002, $34.95 |
| RETAIL OUTLET | (CED) Movies Unlimited Inc., A3-3002, $29.95 |

| | |
|---:|:---|
| TITLE | **...AND JUSTICE FOR ALL** |
| TYPE | Motion Picture (R) |
| FORMAT | Laser CLV, CED |
| DISTRIBUTOR | RCA Videodiscs, 03012, $29.98 |
| RETAIL OUTLET | (CLV) UltraVideo, VLD1020, $29.95 |

| | |
|---:|:---|
| TITLE | **ANDERSON TAPES, THE** |
| TYPE | Motion Picture |
| FORMAT | Laser CLV |
| RETAIL OUTLET | UltraVideo, VLD1025, $29.95 |

| | |
|---:|:---|
| TITLE | **ANGEL OF H.E.A.T.** |
| TYPE | Motion Picture (R) |
| FORMAT | Laser CLV, CED |
| RUNNING TIME | 93 minutes |
| PRODUCER | Warner Home Video |
| DISTRIBUTOR | Pioneer Video, Inc., VL4009, $29.95 |
| RETAIL OUTLET | (CED) Movies Unlimited Inc., A3-4009, $29.95 |

| | |
|---:|:---|
| TITLE | **ANGELS WITH DIRTY FACES** |
| TYPE | Motion Picture |
| FORMAT | CED |
| RETAIL OUTLET | Movies Unlimited Inc., A1-4588, $29.98 |

| | |
|---:|:---|
| TITLE | **ANIMAL CRACKERS** |
| TYPE | Motion Picture, Comedy |
| FORMAT | CED |
| DISTRIBUTOR | RCA Videodiscs, 03318, $19.98 |

| | |
|---:|:---|
| TITLE | **ANIMAL HOUSE** |
| TYPE | Motion Picture, Comedy (R) |
| FORMAT | Laser CLV |
| RUNNING TIME | 109 minutes |
| PRODUCER | Warner Home Video |
| DISTRIBUTOR | Pioneer Video, Inc., 16-007, $29.95 |

| | |
|---:|:---|
| TITLE | **ANNE BYRD'S COOKERY** |
| TYPE | Instructional |
| FORMAT | Laser CAV |
| RETAIL OUTLET | UltraVideo, I800-001, $49.95 |

| | |
|---:|:---|
| TITLE | **ANNIE** |
| TYPE | Motion Picture, Music (PG) |
| FORMAT | Laser CLV, CED (2 discs) |
| FEATURES | Stereo, CX Encoded |
| PRODUCER | RCA/Columbia Pictures Home Video |
| DISTRIBUTOR | Pioneer Video, Inc., VLD1030, $39.95 |
| | RCA Videodiscs, 13033, $39.98 |

| | |
|---:|:---|
| TITLE | **ANNIE HALL** |
| TYPE | Motion Picture, Comedy (PG) |
| FORMAT | Laser CLV, CED |
| RUNNING TIME | 93 minutes |
| PRODUCER | Warner Home Video |
| DISTRIBUTOR | Pioneer Video, Inc., 4518-80, $29.98 |
| | RCA Videodiscs, 01421, $29.98 |

| | |
|---:|:---|
| TITLE | **ANY WHICH WAY YOU CAN** |
| TYPE | Motion Picture, Comedy (R) |
| FORMAT | Laser CLV, CED |
| RUNNING TIME | 116 minutes |
| PRODUCER | Warner Home Video |
| DISTRIBUTOR | Pioneer Video, Inc., 11077LV, $29.98 |
| | RCA Videodiscs, 03144, $29.98 |

| | |
|---:|:---|
| TITLE | **APACHE** |
| TYPE | Motion Picture |
| FORMAT | CED |
| RETAIL OUTLET | Movies Unlimited Inc., A1-4504, $29.98 |

| | |
|---:|:---|
| TITLE | **APARTMENT, THE** |
| TYPE | Motion Picture, Comedy |
| FORMAT | CED |
| DISTRIBUTOR | RCA Videodiscs, 01461, $24.98 |

| | |
|---:|:---|
| TITLE | **APOCALYPSE NOW** |
| TYPE | Motion Picture (R) |
| FORMAT | Laser CLV, CED (2 discs) |
| FEATURES | Stereo, CX Encoded, |
| RUNNING TIME | 153 minutes |
| PRODUCER | Warner Home Video |
| DISTRIBUTOR | Pioneer Video, Inc., LV2306, $39.95 |
| | RCA Videodiscs, 00667, $39.98 |

| | |
|---:|:---|
| TITLE | **APOLLO** |
| DESCRIPTION | Approximately 10,000 NASA photos of the Apollo 11,12,14,15,16, and 17 missions featuring astronaut training, spacecraft preparation, launch, mission, and recovery. Coverage of the "on-board" mission photography includes almost everything shot by the astronauts during each mission; also movie highlights of each mission. |
| TYPE | Technical, Instructional, Archival |
| FORMAT | Laser CAV |
| LEVEL | 1 |
| DEVELOPER | Drew University/Center for Aerospace Education in cooperation with NASA |
| CONTACT FOR MORE INFO. | Patrick Binnes, (201) 377-0302 |
| FEATURES | Chapters, Picture Stops, Stereo |
| PRODUCER | Video Vision Associates, Ltd. |
| DISTRIBUTOR | Video Vision Associates, Ltd. |
| AVAILABILITY | Sale |
| ORDER NO. | SPACE DISC 2 |
| SUGGESTED PRICE | $320 |

| | |
|---:|:---|
| TITLE | **APPLE DUMPLING GANG, THE** |
| TYPE | Motion Picture, Comedy (G) |
| FORMAT | CED |
| PRODUCER | Walt Disney |
| DISTRIBUTOR | RCA Videodiscs, 00712, $24.98 |

| | |
|---:|:---|
| TITLE | **APRIL WINE** |
| TYPE | Music |
| FORMAT | Laser CLV |
| FEATURES | Stereo, CX Encoded |
| RUNNING TIME | 58 minutes |
| PRODUCER | Pioneer Artists |
| DISTRIBUTOR | Pioneer Video, Inc., PA-82-025, $24.95 |

| | |
|---:|:---|
| TITLE | **ARIS KNITWEAR** |
| DESCRIPTION | A point-of-purchase disc that allows shoppers to see the full range of ARIS knitwear products, and select the product best suited to their clothing needs. |
| TYPE | Catalog, Point-of-Purchase, Advertising, Promotional |
| RELEASE DATE | 1983 |
| FORMAT | Laser CAV |
| LEVEL | 2 |
| DEVELOPER | Exhibit Technology, Inc./ARIS Knitwear |
| CONTACT FOR MORE INFO. | Don Cochrane, President - Marketing |
| FEATURES | Chapters, Picture Stops, Digital Dump-Sony, Dual Audio, Sound Over Still |
| PRODUCER | John Rafalski, Exhibit Technology |
| DISTRIBUTOR | ARIS Knitwear - Bloomingdale's |
| AVAILABILITY | All major department stores in the New York metropolitan area will have kiosks in their stores during the 1983 holiday season. |
| SPECIAL CONDITIONS | Available for viewing as part of Exhibit Technology's product line sampling. |

| | |
|---:|:---|
| TITLE | **ARMY COMMUNICATIVE TECHNOLOGY OFFICE (ACTO) VIDEODISC PROTOTYPES** |
| DESCRIPTION | This is a series of ten discs on training and maintenance applications. |
| TYPE | Research and Development, Technical, Instructional, In-house Training, Catalog, Maintenance and Paper Conversions |
| RELEASE DATE | 1982, 1983 |
| FORMAT | Laser CAV |
| LEVEL | 1,2,3 |
| SYSTEM DESCRIPTION | Apple II or SMC 70 computers, Pioneer or Sony videodisc players, touch panel or light pen |
| DEVELOPER | Computer Sciences Corporation |
| CONTACT FOR MORE INFO. | J.G. Palmer, Computer Sciences Corporation P.O. Box M, Ft Eustis, VA 23662 |
| FEATURES | Digital Dump-Pioneer, Dual Audio |
| PRODUCER | Computer Sciences Corporation |
| DISTRIBUTO | ACTO |
| AVAILABILITY | Computer Sciences Corporation can make turnkey systems at several locations. |

| | |
|---:|:---|
| TITLE | **ARSENIC AND OLD LACE** |
| TYPE | Motion Picture |
| FORMAT | CED |
| DISTRIBUTOR | RCA Videodiscs, 03402, $24.98 |

| | |
|---:|:---|
| TITLE | **ART HISTORY RETRIEVAL** |
| DESCRIPTION | One thousand black and white glossy photographs of woodcuts and engravings by the artists Duerer and Raimondi were photographed onto 16 mm film, then transferred to disc. Each image was identified by relevant variables. Users can select all images in a category based on identifying variables, or select a subset of images by intersecting more than one variable. A computer controls the selection and commands the videodisc to randomly access and display the image and all associated information. |
| TYPE | Art, Research and Development, Archival |
| FORMAT | Thomson-CSF transmissive |
| LEVEL | 3 |
| SYSTEM DESCRIPTION | Thomson-CSF TTV 3620 player, HP2000 minicomputer and Apple microcomputer, Sony monitor, and Apple/Thomson driver software developed at the CAI lab for an IEEE 488 board from California Computer Systems. |
| CONTACT FOR MORE INFO. | Dr. Joan Sustik Huntley, (319) 353-3170 |
| PRODUCER | University of Iowa/Computer Assisted Instruction Lab |
| AVAILABILITY | Since this was a research and development effort, neither the disc nor software are publicly available. Interested people may view the program any time at the Weeg Computing Center. |

| | |
|---:|:---|
| TITLE | **ARTHUR** |
| TYPE | Motion Picture, Comedy (PG) |
| FORMAT | Laser CLV, CED |
| RUNNING TIME | 97 minutes |
| PRODUCER | Warner Home Video |
| DISTRIBUTOR | Pioneer Video, Inc., 22020LV, $29.98 |
| | RCA Videodiscs, 03138, $29.98 |

| | |
|---:|:---|
| TITLE | **ASHFORD & SIMPSON** |
| TYPE | Music |
| FORMAT | Laser CLV, CED |
| FEATURES | Stereo, CX Encoded, (CED) Banded |
| RUNNING TIME | 56 minutes |
| PRODUCER | Pioneer Artists |
| DISTRIBUTOR | Pioneer Video, Inc., PA-83-041, $24.95 |
| | RCA Videodiscs, 12142, $24.98 |

| | |
|---:|:---|
| TITLE | **ASIA TRAVEL** |
| DESCRIPTION | Part of the Sony Travel Guide System, which shows prospective travelers highlights of the places they are considering visiting, and provides travel information on printouts. |
| TYPE | Promotional |
| FORMAT | Laser CAV |
| LEVEL | 2 |
| CONTACT FOR MORE INFO. | Sony Corporation of America, (201) 930-1000 |
| FEATURES | Digital Dump-Sony, Stereo, Dual Audio |
| RUNNING TIME | 30 minutes |
| DISTRIBUTOR | Sony Corporation of America |

| | |
|---|---|
| TITLE | **ASSESSMENT OF NEUROMOTOR DYSFUNCTION IN INFANTS** |
| DESCRIPTION | Twenty-seven infants, newborn to 18 months old, provide visual examples of normal and abnormal neuromuscular conditions. Users learn to identify early signs of neuromuscular dysfunction through five parameters: muscle function, reflexes, movement, structure, and gross motor skills. |
| TYPE | Medical, Instructional |
| RELEASE DATE | 1984 |
| FORMAT | Laser CAV |
| LEVEL | 3 |
| SYSTEM DESCRIPTION | IBM PC, Allen UVC, Pioneer 7820-3 (current configuration, will change when marketed) |
| DEVELOPER | Joan Sustik Huntley/Dr. James Blackman/Loretta Lough |
| CONTACT FOR MORE INFO. | Joan Sustik Huntley, (319) 353-7253 |
| PRODUCER | University of Iowa/Weeg Computing Center and the Division of Developmental Disabilities |
| DISTRIBUTOR | Williams & Wilkins |
| AVAILABILITY | Sale |

| | |
|---|---|
| TITLE | **ASTRONOMY** |
| DESCRIPTION | Covers the planets, the sun, and deep sky. Includes still photographs collected from observatories as well as motion footage. |
| TYPE | Technical, Instructional, Archival |
| RELEASE DATE | 3/84 |
| FORMAT | Laser CAV |
| LEVEL | 1 |
| DEVELOPER | Drew University/Center for Aerospace Education in cooperation with NASA |
| CONTACT FOR MORE INFO. | Patrick Binnes, (201) 377-0302 |
| FEATURES | Chapters, Picture Stops, Stereo |
| PRODUCER | Video Vision Associates, Ltd. |
| DISTRIBUTOR | Video Vision Associates, Ltd. |
| AVAILABILITY | Sale |
| ORDER NO. | SPACE DISC 4 |
| SUGGESTED PRICE | $320 |

| | |
|---|---|
| TITLE | **ATLANTIC CITY** |
| TYPE | Motion Picture (R) |
| FORMAT | Laser CLV, CED |
| RUNNING TIME | 104 minutes |
| PRODUCER | Warner Home Video |
| DISTRIBUTOR | Pioneer Video, Inc., LV1460, $29.95<br>RCA Videodiscs, 00674, $24.98 |

| | |
|---|---|
| TITLE | **ATTI (AMERICAN TELEPHONE & TELEGRAPH INTERNATIONAL)** |
| DESCRIPTION | Interactive disc informing viewers of the activities of this marketing arm of American-based AT&T. |
| TYPE | Promotional |
| RELEASE DATE | 1983 |
| FORMAT | Laser CAV |
| LEVEL | 3 |
| PRODUCER | Brien Lee & Company |
| DISTRIBUTOR | ATTI, Basking Ridge, NJ |
| AVAILABILITY | For trade show exhibition, in terminal packaged by Bell Labs. |

| | |
|---:|:---|
| TITLE | **AUTHOR, AUTHOR** |
| TYPE | Motion Picture |
| FORMAT | CED |
| RETAIL OUTLET | Movies Unlimited Inc., A1-1181, $29.98 |

| | |
|---:|:---|
| TITLE | **AUTOBIOGRAPHY OF MISS JANE PITTMAN, THE** |
| TYPE | Motion Picture |
| FORMAT | CED |
| DISTRIBUTOR | RCA Videodiscs, 02023, $19.98 |

| | |
|---:|:---|
| TITLE | **AUTUMN SONATA** |
| TYPE | Motion Picture (PG) |
| FORMAT | Laser CLV |
| RUNNING TIME | 97 minutes |
| PRODUCER | Warner Home Video |
| DISTRIBUTOR | Pioneer Video, Inc., 9021-80, $29.98 |

| | |
|---:|:---|
| TITLE | **BACK ROADS** |
| TYPE | Motion Picture |
| FORMAT | CED |
| RETAIL OUTLET | Movies Unlimited Inc., A2-1038, $24.95 |

| | |
|---:|:---|
| TITLE | **BAD NEWS BEARS, THE** |
| TYPE | Motion Picture (PG) |
| FORMAT | Laser CLV, CED |
| RUNNING TIME | 102 minutes |
| PRODUCER | Warner Home Video |
| DISTRIBUTOR | Pioneer Video, Inc., LV8863, $29.95 |
| | RCA Videodiscs, 00609, $24.98 |

| | |
|---:|:---|
| TITLE | **BAD NEWS BEARS IN BREAKING TRAINING, THE** |
| TYPE | Motion Picture (PG) |
| FORMAT | CED |
| DISTRIBUTOR | RCA Videodiscs, 00682, $24.98 |

| | |
|---:|:---|
| TITLE | **BAFFLED** |
| TYPE | Motion Picture |
| FORMAT | CED |
| RETAIL OUTLET | Movies Unlimited Inc., A1-9033, $29.98 |

| | |
|---:|:---|
| TITLE | **BALLET SERIES** |
| DESCRIPTION | A variety of computer controlled programs are based on this disc, which was filmed at the University of Iowa specifically for this project. The correct pronunciation and spelling accompany each of 190 positions or movements. Drill and test items are randomly generated. |
| TYPE | Dance, Research and Development, Instructional |
| FORMAT | Thomson-CSF transmissive |
| LEVEL | 3 |

| | |
|---|---|
| SYSTEM DESCRIPTION | Thomson-CSF TTV 3620 player; Apple II single disk drive computer; Sony color monitor for videodisc; Sanyo black and white monitor for microcomputer; keyboard game paddle or Symtec light pen; Apple/Thomson driver software developed at the CAI Lab for an IEEE 488 board from California Computer System; video switch developed at the University of Iowa. |
| CONTACT FOR MORE INFO. | Joan Sustik Huntley, (319) 353-3170 |
| PRODUCER | University of Iowa/Computer Assisted Instruction Lab |
| AVAILABILITY | Since this was a research and development effort, neither the disc nor software are publicly available. Interested people may view the program any time at the Weeg Computing Center. |

| | |
|---|---|
| TITLE | **BANANAS** |
| TYPE | Motion Picture, Comedy (R) |
| FORMAT | Laser CLV, CED |
| FEATURES | Stereo |
| RUNNING TIME | 82 minutes |
| PRODUCER | CBS Fox Video |
| DISTRIBUTOR | Pioneer Video, Inc., 4555-80, $34.98 |
| | RCA Videodiscs, 01413, $24.98 |

| | |
|---|---|
| TITLE | **BANG THE DRUM SLOWLY** |
| TYPE | Motion Picture (PG) |
| FORMAT | Laser CLV |
| RUNNING TIME | 98 minutes |
| PRODUCER | Warner Home Video |
| DISTRIBUTOR | Pioneer Video, Inc., LV8732, $29.95 |

| | |
|---|---|
| TITLE | **BARBARELLA** |
| TYPE | Motion Picture (PG) |
| FORMAT | Laser CLV, CED |
| RUNNING TIME | 98 minutes |
| PRODUCER | Warner Home Video |
| DISTRIBUTOR | Pioneer Video, Inc., LV6812, $29.95 |
| | RCA Videodiscs, 00610, $29.98 |

| | |
|---|---|
| TITLE | **BARBAROSA** |
| TYPE | Motion Picture |
| FORMAT | CED |
| RETAIL OUTLET | Movies Unlimited Inc., A1-9048, $29.98 |

| | |
|---|---|
| TITLE | **BAREFOOT CONTESSA** |
| TYPE | Motion Picture |
| FORMAT | CED (2 Discs) |
| PRODUCER | CBS/Fox Video |
| RETAIL OUTLET | Movies Unlimited Inc., A1-4505, $39.98 |

| | |
|---:|:---|
| TITLE | **BAREFOOT IN THE PARK** |
| TYPE | Motion Picture, Comedy (G) |
| FORMAT | Laser CLV, CED |
| RUNNING TIME | 106 minutes |
| PRODUCER | Paramount Home Video |
| DISTRIBUTOR | Pioneer Video, Inc., LV8027, $29.95 |
| | RCA Videodiscs, 03622, $24.98 |

| | |
|---:|:---|
| TITLE | **BASEBALL FUN AND GAMES** |
| TYPE | Motion Picture, Game |
| FORMAT | CED |
| DISTRIBUTOR | RCA Videodiscs, 02112, $19.98 |

| | |
|---:|:---|
| TITLE | **BASEBALL'S HALL OF FAME** |
| TYPE | Motion Picture |
| FORMAT | CED |
| RETAIL OUTLET | Movies Unlimited Inc., A3-E860, $29.95 |

| | |
|---:|:---|
| TITLE | **BASIC AUTOMOTIVE ELECTRONICS** |
| DESCRIPTION | Service training for the Ford video network. |
| TYPE | Technical, Instructional, In-house Training |
| RELEASE DATE | 6/82 |
| LEVEL | 2 |
| CONTACT FOR MORE INFO. | Raymond J. Marx |
| FEATURES | Chapters, Picture Stops, Digital Dump-Sony |
| RUNNING TIME | 27 minutes |
| PRODUCER | Ron Herman |
| DISTRIBUTOR | Ford Video Network |

| | |
|---:|:---|
| TITLE | **BASIC DIESEL PRINCIPLES** |
| DESCRIPTION | Service training for the Ford video network. |
| TYPE | Technical, Instructional, In-house Training |
| RELEASE DATE | 1/83 |
| LEVEL | 2 |
| CONTACT FOR MORE INFO. | Raymond J. Marx |
| FEATURES | Chapters, Picture Stops, Digital Dump-Sony |
| RUNNING TIME | 29 minutes, 40 seconds |
| PRODUCER | Ron Herman |
| DISTRIBUTOR | Ford Video Network |

| | |
|---:|:---|
| TITLE | **BASIC KARATE AND SELF DEFENSE** |
| TYPE | Instructional |
| FORMAT | Laser CAV |
| RETAIL OUTLET | UltraVideo, I230-001, $49.95 |

| | |
|---:|:---|
| TITLE | **BASIC SKILLS EDUCATION FOR THE U.S. ARMY** |
| DESCRIPTION | Interactive videodiscs for adult basic skills education; such as study skills, test-taking skills, test anxiety reduction, spatial orientation, and land navigation skills; using a spatial data management system for information delivery. |

|  |  |
|---|---|
| TYPE | Game, Research and Development, Technical, Instructional |
| FORMAT | Laser CAV |
| LEVEL | 3 |
| FEATURES | Dual Audio |
| PRODUCER | Human Resources Research Organization (HumRRO) |
| AVAILABILITY | Loan |
| SPECIAL CONDITIONS | With permission of the U.S. Army Research Institute |

|  |  |
|---|---|
| TITLE | **BASIC TUMBLING SKILLS** |
| DESCRIPTION | An interactive disc designed for elementary classroom instruction. |
| TYPE | Children, Instructional |
| RELEASE DATE | 1979 |
| FORMAT | Laser CAV |
| LEVEL | 2 |
| CONTACT FOR MORE INFO. | Corporation for Public Broadcasting - KUON-TV, |
| PRODUCER | University of Nebraska/Nebraska Videodisc Design Production Group |
| AVAILABILITY | Not available for release or distribution. |

|  |  |
|---|---|
| TITLE | **BATTLESTAR GALACTICA** |
| TYPE | Motion Picture (PG) |
| FORMAT | Laser CLV, CED |
| RUNNING TIME | 123 minutes |
| PRODUCER | Warner Home Video |
| DISTRIBUTOR | Pioneer Video, Inc., 19-007, $29.95 |
| RETAIL OUTLET | (CED) Movies Unlimited Inc., A5-9007, $29.98 |

|  |  |
|---|---|
| TITLE | **BEACH GIRLS, THE** |
| TYPE | Motion Picture, Comedy (R) |
| FORMAT | Laser CLV, CED |
| RUNNING TIME | 91 minutes |
| PRODUCER | Paramount Home Video |
| DISTRIBUTOR | Pioneer Video, Inc., LV2314, $29.95 |
|  | RCA Videodiscs, 03614, $24.98 |

|  |  |
|---|---|
| TITLE | **BEARS AND I, THE** |
| TYPE | Motion Picture (G) |
| FORMAT | CED |
| PRODUCER | Walt Disney |
| DISTRIBUTOR | RCA Videodiscs, 00704, $24.98 |

|  |  |
|---|---|
| TITLE | **BEAST WITHIN, THE** |
| TYPE | Motion Picture |
| FORMAT | CED |
| RETAIL OUTLET | Movies Unlimited Inc., A2-0172, $29.95 |

|  |  |
|---|---|
| TITLE | **BEASTMASTER, THE** |
| TYPE | Motion Picture |
| FORMAT | Laser CLV, CED |
| RETAIL OUTLET | (CLV) UltraVideo, ML100226, $34.95 |
|  | (CED) Movies Unlimited Inc., A2-0226, $29.95 |

| | |
|---:|:---|
| TITLE | **BEDKNOBS & BROOMSTICKS** |
| TYPE | Motion Picture (G) |
| FORMAT | Laser CLV |
| RUNNING TIME | 117 minutes |
| PRODUCER | Walt Disney |
| DISTRIBUTOR | Pioneer Video, Inc., 42016AS, $34.98 |

| | |
|---:|:---|
| TITLE | **BEING THERE** |
| TYPE | Motion Picture |
| FORMAT | Laser CLV, CED (2 Discs) |
| RETAIL OUTLET | (CLV) UltraVideo, 80-7026, $39.95 |
| | (CED) Movies Unlimited Inc., A1-7026, $39.98 |

| | |
|---:|:---|
| TITLE | **BELL, BOOK & CANDLE** |
| TYPE | Motion Picture |
| FORMAT | CED |
| DISTRIBUTOR | RCA Videodiscs, 03028, $24.98 |

| | |
|---:|:---|
| TITLE | **BELLS ARE RINGING** |
| TYPE | Motion Picture |
| FORMAT | CED |
| RETAIL OUTLET | Movies Unlimited Inc., A2-1025, $29.95 |

| | |
|---:|:---|
| TITLE | **BELLY DANCING** |
| TYPE | Dance, Instructional |
| FORMAT | Laser CAV |
| LEVEL | 1 |
| PRODUCER | Optical Programming Associates |
| DISTRIBUTOR | Pioneer Video, Inc., 37-602, $29.95 |

| | |
|---:|:---|
| TITLE | **BEN HUR** |
| TYPE | Motion Picture |
| FORMAT | Laser CLV, CED (2 Discs) |
| RETAIL OUTLET | (CLV) UltraVideo, ML100226, $34.95 |
| | (CED) Movies Unlimited Inc., A2-0004, $39.95 |

| | |
|---:|:---|
| TITLE | **BENJI** |
| TYPE | Motion Picture |
| FORMAT | CED |
| RETAIL OUTLET | Movies Unlimited Inc., A2-1017, $24.95 |

| | |
|---:|:---|
| TITLE | **BERLIN I** |
| DESCRIPTION | Surrogate travel of West Berlin |
| TYPE | In-house Training |
| FORMAT | Laser CAV |
| LEVEL | 3 |
| DEVELOPER | Interactive Television Company |
| CONTACT FOR MORE INFO. | Dr. Steven Levin, Interactive Television Company |
| RUNNING TIME | 30 minutes |
| AVAILABILITY | Contact Interactive Television Company |

| | |
|---:|:---|
| TITLE | **BERLIN II** |
| DESCRIPTION | Surrogate travel of West Berlin |
| TYPE | In-house Training |
| FORMAT | Laser CAV |
| LEVEL | 3 |
| DEVELOPER | Interactive Television Company |
| CONTACT FOR MORE INFO. | Dr. Steven Levin, Interactive Television Company |
| RUNNING TIME | 30 minutes |
| AVAILABILITY | Contact Interactive Television Company |

| | |
|---:|:---|
| TITLE | **BEST FRIENDS** |
| TYPE | Motion Picture, Comedy (PG) |
| FORMAT | Laser CLV, CED |
| RUNNING TIME | 108 minutes |
| PRODUCER | Warner Home Video |
| DISTRIBUTOR | Pioneer Video, Inc., 11265LV, $29.98 |
| RETAIL OUTLET | (CED) Movies Unlimited Inc., A1-1265, $29.98 |

| | |
|---:|:---|
| TITLE | **BEST LITTLE WHOREHOUSE IN TEXAS, THE** |
| TYPE | Motion Picture, Music, Comedy (R) |
| FORMAT | Laser CLV, CED |
| RUNNING TIME | 114 minutes |
| PRODUCER | MCA Videodisc, Inc. |
| DISTRIBUTOR | Pioneer Video, Inc., 17-008, $32.95 |
| RETAIL OUTLET | (CED) Movies Unlimited Inc., A5-7008, $34.98 |

| | |
|---:|:---|
| TITLE | **BETWEEN THE LINES** |
| TYPE | Motion Picture |
| FORMAT | CED |
| RETAIL OUTLET | Movies Unlimited Inc., A3-5002, $29.95 |

| | |
|---:|:---|
| TITLE | **BIG BAD MAMA** |
| TYPE | Motion Picture (R) |
| FORMAT | CED |
| DISTRIBUTOR | RCA Videodiscs, 00903, $24.98 |

| | |
|---:|:---|
| TITLE | **BIG BLUE MARBLE PRESENTS MY SEVENTEENTH SUMMER** |
| TYPE | Children, Instructional |
| FORMAT | CED |
| DISTRIBUTOR | RCA Videodiscs, 02027, $14.98 |

| | |
|---:|:---|
| TITLE | **BIG FIGHTS, VOL.1 - MUHAMMAD ALI'S GREATEST FIGHTS** |
| TYPE | Motion Picture |
| FORMAT | CED |
| DISTRIBUTOR | RCA Videodiscs, 01701, $19.98 |

| | |
|---:|:---|
| TITLE | **BIG FIGHTS, VOL.2 - HEAVYWEIGHT CHAMPIONS' GREATEST FIGHTS** |
| TYPE | Motion Picture |
| FORMAT | CED |
| DISTRIBUTOR | RCA Videodiscs, 01704, $19.98 |

| | |
|---:|:---|
| TITLE | **BIG FIGHTS, VOL.3 - SUGAR RAY ROBINSON'S GREATEST FIGHTS** |
| TYPE | Motion Picture |
| FORMAT | CED |
| DISTRIBUTOR | RCA Videodiscs, 01706, $19.98 |

| | |
|---:|:---|
| TITLE | **BIG RED ONE, THE** |
| TYPE | Motion Picture |
| FORMAT | CED |
| RETAIL OUTLET | Movies Unlimited Inc., A2-1026, $24.95 |

| | |
|---:|:---|
| TITLE | **BIG SLEEP, THE** |
| TYPE | Motion Picture |
| FORMAT | CED |
| DISTRIBUTOR | RCA Videodiscs, 01429, $24.98 |

| | |
|---:|:---|
| TITLE | **BILLION DOLLAR HOBO, THE** |
| TYPE | Motion Picture |
| FORMAT | CED |
| RETAIL OUTLET | Movies Unlimited Inc., A1-7095, $29.98 |

| | |
|---:|:---|
| TITLE | **BILLY JACK** |
| TYPE | Motion Picture (PG) |
| FORMAT | CED |
| PRODUCER | Warner Home Video |
| DISTRIBUTOR | RCA Videodiscs, 03116, $24.98 |

| | |
|---:|:---|
| TITLE | **BILLY SQUIER LIVE IN THE DARK** |
| TYPE | Music |
| FEATURES | Stereo, CX Encoded |
| RUNNING TIME | 59 minutes |
| PRODUCER | Pioneer Artists |
| DISTRIBUTOR | Pioneer Video, Inc., PA-83-036, $24.95 |

| | |
|---:|:---|
| TITLE | **BINDU VISUALIZATION OF MUSIC** |
| DESCRIPTION | Jazz, classical, blues, and rock music sampler demonstrating a new visual language. |
| TYPE | Music, Game |
| RELEASE DATE | Fall 1984 |
| FORMAT | Laser CAV |
| LEVEL | 0 |
| DEVELOPER | Bindu Productions |
| CONTACT FOR MORE INFO. | Dr. Sanford R. Cohen, (614) 291-2223 |
| FEATURES | Stereo |
| RUNNING TIME | 45 minutes |
| PRODUCER | Bindu Productions |
| DISTRIBUTOR | Not known at this time. |
| AVAILABILITY | Sale |

| | |
|---|---|
| TITLE | **BIO SCI VIDEODISC, THE** |
| DESCRIPTION | A visual data base of 6,000 superb images supporting biological topics from cell chemistry to plant and animal diversity. Sections of motion pictures on frog development, DVA replication, protein synthesis, and molecular dynamics are included. |
| TYPE | Instructional, Archival |
| RELEASE DATE | 10/83 |
| FORMAT | Laser CAV |
| LEVEL | 3 |
| SYSTEM DESCRIPTION | Pioneer 8210 and Sony 1000A players are being used, although disc will operate with all major videodisc players and any number of computers including Apple, TRS80, and IBM PC. |
| DEVELOPER | D. Joseph Clark |
| CONTACT FOR MORE INFO. | D. Joseph Clark, (206) 524-4007 |
| FEATURES | Chapters |
| PRODUCER | D. Joseph Clark |
| DISTRIBUTOR | VIDEODISCOVERY |
| AVAILABILITY | Sale |
| SUGGESTED PRICE | $495 |

| | |
|---|---|
| TITLE | **BIOLOGY I & II** |
| DESCRIPTION | Laboratory simulation for college level |
| TYPE | Instructional |
| FORMAT | Laser CAV |
| LEVEL | 3 |
| CONTACT FOR MORE INFO. | Corporation for Public Broadcasting/Annenberg |
| PRODUCER | University of Nebraska/Nebraska Videodisc Design/Production Group |
| AVAILABILITY | Not available for release or distribution. |

| | |
|---|---|
| TITLE | **BIRDMAN OF ALCATRAZ** |
| TYPE | Motion Picture |
| FORMAT | CED (2 discs) |
| RUNNING TIME | 148 minutes |
| PRODUCER | United Artists |
| DISTRIBUTOR | RCA Videodiscs, 01463, $39.98 |

| | |
|---|---|
| TITLE | **BIRDS, THE** |
| TYPE | Motion Picture |
| FORMAT | CED |
| PRODUCER | Universal Pictures |
| DISTRIBUTOR | RCA Videodiscs, 03317, $19.98 |

| | |
|---|---|
| TITLE | **BLACK HOLE, THE** |
| TYPE | Motion Picture (PG) |
| FORMAT | Laser CLV, CED |
| FEATURES | Stereo, CX Encoded |
| RUNNING TIME | 97 minutes |
| PRODUCER | Warner Home Video |
| DISTRIBUTOR | Pioneer Video, Inc., 42011AS, $34.95 |
| | RCA Videodiscs, 10724, $29.98 |

| | |
|---:|:---|
| TITLE | **BLACK ORPHEUS** |
| TYPE | Motion Picture |
| FORMAT | CED |
| RETAIL OUTLET | Movies Unlimited Inc., A1-7054, $29.98 |

| | |
|---:|:---|
| TITLE | **BLACK STALLION, THE** |
| TYPE | Motion Picture (G) |
| FORMAT | Laser CLV, CED |
| FEATURES | Stereo, CX Encoded |
| PRODUCER | Warner Home Video |
| DISTRIBUTOR | Pioneer Video, Inc., 4503-80, $39.98 |
| | RCA Videodiscs, 01409 (Stereo 11431), $24.98 |

| | |
|---:|:---|
| TITLE | **BLACK STALLION RETURNS, THE** |
| TYPE | Motion Picture |
| FORMAT | CED |
| FEATURES | Stereo |
| DISTRIBUTOR | RCA Videodiscs, 13424, $24.98 |

| | |
|---:|:---|
| TITLE | **BLACK SUNDAY** |
| TYPE | Motion Picture (R) |
| FORMAT | Laser CLV, CED (2 Discs) |
| DISTRIBUTOR | RCA Videodiscs, 00643, $34.98 |
| RETAIL OUTLET | (CLV) UltraVideo, LV88552, $35.95 |

| | |
|---:|:---|
| TITLE | **BLACKJACK: HOW TO PLAY TO WIN** |
| RELEASE DATE | 1982 |
| PRODUCER | Sheldon Renan |
| DISTRIBUTOR | VHD Programs, Inc. |

| | |
|---:|:---|
| TITLE | **BLADE RUNNER** |
| TYPE | Motion Picture (R) |
| FORMAT | Laser CLV, CED |
| FEATURES | Stereo |
| RUNNING TIME | 117 minutes |
| PRODUCER | Embassy Home Entertainment |
| DISTRIBUTOR | Pioneer Video, Inc., 13805, $29.95 |
| RETAIL OUTLET | (CED) Movies Unlimited Inc., A4-3806, $29.95 |

| | |
|---:|:---|
| TITLE | **BLAZING SADDLES** |
| TYPE | Motion Picture, Comedy (R) |
| FORMAT | Laser CLV, CED |
| RUNNING TIME | 90 minutes |
| PRODUCER | Warner Home Video |
| DISTRIBUTOR | Pioneer Video, Inc., 1001LV, $29.98 |
| | RCA Videodiscs, 03103, $29.98 |

| | |
|---:|:---|
| TITLE | **BLONDIE: EAT TO THE BEAT** |
| TYPE | Music |
| FORMAT | CED |
| DISTRIBUTOR | RCA Videodiscs, 02025, $19.98 |

| | |
|---:|:---|
| TITLE | **BLOW OUT** |
| TYPE | Motion Picture (R) |
| FORMAT | Laser CLV, CED |
| FEATURES | Stereo |
| RUNNING TIME | 107 minutes |
| PRODUCER | Vestron Video |
| DISTRIBUTOR | Pioneer Video, Inc., VL4023, $34.95 |
| | RCA Videodiscs, 01903, $24.98 |

| | |
|---:|:---|
| TITLE | **BLUE HAWAII** |
| TYPE | Motion Picture, Music |
| FORMAT | Laser CLV, CED |
| FEATURES | Stereo |
| RUNNING TIME | 101 minutes |
| PRODUCER | Warner Home Video |
| DISTRIBUTOR | Pioneer Video, Inc., 2001-80, $29.98 |
| | RCA Videodiscs, 01506, $24.98 |

| | |
|---:|:---|
| TITLE | **BLUE LAGOON, THE** |
| TYPE | Motion Picture (R) |
| FORMAT | Laser CLV, CED |
| DISTRIBUTOR | RCA Videodiscs, 03003, $29.98 |
| RETAIL OUTLET | (CLV) UltraVideo, VLD2990, $29.95 |

| | |
|---:|:---|
| TITLE | **BLUE MAX, THE** |
| TYPE | Motion Picture |
| FORMAT | CED (2 Discs) |
| RETAIL OUTLET | Movies Unlimited Inc., A1-1062, $39.98 |

| | |
|---:|:---|
| TITLE | **BLUE THUNDER** |
| TYPE | Motion Picture (R) |
| FORMAT | CED |
| FEATURES | Stereo |
| PRODUCER | Columbia Pictures |
| DISTRIBUTOR | RCA Videodiscs, 13052, $29.98 |

| | |
|---:|:---|
| TITLE | **BLUES ALIVE** |
| TYPE | Music |
| FORMAT | CED |
| FEATURES | Stereo, Banded |
| DISTRIBUTOR | RCA Videodiscs, 12145, $24.98 |

|  |  |
|---:|:---|
| TITLE | **BLUES BROTHERS, THE** |
| TYPE | Motion Picture, Comedy, Music (R) |
| FORMAT | Laser CLV, CED (2 discs) |
| FEATURES | Stereo, Banded (CED) |
| RUNNING TIME | 132 minutes |
| PRODUCER | Warner Home Video |
| DISTRIBUTOR | Pioneer Video, Inc., 16-020, $29.95 |
|  | RCA Videodiscs, 13310, $39.98 |

|  |  |
|---:|:---|
| TITLE | **BOAT, THE (DAS BOOT)** |
| TYPE | Motion Picture (R) |
| FORMAT | Laser CLV, CED (2 discs) |
| FEATURES | Stereo, Dubbed in English |
| DISTRIBUTOR | RCA Videodiscs, 13046, $39.98 |
| RETAIL OUTLET | (CLV) UltraVideo, VLD2293, $34.95 |

|  |  |
|---:|:---|
| TITLE | **BOB WELCH & FRIENDS: LIVE AT THE ROXY** |
| TYPE | Music |
| FORMAT | CED |
| FEATURES | Stereo, CX Encoded |
| DISTRIBUTOR | RCA Videodiscs, 12094, $24.98 |

|  |  |
|---:|:---|
| TITLE | **BODY AND SOUL** |
| TYPE | Motion Picture |
| FORMAT | CED |
| RETAIL OUTLET | Movies Unlimited Inc., A2-0229, $29.95 |

|  |  |
|---:|:---|
| TITLE | **BODY HEAT** |
| TYPE | Motion Picture (R) |
| FORMAT | CED |
| DISTRIBUTOR | RCA Videodiscs, 03139, $29.98 |

|  |  |
|---:|:---|
| TITLE | **BOHEME, LA** |
| TYPE | Music |
| FORMAT | Laser CLV |
| FEATURES | Stereo, CX Encoded |
| RUNNING TIME | 116 minutes |
| PRODUCER | Pioneer Artists |
| DISTRIBUTOR | Pioneer Video, Inc., PA-82-028, $49.95 |

|  |  |
|---:|:---|
| TITLE | **BOLERO** |
| TYPE | Motion Picture |
| FORMAT | CED |
| RETAIL OUTLET | Movies Unlimited Inc., A3-5013, $29.98 |

| | |
|---|---|
| TITLE | **BON VOYAGE, CHARLIE BROWN (AND DON'T COME BACK)** |
| TYPE | Motion Picture, Comedy (G) |
| FORMAT | Laser CLV |
| RUNNING TIME | 76 minutes |
| PRODUCER | Paramount Home Video |
| DISTRIBUTOR | Pioneer Video, Inc., LV1158, $29.95 |

| | |
|---|---|
| TITLE | **BONNIE AND CLYDE** |
| TYPE | Motion Picture |
| FORMAT | CED |
| DISTRIBUTOR | RCA Videodiscs, 03117, $29.98 |

| | |
|---|---|
| TITLE | **BONNLAND** |
| DESCRIPTION | Training videodisc for military operations in urban terrain, using the West German village of Bonnland. |
| TYPE | In-house Training |
| FORMAT | Laser CAV |
| LEVEL | 3 |
| DEVELOPER | Interactive Television Company |
| CONTACT FOR MORE INFO. | Dr. Steven Levin, Interactive Television Company |
| RUNNING TIME | 30 minutes |
| AVAILABILITY | Contact Interactive Television Company |

| | |
|---|---|
| TITLE | **BOOGEYMAN, THE** |
| TYPE | Motion Picture |
| FORMAT | CED |
| RETAIL OUTLET | Movies Unlimited Inc., A3-C022, $29.98 |

| | |
|---|---|
| TITLE | **BOOZ-ALLEN & HAMILTON HOME INFORMATION SYSTEM SIMULATOR** |
| DESCRIPTION | Random access interactive full motion and still video support for integrated hardware configuration (personal computer with graphics, hard disk storage, videodisc, and touch-sensitive screen). |
| TYPE | Instructional, Archival, Catalog, Point-of-Purchase, Advertising, Promotional |
| RELEASE DATE | 1982 |
| FORMAT | Laser CAV |
| LEVEL | 3 |
| SYSTEM DESCRIPTION | Proprietary |
| CONTACT FOR MORE INFO. | George Lilly or Mike McLaughlin, Booz-Allen & Hamilton, (212) 639-3039 |
| FEATURES | Chapters, Picture Stops, Dual Audio |
| PRODUCER | Booz-Allen & Hamilton |

| | |
|---|---|
| TITLE | **BORDER, THE** |
| TYPE | Motion Picture (R) |
| FORMAT | Laser CLV |
| RUNNING TIME | 108 minutes |
| PRODUCER | Warner Home Video |
| DISTRIBUTOR | Pioneer Video, Inc., 10-033, $29.95 |

| | |
|---:|:---|
| TITLE | **BORN FREE** |
| TYPE | Motion Picture (G) |
| FORMAT | CED |
| DISTRIBUTOR | RCA Videodiscs, 03020, $24.98 |

| | |
|---:|:---|
| TITLE | **BOSTON** |
| DESCRIPTION | Travel and accommodation for Boston, Massachusetts. |
| TYPE | Promotional |
| FORMAT | Laser CAV |
| LEVEL | 1 |
| CONTACT FOR MORE INFO. | Globalvision |
| PRODUCER | University of Nebraska/Nebraska Videodisc Design/Production Group |
| AVAILABILITY | Not available for release or distribution. |

| | |
|---:|:---|
| TITLE | **BOSTON DISC** |
| DESCRIPTION | Still photos of 70-80 percent of Boston's real estate parcels. |
| TYPE | Technical, Instructional, In-house Training, Archival |
| RELEASE DATE | 6/83 |
| FORMAT | Laser CAV |
| LEVEL | 3 |
| SYSTEM DESCRIPTION | Pioneer VP-1000 player; New Media Graphics DM-1000 interface; DEC VAX-11/780 computer (any other computer may be used) |
| DEVELOPER | City of Boston |
| CONTACT FOR MORE INFO. | Walter R. Crosby |
| FEATURES | Picture Stops |

| | |
|---:|:---|
| TITLE | **BOSTON MEDIA CONSULTANTS - (CLASSIFIED RESEARCH DISC)** |
| TYPE | Music, Dance, Game, Research and Development, Technical |
| FORMAT | CED |
| LEVEL | 3 |

| | |
|---:|:---|
| TITLE | **BOSTON STRANGLER, THE** |
| TYPE | Motion Picture |
| FORMAT | CED |
| RETAIL OUTLET | Movies Unlimited Inc., A1-1015, $29.98 |

| | |
|---:|:---|
| TITLE | **BOXCAR BERTHA** |
| TYPE | Motion Picture |
| FORMAT | CED |
| RETAIL OUTLET | Movies Unlimited Inc., A3-4038, $29.98 |

| | |
|---:|:---|
| TITLE | **BOYS FROM BRAZIL, THE** |
| TYPE | Motion Picture (PG) |
| FORMAT | Laser CLV, CED |
| FEATURES | Stereo |
| RUNNING TIME | 123 minutes |
| PRODUCER | Warner Home Video |
| DISTRIBUTOR | Pioneer Video, Inc., 9002-80, $39.98 |
| | RCA Videodiscs, 00502, $24.98 |

| | |
|---:|:---|
| TITLE | **BOYS IN THE BAND, THE** |
| TYPE | Motion Picture |
| FORMAT | CED |
| RETAIL OUTLET | Movies Unlimited Inc., A1-7017, $29.98 |

| | |
|---:|:---|
| TITLE | **BOYS OF SUMMER, THE** |
| TYPE | Motion Picture |
| FORMAT | CED |
| RETAIL OUTLET | Movies Unlimited Inc., A3-C702, $29.98 |

| | |
|---:|:---|
| TITLE | **BRANNIGAN** |
| TYPE | Motion Picture |
| FORMAT | CED |
| RETAIL OUTLET | Movies Unlimited Inc., A1-4562, $29.98 |

| | |
|---:|:---|
| TITLE | **BREAKFAST AT TIFFANY'S** |
| TYPE | Motion Picture, Comedy |
| FORMAT | Laser CLV, CED |
| RUNNING TIME | 114 minutes |
| PRODUCER | Paramount Home Video |
| DISTRIBUTOR | Pioneer Video, Inc., LV6505, $29.95 |
| | RCA Videodiscs, 00684, $19.98 |

| | |
|---:|:---|
| TITLE | **BREAKHEART PASS** |
| TYPE | Motion Picture |
| FORMAT | CED |
| RETAIL OUTLET | Movies Unlimited Inc., A1-4536, $29.98 |

| | |
|---:|:---|
| TITLE | **BREAKING AWAY** |
| TYPE | Motion Picture |
| FORMAT | CED |
| RETAIL OUTLET | Movies Unlimited Inc., A1-1081, $29.98 |

| | |
|---:|:---|
| TITLE | **BREAKOUT** |
| TYPE | Motion Picture (R) |
| FORMAT | CED |
| DISTRIBUTOR | RCA Videodiscs, 03018, $24.98 |

| | |
|---:|:---|
| TITLE | **BREATHLESS** |
| TYPE | Motion Picture |
| FORMAT | Laser CLV, CED |
| DISTRIBUTOR | RCA Videodiscs, 01913, $24.98 |
| RETAIL OUTLET | (CLV) UltraVideo, VL5017, $34.95 |

| | |
|---:|:---|
| TITLE | **BRIAN'S SONG** |
| TYPE | Motion Picture (G) |
| FORMAT | CED |
| FEATURES | Closed-Captioned |
| PRODUCER | Columbia Pictures |
| DISTRIBUTOR | RCA Videodiscs, 03042, $24.98 |

| | |
|---:|:---|
| TITLE | **BRIDGE ON THE RIVER KWAI, THE** |
| TYPE | Motion Picture |
| FORMAT | Laser CLV, CED (2 discs) |
| PRODUCER | Warner Home Video |
| DISTRIBUTOR | Pioneer Video, Inc., VLD2010, $34.95 |
| | RCA Videodiscs, 03013, $39.98 |

| | |
|---:|:---|
| TITLE | **BRIDGE TOO FAR, A** |
| TYPE | Motion Picture |
| FORMAT | CED (2 Discs) |
| RETAIL OUTLET | Movies Unlimited Inc., A1-4533, $39.98 |

| | |
|---:|:---|
| TITLE | **BRIGADOON** |
| TYPE | Motion Picture, Music, Dance |
| FORMAT | Laser CLV, CED |
| FEATURES | Stereo |
| RUNNING TIME | 108 minutes |
| PRODUCER | MGM/UA Home Video |
| DISTRIBUTOR | Pioneer Video, Inc., ML100040, $34.95 |
| RETAIL OUTLET | (CED) Movies Unlimited Inc., A2-1023, $29.95 |

| | |
|---:|:---|
| TITLE | **BRIMSTONE AND TREACLE** |
| TYPE | Motion Picture |
| FORMAT | CED |
| RETAIL OUTLET | Movies Unlimited Inc., A2-0227, $29.95 |

| | |
|---:|:---|
| TITLE | **BRONCO BILLY** |
| TYPE | Motion Picture (PG) |
| FORMAT | CED |
| PRODUCER | Warner Home Video |
| DISTRIBUTOR | RCA Videodiscs, 03147, $24.98 |

| | |
|---:|:---|
| TITLE | **BRUBAKER** |
| TYPE | Motion Picture (R) |
| FORMAT | Laser CLV, CED (2 discs) |
| FEATURES | Stereo |
| RUNNING TIME | 131 minutes |
| PRODUCER | CBS Fox Video |
| DISTRIBUTOR | Pioneer Video, Inc., 1098-80, $39.98 |
| RETAIL OUTLET | (CED) Movies Unlimited Inc., A1-1098, $39.98 |

| | |
|---:|:---|
| TITLE | **BUCK ROGERS IN THE 25TH CENTURY** |
| TYPE | Motion Picture (PG) |
| FORMAT | Laser CLV |
| FEATURES | Stereo |
| PRODUCER | Warner Home Video |
| DISTRIBUTOR | Pioneer Video, Inc., 13-002, $29.95 |

| | |
|---|---|
| TITLE | **BUDDY BUDDY** |
| TYPE | Motion Picture |
| FORMAT | CED |
| RETAIL OUTLET | Movies Unlimited Inc., A2-1054, $29.95 |

| | |
|---|---|
| TITLE | **BUGS BUNNY/ROAD RUNNER MOVIE, THE** |
| TYPE | Motion Picture (G) |
| FORMAT | CED |
| DISTRIBUTOR | RCA Videodiscs, 03105, $24.98 |

| | |
|---|---|
| TITLE | **BUGSY MALONE** |
| TYPE | Motion Picture, Comedy (G) |
| FORMAT | Laser CLV |
| RUNNING TIME | 94 minutes |
| PRODUCER | Warner Home Video |
| DISTRIBUTOR | Pioneer Video, Inc., LV8898, $29.95 |

| | |
|---|---|
| TITLE | **BULLITT** |
| TYPE | Motion Picture |
| FORMAT | CED |
| DISTRIBUTOR | RCA Videodiscs, 03118, $24.98 |

| | |
|---|---|
| TITLE | **BULLWINKLE AND ROCKY AND FRIENDS, VOL. 1** |
| TYPE | Motion Picture |
| FORMAT | CED |
| FEATURES | Banded |
| DISTRIBUTOR | RCA Videodiscs, 02152, $19.98 |

| | |
|---|---|
| TITLE | **BULLWINKLE AND ROCKY AND FRIENDS, VOL. 2** |
| TYPE | Motion Picture |
| FORMAT | CED |
| FEATURES | Banded |
| DISTRIBUTOR | RCA Videodiscs, 02191, $19.98 |

| | |
|---|---|
| TITLE | **BUS STOP** |
| TYPE | Motion Picture |
| FORMAT | CED |
| RETAIL OUTLET | Movies Unlimited Inc., A1-1031, $29.98 |

| | |
|---|---|
| TITLE | **BUSTIN' LOOSE** |
| TYPE | Motion Picture, Comedy (R) |
| FORMAT | Laser CLV |
| RUNNING TIME | 93 minutes |
| PRODUCER | Warner Home Video |
| DISTRIBUTOR | Pioneer Video, Inc., 16-026, $29.95 |

| | |
|---:|:---|
| TITLE | **BUTCH CASSIDY AND THE SUNDANCE KID** |
| TYPE | Motion Picture, Comedy (PG) |
| FORMAT | Laser CLV, CED |
| FEATURES | Stereo |
| RUNNING TIME | 110 minutes |
| PRODUCER | Warner Home Video |
| DISTRIBUTOR | Pioneer Video, Inc., 1060-80, $29.98 |
| RETAIL OUTLET | (CED) Movies Unlimited Inc., A1-0101, $29.98 |

| | |
|---:|:---|
| TITLE | **BUTTERFLY** |
| TYPE | Motion Picture (R) |
| FORMAT | Laser CLV, CED |
| RUNNING TIME | 105 minutes |
| PRODUCER | Warner Home Video |
| DISTRIBUTOR | Pioneer Video, Inc., VL6007, $34.95 |
| RETAIL OUTLET | (CED) Movies Unlimited Inc., A3-6007, $29.95 |

| | |
|---:|:---|
| TITLE | **BY YOURSELF** |
| DESCRIPTION | Visualized, captioned poems and short stories for hearing-impaired learners. |
| TYPE | Instructional |
| FORMAT | Laser CAV |
| RELEASE DATE | 1980 |
| LEVEL | 2 |
| CONTACT FOR MORE INFO. | Bureau of Education for the Handicapped/University of Nebraska Barkley Center |
| PRODUCER | University of Nebraska/Nebraska Videodisc Design/Production Group |
| AVAILABILITY | Not available for release or distribution. |

| | |
|---:|:---|
| TITLE | **CABARET** |
| TYPE | Motion Picture |
| FORMAT | Laser CLV, CED |
| FEATURES | Stereo |
| RETAIL OUTLET | (CLV) UltraVideo, 80-7035, $34.95 |
| | (CED) Movies Unlimited Inc., A2-1014, $24.95 |

| | |
|---:|:---|
| TITLE | **CADDYSHACK** |
| TYPE | Motion Picture, Comedy (R) |
| FORMAT | Laser CLV, CED |
| RUNNING TIME | 90 minutes |
| PRODUCER | Warner Home Video |
| DISTRIBUTOR | Pioneer Video, Inc., 2005LV, $29.98 |
| | RCA Videodiscs, 03119, $29.98 |

| | |
|---:|:---|
| TITLE | **CAGE AUX FOLLES, LA** |
| TYPE | Motion Picture, Comedy (R) |
| FORMAT | CED |
| DISTRIBUTOR | RCA Videodiscs, 01449, $29.98 |

# Videodiscography

| | |
|---:|:---|
| TITLE | **CAGE AUX FOLLES 2, LA** |
| TYPE | Motion Picture |
| FORMAT | CED |
| RETAIL OUTLET | Movies Unlimited Inc., A1-4551, $29.98 |

| | |
|---:|:---|
| TITLE | **CALEDONIAN DREAMS** |
| TYPE | Music, Adult |
| FORMAT | Laser CAV |
| LEVEL | 1 |
| FEATURES | Stereo, CX Encoded |
| RUNNING TIME | 45 minutes |
| PRODUCER | Pioneer Video Imports |
| DISTRIBUTOR | Pioneer Video, Inc., ME-060US, $27.95 |

| | |
|---:|:---|
| TITLE | **CALIFORNIA SUITE** |
| TYPE | Motion Picture, Comedy (PG) |
| FORMAT | Laser CLV, CED |
| PRODUCER | Warner Home Video |
| DISTRIBUTOR | Pioneer Video, Inc., VLD3010, $29.95 |
| | RCA Videodiscs, 03011, $24.98 |

| | |
|---:|:---|
| TITLE | **CALL FOR FIRE** |
| DESCRIPTION | Tutorial and simulation teaching the skills required to call for artillery fire. Part of a prototype training package. |
| TYPE | Technical, Instructional |
| RELEASE DATE | 6/80 |
| FORMAT | Laser CAV |
| LEVEL | 3 |
| DEVELOPER | WICAT Systems for the U.S. Army Research Institute |
| CONTACT FOR MORE INFO. | Richard M. Cavagnol |
| PRODUCER | WICAT Systems, Inc. |

| | |
|---:|:---|
| TITLE | **CANDID CANDID CAMERA** |
| TYPE | Motion Picture |
| FORMAT | CED |
| RETAIL OUTLET | Movies Unlimited Inc., A3-3020, $29.95 |

| | |
|---:|:---|
| TITLE | **CANDIDATE, THE** |
| TYPE | Motion Picture |
| FORMAT | CED |
| DISTRIBUTOR | RCA Videodiscs, 03120, $24.98 |

| | |
|---:|:---|
| TITLE | **CANDLESHOE** |
| TYPE | Motion Picture (G) |
| FORMAT | CED |
| PRODUCER | Walt Disney |
| DISTRIBUTOR | RCA Videodiscs, 00705, $24.98 |

| | |
|---|---|
| TITLE | **CANNERY ROW** |
| TYPE | Motion Picture |
| FORMAT | CED |
| RETAIL OUTLET | Movies Unlimited Inc., A2-1056, $29.95 |

| | |
|---|---|
| TITLE | **CANNONBALL RUN** |
| TYPE | Motion Picture, Comedy (PG) |
| FORMAT | Laser CLV, CED |
| RUNNING TIME | 96 minutes |
| PRODUCER | Warner Home Video |
| DISTRIBUTOR | Pioneer Video, Inc., VL6001, $34.95 |
| RETAIL OUTLET | (CED) Movies Unlimited Inc., A3-6001, $29.95 |

| | |
|---|---|
| TITLE | **CAPRICORN ONE** |
| TYPE | Motion Picture (PG) |
| FORMAT | Laser CLV, CED |
| FEATURES | (Laser) Stereo, CX Encoded |
| RUNNING TIME | 123 minutes |
| PRODUCER | Warner Home Video |
| DISTRIBUTOR | Pioneer Video, Inc., 9007-80, $39.98 |
| RETAIL OUTLET | (CED) Movies Unlimited Inc., A1-9007, $29.98 |

| | |
|---|---|
| TITLE | **CAPSULE, THE** |
| DESCRIPTION | Fourteen different experiments on disc including surrogate travel, helicopter missile firing simulation, Motorola parts catalog, digital infrared video, and full text document retrieval with images on disc. |
| TYPE | Research and Development, Technical, Instructional, In-house Training, Archival, Catalog |
| FORMAT | Laser CAV |
| LEVEL | 3 |
| SYSTEM DESCRIPTION | IBM PC/XT, Pascal software, Pioneer 7820 player, and Pioneer interface (experiments with machine independent table driven software) |
| DEVELOPER | Dr. Darrell F. Humphrey |
| CONTACT FOR MORE INFO. | Martin Marietta Data Systems/Technology Systems Group, (303) 790-3339 |
| FEATURES | Dual Audio |
| RUNNING TIME | 27 minutes |
| PRODUCER | 3M |

| | |
|---|---|
| TITLE | **CAPTAIN BLOOD** |
| TYPE | Motion Picture |
| FORMAT | CED |
| DISTRIBUTOR | RCA Videodiscs, 03403, $24.98 |

| | |
|---|---|
| TITLE | **CARBON COPY** |
| TYPE | Motion Picture |
| FORMAT | Laser CLV, CED |
| RETAIL OUTLET | (CLV) UltraVideo, E160950, $34.95 |
| | (CED) Movies Unlimited Inc., A4-6096, $29.95 |

| | |
|---:|:---|
| TITLE | **CARDIO-PULMONARY-RESUSCITATION** |
| DESCRIPTION | American Heart Association approved one man method of CPR with simulated rescue requiring student response to save a life. |
| TYPE | Technical, Instructional |
| FORMAT | Laser CAV |
| LEVEL | 3 |
| SYSTEM DESCRIPTION | Pioneer 7820 model 2 player, Apple IIe with 2 drives, touch panel (infrared) monitor, all in a cabinet |
| DEVELOPER | Creativision, Inc. |
| CONTACT FOR MORE INFO. | Mr. Rick Glasby |
| FEATURES | Stereo, Dual Audio, Touch panel monitor |
| PRODUCER | Creativision, Inc. |
| DISTRIBUTOR | Creativision, Inc. |
| AVAILABILITY | Produced for the Navy Electrician's Mate "A" School with contract funds from the U.S. Navy. |

| | |
|---:|:---|
| TITLE | **CARNAL KNOWLEDGE** |
| TYPE | Motion Picture (R) |
| FORMAT | Laser CLV, CED |
| RUNNING TIME | 96 minutes |
| PRODUCER | Warner Home Video |
| DISTRIBUTOR | Pioneer Video, Inc., 4003-80, $29.98 |
| | RCA Videodiscs, 00803, $19.98 |

| | |
|---:|:---|
| TITLE | **CARNY** |
| TYPE | Motion Picture |
| FORMAT | CED |
| PRODUCER | CBS Video |
| RETAIL OUTLET | Movies Unlimited Inc., A1-7028, $29.98 |

| | |
|---:|:---|
| TITLE | **CAROLE KING: ONE TO ONE** |
| TYPE | Music |
| FORMAT | Laser CLV, CED |
| FEATURES | Stereo |
| RETAIL OUTLET | (CLV) UltraVideo, 83-051, $24.95 |
| | (CED) Movies Unlimited Inc., A2-0219, $29.95 |

| | |
|---:|:---|
| TITLE | **CARRIE** |
| TYPE | Motion Picture (R) |
| FORMAT | Laser CLV, CED |
| RUNNING TIME | 98 minutes |
| PRODUCER | Warner Home Video |
| DISTRIBUTOR | Pioneer Video, Inc., 4515-80, $29.98 |
| | RCA Videodiscs, 01448, $24.98 |

| | |
|---:|:---|
| TITLE | **CARTOON CLASSICS VOLUME #1 - CHIP-N-DALE WITH DONALD DUCK** |
| TYPE | Motion Picture |
| FORMAT | Laser CLV |
| RETAIL OUTLET | UltraVideo, 0146AS, $24.95 |

| | |
|---:|:---|
| TITLE | **CARTOON CLASSICS VOLUME #4 - SPORT GOOFY** |
| TYPE | Motion Picture |
| FORMAT | Laser CLV |
| RETAIL OUTLET | UltraVideo, 0165AS, $24.95 |

| | |
|---:|:---|
| TITLE | **CARTOON CLASSICS VOLUME #5 - BEST OF 1931/48** |
| TYPE | Motion Picture |
| FORMAT | Laser CLV |
| RETAIL OUTLET | UltraVideo, 0167AS, $24.95 |

| | |
|---:|:---|
| TITLE | **CASABLANCA** |
| TYPE | Motion Picture |
| FORMAT | Laser CLV, CED |
| FEATURES | Black and white |
| RUNNING TIME | 102 minutes |
| PRODUCER | Warner Home Video |
| DISTRIBUTOR | Pioneer Video, Inc., 4514-80, $39.98 |
| | RCA Videodiscs, 01420, $24.98 |

| | |
|---:|:---|
| TITLE | **CAT BALLOU** |
| TYPE | Motion Picture, Comedy |
| FORMAT | CED |
| DISTRIBUTOR | RCA Videodiscs, 03015, $24.98 |

| | |
|---:|:---|
| TITLE | **CAT ON A HOT TIN ROOF** |
| TYPE | Motion Picture |
| FORMAT | Laser CLV, CED |
| RUNNING TIME | 108 minutes |
| PRODUCER | MGM/UA Home Video |
| DISTRIBUTOR | Pioneer Video, Inc., ML100060, $25.95 |
| RETAIL OUTLET | (CED) Movies Unlimited Inc., A2-1051, $29.95 |

| | |
|---:|:---|
| TITLE | **CAT PEOPLE** |
| TYPE | Motion Picture (R) |
| FORMAT | Laser CLV, CED |
| FEATURES | Stereo, CX Encoded |
| RUNNING TIME | 118 minutes |
| PRODUCER | Warner Home Video |
| DISTRIBUTOR | Pioneer Video, Inc., 11-014, $29.95 |
| RETAIL OUTLET | (CED) Movies Unlimited Inc., A5-1014, $34.98 |

| | |
|---:|:---|
| TITLE | **CATCH-22** |
| TYPE | Motion Picture (R) |
| FORMAT | Laser CLV, CED |
| RUNNING TIME | 119 minutes |
| PRODUCER | Warner Home Video |
| DISTRIBUTOR | Pioneer Video, Inc., LV6924, $29.95 |
| | RCA Videodiscs, 00669, $29.98 |

| | |
|---:|:---|
| TITLE | **CAVEMAN** |
| TYPE | Motion Picture |
| FORMAT | CED |
| RETAIL OUTLET | Movies Unlimited Inc., A1-4543, $29.98 |

| | |
|---:|:---|
| TITLE | **CBS NEWS, JOHN F. KENNEDY** |
| TYPE | Motion Picture |
| FORMAT | CED |
| RETAIL OUTLET | Movies Unlimited Inc., A2-1010, $24.95 |

| | |
|---:|:---|
| TITLE | **CELL, THE** |
| TYPE | Children, Adolescents, Instructional, Archival, Catalog |
| RELEASE DATE | 1/83 |
| FORMAT | Laser CAV |
| LEVEL | 1,2 |
| SYSTEM DESCRIPTION | Two 300 step programs drive two industrial players in interactive mode. Program resides in player's on-board memory permanently. Custom electronics permits some interaction and automatic control of four consumer-type players. |
| DEVELOPER | Exhibits Department/Boston's Museum of Science, in consultation with WGBH, Boston |
| CONTACT FOR MORE INFO. | Larry J. Ralph, (617) 723-2500, ext. 292 |
| FEATURES | Chapters, Dual Audio, Other |
| PRODUCER | Boston's Museum of Science |
| AVAILABILITY | For museum internal use only. |

| | |
|---:|:---|
| TITLE | **CENTERDISK** |
| DESCRIPTION | An overview, a catalog, and an archive of the Center for Advanced Visual Studies at M.I.T. and the works of its artists and fellows. |
| TYPE | Art, Archival, Catalog |
| RELEASE DATE | Fall 1983 |
| FORMAT | Laser CAV, Laser CLV |
| LEVEL | 1, 3 |
| SYSTEM DESCRIPTION | Microcomputer, single laser disc system with keyboard input for use in museums. |
| DEVELOPER | Center for Advanced Visual Studies (CAVS)/Vision Machine Research (VMR)/WGBH |
| CONTACT FOR MORE INFO. | Vin Grabil, CAVS |
| FEATURES | Chapters, Picture Stops, Dual Audio |
| PRODUCER | CAVS/VMR/WGBH coproduction |
| DISTRIBUTOR | CAVS |
| RETAIL OUTLET | CAVS |
| AVAILABILITY | Rent, Loan |
| SPECIAL CONDITIONS | For display in museum or library. |

| | |
|---:|:---|
| TITLE | **CHALLENGE, THE** |
| TYPE | Motion Picture |
| FORMAT | CED |
| RETAIL OUTLET | Movies Unlimited Inc., A1-7137, $29.98 |

| | |
|---:|:---|
| TITLE | **CHALLENGES OF GLAUCOMA, THE** |
| DESCRIPTION | Medical training and information for physicians. |
| TYPE | Medical, Instructional |
| RELEASE DATE | 1981 |
| FORMAT | Laser CAV |
| LEVEL | 2, 3 |
| SYSTEM DESCRIPTION | Originally level 2 program. An external program was added for upgrading to level 3. |
| CONTACT FOR MORE INFO. | Scott Dobbins/Peter Crown |
| FEATURES | Digital Dump-Pioneer, Dual Audio |
| PRODUCER | Scott Dobbins (assoc.)/Peter Crown-Romulus Production |
| DISTRIBUTOR | Merck & Co. |
| AVAILABILITY | Available via Scott Dobbins, (212) 741-138 for demonstration. Contact Merck & Co. for other uses. |

| | |
|---:|:---|
| TITLE | **CHAMP, THE** |
| TYPE | Motion Picture (PG) |
| FORMAT | Laser CLV, CED |
| RUNNING TIME | 119 minutes |
| PRODUCER | MGM/UA Home Video |
| DISTRIBUTOR | Pioneer Video, Inc., ML100034, $34.95 |
| RETAIL OUTLET | (CED) Movies Unlimited Inc., A2-1011, $29.95 |

| | |
|---:|:---|
| TITLE | **CHANGE OF SEASONS, A** |
| TYPE | Motion Picture |
| FORMAT | CED |
| RETAIL OUTLET | Movies Unlimited Inc., A1-1104, $29.98 |

| | |
|---:|:---|
| TITLE | **CHANGELING, THE** |
| TYPE | Motion Picture (R) |
| FORMAT | Laser CLV |
| FEATURES | (CLV) Stereo, CX Encoded |
| RUNNING TIME | 113 minutes |
| PRODUCER | Vestron Video |
| DISTRIBUTOR | Pioneer Video, Inc., VL6006, $34.95 |
| RETAIL OUTLET | (CED) Movies Unlimited Inc., A3-6006, $29.95 |

| | |
|---:|:---|
| TITLE | **CHAPTER TWO** |
| TYPE | Motion Picture, Comedy (PG) |
| FORMAT | Laser CLV |
| PRODUCER | Warner Home Video |
| DISTRIBUTOR | Pioneer Video, Inc., VLD3040, $34.95 |

| | |
|---:|:---|
| TITLE | **CHARGE OF THE LIGHT BRIGADE, THE** |
| TYPE | Motion Picture |
| FORMAT | CED |
| RUNNING TIME | 116 minutes |
| PRODUCER | United Artists |
| DISTRIBUTOR | RCA Videodiscs, 03404, $24.98 |

# Videodiscography

| | |
|---:|:---|
| TITLE | **CHARIOTS OF FIRE** |
| TYPE | Motion Picture (PG) |
| FORMAT | Laser CLV, CED |
| FEATURES | Stereo, CX Encoded |
| RUNNING TIME | 124 minutes |
| PRODUCER | Warner Home Video |
| DISTRIBUTOR | Pioneer Video, Inc., 2004LV, $34.98 |
| | RCA Videodiscs, 13137, $29.98 |

| | |
|---:|:---|
| TITLE | **CHARLIE BROWN FESTIVAL, A** |
| TYPE | Motion Picture (G) |
| FORMAT | CED |
| DISTRIBUTOR | RCA Videodiscs, 01301, $19.98 |

| | |
|---:|:---|
| TITLE | **CHARLIE BROWN FESTIVAL, VOL. 2, A** |
| TYPE | Motion Picture (G) |
| FORMAT | CED |
| DISTRIBUTOR | RCA Videodiscs, 01302, $19.98 |

| | |
|---:|:---|
| TITLE | **CHARLIE BROWN FESTIVAL, VOL. 3, A** |
| TYPE | Motion Picture (G) |
| FORMAT | CED |
| DISTRIBUTOR | RCA Videodiscs, 01303, $19.98 |

| | |
|---:|:---|
| TITLE | **CHARLIE BROWN FESTIVAL, VOL. 4, A** |
| TYPE | Motion Picture (G) |
| FORMAT | CED |
| DISTRIBUTOR | RCA Videodiscs, 01304, $19.98 |

| | |
|---:|:---|
| TITLE | **CHARLIE DANIELS BAND, THE** |
| TYPE | Music |
| FORMAT | Laser CLV, CED |
| FEATURES | Stereo, CX Encoded |
| RUNNING TIME | 131 minutes |
| PRODUCER | CBS Fox Video |
| DISTRIBUTOR | Pioneer Video, Inc., 7105-80, $29.98 |
| RETAIL OUTLET | (CED) Movies Unlimited Inc., A2-1059, $24.95 |

| | |
|---:|:---|
| TITLE | **CHARLOTTE'S WEB** |
| TYPE | Motion Picture, Music (G) |
| FORMAT | Laser CLV, CED |
| RUNNING TIME | 94 minutes |
| PRODUCER | Warner Home Video |
| DISTRIBUTOR | Pioneer Video, Inc., VL8099, $29.95 |
| | RCA Videodiscs, 00624, $24.98 |

| | |
|---:|:---|
| TITLE | **CHARLY** |
| TYPE | Motion Picture (PG) |
| FORMAT | Laser CLV, CED |
| RUNNING TIME | 106 minutes |
| PRODUCER | Warner Home Video |
| DISTRIBUTOR | Pioneer Video, Inc., 8020-80, $34.98 |
| RETAIL OUTLET | (CED) Movies Unlimited Inc., A1-8020, $29.98 |

| | |
|---:|:---|
| TITLE | **CHATTERBOX** |
| TYPE | Motion Picture |
| FORMAT | CED |
| RETAIL OUTLET | Movies Unlimited Inc., A3-3015, $29.95 |

| | |
|---:|:---|
| TITLE | **CHEECH AND CHONG'S NEXT MOVIE** |
| TYPE | Motion Picture, Comedy (R) |
| FORMAT | Laser CLV |
| PRODUCER | Warner Home Video |
| DISTRIBUTOR | Pioneer Video, Inc., 16-021, $29.95 |

| | |
|---:|:---|
| TITLE | **CHEECH AND CHONG'S NICE DREAMS** |
| TYPE | Motion Picture, Comedy (R) |
| FORMAT | Laser CLV |
| PRODUCER | RCA/Columbia Pictures Home Video |
| DISTRIBUTOR | Pioneer Video, Inc., VLD3355, $29.95 |
| | RCA Videodiscs, 03035, $24.98 |

| | |
|---:|:---|
| TITLE | **CHEMISTRY I & II** |
| DESCRIPTION | Laboratory simulation for college level |
| TYPE | Instructional |
| FORMAT | Laser CAV |
| LEVEL | 3 |
| CONTACT FOR MORE INFO. | Corporation for Public Broadcasting/Annenberg |
| PRODUCER | University of Nebraska/Nebraska/Videodisc Design/Production Group |
| AVAILABILITY | Not available for release or distribution. |

| | |
|---:|:---|
| TITLE | **CHICAGO MERCANTILE EXCHANGE** |
| DESCRIPTION | Public access information kiosks dispensing information about the Chicago Mercantile Exchange |
| TYPE | Instructional, Promotional |
| FORMAT | Laser CAV |
| LEVEL | 3 |
| DEVELOPER | Jack Lieb Productions, Inc. |
| FEATURES | Chapters, Picture Stops, Digital Dump-Sony |

| | |
|---:|:---|
| TITLE | **CHICK COREA/GARY BURTON-LIVE IN TOKYO** |
| TYPE | Music |
| FORMAT | Laser CLV |
| FEATURES | Stereo, CX Encoded |
| RUNNING TIME | 58 minutes |
| PRODUCER | Pioneer Artists |
| DISTRIBUTOR | Pioneer Video, Inc., PS-83-037, $24.95 |

| | |
|---:|:---|
| TITLE | **CHINA SYNDROME, THE** |
| TYPE | Motion Picture (PG) |
| FORMAT | Laser CLV, CED |
| PRODUCER | Warner Home Video |
| DISTRIBUTOR | Pioneer Video, Inc., VLD3060, $34.95 |
| | RCA Videodiscs, 03006, $29.98 |

| | |
|---:|:---|
| TITLE | **CHINATOWN** |
| TYPE | Motion Picture (R) |
| FORMAT | Laser CLV, CED (2 discs) |
| RUNNING TIME | 131 minutes |
| PRODUCER | Warner Home Video |
| DISTRIBUTOR | Pioneer Video, Inc., LV8674, $39.95 |
| | RCA Videodiscs, 00613, $34.98 |

| | |
|---:|:---|
| TITLE | **CHIP 'N DALE WITH DONALD DUCK** |
| TYPE | Motion Picture (G) |
| FORMAT | Laser CLV |
| RUNNING TIME | 47 minutes |
| PRODUCER | Walt Disney |
| DISTRIBUTOR | Pioneer Video, Inc., 42146AS, $24.95 |

| | |
|---:|:---|
| TITLE | **CHITTY CHITTY BANG BANG** |
| TYPE | Motion Picture, Comedy (G) |
| FORMAT | Laser CLV, CED |
| FEATURES | (CED) Stereo |
| RUNNING TIME | 142 minutes |
| PRODUCER | CBS Fox Video |
| DISTRIBUTOR | Pioneer Video, Inc., 4557-80, $39.98 |
| RETAIL OUTLET | (CED) Movies Unlimited Inc., A1-4557, $39.98 |

| | |
|---:|:---|
| TITLE | **CHRISTMAS 1983 ELECTRONIC CATALOG** |
| DESCRIPTION | Christmas gift suggestions based on sex, age, price, and life-style. Computer analyzes user input and recommends three gift ideas and a stocking stuffer. |
| TYPE | Catalog, Point-of-Purchase, Advertising |
| RELEASE DATE | 10/83 |
| FORMAT | Laser CAV |
| LEVEL | 3 |
| DEVELOPER | WDC Marketing Associates, Inc. |
| CONTACT FOR MORE INFO. | David Carrington, (205) 945-8951 |
| FEATURES | Digital Dump-Pioneer, Stereo |
| PRODUCER | WDC Marketing Associates, Inc. |
| DISTRIBUTOR | WDC Marketing Associates, Inc. |
| AVAILABILITY | System must be custom designed for individual stores. |

| | |
|---:|:---|
| TITLE | **CHU CHU AND THE PHILLY FLASH** |
| TYPE | Motion Picture |
| FORMAT | CED |
| RETAIL OUTLET | Movies Unlimited Inc., A1-1119, $29.98 |

| | |
|---:|:---|
| TITLE | **CID, EL** |
| TYPE | Motion Picture |
| FORMAT | Laser CLV, CED (2 discs) |
| RUNNING TIME | 189 minutes |
| PRODUCER | Vestron Video |
| DISTRIBUTOR | Pioneer Video, Inc., VL3014, $39.95 |
| RETAIL OUTLET | (CED) Movies Unlimited Inc., A3-3014, $39.95 |

| | |
|---:|:---|
| TITLE | **CINCINNATI KID, THE** |
| TYPE | Motion Picture |
| FORMAT | CED |
| FEATURES | Stereo |
| RETAIL OUTLET | Movies Unlimited Inc., A2-0135, $29.95 |

| | |
|---:|:---|
| TITLE | **CITIBANK** |
| DESCRIPTION | Program to teach customers how to use the automated teller machines and new bank services and products |
| TYPE | Research and Development |
| FORMAT | Laser CAV |
| LEVEL | 2 |
| DEVELOPER | Exhibit Technology, Inc. |
| PRODUCER | Fusion Media, Inc. |

| | |
|---:|:---|
| TITLE | **CITIZEN KANE** |
| TYPE | Motion Picture |
| FORMAT | CED |
| RETAIL OUTLET | Movies Unlimited Inc., 00401, $19.98 |

| | |
|---:|:---|
| TITLE | **CITTA VIVA - GALLIPOLI** |
| DESCRIPTION | Shows the topology, culture, tourist attractions, and sports resources of a medium- sized town. |
| TYPE | Advertising, Promotional |
| RELEASE DATE | May 1983 |
| FORMAT | Laser CAV |
| LEVEL | 3 |
| DEVELOPER | Philips - Eindoven |
| CONTACT FOR MORE INFO. | Roberto Sanguini or Francesco Lotano |
| FEATURES | Chapters, Picture Stops |
| PRODUCER | SEAT - Turin/SARIN - Rome/TECNETRA-Milan |
| AVAILABILITY | Not for distribution. |

| | |
|---:|:---|
| TITLE | **CITY LIGHTS** |
| TYPE | Motion Picture, Comedy |
| FORMAT | CED |
| DISTRIBUTOR | RCA Videodiscs, 02032, $19.98 |

| | |
|---:|:---|
| TITLE | **CLARENCE DARROW** |
| TYPE | Motion Picture |
| FORMAT | CED |
| DISTRIBUTOR | RCA Videodiscs, 02018, $19.98 |

| | |
|---:|:---|
| TITLE | **CLASH OF THE TITANS** |
| TYPE | Motion Picture (PG) |
| FORMAT | Laser CLV, CED |
| FEATURES | (CLV) Stereo, CX Encoded |
| RUNNING TIME | 119 minutes |
| PRODUCER | MGM/UA Home Video |
| DISTRIBUTOR | Pioneer Video, Inc., ML100074, $34.95 |
| RETAIL OUTLET | (CED) Movies Unlimited Inc., A2-1032, $29.95 |

| | |
|---:|:---|
| TITLE | **CLASS** |
| TYPE | Motion Picture |
| FORMAT | CED |
| DISTRIBUTOR | RCA Videodiscs, 01916, $24.98 |

| | |
|---:|:---|
| TITLE | **CLASS OF 1984** |
| TYPE | Motion Picture (R) |
| FORMAT | Laser CLV, CED |
| PRODUCER | Mark Lester |
| DISTRIBUTOR | United Film Distribution Company |
| RETAIL OUTLET | (CLV) UltraVideo, 0146AS, $24.95 |
| | (CED) Movies Unlimited Inc., A3-5022, $29.98 |

| | |
|---:|:---|
| TITLE | **CLASSROOM BEHAVIOR RECORD** |
| DESCRIPTION | This tutorial teaches a systematic procedure for recording normal and abnormal classroom behavior. Users will learn to categorize behavior as one of 36 classifications. The program will emphasize identification of overt behaviors that specify behavioral disorders. |
| TYPE | Instructional |
| FORMAT | Laser CAV |
| LEVEL | 3 |
| SYSTEM DESCRIPTION | 8210 player, Apple II, UP 1000, Allen VMI. |
| DEVELOPER | University of Iowa/Iowa Department of Public Instruction |
| CONTACT FOR MORE INFO. | Dr. Polly Nichols/Gail Fitzgerald, Child Psychiatry Service |
| AVAILABILITY | The system will be field tested in 1984. They anticipate further distribution after the evaluation. Details remain to be negotiated. |

| | |
|---:|:---|
| TITLE | **CLAUDE BOLLING CONCERTO FOR CLASSIC GUITAR AND JAZZ PIANO** |
| TYPE | Music |
| FORMAT | Laser CLV |
| FEATURES | Stereo, CX Encoded |
| RUNNING TIME | 50 minutes |
| PRODUCER | Pioneer Artists |
| DISTRIBUTOR | Pioneer Video, Inc., PA-82-022, $24.95 |

| | |
|---:|:---|
| TITLE | **CLEAN AND INSPECT A GENERATOR** |
| DESCRIPTION | Preventive maintenance system (PMS) performed on an A.C. generator, including brush replacement, brush tension setting, adjusting brush rigging, and measuring insulation resistance. |
| TYPE | Technical, Instructional, Other |
| FORMAT | Laser CAV |
| LEVEL | 3 |

| | |
|---:|:---|
| SYSTEM DESCRIPTION | Pioneer 7820 Model 2, Apple IIe with 2 drives, and touch panel (infrared) monitor, all in a cabinet. |
| DEVELOPER | Creativision, Inc. |
| CONTACT FOR MORE INFO. | Mr. Rick Glasby |
| FEATURES | Stereo, Dual Audio, Touch Panel Monitor |
| PRODUCER | Creativision, Inc. |
| DISTRIBUTOR | Creativision, Inc. |
| AVAILABILITY | Produced for the Navy Electrician's Mate "A" School with contract funds from the U.S. Navy. |

| | |
|---:|:---|
| TITLE | **CLEFT LIP AND PALATE CASE MANAGEMENT** |
| DESCRIPTION | A program for CMS health professionals that describes the types of cleft lip and palate, expectations for growth and development; options for treatment of cleft lip and palate; physical, medical, and psychological aspects of case management in short-term and long-term care. |
| TYPE | Medical, Instructional |
| RELEASE DATE | 1983 |
| FORMAT | Laser CAV |
| LEVEL | 3 |
| SYSTEM DESCRIPTION | Discovision 7820-1 player, Apple II plus microcomputer, and VAI-1 interface card. |
| DEVELOPER | MGT of America, Inc., Talahassee, Florida |
| CONTACT FOR MORE INFO. | Jayne Parker |
| RUNNING TIME | Varies, but approximately 10 hours. |
| AVAILABILITY | Not for sale, but may become available in the future. |

| | |
|---:|:---|
| TITLE | **CLOCKWORK ORANGE, A** |
| TYPE | Motion Picture (R) |
| FORMAT | CED (2 Discs) |
| DISTRIBUTOR | RCA Videodiscs, 03121, $34.98 |

| | |
|---:|:---|
| TITLE | **CLOSE ENCOUNTERS OF THE THIRD KIND** |
| TYPE | Motion Picture (PG) |
| FORMAT | Laser CLV, CED (2 discs) |
| FEATURES | (CLV) Stereo, CX Encoded |
| PRODUCER | Warner Home Video |
| DISTRIBUTOR | Pioneer Video, Inc., VLD3095, $34.95 |
| | RCA Videodiscs, 03004, $39.98 |

| | |
|---:|:---|
| TITLE | **COAL MINER'S DAUGHTER** |
| TYPE | Motion Picture, Music |
| FORMAT | Laser CLV, CED |
| DISTRIBUTOR | RCA Videodiscs, 03308, $29.98 |
| RETAIL OUTLET | (CLV) UltraVideo, 15-005, $29.95 |

| | |
|---:|:---|
| TITLE | **COAST GUARD** |
| DESCRIPTION | Part of a recruiting network used in college placement offices across the country. |
| TYPE | Instructional, Catalog, Promotional |
| FORMAT | Laser CAV |
| LEVEL | 2 |

| | |
|---|---|
| CONTACT FOR MORE INFO. | Aileen McKenna, PREF, the Placement Reference Network, (919) 493-3479 |
| FEATURES | Chapters, Picture Stops |
| RUNNING TIME | 30 minutes |
| AVAILABILITY | PREF has laser disc equipment in college placement offices across the country. They own the equipment and distribute discs describing a company's recruiting programs into the network. The college receives the service at no charge, and the corporation pays a distribution fee. They are planning to open a similar network, the Learning Resources Network, in high school guidance offices with information on colleges. |

| | |
|---|---|
| TITLE | **COAST TO COAST** |
| TYPE | Motion Picture (PG) |
| FORMAT | Laser CLV |
| RUNNING TIME | 95 minutes |
| PRODUCER | Warner Home Video |
| DISTRIBUTOR | Pioneer Video, Inc., LV1342, $29.95 |

| | |
|---|---|
| TITLE | **COLLECTOR, THE** |
| TYPE | Motion Picture |
| FORMAT | CED |
| DISTRIBUTOR | RCA Videodiscs, 03017, $24.98 |

| | |
|---|---|
| TITLE | **COLLEGE FOOTBALL CLASSICS, VOL. 1** |
| TYPE | Motion Picture |
| FORMAT | CED |
| DISTRIBUTOR | RCA Videodiscs, 02011, $14.98 |

| | |
|---|---|
| TITLE | **COLLEGE USE DEMONSTRATION DISC** |
| DESCRIPTION | Side 1 of the disc contains segments from Penn State University and Williams College, an introduction to the system and a discussion of possible production techniques. Side 2 contains the Maryland State Department of Education Functional Reading Program. |
| TYPE | Adolescents, Instructional, Archival, Catalog, Promotional |
| FORMAT | Laser CAV |
| LEVEL | 2 |
| DEVELOPER | Info-Disc Corporation |
| CONTACT FOR MORE INFO. | Donald L. Doty, (301) 424-7950 |
| FEATURES | Chapters, Picture Stops, Dual Audio |
| PRODUCER | Spicer Productions (Side 1)/MITV (Side 2) |
| DISTRIBUTOR | Info-Disc Corporation |
| AVAILABILITY | Sale. Discs and disc players are distributed free of charge to qualified high schools or may be purchased directly through Info-Disc. |
| SUGGESTED PRICE | $100 |

| | |
|---|---|
| TITLE | **COLORATION** |
| DESCRIPTION | L'Oreal program about hair coloration. Side A is for customers. Side B is for hairdressers. |
| TYPE | Instructional, Point-of-Purchase, Promotional |
| RELEASE DATE | 1983 |
| FORMAT | Laser CAV |

| | |
|---:|:---|
| LEVEL | 1 |
| DEVELOPER | CAT Stuttgart- Computer Assisted Televideo GmbH |
| CONTACT FOR MORE INFO. | Maria Schlipf |
| FEATURES | Chapters, Picture Stops, Dual Audio |
| PRODUCER | L'Oreal, GmbH, Karlsruhe, Germany |
| DISTRIBUTOR | L'Oreal |
| AVAILABILITY | Loan |
| SPECIAL CONDITIONS | Has been placed in 200 hair salons in Germany on a test basis. |

| | |
|---:|:---|
| TITLE | **COLUMBIA PICTURES CARTOONS, VOL. 2** |
| TYPE | Motion Picture |
| FORMAT | CED |
| FEATURES | Banded |
| DISTRIBUTOR | RCA Videodiscs, 03039, $19.98 |

| | |
|---:|:---|
| TITLE | **COLUMBIA RIVER PROJECT** |
| DESCRIPTION | Assorted photography, maps, imagery, and data on the Columbia River area in Washington State. Includes registered map photography of 30 different maps as well as integrated surrogate travel. |
| TYPE | Technical |
| FORMAT | Laser CAV |
| LEVEL | 3 |
| DEVELOPER | Interactive Television Company |
| CONTACT FOR MORE INFO. | Dr. Steven Levin, Interactive Television Company |
| RUNNING TIME | 30 minutes |
| AVAILABILITY | Contact Interactive Television Company |

| | |
|---:|:---|
| TITLE | **COMA** |
| TYPE | Motion Picture (PG) |
| FORMAT | Laser CLV, CED |
| RUNNING TIME | 104 minutes |
| PRODUCER | MGM/UA Home Video |
| DISTRIBUTOR | Pioneer Video, Inc., ML100013, $34.95 |
| RETAIL OUTLET | (CED) Movies Unlimited Inc., A2-1021, $29.95 |

| | |
|---:|:---|
| TITLE | **COMANCHEROS, THE** |
| TYPE | Motion Picture |
| FORMAT | CED |
| RETAIL OUTLET | Movies Unlimited Inc., A1-1177, $29.98 |

| | |
|---:|:---|
| TITLE | **COME BACK TO THE FIVE & DIME, JIMMY DEAN, JIMMY DEAN** |
| TYPE | Motion Picture |
| FORMAT | CED |
| RETAIL OUTLET | Movies Unlimited Inc., A4-3336, $29.95 |

| | |
|---:|:---|
| TITLE | **COMEDY TONIGHT** |
| TYPE | Motion Picture |
| FORMAT | CED |
| RETAIL OUTLET | Movies Unlimited Inc., A3-4001, $29.95 |

| | |
|---:|:---|
| TITLE | **COMES A HORSEMAN** |
| TYPE | Motion Picture |
| FORMAT | CED |
| RETAIL OUTLET | Movies Unlimited Inc., A1-4552, $29.98 |

| | |
|---:|:---|
| TITLE | **COMING HOME** |
| TYPE | Motion Picture (R) |
| FORMAT | Laser CLV, CED (2 discs) |
| RUNNING TIME | 127 minutes |
| PRODUCER | Warner Home Video |
| DISTRIBUTOR | Pioneer Video, Inc., 4516-80, $39.98 |
| | RCA Videodiscs, 01402, $34.98 |

| | |
|---:|:---|
| TITLE | **COMPLETE TENNIS FROM THE PROS, VOL.1: STROKES & TECHNIQUES** |
| TYPE | Instructional |
| FORMAT | CED |
| DISTRIBUTOR | RCA Videodiscs, 02014, $19.98 |

| | |
|---:|:---|
| TITLE | **COMPLEAT BEATLES, THE** |
| TYPE | Music |
| FORMAT | Laser CLV, CED |
| FEATURES | Stereo, CX Encoded |
| RUNNING TIME | 119 minutes |
| PRODUCER | MGM/UA Home Video |
| DISTRIBUTOR | Pioneer Video, Inc., ML100166, $34.95 |
| RETAIL OUTLET | (CED) Movies Unlimited Inc., A2-0166, $29.95 |

| | |
|---:|:---|
| TITLE | **COMPREHENSIVE CASE MANAGEMENT OF SPINA BIFIDA** |
| DESCRIPTION | Provides health professionals with the knowledge and skills necessary to provide effective long- term follow-up and case management of spina bifida (myelomeningocele) cases. |
| TYPE | Medical, Instructional |
| RELEASE DATE | 1983 |
| FORMAT | Laser CAV |
| LEVEL | 3 |
| SYSTEM DESCRIPTION | Discovision 7820-1 player, Apple II plus microcomputer, and VAI interface card |
| DEVELOPER | MGT of America, Inc., Tallahassee, FL |
| CONTACT FOR MORE INFO. | Jayne Parker |
| RUNNING TIME | Varies, but approximately 10 hours. |
| PRODUCER | Florida State University, Multi-Media Laboratory |
| AVAILABILITY | Not for sale at this time, but may become available in the future. |

| | |
|---:|:---|
| TITLE | **COMPREHENSIVE JOB SKILLS PRACTICE** |
| DESCRIPTION | Along with four floppy diskettes, this disc provides a realistic simulation of several weeks in the job of a welfare worker in Florida. It is highly interactive, and weaves the problems and solutions of four families requiring assistance into an instructional experience for a new welfare worker. |
| TYPE | Instructional, In-house Training |

| | |
|---:|:---|
| FORMAT | Laser CAV |
| LEVEL | 3 |
| SYSTEM DESCRIPTION | Disc is part of a 160-hour curriculum delivered on Apple II/Pioneer 7820-3/single disk drive system along with ancillary printed reference books. |
| DEVELOPER | University of West Florida/Office for Interactive Technology and Training |
| CONTACT FOR MORE INFO. | Richard Smith; Elizabeth Wright, (413) 256-0444 |
| FEATURES | Dual Audio |
| PRODUCER | University of Nebraska/Nebraska Videodisc Design/Production Group |
| AVAILABILITY | The training program was produced under contract with the Florida Department of Health and Rehabilitative Services, Tallahassee, Florida, and is not available for general distribution. |

| | |
|---:|:---|
| TITLE | **CONAN THE BARBARIAN** |
| TYPE | Motion Picture |
| FORMAT | Laser CLV |
| RETAIL OUTLET | UltraVideo, 13-008, $44.95 |

| | |
|---:|:---|
| TITLE | **CONTINUING MEDICAL EDUCATION LIBRARY** |
| DESCRIPTION | Library of medical simulations on such topics as patient management, differential diagnosis of epigastric pain, gastric carcinoma, and endoscopy techniques. This series has already received accreditation by the AMA. |
| TYPE | Medical, Instructional |
| RELEASE DATE | 1/83 |
| FORMAT | Laser CAV |
| LEVEL | 3 |
| DEVELOPER | WICAT Systems for SmithKline & French Pharmaceuticals |
| CONTACT FOR MORE INFO. | Richard M. Cavagnol |
| PRODUCER | WICAT Systems, Inc. |

| | |
|---:|:---|
| TITLE | **CONVERSATION, THE** |
| TYPE | Motion Picture (PG) |
| FORMAT | Laser CLV |
| FEATURES | CX Encoded |
| RUNNING TIME | 113 minutes |
| PRODUCER | Warner Home Video |
| DISTRIBUTOR | Pioneer Video, Inc., LV2307, $29.95 |

| | |
|---:|:---|
| TITLE | **COUNT OF MONTE-CRISTO, THE** |
| TYPE | Motion Picture |
| FORMAT | CED |
| DISTRIBUTOR | RCA Videodiscs, 00519, $24.98 |

| | |
|---:|:---|
| TITLE | **COUNTRY GIRL, THE** |
| TYPE | Motion Picture |
| FORMAT | CED |
| PRODUCER | Paramount Pictures |
| DISTRIBUTOR | RCA Videodiscs, 00685, $24.98 |

| | |
|---:|:---|
| TITLE | **COURT JESTER, THE** |
| TYPE | Motion Picture |
| FORMAT | Laser CLV |
| RETAIL OUTLET | UltraVideo, LV5512, $29.95 |

| | |
|---:|:---|
| TITLE | **COURT TESTIMONY SKILLS FOR THE EXPERT WITNESS IN CHILD ABUSE/ NEGLECT CASES** |
| DESCRIPTION | Designed to teach physicians, nurses, psychologists, and social workers how to testify as an expert witness in the courtroom. |
| TYPE | Medical, Research and Development, Instructional, In-house Training |
| RELEASE DATE | July 1984 |
| FORMAT | Laser CAV |
| LEVEL | 3 |
| DEVELOPER | University of West Florida/Office for Interactive Technology and Training |
| CONTACT FOR MORE INFO. | Patricia Lynett, (904) 474-2251 |
| FEATURES | Chapters, Dual Audio, Sound Over Still |

| | |
|---:|:---|
| TITLE | **COUSIN, COUSINE** |
| TYPE | Motion Picture |
| FORMAT | CED |
| RETAIL OUTLET | Movies Unlimited Inc., A2-1039, $24.95 |

| | |
|---:|:---|
| TITLE | **CRANBERRY WORLD VISITOR'S CENTER VIDEODISC PROJECT** |
| DESCRIPTION | Segments explaining how cranberries are grown, harvested, processed, and used. |
| TYPE | Instructional, Promotional |
| FORMAT | Laser CAV |
| LEVEL | 2 |
| DEVELOPER | Ocean Spray Cranberries, Inc. |
| FEATURES | Digital Dump-Pioneer |
| PRODUCER | Madeleine L. Butler and Paul M. Raila |
| AVAILABILITY | Not for sale. |

| | |
|---:|:---|
| TITLE | **CREATIVE CAMERA, THE** |
| TYPE | Instructional |
| FORMAT | Laser CAV |
| LEVEL | 1 |
| FEATURES | Dual Audio |
| PRODUCER | Pioneer Video, Inc. |
| DISTRIBUTOR | Pioneer Video, Inc., HP-019, $29.95 |

| | |
|---:|:---|
| TITLE | **CREEPSHOW** |
| TYPE | Motion Picture |
| FORMAT | CED |
| RETAIL OUTLET | Movies Unlimited Inc., A1-1306, $29.98 |

| | |
|---:|:---|
| TITLE | **CROSBY, STILLS AND NASH** |
| TYPE | Music |
| FORMAT | Laser CLV, CED |
| FEATURES | Stereo |
| RETAIL OUTLET | (CLV) UltraVideo, 74-020, $24.95 |
| | (CED) Movies Unlimited Inc., A5-4020, $24.98 |

| | |
|---:|:---|
| TITLE | **CRUISING** |
| TYPE | Motion Picture |
| FORMAT | CED |
| RETAIL OUTLET | Movies Unlimited Inc., A2-1009, $24.95 |

| | |
|---:|:---|
| TITLE | **CUTTER'S WAY** |
| TYPE | Motion Picture |
| FORMAT | CED |
| RETAIL OUTLET | Movies Unlimited Inc., A2-0154, $29.95 |

| | |
|---:|:---|
| TITLE | **D.C. MOTOR MAINTENANCE, PT. 1** |
| DESCRIPTION | D.C. motor types and preventive maintenance, such as commutator inspection, brush seating, and brush tension. |
| TYPE | Technical, Instructional |
| FORMAT | Laser CAV |
| LEVEL | 3 |
| SYSTEM DESCRIPTION | Pioneer 7820 Model 2 player, Apple IIe with 2 drives, touch panel (infrared) monitor, all in one cabinet. |
| DEVELOPER | Creativision, Inc. |
| CONTACT FOR MORE INFO. | Mr. Rick Glasby |
| FEATURES | Stereo, Dual Audio, Touch Panel Monitor |
| PRODUCER | Creativision, Inc. |
| DISTRIBUTOR | Creativision, Inc. |
| AVAILABILITY | Produced for the Navy Electrician's Mate "A" School with contract funds from the U.S. Navy. |

| | |
|---:|:---|
| TITLE | **D.C. MOTOR MAINTENANCE, PT. 2** |
| DESCRIPTION | Troubleshooting of a compound D.C. motor, in a shipboard system, for opens, shorts, and grounds. |
| TYPE | Technical, Instructional |
| FORMAT | Laser CAV |
| LEVEL | 3 |
| SYSTEM DESCRIPTION | Pioneer 7820 Model 2 player, Apple IIe with 2 drives, and touch panel (infrared) monitor, all in one cabinet. |
| DEVELOPER | Creativision, Inc. |
| CONTACT FOR MORE INFO. | Mr. Rick Glasby |
| FEATURES | Stereo, Dual Audio, OT. |
| PRODUCER | Creativision, Inc. |
| DISTRIBUTOR | Creativision, Inc. |
| AVAILABILITY | Produced for the Navy Electrician's Mate "A" School with contract funds from the U.S. Navy. |

| | |
|---:|:---|
| TITLE | **DAMIEN - OMEN II** |
| TYPE | Motion Picture |
| FORMAT | CED |
| RETAIL OUTLET | Movies Unlimited Inc., A1-1087, $29.98 |

| | |
|---:|:---|
| TITLE | **DARBY O'GILL AND THE LITTLE PEOPLE** |
| TYPE | Motion Picture (G) |
| FORMAT | Laser CLV |
| RUNNING TIME | 90 minutes |
| PRODUCER | Walt Disney |
| DISTRIBUTOR | Pioneer Video, Inc., 42038AS, $34.95 |

| | |
|---:|:---|
| TITLE | **DARK VICTORY** |
| TYPE | Motion Picture |
| FORMAT | CED |
| DISTRIBUTOR | RCA Videodiscs, 03405, $19.98 |

| | |
|---:|:---|
| TITLE | **DARTMOUTH COLLEGE ARCHIVAL DISC** |
| DESCRIPTION | Archival footage on Dartmouth presidents and college slide archive. |
| TYPE | Archival |
| FORMAT | Laser CAV |
| LEVEL | 1 |
| DEVELOPER | Prof. William M. Smith |
| CONTACT FOR MORE INFO. | Prof. William M. Smith |
| FEATURES | Picture Stops |
| RUNNING TIME | 25 minutes |
| PRODUCER | Prof. William M. Smith, Dr. Ronald Boehm |
| DISTRIBUTOR | Prof. William M. Smith |

**DAS BOOT.** *See* **BOAT, THE**

| | |
|---:|:---|
| TITLE | **DATA DEMONS** |
| DESCRIPTION | An interactive disc game for AT&T to illustrate a range of special telephone services now being marketed. |
| TYPE | Game, Technical, Instructional, Point-of-Purchase, Advertising, Promotional |
| FORMAT | Sony |
| LEVEL | 3 |
| SYSTEM DESCRIPTION | Sony graphic superimposition on laser disc imagery; Sony computer driven. |
| DEVELOPER | Exhibit Technology, Inc./AT&T |
| CONTACT FOR MORE INFO. | Don Cochrane, President - Marketing |
| FEATURES | Chapters, Picture Stops, Digital Dump-Sony, Dual Audio, Sound Over Still |
| PRODUCER | Mr. John Singer/Exhibit Technology, Inc. |
| DISTRIBUTOR | AT&T |
| AVAILABILITY | Used in AT&T Exhibits and Phone Stores. Only available as part of Exhibit Technology's product line sampling. |

| | |
|---:|:---|
| TITLE | **DAVE MASON LIVE AT PERKINS PALACE** |
| TYPE | Music |
| FORMAT | Laser CLV |
| FEATURES | Stereo, CX Encoded |
| RUNNING TIME | 58 minutes |
| PRODUCER | Pioneer Artists |
| DISTRIBUTOR | Pioneer Video, Inc., PA-82-021, $24.95 |

| | |
|---:|:---|
| TITLE | **DAVY CROCKETT & THE RIVER PIRATES** |
| TYPE | Motion Picture |
| FORMAT | Laser CLV |
| RUNNING TIME | 81 minutes |
| PRODUCER | Warner Home Video |
| DISTRIBUTOR | Pioneer Video, Inc., 42027AS, $34.95 |

| | |
|---:|:---|
| TITLE | **DAY AT THE RACES, A** |
| TYPE | Motion Picture, Comedy |
| FORMAT | Laser CLV, CED |
| FEATURES | Black and white |
| RUNNING TIME | 109 minutes |
| PRODUCER | MGM/UA Home Video |
| DISTRIBUTOR | Pioneer Video, Inc., ML100064, $25.95 |
| RETAIL OUTLET | (CED) Movies Unlimited Inc., A2-1024, $29.95 |

| | |
|---:|:---|
| TITLE | **DAY OF THE DOLPHIN, THE** |
| TYPE | Motion Picture (PG) |
| FORMAT | Laser CLV |
| RUNNING TIME | 104 minutes |
| PRODUCER | Warner Home Video |
| DISTRIBUTOR | Pioneer Video, Inc., 4004-80, $34.98 |

| | |
|---:|:---|
| TITLE | **DAY OF THE LOCUST** |
| TYPE | Motion Picture |
| FORMAT | Laser CLV |
| RETAIL OUTLET | UltraVideo, LV8679, $39.95 |

| | |
|---:|:---|
| TITLE | **DAYS OF HEAVEN** |
| TYPE | Motion Picture (PG) |
| FORMAT | Laser CLV, CED |
| FEATURES | Stereo, CX Encoded |
| RUNNING TIME | 94 minutes |
| PRODUCER | Warner Home Video |
| DISTRIBUTOR | Pioneer Video, Inc., LV8942, $29.95 |
| | RCA Videodiscs, 10659, $24.98 |

| | |
|---:|:---|
| TITLE | **DEAD MEN DON'T WEAR PLAID** |
| TYPE | Motion Picture (PG) |
| FORMAT | Laser CLV |
| FEATURES | Black and white |
| RUNNING TIME | 89 minutes |
| PRODUCER | Warner Home Video |
| DISTRIBUTOR | Pioneer Video, Inc., 16-028, $29.95 |

| | |
|---:|:---|
| TITLE | **DEADLY BLESSING** |
| TYPE | Motion Picture |
| FORMAT | CED |
| RETAIL OUTLET | Movies Unlimited Inc., A4-3216, $29.95 |

| | |
|---:|:---|
| TITLE | **DEAF AWARENESS: LET YOUR FINGERS DO THE TALKING** |
| DESCRIPTION | Disc introduces students to the world of finger spelling and signs. The disc and its accompanying support manual includes both instructional and testing segments on the manual alphabet, simple numbers, word signs, and phrases. |
| TYPE | Instructional |
| FORMAT | Laser CAV |
| LEVEL | 2 |
| DEVELOPER | Alberta Vocational Centre, Calgary, Alberta |
| CONTACT FOR MORE INFO. | Russell Sawchuk at ACCESS Alberta |
| FEATURES | Digital Dump-Sony |
| PRODUCER | ACCESS Alberta and Alberta Vocational Centre |
| DISTRIBUTOR | Marketing Department, ACCESS Alberta |
| AVAILABILITY | Sale |
| SUGGESTED PRICE | $225 |

| | |
|---:|:---|
| TITLE | **DEATH HUNT** |
| TYPE | Motion Picture (R) |
| FORMAT | Laser CLV, CED |
| FEATURES | (CLV) Stereo |
| RUNNING TIME | 97 minutes |
| PRODUCER | Warner Home Video |
| DISTRIBUTOR | Pioneer Video, Inc., 1125-80, $34.98 |
| RETAIL OUTLET | (CED) Movies Unlimited Inc., A1-1125, $29.98 |

| | |
|---:|:---|
| TITLE | **DEATH RACE 2000** |
| TYPE | Motion Picture |
| FORMAT | CED |
| DISTRIBUTOR | RCA Videodiscs, 00905, $24.98 |

| | |
|---:|:---|
| TITLE | **DEATH WISH** |
| TYPE | Motion Picture (R) |
| FORMAT | Laser CLV, CED |
| RUNNING TIME | 93 minutes |
| PRODUCER | Warner Home Video |
| DISTRIBUTOR | Pioneer Video, Inc., LV8774, $29.95 |
| | RCA Videodiscs, 00611, $29.98 |

| | |
|---:|:---|
| TITLE | **DEATH WISH 2** |
| TYPE | Motion Picture |
| FORMAT | Laser CLV, CED |
| RUNNING TIME | 88 minutes |
| PRODUCER | Vestron Video |
| DISTRIBUTOR | Pioneer Video, Inc., VL4017, $34.95 |
| RETAIL OUTLET | (CED) Movies Unlimited Inc., A3-4017, $29.95 |

| | |
|---:|:---|
| TITLE | **DECIMALS AND FRACTIONS** |
| DESCRIPTION | Teaches basic concepts in decimals and fractions. |
| TYPE | Children, Research and Development, Instructional |
| FORMAT | Laser CAV |
| LEVEL | 3 |
| DEVELOPER | Educational Testing Service (ETS) |
| CONTACT FOR MORE INFO. | Isaac Bejar |
| PRODUCER | Frank Capuzzi |
| AVAILABILITY | Not available. This is a research and development project. |

| | |
|---:|:---|
| TITLE | **DEEP, THE** |
| TYPE | Motion Picture (PG) |
| FORMAT | CED |
| DISTRIBUTOR | RCA Videodiscs, 03008, $24.98 |

| | |
|---:|:---|
| TITLE | **DEER HUNTER, THE** |
| TYPE | Motion Picture (R) |
| FORMAT | Laser CLV, CED (2 discs) |
| FEATURES | Stereo, CX Encoded |
| RUNNING TIME | 177 minutes |
| PRODUCER | Warner Home Video |
| DISTRIBUTOR | Pioneer Video, Inc., 10-026, $39.95 |
| | RCA Videodiscs, 13311, $39.98 |

| | |
|---:|:---|
| TITLE | **DEFIANCE** |
| TYPE | Motion Picture |
| FORMAT | CED |
| RETAIL OUTLET | Movies Unlimited Inc., A3-4037, $29.95 |

| | |
|---:|:---|
| TITLE | **DELIVERANCE** |
| TYPE | Motion Picture (R) |
| FORMAT | CED |
| PRODUCER | Warner Home Video |
| DISTRIBUTOR | RCA Videodiscs, 03122, $24.98 |

| | |
|---:|:---|
| TITLE | **DELTA CORRIDOR** |
| DESCRIPTION | Surrogate travel of desert training as well as instruction program on desert familiarization. |
| TYPE | In-house Training |
| FORMAT | Laser CAV |
| LEVEL | 3 |
| DEVELOPER | Interactive Television Company |
| CONTACT FOR MORE INFO. | Dr. Steven Levin, Interactive Television Company |
| RUNNING TIME | 30 minutes |
| AVAILABILITY | Contact Interactive Television Company |

| | |
|---:|:---|
| TITLE | **DEMONSTRATION VIDEODISC (CAT ZURICH)** |
| DESCRIPTION | Demonstration of interactivity, applications, voice over still, quality and design, CAT network, still bank, combination with view data and telecommunication, etc. |

# Videodiscography

| | |
|---|---|
| TYPE | Game, Medical, Research and Development, Instructional, Archival, Point-of-Purchase, Promotional |
| FORMAT | Laser CAV |
| LEVEL | 3 |
| SYSTEM DESCRIPTION | Two Philips industrial VP-835 players controlled by micro with teletex overlays 2 disc sides on-line with 4 audio tracks (German, French, English, and mixed channel for voice over stills). |
| CONTACT FOR MORE INFO. | Hanspanl Schellenberg |
| FEATURES | Stereo, Dual Audio, Sound Over Still |
| PRODUCER | CAT Zurich-Computer Assisted Televideo AG |
| DISTRIBUTOR | CAT Zurich-Computer Assisted Televideo AG |
| AVAILABILITY | Loan. Shown in CAT showrooms and CAT infomobile. |

| | |
|---|---|
| TITLE | **DEMONSTRATION VIDEODISC (WICAT SYSTEMS, INC.)** |
| DESCRIPTION | Disc is a simulation for the Business Services Division of AT&T. |
| TYPE | Research and Development, Technical |
| RELEASE DATE | 7/82 |
| FORMAT | Laser CAV |
| LEVEL | 3 |
| DEVELOPER | WICAT Systems for AT&T |
| CONTACT FOR MORE INFO. | Richard M. Cavagnol |
| PRODUCER | WICAT Systems, Inc. |

| | |
|---|---|
| TITLE | **DEMONSTRATION VIDEODISCS, 1982 AND 1983 (CSC)** |
| DESCRIPTION | Includes samples of several Army training and maintenance/logistics videodisc applications, as well as explanations of Army videodisc efforts. |
| TYPE | Research and Development, Technical, Instructional |
| FORMAT | Laser CAV |
| LEVEL | 3 |
| SYSTEM DESCRIPTION | Pioneer player, Apple computer |
| DEVELOPER | Computer Sciences Corporation |
| CONTACT FOR MORE INFO. | Dr. William A. Stembler |
| FEATURES | Can use any combination of entries on disc for demonstration purposes. |
| PRODUCER | Dr. William A. Stembler |
| AVAILABILITY | Property of U.S. Army |

| | |
|---|---|
| TITLE | **DENTISTRY I ("SUE")** |
| DESCRIPTION | Major oral surgery |
| TYPE | Medical, Instructional |
| LEVEL | 3 |
| FORMAT | Laser CAV |
| CONTACT FOR MORE INFO. | National Library of Medicine/University of Nebraska Dentistry |
| PRODUCER | University of Nebraska/Nebraska Videodisc Design/Production Group |
| AVAILABILITY | Not available for release or distribution. |

| | |
|---|---|
| TITLE | **DENTISTRY II ("JACKIE")** |
| DESCRIPTION | Major oral surgery |
| TYPE | Medical, Instructional |
| LEVEL | 3 |

| | |
|---|---|
| FORMAT | Laser CAV |
| CONTACT FOR MORE INFO. | National Library of Medicine/University of Nebraska Dentistry |
| PRODUCER | University of Nebraska/Nebraska Videodisc Design/Production Group |
| AVAILABILITY | Not available for release or distribution. |

| | |
|---|---|
| TITLE | **DENTISTRY III** |
| DESCRIPTION | Impacted incisor |
| TYPE | Medical, Instructional |
| LEVEL | 3 |
| FORMAT | Laser CAV |
| CONTACT FOR MORE INFO. | National Library of Medicine/University of Nebraska Dentistry |
| PRODUCER | University of Nebraska/Nebraska Videodisc Design/Production Group |
| AVAILABILITY | Not available for release or distribution. |

| | |
|---|---|
| TITLE | **DENTISTRY IV** |
| DESCRIPTION | Dental diagnosis |
| TYPE | Medical, Instructional |
| LEVEL | 3 |
| FORMAT | Laser CAV |
| CONTACT FOR MORE INFO. | National Library of Medicine/University of Nebraska Dentistry |
| PRODUCER | University of Nebraska/Nebraska Videodisc Design/Production Group |
| AVAILABILITY | Not available for release or distribution. |

| | |
|---|---|
| TITLE | **DETECTIVE, THE** |
| TYPE | Motion Picture |
| FORMAT | CED |
| RETAIL OUTLET | Movies Unlimited Inc., A1-1018, $29.98 |

| | |
|---|---|
| TITLE | **DETERMINING FINANCIAL ELIGIBILITY** |
| DESCRIPTION | Along with 13 floppy diskettes, this videodisc emphasizes budget calculations, and instructs welfare workers on the procedures for determining the dollar amount of welfare checks for dependent children in Florida. |
| TYPE | Instructional, In-house Training |
| RELEASE DATE | 10/17/83 |
| FORMAT | Laser CAV |
| LEVEL | 3 |
| SYSTEM DESCRIPTION | Disc is part of a 160-hour curriculum delivered on Apple/Pioneer 7820-3/single disc drive system along with ancillary printed reference books |
| DEVELOPER | University of West Florida/Office for Interactive Technology and Training |
| CONTACT FOR MORE INFO. | Richard Smith; Elizabeth Wright, (413) 256-0444 |
| FEATURES | Dual Audio |
| PRODUCER | University of Nebraska/Nebraska Videodisc Design/Production Group |
| AVAILABILITY | Training program was produced under contract with the Florida Department of Health and Rehabilitative Services, Tallahassee, Florida, and is not available for general distribution. |

| | |
|---|---|
| TITLE | **DEVELOPMENT OF LIVING THINGS** |
| DESCRIPTION | This is the first instructional videodisc ever produced. Uses a sophisticated instructional strategy involving captivating motion sequences and computer-generated animation and still frames for concept presentation and questioning. Ths disc is on developmental biology for high school and college. |
| TYPE | Research and Development, Instructional |
| RELEASE DATE | 5/78 |
| FORMAT | Laser CAV |
| LEVEL | 3 |
| DEVELOPER | WICAT Systems, Inc. with McGraw-Hill, Inc. |
| CONTACT FOR MORE INFO. | Richard M. Cavagnol |
| PRODUCER | WICAT Systems, Inc. |

| | |
|---|---|
| TITLE | **DEXY'S MIDNIGHT RUNNERS** |
| TYPE | Music |
| FORMAT | CED |
| DISTRIBUTOR | RCA Videodiscs, 12218, $19.98 |

| | |
|---|---|
| TITLE | **DIAGNOSIS & TREATMENT OF BIRTH DEFECTS** |
| DESCRIPTION | In-service training for nurses |
| TYPE | Medical, In-house Training |
| LEVEL | 3 |
| CONTACT FOR MORE INFO. | Florida State University |
| PRODUCER | University of Nebraska/Nebraska Videodisc Design/Production Group |
| AVAILABILITY | Not available for release or distribution. |

| | |
|---|---|
| TITLE | **DIAGNOSIS & TREATMENT OF SPINA BIFIDA** |
| DESCRIPTION | In-service training for nurses |
| TYPE | Medical, In-house Training |
| LEVEL | 3 |
| CONTACT FOR MORE INFO. | Florida State University |
| PRODUCER | University of Nebraska/Nebraska Videodisc Design/Production Group |
| AVAILABILITY | Not available for release or distribution. |

| | |
|---|---|
| TITLE | **DIAGNOSTIC CHALLENGES** |
| TYPE | Medical, Instructional |
| RELEASE DATE | 6/80 |
| FORMAT | Laser CAV |
| LEVEL | 3 |
| DEVELOPER | WICAT Systems for SmithKline & French Pharmaceuticals |
| CONTACT FOR MORE INFO. | Richard M. Cavagnol |
| PRODUCER | WICAT Systems, Inc. |

| | |
|---|---|
| TITLE | **DIAL "M" FOR MURDER** |
| TYPE | Motion Picture |
| FORMAT | CED |
| DISTRIBUTOR | RCA Videodiscs, 03157, $24.98 |

| | |
|---:|:---|
| TITLE | **DIAMONDS ARE FOREVER** |
| TYPE | Motion Picture (PG) |
| FORMAT | Laser CLV, CED |
| FEATURES | (CLV) Stereo |
| RUNNING TIME | 119 minutes |
| PRODUCER | CBS Fox Video |
| DISTRIBUTOR | Pioneer Video, Inc., 4605-80, $34.98 |
| | RCA Videodiscs, 01423, $29.98 |

| | |
|---:|:---|
| TITLE | **DIANA ROSS IN CONCERT** |
| TYPE | Music |
| FORMAT | CED |
| FEATURES | Stereo, CX Encoded |
| DISTRIBUTOR | RCA Videodiscs, 12119, $24.98 |

| | |
|---:|:---|
| TITLE | **DIARY OF ANN FRANK** |
| TYPE | Motion Picture |
| FORMAT | CED (2 Discs) |
| RETAIL OUTLET | Movies Unlimited Inc., A1-1074, $39.98 |

| | |
|---:|:---|
| TITLE | **DICK CAVETT'S HOCUS POCUS, IT'S MAGIC** |
| TYPE | Motion Picture (G) |
| FORMAT | Laser CLV |
| RUNNING TIME | 84 minutes |
| PRODUCER | Warner Home Video |
| DISTRIBUTOR | Pioneer Video, Inc., VL3006, $34.95 |

| | |
|---:|:---|
| TITLE | **DIGITAL VIDEO CORPORATION - (CONFIDENTIAL)** |
| TYPE | Research and Development |
| FORMAT | Laser CAV |
| LEVEL | 3 |
| FEATURES | Chapters, Picture Stops, Stereo |
| AVAILABILITY | Digital Video Corporation has completed six proprietary videodisc projects that are not for resale. |

| | |
|---:|:---|
| TITLE | **DINER** |
| TYPE | Motion Picture (R) |
| FORMAT | Laser CLV, CED |
| RUNNING TIME | 110 minutes |
| PRODUCER | MGM/UA Home Video |
| DISTRIBUTOR | Pioneer Video, Inc., ML100164, $34.95 |
| RETAIL OUTLET | (CED) Movies Unlimited Inc., A2-1063, $29.95 |

| | |
|---:|:---|
| TITLE | **DINNER FOR THE BOSS** |
| DESCRIPTION | A broadcast Julia Child episode reformatted for videodisc demonstration. |
| TYPE | Instructional |
| RELEASE DATE | 9/82 |
| FORMAT | Laser CAV |
| LEVEL | 2 |

| | |
|---|---|
| DEVELOPER | WGBH/Sony |
| CONTACT FOR MORE INFO. | Rob Lippincott, Director of Videodisc Projects, WGBH |
| FEATURES | Digital Dump-Sony |
| RUNNING TIME | 29 minutes |
| PRODUCER | WGBH |
| DISTRIBUTOR | WGBH |
| AVAILABILITY | Not for sale. Will barter for similar level 2 demonstration projects. |

| | |
|---|---|
| TITLE | **DIRT BAND, THE** |
| TYPE | Music |
| FORMAT | Laser CLV |
| FEATURES | Stereo, CX Encoded |
| RUNNING TIME | 58 minutes |
| PRODUCER | Pioneer Artists |
| DISTRIBUTOR | Pioneer Video, Inc., PA-83-034, $24.95 |

| | |
|---|---|
| TITLE | **DIRTY DOZEN, THE** |
| TYPE | Motion Picture |
| FORMAT | CED (2 discs) |
| FEATURES | Stereo |
| RETAIL OUTLET | Movies Unlimited Inc., A2-0008, $39.95 |

| | |
|---|---|
| TITLE | **DIRTY HARRY** |
| TYPE | Motion Picture (R) |
| FORMAT | Laser CLV, CED |
| RUNNING TIME | 103 minutes |
| PRODUCER | Warner Home Video |
| DISTRIBUTOR | Pioneer Video, Inc., 1019LV, $29.98 |
| | RCA Videodiscs, 03104, $24.98 |

| | |
|---|---|
| TITLE | **DISC SUPPLEMENT TO JOURNAL, CELL MOTILITY** |
| DESCRIPTION | Photomicrographic material correlated with 1983 volume of "Cell Motility." |
| TYPE | Technical |
| RELEASE DATE | 12/83 |
| FORMAT | Laser CAV |
| LEVEL | 1 |
| DEVELOPER | Prof. William M. Smith |
| CONTACT FOR MORE INFO. | Prof. William M. Smith |
| FEATURES | Picture Stops, Digital Dump-Sony |
| RUNNING TIME | 29 minutes |
| PRODUCER | Prof. William M. Smith, Dr. Ronald Boehm |
| DISTRIBUTOR | Alan R. Liss, Inc., NYC Publisher |

| | |
|---|---|
| TITLE | **DISCOVERING AMERICA** |
| DESCRIPTION | Interactive videodisc game for foreign businessmen and their families locating in the U.S. |
| TYPE | In-house Training |
| RELEASE DATE | 1/84 |

| | |
|---:|:---|
| TITLE | **DISNEY CARTOON PARADE, VOL. 1** |
| TYPE | Motion Picture (G) |
| FORMAT | CED |
| DISTRIBUTOR | RCA Videodiscs, 00709, $19.98 |

| | |
|---:|:---|
| TITLE | **DISNEY CARTOON PARADE, VOL. 2** |
| TYPE | Motion Picture (G) |
| FORMAT | CED |
| PRODUCER | Walt Disney |
| DISTRIBUTOR | RCA Videodiscs, 00715, $19.98 |

| | |
|---:|:---|
| TITLE | **DISNEY CARTOON PARADE, VOL. 3** |
| TYPE | Motion Picture (G) |
| FORMAT | CED |
| PRODUCER | Walt Disney |
| DISTRIBUTOR | RCA Videodiscs, 00717, $19.98 |

| | |
|---:|:---|
| TITLE | **DISNEY CARTOON PARADE, VOL. 4** |
| TYPE | Motion Picture (G) |
| FORMAT | CED |
| PRODUCER | Walt Disney |
| DISTRIBUTOR | RCA Videodiscs, 00727, $19.98 |

| | |
|---:|:---|
| TITLE | **DISNEY CARTOON PARADE, VOL. 5** |
| TYPE | Motion Picture (G) |
| FORMAT | CED |
| DISTRIBUTOR | RCA Videodiscs, 00729, $19.98 |

| | |
|---:|:---|
| TITLE | **DISNEYDISC OF MYSTERY AND MAGIC** |
| TYPE | Game, Children, Instructional |
| RELEASE DATE | July 1984 |
| FORMAT | CED |
| LEVEL | 1 |
| DEVELOPER | Walt Disney Telecommunications |
| CONTACT FOR MORE INFO. | Diane Smook, (212) 930-4767 |
| FEATURES | Chapters, Picture Stops, Dual Audio |
| PRODUCER | Walt Disney Telecommunications |
| DISTRIBUTOR | RCA Videodiscs, 25007 |

| | |
|---:|:---|
| TITLE | **DIVA** |
| TYPE | Motion Picture (R) |
| FORMAT | Laser CLV, CED |
| FEATURES | English subtitles |
| RETAIL OUTLET | (CLV) UltraVideo, ML100183, $34.95 |
| | (CED) Movies Unlimited Inc., A2-0183, $29.95 |

| | |
|---:|:---|
| TITLE | **DIVE TO ANOTHER WORLD** |
| DESCRIPTION | An adventure dive in the "Alvin" submarine to the hydrothermal vent life 2 1/2 miles deep in the Pacific Ocean. |
| TYPE | Instructional |

| | |
|---:|:---|
| RELEASE DATE | 12/83 |
| FORMAT | Laser CAV, CED |
| LEVEL | 1 |
| DEVELOPER | Laser 7 Group |
| CONTACT FOR MORE INFO. | Phil Rapp, (619) 453-0622 |
| FEATURES | Chapters, Picture Stops, Digital Dump-Sony, Digital Dump-Pioneer, Stereo, Dual Audio, CX Encoded |
| RUNNING TIME | 51 minutes |
| PRODUCER | Laser 7 Group |
| DISTRIBUTOR | In negotiation. |

| | |
|---:|:---|
| TITLE | **DIVINE MADNESS** |
| TYPE | Motion Picture, Music, Comedy (R) |
| FORMAT | CED |
| FEATURES | Stereo |
| RUNNING TIME | 94 minutes |
| PRODUCER | Warner Home Video |
| DISTRIBUTOR | RCA Videodiscs, 13148, $24.98 |

| | |
|---:|:---|
| TITLE | **DOCTOR ZHIVAGO** |
| TYPE | Motion Picture (PG) |
| FORMAT | Laser CLV, CED (2 discs) |
| FEATURES | (CLV) Stereo |
| RUNNING TIME | 133 minutes |
| PRODUCER | MGM/UA Home Video |
| DISTRIBUTOR | Pioneer Video, Inc., ML100003, $39.95 |
| RETAIL OUTLET | (CED) Movies Unlimited Inc., A2-1000, $39.95 |

| | |
|---:|:---|
| TITLE | **DODGE CITY** |
| TYPE | Motion Picture |
| FORMAT | CED |
| RETAIL OUTLET | Movies Unlimited Inc., A1-4625, $29.98 |

| | |
|---:|:---|
| TITLE | **DOG DAY AFTERNOON** |
| TYPE | Motion Picture |
| FORMAT | CED |
| DISTRIBUTOR | RCA Videodiscs, 03123, $29.98 |

| | |
|---:|:---|
| TITLE | **DOGS OF WAR, THE** |
| TYPE | Motion Picture |
| FORMAT | CED |
| RETAIL OUTLET | Movies Unlimited Inc., A1-4539, $29.98 |

| | |
|---:|:---|
| TITLE | **DOLL'S HOUSE, A** |
| TYPE | Motion Picture |
| FORMAT | CED |
| DISTRIBUTOR | RCA Videodiscs, 02024, $19.98 |

| | |
|---:|:---|
| TITLE | **DOLLY PARTON** |
| TYPE | Music |
| FORMAT | CED |
| FEATURES | Stereo, Banded |
| DISTRIBUTOR | RCA Videodiscs, 12130, $19.98 |

| | |
|---:|:---|
| TITLE | **DOMINO PRINCIPLE, THE** |
| TYPE | Motion Picture |
| FORMAT | CED |
| RETAIL OUTLET | Movies Unlimited Inc., A1-9008, $29.98 |

| | |
|---:|:---|
| TITLE | **DON KIRSHNER'S ROCK CONCERT, VOL.1** |
| TYPE | Music |
| FORMAT | CED |
| DISTRIBUTOR | RCA Videodiscs, 02029, $19.98 |

| | |
|---:|:---|
| TITLE | **DON'T LOOK NOW** |
| TYPE | Motion Picture (R) |
| FORMAT | Laser CLV, CED |
| RUNNING TIME | 110 minutes |
| PRODUCER | Paramount Home Video |
| DISTRIBUTOR | Pioneer Video, Inc., LV8704, $29.95 |
| | RCA Videodiscs, 00679, $29.98 |

| | |
|---:|:---|
| TITLE | **DOOBIE BROTHERS LIVE, THE** |
| TYPE | Music |
| FORMAT | CED |
| FEATURES | Stereo, CX Encoded |
| DISTRIBUTOR | RCA Videodiscs, 12121, $24.98 |

| | |
|---:|:---|
| TITLE | **DORADO, EL** |
| TYPE | Motion Picture |
| FORMAT | CED (2 Discs) |
| DISTRIBUTOR | RCA Videodiscs, 03603, $39.98 |

| | |
|---:|:---|
| TITLE | **DOT AND THE KANGAROO** |
| TYPE | Motion Picture |
| FORMAT | CED |
| RETAIL OUTLET | Movies Unlimited Inc., A1-6112, $29.98 |

| | |
|---:|:---|
| TITLE | **DOWNHILL RACER** |
| TYPE | Motion Picture (PG) |
| FORMAT | Laser CLV |
| RUNNING TIME | 102 minutes |
| PRODUCER | Warner Home Video |
| DISTRIBUTOR | Pioneer Video, Inc., LV6910, $29.95 |

| | |
|---:|:---|
| TITLE | **DR. CRYPTON'S BRAIN BUSTERS** |
| PRODUCER | Sheldon Renan, Peter Bloch |
| DISTRIBUTOR | RCA Selectavision Videodiscs |

# Videodiscography

| | |
|---|---|
| TITLE | **DR. DETROIT** |
| TYPE | Motion Picture, Comedy (R) |
| FORMAT | Laser CLV |
| RETAIL OUTLET | UltraVideo, 40-001, $29.95 |

| | |
|---|---|
| TITLE | **DR. DOOLITTLE** |
| TYPE | Motion Picture |
| FORMAT | CED (2 discs) |
| FEATURES | Stereo |
| RETAIL OUTLET | Movies Unlimited Inc., A1-1025, $39.98 |

| | |
|---|---|
| TITLE | **DR. NO** |
| TYPE | Motion Picture (PG) |
| FORMAT | Laser CLV, CED |
| RUNNING TIME | 111 minutes |
| PRODUCER | Warner Home Video |
| DISTRIBUTOR | Pioneer Video, Inc., 4524-80, $34.98 |
| | RCA Videodiscs, 01426, $29.98 |

| | |
|---|---|
| TITLE | **DR. SEUSS VIDEO FESTIVAL** |
| TYPE | Motion Picture |
| FORMAT | CED |
| RETAIL OUTLET | Movies Unlimited Inc., A2-1061, $29.95 |

| | |
|---|---|
| TITLE | **DR. SPOCK - CARING FOR YOUR NEWBORN** |
| TYPE | Instructional |
| FORMAT | CED |
| DISTRIBUTOR | RCA Videodiscs, 02016, $19.98 |

| | |
|---|---|
| TITLE | **DR. STRANGELOVE** |
| TYPE | Motion Picture, Comedy |
| FORMAT | CED |
| DISTRIBUTOR | RCA Videodiscs, 03036, $24.98 |

| | |
|---|---|
| TITLE | **DRACULA (1979)** |
| TYPE | Motion Picture (R) |
| FORMAT | Laser CLV, CED |
| FEATURES | (CLV) Stereo |
| RUNNING TIME | 109 minutes |
| PRODUCER | Warner Home Video |
| DISTRIBUTOR | Pioneer Video, Inc., 11-011, $29.95 |
| | RCA Videodiscs, 03314, $19.98 |

| | |
|---|---|
| TITLE | **DRAGONSLAYER** |
| TYPE | Motion Picture (PG) |
| FORMAT | Laser CLV, CED |
| FEATURES | Stereo, CX Encoded |
| RUNNING TIME | 110 minutes |
| PRODUCER | Warner Home Video |
| DISTRIBUTOR | Pioneer Video, Inc., LV1367, $29.95 |
| | RCA Videodiscs, 10675, $29.98 |

| | |
|---:|:---|
| TITLE | **DRESSED TO KILL** |
| TYPE | Motion Picture (R) |
| FORMAT | Laser CLV, CED |
| DISTRIBUTOR | RCA Videodiscs, 01901, $24.98 |
| RETAIL OUTLET | (CLV) UltraVideo, VL4050, $34.95 |

| | |
|---:|:---|
| TITLE | **DUCHESS AND THE DIRTWATER FOX** |
| TYPE | Motion Picture |
| FORMAT | CED |
| RETAIL OUTLET | Movies Unlimited Inc., A1-1059, $29.98 |

| | |
|---:|:---|
| TITLE | **DUCK SOUP** |
| TYPE | Motion Picture, Comedy |
| FORMAT | CED |
| DISTRIBUTOR | RCA Videodiscs, 03316, $19.98 |

| | |
|---:|:---|
| TITLE | **DUMBO** |
| TYPE | Motion Picture (G) |
| FORMAT | Laser CLV, CED |
| RUNNING TIME | 63 minutes |
| PRODUCER | Warner Home Video |
| DISTRIBUTOR | Pioneer Video, Inc., 42024AS, $34.95 |
| | RCA Videodiscs, 00723, $24.98 |

| | |
|---:|:---|
| TITLE | **DUNDER KLUMPEN** |
| TYPE | Motion Picture |
| FORMAT | Laser CLV, CED |
| RUNNING TIME | 85 minutes |
| PRODUCER | Vestron Video |
| DISTRIBUTOR | Pioneer Video, Inc., VL3012, $34.95 |
| RETAIL OUTLET | (CED) Movies Unlimited Inc., A3-3012, $29.95 |

| | |
|---:|:---|
| TITLE | **DURAN DURAN** |
| TYPE | Music |
| FORMAT | Laser CLV, CED |
| FEATURES | Stereo, CX Encoded, (CED) Banded |
| RUNNING TIME | 55 minutes |
| PRODUCER | Pioneer Artists |
| DISTRIBUTOR | Pioneer Video, Inc., PA-83-044, $24.95 |
| | RCA Videodiscs, 12172, $19.98 |

| | |
|---:|:---|
| TITLE | **DVORAK 9** |
| TYPE | Music |
| FORMAT | Laser CLV |
| FEATURES | Stereo |
| RETAIL OUTLET | UltraVideo, MC036, $27.95 |

| | |
|---:|:---|
| TITLE | **DVORAK'S SLAVIC DANCE** |
| TYPE | Music |
| FORMAT | Laser CLV |

| | |
|---|---|
| FEATURES | Stereo, CX Encoded |
| RUNNING TIME | 71 minutes |
| PRODUCER | Pioneer Video Imports |
| DISTRIBUTOR | Pioneer Video, Inc., MC-037US, $29.95 |

| | |
|---|---|
| TITLE | **E-3 FLOATING DECK PULSER** |
| DESCRIPTION | Maintenance-aiding and training program for the E-3 Floating Deck Pulser (FDP). |
| TYPE | Research and Development, Instructional, In-house Training |
| FORMAT | Laser CAV |
| LEVEL | 3 |
| SYSTEM DESCRIPTION | The E-3 FDP Interactive Videodisc System is an stand-alone training and maintenance-aiding system, designed for individual use and housed in a carrel. |
| DEVELOPER | Westinghouse Technical Training Operations |
| CONTACT FOR MORE INFO. | Steve G. Kukla, (301) 995-5482 |
| FEATURES | Text/graphics Overlays, Touch Panel |
| PRODUCER | Westinghouse Multi-Media Production Center |
| AVAILABILITY | Not available. |

| | |
|---|---|
| TITLE | **EARTH, WIND AND FIRE** |
| TYPE | Music |
| FORMAT | Laser CLV, CED |
| FEATURES | Stereo, CX Encoded |
| RUNNING TIME | 58 minutes |
| PRODUCER | Pioneer Artists |
| DISTRIBUTOR | Pioneer Video, Inc., PA-83-039, $24.95 |
| RETAIL OUTLET | (CED) Movies Unlimited Inc., A3-2006, $24.95 |

| | |
|---|---|
| TITLE | **EARTHLING, THE** |
| TYPE | Motion Picture |
| FORMAT | CED |
| RETAIL OUTLET | Movies Unlimited Inc., A3-4029, $29.95 |

| | |
|---|---|
| TITLE | **EAST OF EDEN** |
| TYPE | Motion Picture |
| FORMAT | CED |
| DISTRIBUTOR | RCA Videodiscs, 03124, $24.98 |

| | |
|---|---|
| TITLE | **EASY MONEY** |
| TYPE | Motion Picture, Comedy |
| FORMAT | CED |
| DISTRIBUTOR | RCA Videodiscs, 01917, $24.98 |

| | |
|---|---|
| TITLE | **EASY RIDER** |
| TYPE | Motion Picture (R) |
| FORMAT | Laser CLV, CED |
| DISTRIBUTOR | RCA Videodiscs, 03005, $24.98 |
| RETAIL OUTLET | (CLV) UltraVideo, VLD3140, $29.95 |

| | |
|---:|:---|
| TITLE | **EDDIE MACON'S RUN** |
| TYPE | Motion Picture (PG) |
| FORMAT | Laser CLV, CED |
| RUNNING TIME | 95 minutes |
| PRODUCER | MCA Videodisc, Inc. |
| DISTRIBUTOR | Pioneer Video, Inc., 12-025, $29.95 |
| RETAIL OUTLET | (CED) Movies Unlimited Inc., A5-2025, $29.98 |

---

| | |
|---:|:---|
| TITLE | **EDDIE MURPHY - DELIRIOUS** |
| TYPE | Comedy |
| FORMAT | Laser CLV |
| RETAIL OUTLET | UltraVideo, LV2323, $29.95 |

---

| | |
|---:|:---|
| TITLE | **EDITING SIMULATION DISC #1** |
| DESCRIPTION | This system simulates an editing system. The videodisc contains typical raw footage of a local event. Three videodisc players under simultaneous control will allow users to select segments, construct an edit list, modify the edit list, and view immediate playback of editing decisions. |
| TYPE | Instructional |
| FORMAT | Laser CAV |
| LEVEL | 3 |
| SYSTEM DESCRIPTION | 3 Sony players, Apple II computer, 3 Allen VMI interfaces |
| DEVELOPER | University of Iowa/Weeg Computing Center and Department of Film and Broadcasting |
| CONTACT FOR MORE INFO. | Dr. Joan Sustik Huntley |
| PRODUCER | Franklin Miller, Department of Film and Broadcasting |

---

| | |
|---:|:---|
| TITLE | **EEMT COMPOSITE NO. 1** |
| DESCRIPTION | Still frame simulations of electronic systems used in Electronics Equipment Maintenance Trainers (EEMT). |
| TYPE | Research and Development, Technical, Instructional, In-house Training |
| FORMAT | Laser CAV |
| LEVEL | 3 |
| DEVELOPER | Naval Personnel Research & Development Center |
| CONTACT FOR MORE INFO. | George F. Lahey, (619) 225-7122 |
| FEATURES | Still frames only |
| PRODUCER | U.S. Navy |
| AVAILABILITY | For government use only. |

---

**EL CID.** See **CID, EL**

---

**EL DORADO.** See **DORADO, EL**

---

| | |
|---:|:---|
| TITLE | **ELECTRIC HORSEMAN, THE** |
| TYPE | Motion Picture (PG) |
| FORMAT | Laser CLV, CED |
| RUNNING TIME | 120 minutes |
| PRODUCER | Warner Home Video |
| DISTRIBUTOR | Pioneer Video, Inc., 10-025, $29.95<br>RCA Videodiscs, 03312, $29.98 |

| | |
|---:|:---|
| TITLE | **ELECTRIC LIGHT ORCHESTRA** |
| TYPE | Music |
| FORMAT | CED |
| RETAIL OUTLET | Movies Unlimited Inc., A2-1006, $24.95 |

| | |
|---:|:---|
| TITLE | **ELECTRONIC IGNITION SYSTEM** |
| DESCRIPTION | Part of a large computer-based instructional system dealing with electronic ignition systems for Ford Motor Company. This project was implemented on the PLATO computer system. |
| TYPE | Technical, Instructional |
| RELEASE DATE | 10/79 |
| FORMAT | Laser CAV |
| LEVEL | 3 |
| DEVELOPER | WICAT Systems, Inc. for Ford Motor Company |
| CONTACT FOR MORE INFO. | Richard M. Cavagnol |
| PRODUCER | WICAT Systems, Inc. |

| | |
|---:|:---|
| TITLE | **ELEPHANT MAN, THE** |
| TYPE | Motion Picture (PG) |
| FORMAT | Laser CLV, CED (2 discs) |
| FEATURES | Stereo, Black and white |
| RUNNING TIME | 124 minutes |
| PRODUCER | Warner Home Video |
| DISTRIBUTOR | Pioneer Video, Inc., LV1347, $35.95 |
| | RCA Videodiscs, 00642, $34.98 |

| | |
|---:|:---|
| TITLE | **ELEPHANT PARTS** |
| TYPE | Music, Comedy |
| FORMAT | Laser CAV |
| LEVEL | 1 |
| FEATURES | Stereo, CX Encoded |
| RUNNING TIME | 60 minutes |
| PRODUCER | Pioneer Artists |
| DISTRIBUTOR | Pioneer Video, Inc., PA-81-005, $24.95 |

| | |
|---:|:---|
| TITLE | **ELMER GANTRY** |
| TYPE | Motion Picture |
| FORMAT | CED (2 Discs) |
| RETAIL OUTLET | Movies Unlimited Inc., A1-4582, $39.98 |

| | |
|---:|:---|
| TITLE | **ELTON JOHN: VISIONS** |
| TYPE | Music |
| FORMAT | Laser CLV, CED |
| FEATURES | Stereo, CX Encoded |
| RUNNING TIME | 45 minutes |
| PRODUCER | Embassy Home Entertainment |
| DISTRIBUTOR | Pioneer Video, Inc., 120150, $29.95 |
| RETAIL OUTLET | (CED) Movies Unlimited Inc., A4-2016, $21.98 |

| | |
|---:|:---|
| TITLE | **ELVIS - ALOHA FROM HAWAII** |
| TYPE | Music |
| FORMAT | CED |
| DISTRIBUTOR | RCA Videodiscs, 02055, $24.98 |

| | |
|---:|:---|
| TITLE | **ELVIS - HIS 1968 COMEBACK SPECIAL** |
| TYPE | Music |
| FORMAT | CED |
| DISTRIBUTOR | RCA Videodiscs, 02056, $24.98 |

| | |
|---:|:---|
| TITLE | **ELVIS ON TOUR** |
| TYPE | Music |
| FORMAT | CED |
| RETAIL OUTLET | Movies Unlimited Inc., A2-0153, $29.95 |

| | |
|---:|:---|
| TITLE | **EMERGENCY FIRST AID** |
| RELEASE DATE | 1982 |
| PRODUCER | Sheldon Renan |
| DISTRIBUTOR | VHD Programs, Inc. |

| | |
|---:|:---|
| TITLE | **EMILY** |
| TYPE | Motion Picture |
| FORMAT | CED |
| RETAIL OUTLET | Movies Unlimited Inc., A2-0204, $29.95 |

| | |
|---:|:---|
| TITLE | **EMMANUELLE IN BANGKOK** |
| TYPE | Motion Picture |
| FORMAT | CED |
| RETAIL OUTLET | Movies Unlimited Inc., A3-E970, $29.95 |

| | |
|---:|:---|
| TITLE | **EMMANUELLE, THE JOYS OF A WOMAN** |
| TYPE | Motion Picture, Adult (X) |
| FORMAT | Laser CLV |
| RUNNING TIME | 92 minutes |
| PRODUCER | Paramount Home Video |
| DISTRIBUTOR | Pioneer Video, Inc., LV8890, $29.95 |

| | |
|---:|:---|
| TITLE | **ENDANGERED SPECIES** |
| TYPE | Motion Picture |
| FORMAT | CED |
| RETAIL OUTLET | Movies Unlimited Inc., A2-0217, $29.95 |

| | |
|---:|:---|
| TITLE | **ENDLESS LOVE** |
| TYPE | Motion Picture (R) |
| FORMAT | CED |
| DISTRIBUTOR | RCA Videodiscs, 02114, $24.98 |

# Videodiscography

| | |
|---:|:---|
| TITLE | **ENFORCER, THE** |
| TYPE | Motion Picture |
| FORMAT | CED |
| DISTRIBUTOR | RCA Videodiscs, 03149, $29.98 |

| | |
|---:|:---|
| TITLE | **ENTER THE DRAGON** |
| TYPE | Motion Picture (R) |
| FORMAT | Laser CLV, CED |
| RUNNING TIME | 90 minutes |
| PRODUCER | Warner Home Video |
| DISTRIBUTOR | Pioneer Video, Inc., 61006LV, $29.98 |
| | RCA Videodiscs, 03125, $29.98 |

| | |
|---:|:---|
| TITLE | **ENTER THE NINJA** |
| TYPE | Motion Picture |
| FORMAT | CED |
| RETAIL OUTLET | Movies Unlimited Inc., A2-0186, $29.95 |

| | |
|---:|:---|
| TITLE | **ENTERTAINMENT GAME, THE** |
| DESCRIPTION | Movie and television trivia with Paramount Pictures film clips. |
| TYPE | Game |
| RELEASE DATE | 4/84 |
| FORMAT | CED |
| LEVEL | 1 |
| DEVELOPER | Paramount Home Video |
| CONTACT FOR MORE INFO. | Diane Smook, (212) 930-4767 |
| FEATURES | Chapters, Picture Stops, Dual Audio, |
| PRODUCER | Paramount Home Video |
| DISTRIBUTOR | RCA Videodiscs, 25005 |

| | |
|---:|:---|
| TITLE | **EPCOT CENTER VIDEODISCS (WALT DISNEY WORLD - FLORIDA)** |
| DESCRIPTION | A series of videodiscs providing 83 different programs for Epcot Center visitors. |
| TYPE | Music, Game, Children, Adolescents, Instructional, Promotional |
| RELEASE DATE | 10/82 |
| FORMAT | Laser CAV |
| LEVEL | 3 |
| SYSTEM DESCRIPTION | The videodiscs operate on a variety of systems, from simple "push-to-start" configurations to a 16-player, 8-guest station touch-screen interface. |
| DEVELOPER | WED Enterprises, 1401 Flower St. Glendale CA 91201 |
| CONTACT FOR MORE INFO. | Ken Christie, Videodisc Designer, Show Design, Department 410 |
| AVAILABILITY | For visitors to EPCOT, not for sale. |

| | |
|---:|:---|
| TITLE | **EROTICISE** |
| TYPE | Adult |
| FORMAT | CED |
| FEATURES | Stereo |
| RETAIL OUTLET | Movies Unlimited Inc., A3-3018, $29.95 |

| | |
|---:|:---|
| TITLE | **ESCAPE FROM ALCATRAZ** |
| TYPE | Motion Picture (PG) |
| FORMAT | Laser CLV, CED |
| RUNNING TIME | 112 minutes |
| PRODUCER | Warner Home Video |
| DISTRIBUTOR | Pioneer Video, Inc., LV1256, $29.95 |
| | RCA Videodiscs, 00635, $29.98 |

| | |
|---:|:---|
| TITLE | **ESCAPE FROM NEW YORK** |
| TYPE | Motion Picture (R) |
| FORMAT | Laser CLV, CED |
| FEATURES | (CLV) Stereo |
| RUNNING TIME | 106 minutes |
| PRODUCER | Embassy Home Entertainment |
| DISTRIBUTOR | Pioneer Video, Inc., 16025, $29.95 |
| | RCA Videodiscs, 00809, $24.98 |

| | |
|---:|:---|
| TITLE | **ESCAPE TO ATHENA** |
| TYPE | Motion Picture |
| FORMAT | CED |
| RETAIL OUTLET | Movies Unlimited Inc., A1-9019, $29.98 |

| | |
|---:|:---|
| TITLE | **ESCAPE TO WITCH MOUNTAIN** |
| TYPE | Motion Picture (G) |
| FORMAT | Laser CLV, CED |
| RUNNING TIME | 94 minutes |
| PRODUCER | Warner Home Video |
| DISTRIBUTOR | Pioneer Video, Inc., 13AS, $34.95 |
| | RCA Videodiscs, 00711, $24.98 |

| | |
|---:|:---|
| TITLE | **ESTABLISHING TECHNICAL ELIGIBILITY** |
| DESCRIPTION | Along with 14 floppy diskettes, this videodisc is designed to teach the technical requirements, such as age and citizenship, for receiving welfare for dependent children in Florida. |
| TYPE | Instructional, In-house Training |
| RELEASE DATE | 2/28/83 |
| FORMAT | Laser CAV |
| LEVEL | 3 |
| SYSTEM DESCRIPTION | Disc is part of a 160-hour curriculum delivered on Apple II/Pioneer 7820-3/single disk drive system along with ancillary printed reference books. |
| DEVELOPER | University of West Florida/Office for Interactive Technology and Training |
| CONTACT FOR MORE INFO. | Richard Smith; Elizabeth Wright, (413) 256-0444 |
| FEATURES | Dual Audio |
| PRODUCER | University of Nebraska/Nebraska Videodisc Design/Production Group |
| AVAILABILITY | Training program produced under contract with the Florida Department of Health and Rehabilitative Services, Tallahassee, Florida, and is not available for general distribution. |

| | |
|---:|:---|
| TITLE | **EUBIE!** |
| TYPE | Motion Picture, Music |
| FORMAT | CED |
| FEATURES | Stereo, CX Encoded |
| DISTRIBUTOR | RCA Videodiscs, 12076, $24.98 |

| | |
|---:|:---|
| TITLE | **EURYTHMICS SWEET DREAMS (THE VIDEO ALBUM)** |
| TYPE | Music |
| FORMAT | CED |
| FEATURES | Stereo |
| PRODUCER | RCA |
| DISTRIBUTOR | RCA Videodiscs, 12185, $19.98 |

| | |
|---:|:---|
| TITLE | **EVE, THE** |
| TYPE | Music |
| FORMAT | Laser CAV |
| LEVEL | 1 |
| FEATURES | Stereo |
| PRODUCER | Pioneer Video Imports |
| DISTRIBUTOR | Pioneer Video, Inc., MJ-001, $27.95 |

| | |
|---:|:---|
| TITLE | **EVENING WITH RAY CHARLES, AN** |
| TYPE | Music |
| FORMAT | Laser CAV |
| LEVEL | 1 |
| FEATURES | Stereo, CX Encoded |
| PRODUCER | Warner Home Video |
| DISTRIBUTOR | Pioneer Video, Inc., 74-612, $24.95 |

| | |
|---:|:---|
| TITLE | **EVENING WITH ROBIN WILLIAMS, AN** |
| TYPE | Comedy |
| FORMAT | CED |
| DISTRIBUTOR | RCA Videodiscs, 03611, $24.98 |

| | |
|---:|:---|
| TITLE | **EVENING WITH THE ROYAL BALLET - FONTEYN & NUREYEV** |
| TYPE | Dance |
| FORMAT | CED |
| DISTRIBUTOR | RCA Videodiscs, 02041, $24.98 |

| | |
|---:|:---|
| TITLE | **EVERY WHICH WAY BUT LOOSE** |
| TYPE | Motion Picture (PG) |
| FORMAT | CED |
| DISTRIBUTOR | RCA Videodiscs, 03107, $29.98 |

| | |
|---:|:---|
| TITLE | **EVERYTHING YOU ALWAYS WANTED TO KNOW ABOUT SEX, BUT WERE AFRAID TO ASK** |
| TYPE | Motion Picture, Comedy (R) |
| FORMAT | Laser CLV, CED |
| DISTRIBUTOR | RCA Videodiscs, 01444, $24.98 |
| RETAIL OUTLET | (CLV) UltraVideo, 80-4598, $34.95 |

| | |
|---:|:---|
| TITLE | **EVIDENCE & OBJECTIONS** |
| DESCRIPTION | Trial law for first year law students. |
| TYPE | Instructional |
| FORMAT | Laser CAV |
| LEVEL | 3 |
| CONTACT FOR MORE INFO. | University of Nebraska at Lincoln Law School |
| PRODUCER | University of Nebraska/Nebraska Videodisc Design/Production Group |
| AVAILABILITY | Not available for release or distribution. |

| | |
|---:|:---|
| TITLE | **EVILSPEAK** |
| TYPE | Motion Picture |
| FORMAT | CED |
| RETAIL OUTLET | Movies Unlimited Inc., A1-6127, $29.98 |

| | |
|---:|:---|
| TITLE | **EXCALIBUR** |
| TYPE | Motion Picture |
| FORMAT | CED (2 Discs) |
| DISTRIBUTOR | RCA Videodiscs, 03145, $39.98 |

| | |
|---:|:---|
| TITLE | **EXODUS** |
| TYPE | Motion Picture |
| FORMAT | CED (2 Discs) |
| RETAIL OUTLET | Movies Unlimited Inc., A1-4544, $39.98 |

| | |
|---:|:---|
| TITLE | **EXORCIST, THE** |
| TYPE | Motion Picture (R) |
| FORMAT | Laser CLV, CED |
| PRODUCER | Warner Home Video |
| DISTRIBUTOR | Pioneer Video, Inc., 1007LV, $29.98 |
| | RCA Videodiscs, 03126, $29.98 |

| | |
|---:|:---|
| TITLE | **EXPLORING LANGUAGE** |
| DESCRIPTION | Disc contains a television series that deals with composition and language as a social and political issue. |
| TYPE | Instructional |
| FORMAT | Laser CAV |
| LEVEL | 1 |
| DEVELOPER | Maryland Center for Public Broadcasting/International University Consortium |
| CONTACT FOR MORE INFO. | Ralph France (301) 356-5600 |
| FEATURES | Chapters, Picture Stops |
| RUNNING TIME | 28 minutes, 50 seconds |
| PRODUCER | Maryland Center for Public Broadcasting |
| DISTRIBUTOR | Maryland Center for Public Broadcasting/International University Consortium |
| AVAILABILITY | Limited quantity available for sale to member institutions of the International University Consortium. |

| | |
|---|---|
| TITLE | **EXTERMINATOR, THE** |
| TYPE | Motion Picture |
| FORMAT | CED |
| RETAIL OUTLET | Movies Unlimited Inc., A4-0026, $29.95 |

| | |
|---|---|
| TITLE | **EXTRA MILE, THE** |
| DESCRIPTION | Produced for TNT Canada, a multimodal transportation company based in Canada. Administrative and management training. The goal is to make service employees more customer-oriented in their work style, to give them better problem-solving skills. The needs of 15 interdependent job areas are addressed, spread across 12 operating divisions of the company. |
| TYPE | Technical, Instructional, In-house Training |
| RELEASE DATE | 9/83 |
| FORMAT | Laser CAV |
| LEVEL | 3 |
| SYSTEM DESCRIPTION | Three videodiscs, two diskettes, delivering 25 hours of training; designed to run on an Apple II or IIe with the Allen Communications interface, using the Pioneer Model III industrial player. |
| DEVELOPER | Matrix Information Systems |
| CONTACT FOR MORE INFO. | Michael North, Director, (213) 391-0243 |
| FEATURES | Dual Audio, Sound Over Still, Printer and Modem Drivers |
| PRODUCER | Matrix Information Systems |
| DISTRIBUTOR | Matrix Information Systems |
| AVAILABILITY | By special arrangement. Not available to transportation companies. |

| | |
|---|---|
| TITLE | **EYE FOR AN EYE, AN** |
| TYPE | Motion Picture |
| FORMAT | CED |
| RETAIL OUTLET | Movies Unlimited Inc., A4-6016, $29.95 |

| | |
|---|---|
| TITLE | **EYE OF THE NEEDLE** |
| TYPE | Motion Picture |
| FORMAT | CED |
| RETAIL OUTLET | Movies Unlimited Inc., A1-4581, $29.98 |

| | |
|---|---|
| TITLE | **EYES OF LAURA MARS, THE** |
| TYPE | Motion Picture (R) |
| FORMAT | CED |
| DISTRIBUTOR | RCA Videodiscs, 03007, $24.98 |

| | |
|---|---|
| TITLE | **EYEWITNESS** |
| TYPE | Motion Picture |
| FORMAT | CED |
| RETAIL OUTLET | Movies Unlimited Inc., A1-1116, $29.98 |

| | |
|---|---|
| TITLE | **F.I.S.T.** |
| TYPE | Motion Picture |
| FORMAT | CED (2 Discs) |
| RETAIL OUTLET | Movies Unlimited Inc., A1-4520, $39.98 |

| | |
|---|---|
| TITLE | **FAIR EMPLOYMENT PRACTICE** |
| DESCRIPTION | Dramatized EEOC guidelines concerning recruitment, placement, selection, promotion, transfer, discipline and discharge, and sexual harassment. Interactive questions with review (true and false as well as multiple choice). |
| TYPE | Instructional |
| RELEASE DATE | 2/81 |
| FORMAT | Laser CAV |
| LEVEL | 2 |
| DEVELOPER | BNA Communications Inc. |
| CONTACT FOR MORE INFO. | Emily Galloway or Cliff Witt, (301) 948-0540 |
| FEATURES | Chapters, Picture Stops, Digital Dump-Pioneer |
| PRODUCER | BNA Communications Inc. |
| DISTRIBUTOR | BNA Communications Inc. |
| AVAILABILITY | Sale, rent. |
| ORDER NO. | FEF-08 |
| SUGGESTED PRICE | $3000 (rental prices vary with term of rental). |

| | |
|---|---|
| TITLE | **FAME** |
| TYPE | Motion Picture, Music, Dance (R) |
| FORMAT | Laser CLV, CED (2 discs) |
| FEATURES | Stereo, CX Encoded |
| RUNNING TIME | 133 minutes |
| PRODUCER | MGM/UA Home Video |
| DISTRIBUTOR | Pioneer Video, Inc., ML100027, $39.95 |
| RETAIL OUTLET | (CED) Movies Unlimited Inc., A2-1008, $39.95 |

| | |
|---|---|
| TITLE | **FAMILY ENTERTAINMENT PLAYHOUSE, VOL. 2** |
| TYPE | Motion Picture (G) |
| FORMAT | CED |
| DISTRIBUTOR | RCA Videodiscs, 02065, $14.98 |

| | |
|---|---|
| TITLE | **FAN, THE** |
| TYPE | Motion Picture (R) |
| FORMAT | Laser CLV |
| FEATURES | CX Encoded |
| RUNNING TIME | 95 minutes |
| PRODUCER | Warner Home Video |
| DISTRIBUTOR | Pioneer Video, Inc., LV1469, $29.95 |

| | |
|---|---|
| TITLE | **FANTASTIC VOYAGE** |
| TYPE | Motion Picture |
| FORMAT | CED |
| RETAIL OUTLET | Movies Unlimited Inc., A1-0102, $29.98 |

| | |
|---|---|
| TITLE | **FAREWELL, MY LOVELY** |
| TYPE | Motion Picture (R) |
| FORMAT | CED |
| DISTRIBUTOR | RCA Videodiscs, 00505, $24.98 |

| | |
|---:|:---|
| TITLE | **FAST TIMES AT RIDGEMONT HIGH** |
| TYPE | Motion Picture, Comedy (R) |
| FORMAT | Laser CLV |
| RUNNING TIME | 90 minutes |
| PRODUCER | MCA Videodisc, Inc. |
| DISTRIBUTOR | Pioneer Video, Inc., 16-029, $29.95 |
| RETAIL OUTLET | (CED) Movies Unlimited Inc., A5-6029, $34.98 |

| | |
|---:|:---|
| TITLE | **FIDDLER ON THE ROOF** |
| TYPE | Motion Picture, Music (G) |
| FORMAT | Laser CLV, CED (2 discs) |
| FEATURES | (CLV) Stereo, CX Encoded |
| RUNNING TIME | 169 minutes |
| PRODUCER | Warner Home Video |
| DISTRIBUTOR | Pioneer Video, Inc., 4524-80, $49.98 |
| | RCA Videodiscs, 01404, $39.98 |

| | |
|---:|:---|
| TITLE | **FILLE MAL GARDEE, LA** |
| TYPE | Music, Dance |
| FORMAT | Laser CLV |
| FEATURES | Stereo, CX Encoded |
| RUNNING TIME | 100 minutes |
| PRODUCER | Pioneer Artists |
| DISTRIBUTOR | Pioneer Video, Inc., PA-81-007, $34.95 |

| | |
|---:|:---|
| TITLE | **FINAL CONFLICT, THE** |
| TYPE | Motion Picture |
| FORMAT | CED |
| RETAIL OUTLET | Movies Unlimited Inc., A1-1115, $29.98 |

| | |
|---:|:---|
| TITLE | **FINAL COUNTDOWN, THE** |
| TYPE | Motion Picture |
| FORMAT | CED |
| PRODUCER | Richard R. St. Johns |
| RETAIL OUTLET | Movies Unlimited Inc., A3-4047, $29.98 |

| | |
|---:|:---|
| TITLE | **FINAL EXAM** |
| TYPE | Motion Picture |
| FORMAT | CED |
| RETAIL OUTLET | Movies Unlimited Inc., A4-6168, $29.95 |

| | |
|---:|:---|
| TITLE | **FIREFOX** |
| TYPE | Motion Picture (PG) |
| FORMAT | Laser CLV, CED |
| FEATURES | Stereo |
| RUNNING TIME | 105 minutes |
| PRODUCER | Warner Home Video |
| DISTRIBUTOR | Pioneer Video, Inc., 11219LV, $39.98 |
| RETAIL OUTLET | (CED) Movies Unlimited Inc., A1-1219, $39.98 |

|               |                                                          |
|--------------:|----------------------------------------------------------|
| TITLE         | **FIRST BARRY MANILOW SPECIAL, THE**                     |
| TYPE          | Music                                                    |
| FEATURES      | Stereo                                                   |
| FORMAT        | Laser CLV, CED                                           |
| RUNNING TIME  | 52 minutes                                               |
| PRODUCER      | MGM/UA Home Video                                        |
| DISTRIBUTOR   | Pioneer Video, Inc., ML100148, $25.95                    |
| RETAIL OUTLET | (CED) Movies Unlimited Inc., A2-1058, $29.95             |

|             |                                     |
|------------:|-------------------------------------|
| TITLE       | **FIRST BLOOD**                     |
| TYPE        | Motion Picture                      |
| FORMAT      | CED                                 |
| FEATURES    | Stereo                              |
| DISTRIBUTOR | RCA Videodiscs, 12143, $29.98       |

|               |                                          |
|--------------:|------------------------------------------|
| TITLE         | **FIRST MONDAY IN OCTOBER**              |
| TYPE          | Motion Picture (R)                       |
| FORMAT        | Laser CLV                                |
| RUNNING TIME  | 99 minutes                               |
| PRODUCER      | Warner Home Video                        |
| DISTRIBUTOR   | Pioneer Video, Inc., LV1408, $29.95      |

|             |                                       |
|------------:|---------------------------------------|
| TITLE       | **FIRST NATIONAL KIDISC, THE**        |
| TYPE        | Game, Children, Instructional         |
| FORMAT      | Laser CAV                             |
| LEVEL       | 1                                     |
| PRODUCER    | Optical Programming Associates        |
| DISTRIBUTOR | Pioneer Video, Inc., 64-619, $19.95   |

|                         |                                                                                                                        |
|------------------------:|------------------------------------------------------------------------------------------------------------------------|
| TITLE                   | **FIRST SIXTY SECONDS, THE**                                                                                           |
| DESCRIPTION             | A custom sales-training disc produced for Ford Motor Company. Teaches salespeople how to be more effective with the woman customer. |
| TYPE                    | Instructional, In-house Training                                                                                       |
| LEVEL                   | 2                                                                                                                      |
| DEVELOPER               | IT, Inc.                                                                                                               |
| CONTACT FOR MORE INFO.  | Gary Quinlin, President, or Constance Buffalo, Vice President for Production, (800) 328-7937                           |

|             |                                |
|------------:|--------------------------------|
| TITLE       | **FISTFUL OF DOLLARS, A**      |
| TYPE        | Motion Picture                 |
| FORMAT      | CED                            |
| DISTRIBUTOR | RCA Videodiscs, 01415, $29.98  |

|               |                                            |
|--------------:|--------------------------------------------|
| TITLE         | **FLAMING STAR**                           |
| TYPE          | Motion Picture                             |
| FORMAT        | CED                                        |
| RETAIL OUTLET | Movies Unlimited Inc., A1-1173, $29.98     |

| | |
|---:|:---|
| TITLE | **FLASH GORDON** |
| TYPE | Motion Picture (PG) |
| FORMAT | Laser CLV, CED |
| FEATURES | Stereo |
| RUNNING TIME | 112 minutes |
| PRODUCER | Warner Home Video |
| DISTRIBUTOR | Pioneer Video, Inc., 13-006, $29.95 |
| RETAIL OUTLET | (CED) Movies Unlimited Inc., A5-3006, $34.98 |

| | |
|---:|:---|
| TITLE | **FLASHDANCE** |
| TYPE | Motion Picture (R) |
| FORMAT | Laser CLV |
| FEATURES | Stereo |
| RETAIL OUTLET | UltraVideo, LV1454, $29.95 |

| | |
|---:|:---|
| TITLE | **FLEETWOOD MAC** |
| TYPE | Music |
| FORMAT | Laser CAV, CED |
| LEVEL | (CAV) 1 |
| FEATURES | (CAV) Stereo |
| PRODUCER | Warner Home Video |
| DISTRIBUTOR | Pioneer Video, Inc., 74-001, $24.95 |
| | RCA Videodiscs, 02046, $19.98 |

| | |
|---:|:---|
| TITLE | **FLEETWOOD MAC IN CONCERT MIRAGE TOUR '82** |
| TYPE | Music |
| FORMAT | Laser CLV, CED |
| FEATURES | Stereo, CX Encoded |
| RUNNING TIME | 57 minutes |
| PRODUCER | Pioneer Artists |
| DISTRIBUTOR | Pioneer Video, Inc., PA-83-048, $24.95 |
| | RCA Videodiscs, 12139, $24.98 |

| | |
|---:|:---|
| TITLE | **FLIGHT 505** |
| DESCRIPTION | A program intended to teach English as a foreign language. It simulates conversations in a visit to the USA by a Japanese businessman. |
| TYPE | Instructional, In-house Training |
| RELEASE DATE | 4/84 |
| FORMAT | Laser CAV |
| LEVEL | 3 |
| DEVELOPER | David O. McKay Institute and British Broadcasting Corporation |
| CONTACT FOR MORE INFO. | Larrie E. Gale |
| FEATURES | Dual Audio |
| RUNNING TIME | 180 minutes |
| PRODUCER | British Broadcasting Corporation/Brigham Young University (co-producers) |
| DISTRIBUTOR | British Broadcasting Corporation |
| AVAILABILITY | Sale |

| | |
|---|---|
| TITLE | **FLYING DEUCES, THE** |
| TYPE | Motion Picture |
| FORMAT | Laser CLV, CED |
| RETAIL OUTLET | (CLV) UltraVideo, VL2009, $34.95 |
| | (CED) Movies Unlimited Inc., A3-2009, $29.98 |

| | |
|---|---|
| TITLE | **FOG, THE** |
| TYPE | Motion Picture (R) |
| FORMAT | Laser CLV, CED |
| RUNNING TIME | 94 minutes |
| PRODUCER | Warner Home Video |
| DISTRIBUTOR | Pioneer Video, Inc., 4067-80, $29.98 |
| | RCA Videodiscs, 00805, $24.98 |

| | |
|---|---|
| TITLE | **FOOTBALL FOLLIES AND SENSATIONAL SIXTIES** |
| TYPE | Motion Picture |
| FORMAT | Laser CAV |
| LEVEL | 1 |
| RUNNING TIME | 46 minutes |
| PRODUCER | Warner Home Video |
| DISTRIBUTOR | Pioneer Video, Inc., LV001, $24.95 |

| | |
|---|---|
| TITLE | **FOOTLIGHT PARADE** |
| TYPE | Motion Picture, Music |
| FORMAT | CED |
| DISTRIBUTOR | RCA Videodiscs, 03406, $24.98 |

| | |
|---|---|
| TITLE | **FOR A FEW DOLLARS MORE** |
| TYPE | Motion Picture |
| FORMAT | CED (2 Discs) |
| DISTRIBUTOR | RCA Videodiscs, 01467, $34.98 |

| | |
|---|---|
| TITLE | **FOR YOUR EYES ONLY** |
| TYPE | Motion Picture (PG) |
| FORMAT | Laser CLV, CED (2 Discs) |
| FEATURES | Stereo, CX Encoded |
| RUNNING TIME | 128 minutes |
| PRODUCER | Warner Home Video |
| DISTRIBUTOR | Pioneer Video, Inc., 4568-80, $39.98 |
| RETAIL OUTLET | (CED) Movies Unlimited Inc., A2-0180, $39.95 |

| | |
|---|---|
| TITLE | **FORBIDDEN PLANET** |
| TYPE | Motion Picture |
| FORMAT | Laser CLV, CED |
| FEATURES | Stereo |
| RUNNING TIME | 110 minutes |
| PRODUCER | MGM/UA Home Video |
| DISTRIBUTOR | Pioneer Video, Inc., ML100041, $25.95 |
| RETAIL OUTLET | (CED) Movies Unlimited Inc., A2-1013, $29.95 |

| | |
|---:|:---|
| TITLE | **FORCE 10 FROM NAVARONE** |
| TYPE | Motion Picture |
| FORMAT | Laser CLV |
| RETAIL OUTLET | UltraVideo, VL4051, $34.95 |

| | |
|---:|:---|
| TITLE | **FORCED VENGEANCE** |
| TYPE | Motion Picture (R) |
| FORMAT | CED |
| DISTRIBUTOR | MGM/UA Entertainment Co. |
| RETAIL OUTLET | Movies Unlimited Inc., A2-0189, $29.95 |

| | |
|---:|:---|
| TITLE | **FORD MOTOR VIDEO NETWORK (15 TOTAL DISC SIDES)** |
| DESCRIPTION | Technical and sales training. |
| TYPE | Technical, Instructional, In-house Training, Archival |
| RELEASE DATE | 1981-83 |
| FORMAT | Laser CAV |
| LEVEL | 0,1,2 |
| DEVELOPER | Full Circle Communications, Inc. |
| FEATURES | Digital Dump-Sony, Dual Audio |
| RUNNING TIME | 20-30 minutes |
| PRODUCER | Full Circle Communications, Inc. |
| DISTRIBUTOR | Ford Motor Video Network |
| AVAILABILITY | Ford network only. |

| | |
|---:|:---|
| TITLE | **FOREPLAY** |
| TYPE | Motion Picture |
| FORMAT | CED |
| RETAIL OUTLET | Movies Unlimited Inc., A3-3022, $29.98 |

| | |
|---:|:---|
| TITLE | **FORMULA, THE** |
| TYPE | Motion Picture |
| FORMAT | CED |
| RETAIL OUTLET | Movies Unlimited Inc., A2-1012, $29.95 |

| | |
|---:|:---|
| TITLE | **FORT APACHE, THE BRONX** |
| TYPE | Motion Picture (R) |
| FORMAT | Laser CLV, CED |
| RUNNING TIME | 120 minutes |
| PRODUCER | Warner Home Video |
| DISTRIBUTOR | Pioneer Video, Inc., LV1116, $29.95 |
| RETAIL OUTLET | (CED) Movies Unlimited Inc., A3-6000, $29.95 |

| | |
|---:|:---|
| TITLE | **FORT SILL VIDEODISC COURSEWARE PACKAGE** |
| DESCRIPTION | Eight computer-controlled videodisc lessons on the theory and fault diagnosis of AN/VRD-12 series FM radio sets. |
| TYPE | Technical, Instructional |
| RELEASE DATE | 8/81 |
| FORMAT | Laser CAV |
| LEVEL | 3 |
| DEVELOPER | WICAT Systems, Inc. for the U.S. Army (TDI) |
| CONTACT FOR MORE INFO. | Richard M. Cavagnol |
| PRODUCER | Ft. Sill Television facility with consultation from WICAT Systems, Inc. |

| | |
|---:|:---|
| TITLE | **42ND STREET** |
| TYPE | Motion Picture, Music, Dance |
| FORMAT | Laser CLV, CED |
| FEATURES | Black and white |
| DISTRIBUTOR | RCA Videodiscs, 01445, $24.98 |
| RETAIL OUTLET | (CLV) UltraVideo, 80-4502, $29.95 |

| | |
|---:|:---|
| TITLE | **48 HOURS** |
| TYPE | Motion Picture (R) |
| FORMAT | Laser CLV, CED |
| FEATURES | Stereo, CX Encoded |
| RUNNING TIME | 97 minutes |
| PRODUCER | Paramount Home Video |
| DISTRIBUTOR | Pioneer Video, Inc., LV1139, $29.95 |
| | RCA Videodiscs, 13612, $29.98 |

| | |
|---:|:---|
| TITLE | **FOUL PLAY** |
| TYPE | Motion Picture, Comedy (PG) |
| FORMAT | Laser CLV |
| RUNNING TIME | 118 minutes |
| PRODUCER | Warner Home Video |
| DISTRIBUTOR | Pioneer Video, Inc., LV1116, $29.95 |
| | RCA Videodiscs, 00603, $29.98 |

| | |
|---:|:---|
| TITLE | **FOUR FRIENDS** |
| TYPE | Motion Picture (R) |
| FORMAT | CED |
| PRODUCER | Filmways Pictures, Inc. |
| RETAIL OUTLET | Movies Unlimited Inc., A3-4019, $29.95 |

| | |
|---:|:---|
| TITLE | **FOUR MUSKETEERS, THE** |
| TYPE | Motion Picture (PG) |
| FORMAT | CED |
| DISTRIBUTOR | RCA Videodiscs, 02104, $24.98 |

| | |
|---:|:---|
| TITLE | **FOUR SEASONS, THE** |
| TYPE | Motion Picture (PG) |
| FORMAT | Laser CLV, CED |
| RUNNING TIME | 108 minutes |
| PRODUCER | Warner Home Video |
| DISTRIBUTOR | Pioneer Video, Inc., 16-025, $29.95 |
| | RCA Videodiscs, 03309, $29.98 |

| | |
|---:|:---|
| TITLE | **FRANKENSTEIN** |
| TYPE | Motion Picture |
| FORMAT | CED |
| PRODUCER | Universal Pictures |
| DISTRIBUTOR | RCA Videodiscs, 03313, $19.98 |

# Videodiscography

| | |
|---|---|
| TITLE | **FREE TO BE YOU AND ME/MANDY'S GRANDMOTHER** |
| TYPE | Children |
| FORMAT | CED |
| DISTRIBUTOR | RCA Videodiscs, 02110, $24.98 |

| | |
|---|---|
| TITLE | **FRENCH CONNECTION, THE** |
| TYPE | Motion Picture (R) |
| FORMAT | Laser CLV, CED |
| FEATURES | Stereo |
| RUNNING TIME | 104 minutes |
| PRODUCER | Warner Home Video |
| DISTRIBUTOR | Pioneer Video, Inc., 1009-80, $29.98 |
| RETAIL OUTLET | (CED) Movies Unlimited Inc., A1-0103, $29.98 |

| | |
|---|---|
| TITLE | **FRENCH LIEUTENANT'S WOMAN, THE** |
| TYPE | Motion Picture (R) |
| FORMAT | Laser CLV, CED |
| RUNNING TIME | 124 minutes |
| PRODUCER | Warner Home Video |
| DISTRIBUTOR | Pioneer Video, Inc. 4586-80, $39.98 |
| RETAIL OUTLET | (CED) Movies Unlimited Inc., A2-0181, $29.95 |

| | |
|---|---|
| TITLE | **FRIDAY THE 13TH** |
| TYPE | Motion Picture (R) |
| FORMAT | Laser CLV, CED |
| RUNNING TIME | 95 minutes |
| PRODUCER | Warner Home Video |
| DISTRIBUTOR | Pioneer Video, Inc., LV1395, $29.95 |
| | RCA Videodiscs, 00653, $29.98 |

| | |
|---|---|
| TITLE | **FRIDAY THE 13TH, PART 2** |
| TYPE | Motion Picture (R) |
| FORMAT | Laser CLV, CED |
| RUNNING TIME | 87 minutes |
| PRODUCER | Warner Home Video |
| DISTRIBUTOR | Pioneer Video, Inc., LV1457, $29.95 |
| | RCA Videodiscs, 03618, $29.98 |

| | |
|---|---|
| TITLE | **FRIDAY THE 13TH, PART 3** |
| TYPE | Motion Picture (R) |
| FORMAT | Laser CLV |
| RUNNING TIME | 96 minutes |
| PRODUCER | Paramount Home Video |
| DISTRIBUTOR | Pioneer Video, Inc., LV1539, $29.95 |

| | |
|---|---|
| TITLE | **FROG PRINCE, THE** |
| TYPE | Motion Picture (G) |
| FORMAT | Laser CLV |
| RUNNING TIME | 50 minutes |
| PRODUCER | Walt Disney |
| DISTRIBUTOR | Pioneer Video, Inc., 42805AS, $24.95 |

| | |
|---:|:---|
| TITLE | **FROM RUSSIA WITH LOVE** |
| TYPE | Motion Picture (PG) |
| FORMAT | Laser CLV, CED |
| RUNNING TIME | 118 minutes |
| PRODUCER | Warner Home Video |
| DISTRIBUTOR | Pioneer Video, Inc., 4566-80, $34.98 |
| | RCA Videodiscs, 01412, $29.98 |

| | |
|---:|:---|
| TITLE | **FROM THE NEW WORLD SYMPHONY NO. 9 IN E MINOR, OPUS 95** |
| TYPE | Music |
| FORMAT | Laser CLV |
| FEATURES | Stereo, CX Encoded |
| PRODUCER | Pioneer Video Imports |
| DISTRIBUTOR | Pioneer Video, Inc., MC-036, $27.95 |

| | |
|---:|:---|
| TITLE | **FUGITIVE THE FINAL EPISODE, THE** |
| TYPE | Motion Picture |
| FORMAT | CED |
| DISTRIBUTOR | RCA Videodiscs, 02030, $19.98 |

| | |
|---:|:---|
| TITLE | **FULCRUM CENTRAL AMERICA MAPS** |
| DESCRIPTION | Maps of Central America |
| TYPE | Archival |
| FORMAT | Laser CLV |
| LEVEL | 3 |
| DEVELOPER | Interactive Television Company |
| CONTACT FOR MORE INFO. | Dr. Steven Levin, Interactive Television Company |
| RUNNING TIME | 30 minutes |
| AVAILABILITY | Contact Interactive Television Company |

| | |
|---:|:---|
| TITLE | **FUN & GAMES** |
| TYPE | Comedy, Music, Dance, Game, Children, Instructional |
| FORMAT | Laser CAV |
| LEVEL | 1 |
| FEATURES | Chapters, Picture Stops, Stereo, Dual Audio |
| PRODUCER | Optical Programming Associates |
| DISTRIBUTOR | Pioneer Video, Inc., 37-601, $29.95 |

| | |
|---:|:---|
| TITLE | **FUN IN ACAPULCO** |
| TYPE | Motion Picture, Music |
| FORMAT | CED |
| DISTRIBUTOR | RCA Videodiscs, 01505, $24.98 |

| | |
|---:|:---|
| TITLE | **FUNNY GIRL** |
| TYPE | Motion Picture, Music |
| FORMAT | CED (2 Discs) |
| FEATURES | Stereo, Banded |
| DISTRIBUTOR | RCA Videodiscs, 13045, $39.98 |

| | |
|---:|:---|
| TITLE | **FUNNY THING HAPPENED ON THE WAY TO THE FORUM, A** |
| TYPE | Motion Picture, Comedy, Music |
| FORMAT | CED |
| RUNNING TIME | 99 minutes |
| PRODUCER | United Artists |
| DISTRIBUTOR | RCA Videodiscs, 01468, $24.98 |

| | |
|---:|:---|
| TITLE | **FUTUREWORLD** |
| TYPE | Motion Picture |
| FORMAT | CED |
| RETAIL OUTLET | Movies Unlimited Inc., A3-4053, $29.98 |

| | |
|---:|:---|
| TITLE | **G.I. BLUES** |
| TYPE | Motion Picture, Music |
| FORMAT | CED |
| DISTRIBUTOR | RCA Videodiscs, 01502, $24.98 |

| | |
|---:|:---|
| TITLE | **GALAXINA** |
| TYPE | Motion Picture (R) |
| FORMAT | Laser CLV |
| RUNNING TIME | 85 minutes |
| PRODUCER | Warner Home Video |
| DISTRIBUTOR | Pioneer Video, Inc., 13-007, $29.95 |

| | |
|---:|:---|
| TITLE | **GALLAGHER - SOME UNCENSORED EVENING** |
| TYPE | Comedy (Adult) |
| FORMAT | Laser CLV |
| FEATURES | Stereo, CX Encoded |
| RUNNING TIME | 57 minutes |
| PRODUCER | Pioneer Artists |
| DISTRIBUTOR | Pioneer Video, Inc., PA-83-046, $24.95 |

| | |
|---:|:---|
| TITLE | **GALLIPOLI** |
| TYPE | Motion Picture (PG) |
| FORMAT | Laser CLV |
| RUNNING TIME | 11 minutes |
| PRODUCER | Warner Home Video |
| DISTRIBUTOR | Pioneer Video, Inc., LV1504, $29.95 |

| | |
|---:|:---|
| TITLE | **GAMBLER, THE** |
| TYPE | Motion Picture (R) |
| FORMAT | Laser CLV |
| RUNNING TIME | 111 minutes |
| PRODUCER | Warner Home Video |
| DISTRIBUTOR | Pioneer Video, Inc., LV8678, $29.95 |

| | |
|---:|:---|
| TITLE | **GAME OF DEATH, THE** |
| TYPE | Motion Picture |
| FORMAT | CED |
| RETAIL OUTLET | Movies Unlimited Inc., A1-6124, $29.98 |

| | |
|---:|:---|
| TITLE | **GANDHI** |
| TYPE | Motion Picture |
| FORMAT | CED (2 Discs) |
| FEATURES | Stereo |
| PRODUCER | Columbia Pictures |
| DISTRIBUTOR | RCA Videodiscs, 13051, $39.98 |

| | |
|---:|:---|
| TITLE | **GARDENING AT HOME** |
| TYPE | Instructional |
| FORMAT | Laser CAV |
| LEVEL | 1 |
| PRODUCER | Xerox Information Resources Group |
| DISTRIBUTOR | Pioneer Video, Inc., 11-110, $29.95 |

| | |
|---:|:---|
| TITLE | **GAS PUMP GIRLS** |
| TYPE | Motion Picture, Comedy |
| FORMAT | Laser CLV, CED |
| RUNNING TIME | 102 minutes |
| PRODUCER | Vestron Video |
| DISTRIBUTOR | Pioneer Video, Inc., WL5502, $29.95 |
| RETAIL OUTLET | (CED) Movies Unlimited Inc., A3-5502, $24.95 |

| | |
|---:|:---|
| TITLE | **GATHERING AND VERIFYING INFORMATION** |
| DESCRIPTION | This experimental videodisc was designed to test the combination of an Apple computer, a computer graphics system, and an intelligent video player as an interactive, training delivery system. The courseware was compared with favorable results to a course on the same subject presented in XLS's traditional format. |
| TYPE | Research and Development |
| FORMAT | Laser CAV |
| SYSTEM DESCRIPTION | The system interfaces a Sony player, an Apple IIe, and a Symtec PGS. |
| DEVELOPER | The Glyn Group, Inc. and Designware |
| CONTACT FOR MORE INFO. | J.M. McCarthy, The Glyn Group, (212) 255-5156 |
| PRODUCER | Xerox Learning Systems |
| AVAILABILITY | Not available |

| | |
|---:|:---|
| TITLE | **GATOR** |
| TYPE | Motion Picture |
| FORMAT | CED |
| RETAIL OUTLET | Movies Unlimited Inc., A1-4617, $29.98 |

| | |
|---:|:---|
| TITLE | **GAUNTLET, THE** |
| TYPE | Motion Picture |
| FORMAT | CED |
| DISTRIBUTOR | RCA Videodiscs, 03150, $29.98 |

| | |
|---:|:---|
| TITLE | **GENERAL MOTORS OF CANADA** |
| DESCRIPTION | Ten releases throughout 1984 for sales and service training. |
| TYPE | Research and Development, Technical, Instructional, In-house Training, Archival, Point-of-Purchase |

| | |
|---|---|
| FORMAT | Laser CAV |
| LEVEL | 2, 3 |
| DEVELOPER | Michael J. Petro Ltd. |
| CONTACT FOR MORE INFO. | Michael J. Petro, (519) 258-9008 |
| FEATURES | Digital Dump-Pioneer, Dual Audio |
| PRODUCER | Michael J. Petro Ltd. |
| DISTRIBUTOR | General Motors of Canada Videodisc Network |
| AVAILABILITY | Available to subscribed GMC of Canada Videodisc Network Dealers, and to bona fide Canadian technical schools. |

| | |
|---|---|
| TITLE | **GENTLEMAN JIM** |
| TYPE | Motion Picture |
| FORMAT | CED |
| RUNNING TIME | 104 minutes |
| PRODUCER | United Artists |
| DISTRIBUTOR | RCA Videodiscs, 03407, $24.98 |

| | |
|---|---|
| TITLE | **GENTLEMEN PREFER BLONDES** |
| TYPE | Motion Picture |
| FORMAT | CED |
| RETAIL OUTLET | Movies Unlimited Inc., A1-1019, $29.98 |

| | |
|---|---|
| TITLE | **GEORGE CARLIN AND STEVE MARTIN HOST SATURDAY NIGHT LIVE, VOL. 1** |
| TYPE | Comedy |
| FORMAT | CED |
| DISTRIBUTOR | RCA Videodiscs, 01204, $19.98 |

| | |
|---|---|
| TITLE | **GEOSCIENCE** |
| DESCRIPTION | Covers the geology of the solar system. Uses U2 high-altitude photography, spacecraft photography, LANDSAT coverage, radar studies, and advanced analysis and mapping techniques. |
| TYPE | Technical, Instructional, Archival |
| FORMAT | Laser CAV |
| LEVEL | 1 |
| DEVELOPER | Drew University/Center for Aerospace Education in cooperation with NASA |
| CONTACT FOR MORE INFO. | Patrick Binnes, (201) 377-0302 |
| FEATURES | Chapters, Picture Stops, Stereo |
| PRODUCER | Video Vision Associates, Ltd. |
| DISTRIBUTOR | Video Vision Associates, Ltd. |
| AVAILABILITY | Sale |
| ORDER NO. | SPACE DISC 6 |
| SUGGESTED PRICE | $320 |

| | |
|---|---|
| TITLE | **GETTING OF WISDOM, THE** |
| TYPE | Motion Picture |
| FORMAT | CED |
| RETAIL OUTLET | Movies Unlimited Inc., A1-7048, $29.98 |

| | |
|---:|:---|
| TITLE | **GHOST STORY** |
| TYPE | Motion Picture (R) |
| FORMAT | Laser CLV |
| RUNNING TIME | 111 minutes |
| PRODUCER | Warner Home Video |
| DISTRIBUTOR | Pioneer Video, Inc., 11-013, $29.95 |

| | |
|---:|:---|
| TITLE | **GIGI** |
| TYPE | Motion Picture |
| FORMAT | Laser CLV, CED |
| FEATURES | Stereo |
| RETAIL OUTLET | (CLV) UltraVideo, ML100050, $34.95 (CED) Movies Unlimited Inc., A2-0050, $29.95 |

| | |
|---:|:---|
| TITLE | **GILDA** |
| TYPE | Motion Picture, Comedy |
| FORMAT | CED |
| DISTRIBUTOR | RCA Videodiscs, 03038, $24.98 |

| | |
|---:|:---|
| TITLE | **GIMME SHELTER** |
| TYPE | Music (PG) |
| FORMAT | CED |
| DISTRIBUTOR | RCA Videodiscs, 11806 (Stereo, Banded), 01801, $19.98 |

| | |
|---:|:---|
| TITLE | **GIRL GROUPS - THE STORY OF A SOUND** |
| TYPE | Motion Picture, Music |
| FORMAT | Laser CLV |
| RETAIL OUTLET | UltraVideo, ML100194, $34.95 |

| | |
|---:|:---|
| TITLE | **GISELLE - NUREYEV** |
| TYPE | Dance |
| FORMAT | CED |
| DISTRIBUTOR | RCA Videodiscs, 00517, $24.98 |

| | |
|---:|:---|
| TITLE | **GLORIA** |
| TYPE | Motion Picture (PG) |
| FORMAT | Laser CLV |
| PRODUCER | Warner Home Video |
| DISTRIBUTOR | Pioneer Video, Inc., VLD3160, $34.95 |

| | |
|---:|:---|
| TITLE | **GO FOR THE GREEN** |
| DESCRIPTION | A simulated game of golf played on six of the best holes in the United States. |
| TYPE | Game, In-house Training, Archival |
| FORMAT | Laser CAV |
| LEVEL | 2 |
| DEVELOPER | Joseph E. Barrett and Frederick E. Schoen |
| CONTACT FOR MORE INFO. | Barrett & Associates, (313) 342-9175 |
| FEATURES | Chapters, Picture Stops, Digital Dump-Sony, CX Encoded |
| PRODUCER | Barrett & Associates |
| DISTRIBUTOR | Ford Motor Company |
| AVAILABILITY | Available only through FMC. |

| | |
|---:|:---|
| TITLE | **GO TELL THE SPARTANS** |
| TYPE | Motion Picture (R) |
| FORMAT | Laser CLV, CED |
| RUNNING TIME | 114 minutes |
| PRODUCER | Vestron Video |
| DISTRIBUTOR | Pioneer Video, Inc., VL5000, $34.95 |
| RETAIL OUTLET | (CED) Movies Unlimited Inc., A3-5000, $29.95 |

| | |
|---:|:---|
| TITLE | **GODFATHER, THE** |
| TYPE | Motion Picture (R) |
| FORMAT | Laser CLV, CED (2 Discs) |
| RUNNING TIME | 171 minutes |
| PRODUCER | Warner Home Video |
| DISTRIBUTOR | Pioneer Video, Inc., LV8049, $39.95 |
| | RCA Videodiscs, 00604, $39.98 |

| | |
|---:|:---|
| TITLE | **GODFATHER, PART 2, THE** |
| TYPE | Motion Picture (R) |
| FORMAT | Laser CLV, CED (2 Discs) |
| FEATURES | CX Encoded |
| RUNNING TIME | 200 minutes |
| PRODUCER | Warner Home Video |
| DISTRIBUTOR | Pioneer Video, Inc., LV8459, $39.95 |
| | RCA Videodiscs, 00644, $39.98 |

| | |
|---:|:---|
| TITLE | **GODZILLA** |
| TYPE | Motion Picture |
| FORMAT | Laser CLV, CED |
| RUNNING TIME | 80 minutes |
| PRODUCER | Vestron Video |
| DISTRIBUTOR | Pioneer Video, Inc., VL3010, $34.95 |
| RETAIL OUTLET | (CED) Movies Unlimited Inc., A3-3010, $29.95 |

| | |
|---:|:---|
| TITLE | **GOIN' SOUTH** |
| TYPE | Motion Picture |
| FORMAT | Laser CLV |
| RETAIL OUTLET | UltraVideo, LV1133, $29.95 |

| | |
|---:|:---|
| TITLE | **GOING METRIC** |
| DESCRIPTION | An interactive disc designed for the consumer optical player. |
| TYPE | Instructional |
| RELEASE DATE | 1979 |
| FORMAT | Laser CAV |
| LEVEL | 2 |
| CONTACT FOR MORE INFO. | Corporation for Public Broadcasting, KUON-TV, University of Nebraska KUON-TV |
| PRODUCER | University of Nebraska/Nebraska Videodisc Design/Production Group |
| AVAILABILITY | Not available for release or distribution. |

| | |
|---:|:---|
| TITLE | **GOLD BUG, THE/RODEO RED AND THE RUNAWAY** |
| TYPE | Motion Picture |
| FORMAT | CED |
| DISTRIBUTOR | RCA Videodiscs, 02021, $14.98 |

| | |
|---:|:---|
| TITLE | **GOLD DIGGERS OF 1933** |
| TYPE | Motion Picture, Music |
| FORMAT | CED |
| RUNNING TIME | 98 minutes |
| PRODUCER | United Artists |
| DISTRIBUTOR | RCA Videodiscs, 03408, $24.98 |

| | |
|---:|:---|
| TITLE | **GOLDEN AGE OF COLLEGE FOOTBALL, THE** |
| TYPE | Motion Picture |
| FORMAT | CED |
| RETAIL OUTLET | Movies Unlimited Inc., A1-3706, $29.98 |

| | |
|---:|:---|
| TITLE | **GOLDEN DECADE OF COLLEGE FOOTBALL, 1970-1979, A** |
| TYPE | Motion Picture |
| FORMAT | Laser CLV |
| RUNNING TIME | 90 minutes |
| PRODUCER | Warner Home Video |
| DISTRIBUTOR | Pioneer Video, Inc., 3701-80, $29.98 |

| | |
|---:|:---|
| TITLE | **GOLDFINGER** |
| TYPE | Motion Picture (PG) |
| FORMAT | Laser CLV, CED |
| RUNNING TIME | 108 minutes |
| PRODUCER | Warner Home Video |
| DISTRIBUTOR | Pioneer Video, Inc., 4595-80, $34.98 |
| | RCA Videodiscs, 01410, $29.98 |

| | |
|---:|:---|
| TITLE | **GOLDILOCKS AND THE THREE BEARS** |
| TYPE | Motion Picture |
| FORMAT | Laser CLV |
| FEATURES | Closed captioned |
| RETAIL OUTLET | UltraVideo, 80-6386, $29.95 |

| | |
|---:|:---|
| TITLE | **GOOD GUYS WEAR BLACK** |
| TYPE | Motion Picture (R) |
| FORMAT | Laser CLV, CED |
| RUNNING TIME | 96 minutes |
| PRODUCER | Vestron Video |
| DISTRIBUTOR | Pioneer Video, Inc., VL6002, $34.95 |
| RETAIL OUTLET | (CED) Movies Unlimited Inc., A3-6002, $39.98 |

| | |
|---:|:---|
| TITLE | **GOOD, THE BAD AND THE UGLY, THE** |
| TYPE | Motion Picture |
| FORMAT | Laser CLV, CED (2 Discs) |
| RUNNING TIME | 161 minutes |

|  |  |
|---|---|
| PRODUCER | Warner Home Video |
| DISTRIBUTOR | Pioneer Video, Inc., 4545-80, $39.98 |
|  | RCA Videodiscs, 01416, $34.98 |

|  |  |
|---|---|
| TITLE | **GOODBYE GIRL, THE** |
| TYPE | Motion Picture, Comedy (PG) |
| FORMAT | Laser CLV, CED |
| RUNNING TIME | 110 minutes |
| PRODUCER | MGM/UA Home Video |
| DISTRIBUTOR | Pioneer Video, Inc., ML100069, $34.95 |
| RETAIL OUTLET | (CED) Movies Unlimited Inc., A2-1031, $29.95 |

|  |  |
|---|---|
| TITLE | **GOODBYE, COLUMBUS** |
| TYPE | Motion Picture, Comedy (R) |
| FORMAT | Laser CLV, CED |
| PRODUCER | Paramount Home Video, Inc. |
| DISTRIBUTOR | Pioneer Video, Inc., LV5472, $29.95 |
|  | RCA Videodiscs, 00645, $24.98 |

|  |  |
|---|---|
| TITLE | **GOSPEL** |
| TYPE | Motion Picture, Music |
| FORMAT | CED |
| FEATURES | Stereo |
| DISTRIBUTOR | RCA Videodiscs, 12186, $19.98 |

|  |  |
|---|---|
| TITLE | **GRACE JONES: ONE MAN SHOW** |
| TYPE | Music |
| FORMAT | CED |
| RETAIL OUTLET | Movies Unlimited Inc., A3-2004, $24.95 |

|  |  |
|---|---|
| TITLE | **GRADUATE, THE** |
| TYPE | Motion Picture (PG) |
| FORMAT | Laser CLV, CED |
| FEATURES | (CLV) Stereo |
| RUNNING TIME | 106 minutes |
| PRODUCER | Warner Home Video |
| DISTRIBUTOR | Pioneer Video, Inc., 4006-80, $34.98 |
|  | RCA Videodiscs, 00801, $19.98 |

|  |  |
|---|---|
| TITLE | **GRAPES OF WRATH, THE** |
| TYPE | Motion Picture |
| FORMAT | CED (2 Discs) |
| RETAIL OUTLET | Movies Unlimited Inc., A1-1024, $39.98 |

|  |  |
|---|---|
| TITLE | **GRATEFUL DEAD IN CONCERT** |
| TYPE | Music |
| FORMAT | CED |
| DISTRIBUTOR | RCA Videodiscs, 02010, $24.98 |

| | |
|---|---|
| TITLE | **GRATEFUL DEAD/DEAD AHEAD** |
| TYPE | Music |
| FORMAT | Laser CLV |
| FEATURES | Stereo, CX Encoded |
| RUNNING TIME | 114 minutes |
| PRODUCER | Pioneer Artists |
| DISTRIBUTOR | Pioneer Video, Inc., PA-82-010, $29.95 |

| | |
|---|---|
| TITLE | **GREASE** |
| TYPE | Motion Picture, Music (PG) |
| FORMAT | Laser CLV, CED |
| FEATURES | Stereo, CX Encoded |
| RUNNING TIME | 110 minutes |
| PRODUCER | Warner Home Video |
| DISTRIBUTOR | Pioneer Video, Inc., LV1108, $29.95 |
| | RCA Videodiscs, 00601, $29.98 |

| | |
|---|---|
| TITLE | **GREASE 2** |
| TYPE | Motion Picture, Music (PG) |
| FORMAT | Laser CLV, CED |
| FEATURES | Stereo, CX Encoded |
| RUNNING TIME | 115 minutes |
| PRODUCER | Warner Home Video |
| DISTRIBUTOR | Pioneer Video, Inc., LV1193, $29.95 |
| | RCA Videodiscs, 13615, $29.98 |

| | |
|---|---|
| TITLE | **GREAT CARUSO, THE** |
| TYPE | Motion Picture |
| FORMAT | CED |
| RETAIL OUTLET | Movies Unlimited Inc., A2-1030, $29.95 |

| | |
|---|---|
| TITLE | **GREAT CITIES: LONDON, ROME, DUBLIN, ATHENS** |
| TYPE | Instructional |
| FORMAT | CED |
| DISTRIBUTOR | RCA Videodiscs, 02052, $14.98 |

| | |
|---|---|
| TITLE | **GREAT DICTATOR, THE** |
| TYPE | Motion Picture |
| FORMAT | CED |
| DISTRIBUTOR | RCA Videodiscs, 02083, $19.98 |

| | |
|---|---|
| TITLE | **GREAT ESCAPE, THE** |
| TYPE | Motion Picture |
| FORMAT | Laser CLV, CED |
| RUNNING TIME | 170 minutes |
| PRODUCER | CBS Fox Video |
| DISTRIBUTOR | Pioneer Video, Inc., 4558-80, $39.98 |
| | RCA Videodiscs, 01406, $39.98 |

| | |
|---|---|
| TITLE | **GREAT GATSBY, THE** |
| TYPE | Motion Picture (PG) |
| FORMAT | Laser CLV, CED (2 Discs) |
| RUNNING TIME | 146 minutes |
| PRODUCER | Warner Home Video |
| DISTRIBUTOR | Pioneer Video, Inc., LV8469, $35.95 |
| | RCA Videodiscs, 00646, $34.98 |

| | |
|---|---|
| TITLE | **GREAT LOCOMOTIVE CHASE, THE** |
| TYPE | Motion Picture (G) |
| FORMAT | CED |
| PRODUCER | Walt Disney |
| DISTRIBUTOR | RCA Videodiscs, 00707, $24.98 |

| | |
|---|---|
| TITLE | **GREAT MUPPET CAPER, THE** |
| TYPE | Motion Picture, Comedy (G) |
| FORMAT | Laser CLV, CED |
| FEATURES | Stereo, CX Encoded |
| RUNNING TIME | 98 minutes |
| PRODUCER | Warner Home Video |
| DISTRIBUTOR | Pioneer Video, Inc., 9035-80, $34.98 |
| | RCA Videodiscs, 10526 (Stereo, Banded), 00522, $24.98 |

| | |
|---|---|
| TITLE | **GREAT SANTINI, THE** |
| TYPE | Motion Picture |
| FORMAT | CED |
| DISTRIBUTOR | RCA Videodiscs, 03127, $24.98 |

| | |
|---|---|
| TITLE | **GREAT SCOUT & CATHOUSE THURSDAY** |
| TYPE | Motion Picture |
| FORMAT | CED |
| RETAIL OUTLET | Movies Unlimited Inc., A3-4032, $29.95 |

| | |
|---|---|
| TITLE | **GREAT SPACE COASTER SUPERSHOW** |
| TYPE | Motion Picture |
| FORMAT | CED |
| RETAIL OUTLET | Movies Unlimited Inc., A2-0158, $29.95 |

| | |
|---|---|
| TITLE | **GREAT TRAIN ROBBERY, THE** |
| TYPE | Motion Picture (PG) |
| FORMAT | Laser CLV, CED |
| FEATURES | Stereo, CX Encoded |
| RUNNING TIME | 111 minutes |
| PRODUCER | Warner Home Video |
| DISTRIBUTOR | Pioneer Video, Inc., 4531-80, $34.98 |
| RETAIL OUTLET | (CED) Movies Unlimited Inc., A1-4531, $29.98 |

| | |
|---:|:---|
| TITLE | **GREATEST ADVENTURE, THE** |
| TYPE | Motion Picture |
| FORMAT | Laser CLV, CED |
| RUNNING TIME | 54 minutes |
| PRODUCER | Vestron Video |
| DISTRIBUTOR | Pioneer Video, Inc., VL2005, $29.95 |
| RETAIL OUTLET | (CED) Movies Unlimited Inc., A3-2005, $24.95 |

| | |
|---:|:---|
| TITLE | **GREATEST FIGHTS OF THE '70'S, THE** |
| TYPE | Motion Picture |
| FORMAT | CED |
| RETAIL OUTLET | Movies Unlimited Inc., A2-1046, $24.95 |

| | |
|---:|:---|
| TITLE | **GREATEST SHOW ON EARTH, THE** |
| TYPE | Motion Picture |
| FORMAT | Laser CLV, CED (2 Discs) |
| RUNNING TIME | 149 minutes |
| PRODUCER | Warner Home Video |
| DISTRIBUTOR | Pioneer Video, Inc., LV6617, $35.95 |
| | RCA Videodiscs, 00614, $34.98 |

| | |
|---:|:---|
| TITLE | **GREEN BERETS, THE** |
| TYPE | Motion Picture (G) |
| FORMAT | CED (2 Discs) |
| PRODUCER | Warner Home Video |
| DISTRIBUTOR | RCA Videodiscs, 03128, $34.98 |

| | |
|---:|:---|
| TITLE | **GREENDAY VIDEO, VOL. 1 NO.4** |
| DESCRIPTION | An interactive system and disc placed at point of sale; program material is gardening-related and includes brief "how to" segments, cataloging of tools and plant material, and product demonstrations. Most segments are manufacturer-sponsored. |
| TYPE | Research and Development, Instructional Catalog, Point-of-Purchase, Advertising, Promotional |
| FORMAT | Laser CAV |
| LEVEL | 3 |
| SYSTEM DESCRIPTION | System: Sony LDP-1000 player, SMC-70, superimposer, cache disc, KX-1211 HG monitor, custom keypad. Purpose to provide consumers with generic and specific product information at the point of sale. Technical information developed computer program to store and report user data. Special features accompanied by printed program guide. |
| DEVELOPER | VCM Systems, Inc. |
| CONTACT FOR MORE INFO. | Thomas Hedges, Executive Producer |
| PRODUCER | VCM Systems, Inc. |

| | |
|---:|:---|
| TITLE | **GROVER WASHINGTON, JR. IN CONCERT** |
| TYPE | Music |
| FORMAT | Laser CAV, CED (Stereo) |
| LEVEL | 1 |
| FEATURES | Stereo, CX Encoded |
| RUNNING TIME | 53 minutes |

| | |
|---:|:---|
| PRODUCER | Pioneer Artists |
| DISTRIBUTOR | Pioneer Video, Inc., PA-82-011, $24.95 |
| RETAIL OUTLET | (CED) Movies Unlimited Inc., A2-0203, $29.95 |

| | |
|---:|:---|
| TITLE | **GULLIVER'S TRAVELS** |
| TYPE | Motion Picture |
| FORMAT | CED |
| RETAIL OUTLET | Movies Unlimited Inc., A3-2011, $29.98 |

| | |
|---:|:---|
| TITLE | **GUMBY ADVENTURE, A** |
| TYPE | Motion Picture |
| FORMAT | CED |
| RETAIL OUTLET | Movies Unlimited Inc., A2-1068, $29.95 |

| | |
|---:|:---|
| TITLE | **GUNFIGHT AT THE O.K. CORRAL** |
| TYPE | Motion Picture |
| FORMAT | Laser CLV, CED |
| RUNNING TIME | 117 minutes |
| PRODUCER | Paramount Home Video |
| DISTRIBUTOR | Pioneer Video, Inc., LV6218, $29.95 |
| | RCA Videodiscs, 00687, $24.98 |

| | |
|---:|:---|
| TITLE | **GUNS OF NAVARONE, THE** |
| TYPE | Motion Picture |
| FORMAT | Laser CLV, CED (2 Discs) |
| PRODUCER | Warner Home Video |
| DISTRIBUTOR | Pioneer Video, Inc., LVD2010, $34.95 |
| | RCA Videodiscs, 03014, $39.98 |

| | |
|---:|:---|
| TITLE | **GUS** |
| TYPE | Motion Picture (G) |
| FORMAT | Laser CLV, CED |
| RUNNING TIME | 96 minutes |
| PRODUCER | Walt Disney |
| DISTRIBUTOR | Pioneer Video, Inc., 42029AS, $34.95 |
| | RCA Videodiscs, 00719, $24.98 |

| | |
|---:|:---|
| TITLE | **GUYS AND DOLLS** |
| TYPE | Motion Picture |
| FORMAT | Laser CLV, CED (2 Discs) |
| RETAIL OUTLET | (CLV) UltraVideo, 80-7039, $39.95 (CED) |
| | Movies Unlimited Inc., A2-1029, $39.95 |

| | |
|---:|:---|
| TITLE | **GYNECOLOGY PATIENT EDUCATION** |
| DESCRIPTION | Follows an actual patient before, during and following radiation therapy for cervical cancer. It also provides explanations by physicians on the placement of radiation implements and the process of radiation relative to the cancer. |
| TYPE | Medical, Research and Development, Instructional |
| FORMAT | Thomson-CSF Transmissive |
| LEVEL | 3 |

|   |   |
|---|---|
| SYSTEM DESCRIPTION | Thomson-CSF TTV 3620 player, Apple II single disk drive computer, Sony color monitor, Apple/Thomson driver software developed at the CAI Lab for an IEEE 488 board from California Computer Systems. |
| CONTACT FOR MORE INFO. | Joan Sustik Huntley, (319) 353-3170 |
| RUNNING TIME | 27 minutes |
| PRODUCER | University of Iowa/Computer Assisted Instruction Lab |
| AVAILABILITY | Since this was a research and development effort, neither the disc nor software are publicly available. Interested people may view the program any time at the Weeg Computing Center. |

---

|   |   |
|---|---|
| TITLE | **HAIR** |
| TYPE | Motion Picture, Music, Dance |
| FORMAT | Laser CLV, CED |
| FEATURES | Stereo, CX Encoded |
| DISTRIBUTOR | RCA Videodiscs, 11440, $24.98 |
| RETAIL OUTLET | UltraVideo, 80-4593, $34.95 |

---

|   |   |
|---|---|
| TITLE | **HALL & OATES ROCK AND SOUL LIVE** |
| TYPE | Music |
| FORMAT | CED |
| FEATURES | Stereo, Banded |
| DISTRIBUTOR | RCA Videodiscs, 12214, $19.98 |

---

|   |   |
|---|---|
| TITLE | **HALLOWEEN 2** |
| TYPE | Motion Picture (R) |
| FORMAT | Laser CLV |
| FEATURES | Stereo, CX Encoded |
| RUNNING TIME | 92 minutes |
| PRODUCER | MCA Videodisc, Inc. |
| DISTRIBUTOR | Pioneer Video, Inc., 11-019, $32.95 |

---

|   |   |
|---|---|
| TITLE | **HALLOWEEN 3** |
| TYPE | Motion Picture (R) |
| FORMAT | Laser CLV |
| RETAIL OUTLET | UltraVideo, 11-020, $32.95 |

---

|   |   |
|---|---|
| TITLE | **HAMLET** |
| TYPE | Motion Picture |
| FORMAT | CED (2 Discs) |
| DISTRIBUTOR | RCA Videodiscs, 01004, $34.98 |

---

|   |   |
|---|---|
| TITLE | **HANDS-ON WORD PROCESSING** |
| DESCRIPTION | Introduces word processing functions and compares Microsoft Word, Wordstar, Easywriter II, Applewriter II, and Bank Street Writer. |
| TYPE | Point-of-Purchase, Promotional |
| FORMAT | Laser CAV |
| LEVEL | 3 |
| SYSTEM DESCRIPTION | Point-of-purchase merchandising consists of player, control panel, monitor, and disc in 25" cube. |
| DEVELOPER | Intermedia |
| CONTACT FOR MORE INFO. | Susan Hoffman or Stan Barnes, (206) 282-7262 |
| DISTRIBUTOR | Intermedia |

| | |
|---|---|
| TITLE | **HANG 'EM HIGH** |
| TYPE | Motion Picture |
| FORMAT | Laser CLV, CED |
| RUNNING TIME | 114 minutes |
| PRODUCER | CBS Fox Video |
| DISTRIBUTOR | Pioneer Video, Inc., 4628-80, $34.98 |
| | RCA Videodiscs, 01470, $19.98 |

| | |
|---|---|
| TITLE | **HANKY PANKY** |
| TYPE | Motion Picture, Comedy |
| FORMAT | CED |
| DISTRIBUTOR | RCA Videodiscs, 03068, $29.95 |

| | |
|---|---|
| TITLE | **HANSEL & GRETEL** |
| TYPE | Motion Picture (G) |
| FORMAT | CED |
| DISTRIBUTOR | RCA Videodiscs, 02105, $14.98 |

| | |
|---|---|
| TITLE | **HANSEN'S DISEASE** |
| DESCRIPTION | Clinical motion and still-frame footage of Hansen's Disease (leprosy), including epidemiology, diagnosis, treatment, and research aspects. This is a prototype disc that will lead to an intenational archival collection and medical teaching source of all suitable visual material on leprosy. Lessons are authored with Pilot so that they can be easily adapted for local use. Also, use of a data base management system will allow flexibility in searching through the entire collection, with cross- reference to 35 mm and videotape formats. Application testing and field trails will be carried out at 14 locations in the U.S. and 50 hospitals and medical schools in India. |
| TYPE | Medical, Research and Development, Technical, Instructional, In-house Training, Archival |
| FORMAT | Laser CAV |
| LEVEL | 3 |
| SYSTEM DESCRIPTION | Sony LPD-1000 player interfaced via Allen VMI to Apple IIe with authoring in Apple SuperPilot. Record keeping with SuperPilot log and DB Master (Version IV). |
| DEVELOPER | U.S. Public Health Service |
| CONTACT FOR MORE INFO. | Dr. Richard J. O'Connor, (504) 642-7771, ext. 281 |
| RUNNING TIME | 27 minutes 46 seconds |
| PRODUCER | U.S. Public Health Service, National Hansen's Disease Center |
| DISTRIBUTOR | Same as producer |
| AVAILABILITY | Loaned for educational research studies only. |

| | |
|---|---|
| TITLE | **HAPPY BIRTHDAY TO ME** |
| TYPE | Motion Picture (R) |
| FORMAT | Laser CLV |
| PRODUCER | Warner Home Video |
| DISTRIBUTOR | Pioneer Video, Inc., VLD3175, $29.95 |

| | |
|---|---|
| TITLE | **HAPPY HOOKER GOES HOLLYWOOD, THE** |
| TYPE | Motion Picture, Adult (R) |
| FORMAT | Laser CLV |
| RUNNING TIME | 85 minutes |
| PRODUCER | Warner Home Video |
| DISTRIBUTOR | Pioneer Video, Inc., 28-009, $29.95 |

| | |
|---|---|
| TITLE | **HAPPY HOOKER GOES TO WASHINGTON, THE** |
| TYPE | Motion Picture, Adult |
| FORMAT | CED |
| RETAIL OUTLET | Movies Unlimited Inc., A3-4010, $29.95 |

| | |
|---|---|
| TITLE | **HAPPY HOOKER, THE** |
| TYPE | Motion Picture, Adult (R) |
| FORMAT | Laser CLV, CED |
| PRODUCER | Warner Home Video |
| DISTRIBUTOR | Pioneer Video, Inc., WL5503, $29.95 |
| RETAIL OUTLET | (CED) Movies Unlimited Inc., A3-5503, $24.95 |

| | |
|---|---|
| TITLE | **HARD COUNTRY, THE** |
| TYPE | Motion Picture |
| FORMAT | CED |
| RETAIL OUTLET | Movies Unlimited Inc., A1-9030, $29.98 |

| | |
|---|---|
| TITLE | **HARDER THEY COME, THE** |
| TYPE | Motion Picture, Music (R) |
| FORMAT | CED |
| DISTRIBUTOR | RCA Videodiscs, 00902, $19.98 |

| | |
|---|---|
| TITLE | **HAROLD AND MAUDE** |
| TYPE | Motion Picture, Comedy (PG) |
| FORMAT | Laser CLV, CED |
| RUNNING TIME | 91 minutes |
| PRODUCER | Warner Home Video |
| DISTRIBUTOR | Pioneer Video, Inc., LV8042, $29.95 |
| | RCA Videodiscs, 00615, $29.98 |

| | |
|---|---|
| TITLE | **HARPER VALLEY P.T.A.** |
| TYPE | Motion Picture |
| FORMAT | CED |
| RETAIL OUTLET | Movies Unlimited Inc., A3-4043, $29.95 |

| | |
|---|---|
| TITLE | **HARRY CHAPIN IN CONCERT** |
| TYPE | Music |
| FORMAT | CED |
| RETAIL OUTLET | Movies Unlimited Inc., A1-7096, $29.98 |

| | |
|---|---|
| TITLE | **HARRY CHAPIN, THE FINAL CONCERT** |
| TYPE | Music |
| FORMAT | Laser CLV |

| | |
|---:|:---|
| FEATURES | Stereo |
| RUNNING TIME | 89 minutes |
| PRODUCER | CBS Fox Video |
| DISTRIBUTOR | Pioneer Video, Inc., 7096-80, $34.98 |

| | |
|---:|:---|
| TITLE | **HATARI!** |
| TYPE | Motion Picture |
| FORMAT | Laser CLV |
| RUNNING TIME | 158 minutes |
| PRODUCER | Paramount Home Video |
| DISTRIBUTOR | Pioneer Video, Inc., LV6629-2, $39.95 |

| | |
|---:|:---|
| TITLE | **HAWAII** |
| TYPE | Motion Picture |
| FORMAT | CED (2 Discs) |
| RETAIL OUTLET | Movies Unlimited Inc., A1-4549, $39.98 |

| | |
|---:|:---|
| TITLE | **HAWAII TRAVEL** |
| DESCRIPTION | Part of the Sony Travel Guide System, which shows prospective travelers highlights of the place they are considering visiting, and provides travel information on printouts. |
| TYPE | Promotional |
| FORMAT | Laser CAV |
| LEVEL | 2 |
| CONTACT FOR MORE INFO. | Sony Corporation of America, (201) 930-1000 |
| FEATURES | Digital Dump-Sony, Stereo, Dual Audio |
| RUNNING TIME | 30 minutes |
| DISTRIBUTOR | Sony Corporation of America |

| | |
|---:|:---|
| TITLE | **HAWK TROUBLESHOOTING COURSE** |
| DESCRIPTION | A three-sided videodisc package containing 11 CAI lessons on the HAWK missile system electronic circuits and fault isolation during Integrated system checks. The package also includes a unit of troubleshooting aids. |
| TYPE | Technical, Instructional |
| FORMAT | Laser CAV |
| LEVEL | 3 |
| SYSTEM DESCRIPTION | To support the courseware, WICAT designed and manufactured a computer-controlled videodisc training system called the Distributed Instructional System (DIS). |
| DEVELOPER | WICAT Systems, Inc. for DARPA |
| CONTACT FOR MORE INFO. | Richard M. Cavagnol |
| PRODUCER | WICAT Systems, Inc. for DARPA |

| | |
|---:|:---|
| TITLE | **HE KNOWS YOU'RE ALONE** |
| TYPE | Motion Picture (R) |
| FORMAT | CED |
| PRODUCER | MGM/UNITED ARTISTS |
| RETAIL OUTLET | Movies Unlimited Inc., A2-0220, $29.95 |

| | |
|---:|:---|
| TITLE | **HE-MAN AND THE MASTERS OF THE UNIVERSE, VOL. 1** |
| TYPE | Motion Picture |
| FORMAT | CED |
| PRODUCER | Filmation |
| DISTRIBUTOR | RCA Videodiscs, 02174, $19.98 |

| | |
|---:|:---|
| TITLE | **HE-MAN AND THE MASTERS OF THE UNIVERSE, VOL. 2** |
| TYPE | Motion Picture |
| FORMAT | CED |
| PRODUCER | Filmation |
| DISTRIBUTOR | RCA Videodiscs, 02175, $19.98 |

| | |
|---:|:---|
| TITLE | **HEARTACHES** |
| TYPE | Motion Picture |
| FORMAT | Laser CLV, CED |
| RUNNING TIME | 93 minutes |
| PRODUCER | Vestron Video |
| DISTRIBUTOR | Pioneer Video, Inc., VL4024, $34.95 |
| RETAIL OUTLET | (CED) Movies Unlimited Inc., A3-4024, $29.95 |

| | |
|---:|:---|
| TITLE | **HEARTBREAK KID, THE** |
| TYPE | Motion Picture |
| FORMAT | CED |
| RETAIL OUTLET | Movies Unlimited Inc., A1-1083, $29.98 |

| | |
|---:|:---|
| TITLE | **HEAVEN CAN WAIT** |
| TYPE | Motion Picture (PG) |
| FORMAT | Laser CLV, CED |
| RUNNING TIME | 101 minutes |
| PRODUCER | Warner Home Video |
| DISTRIBUTOR | Pioneer Video, Inc., LV1109, $29.95 |
| | RCA Videodiscs, 00605, $29.98 |

| | |
|---:|:---|
| TITLE | **HEIDI** |
| TYPE | Motion Picture (G) |
| FORMAT | CED |
| DISTRIBUTOR | RCA Videodiscs, 01203, $19.98 |

| | |
|---:|:---|
| TITLE | **HELL IN THE PACIFIC** |
| TYPE | Motion Picture |
| FORMAT | CED |
| RETAIL OUTLET | Movies Unlimited Inc., A1-8028, $29.98 |

| | |
|---:|:---|
| TITLE | **HELLO DOLLY** |
| TYPE | Motion Picture, Music, Comedy (G) |
| FORMAT | Laser CLV, CED (2 Discs) |
| FEATURES | Stereo, CX Encoded |
| RUNNING TIME | 148 minutes |
| PRODUCER | Warner Home Video |
| DISTRIBUTOR | Pioneer Video, Inc., 1001-80, $39.98 |
| RETAIL OUTLET | (CED) Movies Unlimited Inc., A1-0113, $39.98 |

| | |
|---:|:---|
| TITLE | **HENRY V** |
| TYPE | Motion Picture |
| FORMAT | CED (2 Discs) |
| DISTRIBUTOR | RCA Videodiscs, 01003, $34.98 |

| | |
|---:|:---|
| TITLE | **HERBIE RIDES AGAIN** |
| TYPE | Motion Picture (G) |
| FORMAT | Laser CLV, CED |
| RUNNING TIME | 88 minutes |
| PRODUCER | Walt Disney |
| DISTRIBUTOR | Pioneer Video, Inc., 42042AS, $34.95 |
| | RCA Videodiscs, 00728, $24.98 |

| | |
|---:|:---|
| TITLE | **HERCULES** |
| TYPE | Motion Picture |
| FORMAT | CED |
| RETAIL OUTLET | Movies Unlimited Inc., A4-0736, $29.95 |

| | |
|---:|:---|
| TITLE | **HERE IT IS, BURLESQUE** |
| TYPE | Music, Comedy, Dance, Adult |
| FORMAT | Laser CLV, CED |
| RUNNING TIME | 88 minutes |
| PRODUCER | Warner Home Video |
| DISTRIBUTOR | Pioneer Video, Inc., VL3004, $34.95 |
| RETAIL OUTLET | (CED) Movies Unlimited Inc., A3-3004, $29.95 |

| | |
|---:|:---|
| TITLE | **HERITAGE OF THE BIBLE: THE LAW AND THE PROPHETS** |
| TYPE | Motion Picture, Art |
| FORMAT | CED |
| DISTRIBUTOR | RCA Videodiscs, 01212, $14.98 |

| | |
|---:|:---|
| TITLE | **HEROES** |
| TYPE | Motion Picture |
| FORMAT | Laser CLV |
| RETAIL OUTLET | UltraVideo, 10-012, $24.95 |

| | |
|---:|:---|
| TITLE | **HEY, CINDERELLA** |
| TYPE | Motion Picture |
| FORMAT | Laser CLV |
| RETAIL OUTLET | UltraVideo, 0806AS, $24.95 |

| | |
|---:|:---|
| TITLE | **HIGH ANXIETY** |
| TYPE | Motion Picture |
| FORMAT | CED |
| RETAIL OUTLET | Movies Unlimited Inc., A1-1107, $29.98 |

| | |
|---:|:---|
| TITLE | **HIGH COUNTRY** |
| TYPE | Motion Picture |
| FORMAT | CED |
| RETAIL OUTLET | Movies Unlimited Inc., A3-5009, $29.95 |

| | |
|---:|:---|
| TITLE | **HIGH NOON** |
| TYPE | Motion Picture |
| FORMAT | CED |
| DISTRIBUTOR | RCA Videodiscs, 00301, $19.98 |

| | |
|---:|:---|
| TITLE | **HIGH POWER ILLUMINATOR RAM RADAR DEMO** |
| DESCRIPTION | Training on operational and test procedures. |
| TYPE | Technical, Instructional, Promotional |
| FORMAT | Laser CAV |
| LEVEL | 3 |
| DEVELOPER | D. Gelotti/E. Dextraze |
| CONTACT FOR MORE INFO. | Ed Dextraze, (617) 272-9300 |
| FEATURES | Sound Over Still |
| PRODUCER | Raytheon Service Company |
| AVAILABILITY | Not for sale. |

| | |
|---:|:---|
| TITLE | **HIGH ROAD TO CHINA** |
| TYPE | Motion Picture (PG) |
| FORMAT | Laser CLV, CED |
| RUNNING TIME | 105 minutes |
| PRODUCER | Warner Home Video |
| DISTRIBUTOR | Pioneer Video, Inc., 11309LV, $29.98 |
| RETAIL OUTLET | (CED) Movies Unlimited Inc., A1-1309, $29.98 |

| | |
|---:|:---|
| TITLE | **HIGH SIERRA** |
| TYPE | Motion Picture |
| FORMAT | Laser CLV, CED |
| FEATURES | Black and white |
| RUNNING TIME | 100 minutes |
| PRODUCER | CBS Fox Video |
| DISTRIBUTOR | Pioneer Video, Inc., 4629-80, $34.98 |
| | RCA Videodiscs, 03409, $19.98 |

| | |
|---:|:---|
| TITLE | **HISTORY DISQUIZ, THE** |
| FORMAT | Laser CAV |
| LEVEL | 1 |
| PRODUCER | Optical Programming Associates |
| DISTRIBUTOR | Pioneer Video, Inc., 37-069, $29.95 |

| | |
|---:|:---|
| TITLE | **HISTORY OF THE WORLD, PART 1** |
| TYPE | Motion Picture, Comedy (R) |
| FORMAT | Laser CLV, CED |
| FEATURES | Stereo, CX Encoded |
| RUNNING TIME | 93 minutes |
| PRODUCER | Warner Home Video |
| DISTRIBUTOR | Pioneer Video, Inc., 1114-80, $34.98 |
| | RCA Videodiscs, 02063, $29.98 |

| | |
|---:|:---|
| TITLE | **HOBBIT, THE** |
| TYPE | Motion Picture |
| FORMAT | CED |
| DISTRIBUTOR | RCA Videodiscs, 01705, $24.98 |

| | |
|---:|:---|
| TITLE | **HOCUS POCUS - IT'S MAGIC** |
| TYPE | Motion Picture |
| FORMAT | Laser CLV, CED |
| RETAIL OUTLET | (CLV) UltraVideo, VL3006, $34.95 |
| | (CED) Movies Unlimited Inc., A3-3006, $29.95 |

| | |
|---:|:---|
| TITLE | **HOLOCAUST** |
| TYPE | Motion Picture |
| FORMAT | CED (4 Discs) |
| DISTRIBUTOR | RCA Videodiscs, 02130, $79.98 |

| | |
|---:|:---|
| TITLE | **HOMBRE** |
| TYPE | Motion Picture |
| FORMAT | CED |
| RETAIL OUTLET | Movies Unlimited Inc., A1-1012, $29.98 |

| | |
|---:|:---|
| TITLE | **HOMEWORK** |
| TYPE | Motion Picture (R) |
| FORMAT | Laser CLV |
| RUNNING TIME | 90 minutes |
| PRODUCER | MCA Videodisc, Inc. |
| DISTRIBUTOR | Pioneer Video, Inc., 16-030, $29.95 |

| | |
|---:|:---|
| TITLE | **HONEYSUCKLE ROSE** |
| TYPE | Motion Picture, Music (PG) |
| FORMAT | Laser CLV |
| FEATURES | Stereo, CX Encoded |
| RUNNING TIME | 120 minutes |
| PRODUCER | Warner Home Video |
| DISTRIBUTOR | Pioneer Video, Inc., 1043LV, $29.98 |

| | |
|---:|:---|
| TITLE | **HOOPER** |
| TYPE | Motion Picture (PG) |
| FORMAT | CED |
| DISTRIBUTOR | RCA Videodiscs, 03111, $29.98 |

| | |
|---:|:---|
| TITLE | **HOPSCOTCH** |
| TYPE | Motion Picture |
| FORMAT | CED |
| RETAIL OUTLET | Movies Unlimited Inc., A4-0756, $29.95 |

| | |
|---:|:---|
| TITLE | **HOROWITZ** |
| TYPE | Music |
| FORMAT | Laser CLV |
| FEATURES | Stereo, CX Encoded |
| RUNNING TIME | 116 minutes |
| PRODUCER | Pioneer Artists |
| DISTRIBUTOR | Pioneer Video, Inc., PA-82-031, $29.95 |

| | |
|---|---|
| TITLE | **HORSE SOLDIERS, THE** |
| TYPE | Motion Picture |
| FORMAT | CED |
| DISTRIBUTOR | RCA Videodiscs, 01438, $29.98 |

| | |
|---|---|
| TITLE | **HOUSE CALLS** |
| TYPE | Motion Picture (PG) |
| FORMAT | Laser CLV |
| RUNNING TIME | 98 minutes |
| PRODUCER | Warner Home Video |
| DISTRIBUTOR | Pioneer Video, 16-006, $29.95 |

| | |
|---|---|
| TITLE | **HOUSE OF SORORITY ROW, THE** |
| TYPE | Motion Picture |
| FORMAT | Laser CLV |
| RETAIL OUTLET | UltraVideo, VL4059, $34.95 |

| | |
|---|---|
| TITLE | **HOW TO BEAT THE HIGH COST OF LIVING** |
| TYPE | Motion Picture |
| FORMAT | CED |
| RETAIL OUTLET | Movies Unlimited Inc., A3-4031, $29.98 |

| | |
|---|---|
| TITLE | **HOW TO COPE WITH...** |
| TYPE | Instructional |
| FORMAT | Laser CLV |
| RETAIL OUTLET | UltraVideo, I500-001, $39.95 |

| | |
|---|---|
| TITLE | **HOW TO MARRY A MILLIONAIRE** |
| TYPE | Motion Picture |
| FORMAT | CED |
| RETAIL OUTLET | Movies Unlimited Inc., A1-1023, $29.98 |

| | |
|---|---|
| TITLE | **HOW TO WATCH PRO FOOTBALL** |
| FORMAT | Laser CAV |
| LEVEL | 1 |
| FEATURES | Dual Audio |
| PRODUCER | Optical Programming Associates |
| DISTRIBUTOR | Pioneer Video, Inc., 86-515, $24.95 |

| | |
|---|---|
| TITLE | **HOW YOUR HEART AND CIRCULATORY SYSTEM WORK** |
| DESCRIPTION | A section of the active health curriculum as taught in the British Columbia school system, grades 4 through 7. |
| TYPE | In-house Training, Adolescents, Instructional |
| FORMAT | Laser CAV |
| LEVEL | 2, 3 |
| SYSTEM DESCRIPTION | Totally interactive, using a Discovision 7420 player with a mixture of motion picture footage and 550 interactive stills. |
| DEVELOPER | Dr. Glenn Kirchner, Faculty of Education, Simon Fraser University |
| CONTACT FOR MORE INFO. | Chris Hildred, Manager for Photo Services, Instructional Media Center |

# Videodiscography

| | |
|---:|:---|
| FEATURES | Chapters, Picture Stops, Digital Dump-Pioneer |
| PRODUCER | Simon Fraser University |
| DISTRIBUTOR | Simon Fraser University |
| AVAILABILITY | Sale or rent |

| | |
|---:|:---|
| TITLE | **HOWLING, THE** |
| TYPE | Motion Picture (R) |
| FORMAT | Laser CLV, CED |
| DISTRIBUTOR | RCA Videodiscs, 00810, $24.98 |
| RETAIL OUTLET | (CLV) UltraVideo, E161550, $34.95 |

| | |
|---:|:---|
| TITLE | **HUCKLEBERRY FINN** |
| TYPE | Motion Picture |
| FORMAT | CED |
| RETAIL OUTLET | Movies Unlimited Inc., A3-E909, $29.95 |

| | |
|---:|:---|
| TITLE | **HUD** |
| TYPE | Motion Picture |
| FORMAT | Laser CLV, CED |
| DISTRIBUTOR | RCA Videodiscs, 00625, $24.98 |
| RETAIL OUTLET | (CLV) UltraVideo, LV6630, $29.95 |

| | |
|---:|:---|
| TITLE | **HUMAN GENETICS TRAINING FOR NURSES: PART I** |
| DESCRIPTION | Basic training in human genetics, focusing on applications for nurses. |
| TYPE | Medical, Instructional |
| FORMAT | Laser CAV |
| LEVEL | 3 |
| SYSTEM DESCRIPTION | Pioneer 7820-I player, Apple II plus, and a Coloney Interface. |
| DEVELOPER | Florida State University/Center for Educational Technology |
| CONTACT FOR MORE INFO. | Dr. Brent Hewlett, (904) 644-4720 or Jayne Parker |
| FEATURES | Dual Audio |
| RUNNING TIME | 6 hours |
| PRODUCER | Florida State University, Multi-Media Laboratory |
| DISTRIBUTOR | Florida Department of Health and Rehabilitative Services, Children's Medical Services |
| AVAILABILITY | Not available. |

| | |
|---:|:---|
| TITLE | **HUNCHBACK OF NOTRE DAME, THE** |
| TYPE | Motion Picture |
| FORMAT | CED |
| DISTRIBUTOR | RCA Videodiscs, 00402, $19.98 |

| | |
|---:|:---|
| TITLE | **HUNTER, THE** |
| TYPE | Motion Picture (PG) |
| FORMAT | Laser CLV |
| RUNNING TIME | 97 minutes |
| PRODUCER | Warner Home Video |
| DISTRIBUTOR | Pioneer Video, Inc., LV1192, $29.95 |

| | |
|---:|:---|
| TITLE | **HUSTLER, THE** |
| TYPE | Motion Picture |
| FORMAT | CED (2 Discs) |
| RETAIL OUTLET | Movies Unlimited Inc., A1-1006, $39.98 |

| | |
|---:|:---|
| TITLE | **I AM A FUGITIVE FROM A CHAIN GANG** |
| TYPE | Motion Picture |
| FORMAT | CED |
| PRODUCER | United Artists |
| DISTRIBUTOR | RCA Videodiscs, 03410, $24.98 |

| | |
|---:|:---|
| TITLE | **I LOVE YOU** |
| TYPE | Motion Picture |
| FORMAT | CED |
| RETAIL OUTLET | Movies Unlimited Inc., A2-0209, $29.95 |

| | |
|---:|:---|
| TITLE | **I OUGHT TO BE IN PICTURES** |
| TYPE | Motion Picture |
| FORMAT | CED |
| RETAIL OUTLET | Movies Unlimited Inc., A1-1150, $29.98 |

| | |
|---:|:---|
| TITLE | **I SPIT ON YOUR GRAVE** |
| TYPE | Motion Picture |
| FORMAT | Laser CLV, CED |
| FEATURES | Closed captioned |
| RETAIL OUTLET | (CLV) UltraVideo, ZL016, $34.95 |
| | (CED) Movies Unlimited Inc., A3-C016, $29.95 |

| | |
|---:|:---|
| TITLE | **I, THE JURY** |
| TYPE | Motion Picture |
| FORMAT | CED |
| RETAIL OUTLET | Movies Unlimited Inc., A1-1186, $29.98 |

| | |
|---:|:---|
| TITLE | **I'M DANCING AS FAST AS I CAN** |
| TYPE | Motion Picture (R) |
| FORMAT | Laser CLV |
| RUNNING TIME | 107 minutes |
| PRODUCER | Warner Home Video |
| DISTRIBUTOR | Pioneer Video, Inc., LV1295, $29.95 |

| | |
|---:|:---|
| TITLE | **IBM - (CONFIDENTIAL)** |
| TYPE | Technical, Instructional, In-house Training, Point-of-Purchase |
| FORMAT | Laser CAV |
| LEVEL | 3 |
| DEVELOPER | Paul Frauenhoffer, IBM, (305) 241-7632 |

| | |
|---:|:---|
| TITLE | **IBM 1750 TELEPHONE SYSTEM** |
| DESCRIPTION | Training videodisc, using live animation to instruct in-house users in the operation of IBM's 1750 telephone system. |

| | |
|---|---|
| TYPE | Technical, Instructional, In-house Training |
| RELEASE DATE | 1981 |
| FORMAT | Laser CAV |
| LEVEL | 3 |
| DEVELOPER | WICAT Systems, Inc. for IBM |
| CONTACT FOR MORE INFO. | Richard M. Cavagnol |
| PRODUCER | WICAT Systems, Inc. |

| | |
|---|---|
| TITLE | **IBM HERITAGE** |
| DESCRIPTION | Provides IBM employees with a summary history of their company's business and social impact. |
| RELEASE DATE | 1981 |
| TYPE | Instructional, In-house Training, Archival |
| FORMAT | Laser CAV |
| LEVEL | 2 |
| DEVELOPER | Exhibit Technology, Inc., Ken Valley and Don Cochrane |
| CONTACT FOR MORE INFO. | Don Cochrane, President - Marketing |
| FEATURES | Chapters, Picture Stops, Digital Dump-Pioneer |
| PRODUCER | Ken Valley |
| AVAILABILITY | Available as a company product sample for tutorial and prototype demonstrations. |

| | |
|---|---|
| TITLE | **IBM OFFICE SYSTEMS** |
| DESCRIPTION | Two-sided marketing videodisc, which introduces the entire line of IBM Office Systems products. |
| TYPE | Technical, Instructional, In-house Training |
| RELEASE DATE | June 1982 |
| FORMAT | Laser CAV |
| LEVEL | 3 |
| DEVELOPER | WICAT Systems, Inc. for IBM |
| CONTACT FOR MORE INFO. | Richard M. Cavagnol |
| PRODUCER | WICAT Systems, Inc. |

| | |
|---|---|
| TITLE | **IBM PC EXPERIENCE** |
| DESCRIPTION | Trains people to use the IBM Personal Computer keyboard, DOS, and Wordstar and Visicalc programs. Users can work on the computer while viewing the videodisc. |
| TYPE | Instructional, In-house Training |
| FORMAT | Laser CAV |
| LEVEL | 3 |
| SYSTEM DESCRIPTION | Player is controlled by IBM PC over RS232. Features digitally synthesized voice over still frames produced by voice synthesis from the PC. |
| DEVELOPER | Interactive Image Technologies for Interactive Research Corporation |
| CONTACT FOR MORE INFO. | Patrick Lee/Joseph Koenig at Interactive Image or Allen Lee Adkins at Interactive Research |
| FEATURES | Picture Stops, Sound Over Still |
| PRODUCER | Patrick Lee |
| DISTRIBUTOR | Interactive Research Corporation |
| AVAILABILITY | Sale. Qualified customers have a 20-day evaluation period. |
| ORDER NO. | IBM PC-1 |
| SUGGESTED PRICE | $1880 |

| | |
|---:|:---|
| TITLE | **IF YOU COULD SEE WHAT I HEAR** |
| TYPE | Motion Picture |
| FORMAT | Laser CLV, CED |
| RUNNING TIME | 100 minutes |
| PRODUCER | Vestron Video |
| DISTRIBUTOR | Pioneer Video, Inc., 5014, $34.95 |
| RETAIL OUTLET | (CED) Movies Unlimited Inc., A3-5014, $29.95 |

| | |
|---:|:---|
| TITLE | **IMPLEMENTING ADDITIONAL PROGRAMS AND POLICIES** |
| DESCRIPTION | Along with 12 floppy diskettes, this videodisc is designed to teach Florida's welfare workers about various state programs such as Child Support Enforcement so that they can be better prepared to respond to the client's needs. It contains job-simulated activities that allow the learner to function as a case worker. Other practices are presented as a radio drama (audio 2), a television game show (motion), and a press interview (motion, still and audio 2). |
| TYPE | Instructional, In-house Training |
| FORMAT | Laser CAV |
| LEVEL | 3 |
| SYSTEM DESCRIPTION | Disc is part of 160-hour curriculum delivered on an Apple II/Pioneer 7820-3/single disc drive system along with ancillary printed reference books. |
| DEVELOPER | University of West Florida/Office for Interactive Technology and Training |
| CONTACT FOR MORE INFO. | Richard Smith; Elizabeth Wright, (413) 256-0444 |
| FEATURES | Dual Audio |
| PRODUCER | University of Nebraska/Nebraska Videodisc Design/Production Group |
| AVAILABILITY | Training program produced under contract with the Florida Department of Health and Rehabilitative Services, Tallahassee, Florida, and is not available for general distribution. |

| | |
|---:|:---|
| TITLE | **IMPROPER CHANNELS** |
| TYPE | Motion Picture, Comedy |
| FORMAT | Laser CLV, CED |
| RUNNING TIME | 91 minutes |
| PRODUCER | Vestron Video |
| DISTRIBUTOR | Pioneer Video, Inc., VL5011, $34.95 |
| RETAIL OUTLET | (CED) Movies Unlimited Inc., A3-5011, $29.95 |

| | |
|---:|:---|
| TITLE | **IN PRAISE OF OLDER WOMEN** |
| TYPE | Motion Picture, Adult (R) |
| FORMAT | CED |
| DISTRIBUTOR | RCA Videodiscs, 00811, $19.98 |

| | |
|---:|:---|
| TITLE | **IN THE HEAT OF THE NIGHT** |
| TYPE | Motion Picture |
| FORMAT | CED |
| DISTRIBUTOR | RCA Videodiscs, 01478, $29.98 |

| | |
|---:|:---|
| TITLE | **IN-LAWS, THE** |
| TYPE | Motion Picture, Comedy (PG) |
| FORMAT | CED |
| DISTRIBUTOR | RCA Videodiscs, 03108, $24.98 |

| | |
|---:|:---|
| TITLE | **INCUBUS, THE** |
| TYPE | Motion Picture |
| FORMAT | Laser CLV, CED |
| RUNNING TIME | 90 minutes |
| PRODUCER | Vestron Video |
| DISTRIBUTOR | Pioneer Video, Inc., VL4016, $34.95 |
| RETAIL OUTLET | (CED) Movies Unlimited Inc., A3-4016, $29.95 |

| | |
|---:|:---|
| TITLE | **INCREDIBLE SHRINKING WOMAN, THE** |
| TYPE | Motion Picture |
| FORMAT | Laser CLV |
| RETAIL OUTLET | UltraVideo, 16-023, $29.95 |

| | |
|---:|:---|
| TITLE | **INHERIT THE WIND** |
| TYPE | Motion Picture |
| FORMAT | CED |
| DISTRIBUTOR | RCA Videodiscs, 01474, $24.98 |

| | |
|---:|:---|
| TITLE | **INHERIT THE WIND** |
| TYPE | Motion Picture |
| FORMAT | CED (2 Discs) |
| RETAIL OUTLET | Movies Unlimited Inc., A1-4651, $39.98 |

| | |
|---:|:---|
| TITLE | **INN OF THE SIXTH HAPPINESS** |
| TYPE | Motion Picture |
| FORMAT | CED (2 Discs) |
| FEATURES | Stereo |
| RETAIL OUTLET | Movies Unlimited Inc., A1-1170, $39.98 |

| | |
|---:|:---|
| TITLE | **INTEGRATED READINESS CENTER** |
| DESCRIPTION | Disc specifically developed for a new product line of the Westinghouse Integrated Logistics Support Division - the Integrated Readiness Center (IRC). The theme of the disc is "Maintenance-Aiding," and it includes marketing segments that introduce the IRC and a technical sample for the electronics warfare pod for the AN/ALQ-131 system. |
| TYPE | Research and Development, Instructional, In-house Training |
| FORMAT | Laser CAV |
| LEVEL | 3 |
| SYSTEM DESCRIPTION | The IRC is a VAX-based automatic test station for Westinghouse products and incorporates a maintenance-aiding console (MAC). The MAC is a stand-alone interactive videodisc station, and features a color RGB monitor with touch panel, a videodisc player, a text-graphics keyer, and microcomputer control with mass storage disks. |
| DEVELOPER | Westinghouse Technical Training Operations |
| CONTACT FOR MORE INFO. | Steve G. Kukla, (301) 995-5482 |
| FEATURES | Text/Graphics Overlays, Touch Panel |
| RUNNING TIME | 30 minutes |
| PRODUCER | Westinghouse Multi-Media Production Center |
| AVAILABILITY | Not available. This disc was inspected by the USAF and has been cleared for public viewing. |

| | |
|---|---|
| TITLE | **INTEGRATING ADMINISTRATIVE POLICIES** |
| DESCRIPTION | Along with 17 floppy diskettes, this videodisc is designed to introduce a new trainee to the administrative rules, policies, and procedures that a welfare worker in Florida must follow in order to correctly determine whether families are eligible to receive aid and assistance for dependent children. Along with other job- simulated activities it contains a futuristic scenario in which elderly case workers in 2025 reminisce about "the way it was," and a contemporary scene in a laundromat in which clients are discussing their experiences with the AFDC Program. |
| TYPE | Instructional, In-house Training |
| FORMAT | Laser CAV |
| LEVEL | 3 |
| SYSTEM DESCRIPTION | Disc is part of 160-hour curriculum delivered on an Apple II/Pioneer 7820-3/single disc drive system along with ancillary printed reference books. |
| DEVELOPER | University of West Florida/Office for Interactive Technology and Training |
| CONTACT FOR MORE INFO. | Richard Smith; Elizabeth Wright, (413) 256-0444 |
| FEATURES | Dual Audio |
| PRODUCER | University of Nebraska/Nebraska Videodisc Design/Production Group |
| AVAILABILITY | Training program produced under contract with the Florida Department of Health and Rehabilitative Services, Tallahasee, Florida, and is not available for general distribution. |

| | |
|---|---|
| TITLE | **INTELLIGENT DISC, THE** |
| DESCRIPTION | Medical education and flight training. |
| TYPE | Medical, Technical, Instructional |
| LEVEL | 2, 3 |
| CONTACT FOR MORE INFO. | Corporation for Public Broadcasting, University of Nebraska, KUON-TV |
| PRODUCER | University of Nebraska/Nebraska Videodisc Design/Production Group |
| AVAILABILITY | Not available for release or distribution. |

| | |
|---|---|
| TITLE | **INTERACTION AND INTERVENTION WITH GRIEVING CLIENTS AND FAMILIES** |
| DESCRIPTION | Simulation program designed to aid pediatric nurses to interact and intervene with chronically and terminally ill clients. |
| TYPE | Medical, In-house Training |
| FORMAT | Laser CAV |
| LEVEL | 3 |
| SYSTEM DESCRIPTION | Coloney Learning Carrel System |
| DEVELOPER | Interactive Video Concepts, Inc. (IVC) |
| CONTACT FOR MORE INFO. | Peg Callahan, IVC/Jayne Parker, Florida Department of Health and Rehabilitative Services/Children's Medical Services |
| RUNNING TIME | Varies, but approximately 8 hours |
| PRODUCER | Interactive Video Concepts, Inc. (IVC) |
| AVAILABILITY | Used for continuing nursing education by Children's Medical Services. |

| | |
|---|---|
| TITLE | **INTERACTIVE ARTS & SCIENCE, INC. - (CONFIDENTIAL)** |
| TYPE | Game, Research and Development |
| FORMAT | CED |
| LEVEL | 1,2,3 |
| SYSTEM DESCRIPTION | (Confidential) |
| DEVELOPER | Interactive Arts & Science (IAS, Inc.) working in concert with a motion picture studio. |
| CONTACT FOR MORE INFO. | Jack D. Cooper, President, IAS, Inc. |

|   |   |
|---|---|
| TITLE | **INTERACTIVE ARTS INTERNATIONAL - (UNANNOUNCED, PROPRIETARY UNTIL DATE OF ANNOUNCEMENT)** |
| DESCRIPTION | Advanced, inexpensive, consumer adventure/ strategy game that will appeal to the whole family. First in a series of entertainment/educational programs from Interactive Arts International. May be played with a remote-control hand unit or with an outboard microcomputer. Acts as level 1 or 3 due to special innovation in programming and disc design/game strategy. |
| TYPE | Game, Children, Adolescents, Archival |
| RELEASE DATE | Second quarter 1984 |
| FORMAT | CED, Laser CAV |
| LEVEL | 1, 3 |
| SYSTEM DESCRIPTION | Details will be announced at time of product introduction. The system will include a cheap, highly functional interface for most home computers, including most Apple models, Atari, Coleco Adam, IBM PC Jr., and Commodore Pet and 64. An interface for new RCA interactive player will also be available. |
| DEVELOPER | Interactive Arts International |
| CONTACT FOR MORE INFO. | Allen Lee Adkins |
| FEATURES | Chapters, Picture Stops, Dual Audio, Sound Over Still |
| PRODUCER | Allen Lee Adkins |
| DISTRIBUTOR | Interactive Arts International |
| RETAIL OUTLET | Vernes Magnavox, Westminster, CA<br>Dennis Webb, owner/proprietor, (714) 898-9561 |
| AVAILABILITY | Sale, Rent |
| SPECIAL CONDITIONS | Distributors have collective advertising and promotional options. Contact Interactive Arts for details. |
| SUGGESTED PRICE | Approximately $49 for level 1 disc game set and $65 for level 3 set. |

|   |   |
|---|---|
| TITLE | **INTERACTIVE JULIA CHILD** |
| DESCRIPTION | Instructions on one full course dinner by Julia Child herself. This interactive disc includes ingredient pages, wine choices, and preparations. |
| TYPE | Instructional |
| FORMAT | Laser CAV |
| LEVEL | 2 |
| FEATURES | Digital Dump-Sony |
| PRODUCER | WGBH Boston |
| DISTRIBUTOR | Sony Corporation of America |
| AVAILABILITY | Sale |
| SUGGESTED PRICE | $29.95 |

|   |   |
|---|---|
| TITLE | **INTERNATIONAL HARVESTER** |
| DESCRIPTION | Training on the mechanical systems of the TD-20 equipment. Used as a demonstration to training directors. |
| TYPE | Instructional, Promotional |
| FORMAT | Laser CAV |
| LEVEL | 3 |
| DEVELOPER | Jack Lieb Productions, Inc. |
| FEATURES | Chapters, Picture Stops, Digital Dump-Sony |

| | |
|---|---|
| TITLE | **INTERVENTION IN CHILD ABUSE AND NEGLECT** |
| DESCRIPTION | Helps CMS health professionals recognize indicators of child abuse and neglect, and their own responsibility to report suspected cases. |
| TYPE | Medical, Instructional |
| FORMAT | Laser CAV |
| LEVEL | 3 |
| SYSTEM DESCRIPTION | Discovision 7820-1 player, Apple II Plus microcomputer, and a VAI-1 interface card. |
| DEVELOPER | MGT of America, Inc., Tallahassee, Florida |
| CONTACT FOR MORE INFO. | Jayne Parker |
| RUNNING TIME | Approximately 6 hours |
| AVAILABILITY | Not for sale, but they may become available in the future. |

| | |
|---|---|
| TITLE | **INTRODUCTION TO COMPUTER LITERACY** |
| DESCRIPTION | Instructional discs designed to give a nontechnical overview of microcomputer technology to industry, business, government, and educational markets. This is part of a library of videodisc programs about microcomputer training. |
| RELEASE DATE | February 1984 |
| TYPE | Instructional |
| FORMAT | Laser CAV |
| LEVEL | 3 |
| SYSTEM DESCRIPTION | Requires an Apple or IBM Personal Computer and any industrial player. |
| DEVELOPER | Jam, Inc. |
| CONTACT FOR MORE INFO. | Michael Yampolsky, Vice President, Research and Development |
| FEATURES | Picture Stops |
| RUNNING TIME | 29 minutes |
| PRODUCER | Jam, Inc. |
| DISTRIBUTOR | Same as producer |
| AVAILABILITY | Sale |
| SUGGESTED PRICE | $1995 |

| | |
|---|---|
| TITLE | **INTRODUCTION TO ECONOMICS** |
| DESCRIPTION | A lesson in basic economic forces to be used with an Apple computer. |
| TYPE | Instructional |
| LEVEL | 3 |
| CONTACT FOR MORE INFO. | Kent Kehrberg, Project Manager, MECC |
| PRODUCER | University of Nebraska/Nebraska Videodisc Design/Production Group |
| AVAILABILITY | Not available for release or distribution. |

| | |
|---|---|
| TITLE | **INTRODUCTION TO THE JOIN SYSTEM** |
| DESCRIPTION | An interactive training disc distributed to over 4,000 Army recruiting offices. Teaches recruiters how to operate a computer operated laser disc station. |
| TYPE | Technical, Instructional |
| RELEASE DATE | February 1983 |
| FORMAT | Laser CAV |
| LEVEL | 3 |
| SYSTEM DESCRIPTION | Sony LV 1000 player controlled by a microcomputer accessed through a custom and ASCII keyboard and a printer for output. Telephone and modem are all in one case. |
| DEVELOPER | C3 |

# Videodiscography

|  |  |
|---|---|
| CONTACT FOR MORE INFO. | Garri Garripoli, Citidisc Productions |
| FEATURES | Chapters, Picture Stops, Digital Dump-Sony |
| RUNNING TIME | 28 minutes, 50 seconds |
| PRODUCER | Citidisc Productions |
| AVAILABILITY | In-house for Army use. |

|  |  |
|---|---|
| TITLE | **INTRODUCTION TO THE LEARNING SYSTEM, AN** |
| DESCRIPTION | Following an overview of interactive technology, step-by step procedures for operating a computer-controlled interactive videodisc system arte detailed. Throughout the presentation, the viewer is given the opportunity to practice using the various functions. |
| TYPE | Instructional, In-house Training |
| FORMAT | Laser CAV |
| LEVEL | 3 |
| SYSTEM DESCRIPTION | Disc is part of 160-hour curriculum delivered on an Apple II/Pioneer 7820-3/single disc drive system along with ancillary printed reference books. |
| DEVELOPER | University of West Florida/Office for Interactive Technology and Training |
| CONTACT FOR MORE INFO. | Richard Smith; Elizabeth Wright, (413) 256-0444 |
| FEATURES | Dual Audio |
| PRODUCER | University of Nebraska/Nebraska Videodisc Design/Production Group |
| AVAILABILITY | Training program produced under contract with the Florida Department of Health and Rehabilitative Services, Tallahassee, Florida, and is not available for general distribution. |

|  |  |
|---|---|
| TITLE | **INVASION OF THE BODY SNATCHERS - ORIGINAL VERSION** |
| TYPE | Motion Picture |
| FORMAT | CED |
| DISTRIBUTOR | RCA Videodiscs, 00302, $19.98 |

|  |  |
|---|---|
| TITLE | **INVESTMENT SERVICES** |
| DESCRIPTION | This is a point-of-purchase videodisc that will be used in regular banks by one of Cutting Edge's clients to promote investment services. |
| TYPE | Point-of-Purchase |
| FORMAT | Laser CAV |
| LEVEL | 3 |
| SYSTEM DESCRIPTION | Pioneer 8210 combined with a Commodore microcomputer |
| FEATURES | Chapters |
| RUNNING TIME | 15 minutes |
| PRODUCER | Judy Bergsma-Owen |

|  |  |
|---|---|
| TITLE | **INVITATION TO THE DANCE** |
| TYPE | Motion Picture |
| FORMAT | CED |
| RETAIL OUTLET | Movies Unlimited Inc., A2-0192, $29.95 |

|  |  |
|---|---|
| TITLE | **IRMA LA DOUCE** |
| TYPE | Motion Picture |
| FORMAT | CED (2 Discs) |
| RETAIL OUTLET | Movies Unlimited Inc., A1-4548, $39.98 |

| | |
|---:|:---|
| TITLE | **ISLAND OF DR. MOREAU** |
| TYPE | Motion Picture |
| FORMAT | CED |
| RETAIL OUTLET | Movies Unlimited Inc., A3-4054, $29.98 |

| | |
|---:|:---|
| TITLE | **ISRAELI BOY: LIFE ON A KIBBUTZ** |
| DESCRIPTION | An experiment in reformatting an existing film. |
| TYPE | Research and Development |
| FORMAT | Laser CAV |
| LEVEL | 2 |
| CONTACT FOR MORE INFO. | Bureau of Educations for the Handicapped/University of Nebraska - Barkley Center |
| PRODUCER | University of Nebraska/Nebraska Videodisc Design/Production Group |
| AVAILABILITY | Not available for release or distribution. |

| | |
|---:|:---|
| TITLE | **IT CAME FROM HOLLYWOOD** |
| TYPE | Motion Picture, Comedy (PG) |
| FORMAT | Laser CLV, CED |
| RUNNING TIME | 80 minutes |
| PRODUCER | Paramount Home Video |
| DISTRIBUTOR | Pioneer Video, Inc., LV1421, $29.95 |
| | RCA Videodiscs, 03616, $24.98 |

| | |
|---:|:---|
| TITLE | **IT'S A MAD, MAD, MAD, MAD, WORLD** |
| TYPE | Motion Picture, Comedy (G) |
| FORMAT | Laser CLV, CED (2 Discs) |
| FEATURES | Stereo |
| RUNNING TIME | 154 minutes |
| PRODUCER | United Artists |
| DISTRIBUTOR | RCA Videodiscs, 11478, $39.98 |
| RETAIL OUTLET | (CLV) UltraVideo, 80-4534, $34.95 |

| | |
|---:|:---|
| TITLE | **IT'S MY TURN** |
| TYPE | Motion Picture (R) |
| FORMAT | Laser CLV |
| PRODUCER | Warner Home Video |
| DISTRIBUTOR | Pioneer Video, Inc., VLD3190, $29.95 |

| | |
|---:|:---|
| TITLE | **ITZHAK PERLMAN** |
| TYPE | Music |
| FEATURES | Stereo, CX Encoded |
| FORMAT | Laser CLV |
| RUNNING TIME | 45 minutes |
| PRODUCER | Pioneer Artists |
| DISTRIBUTOR | Pioneer Video, Inc., PA-83-042, $24.95 |

| | |
|---:|:---|
| TITLE | **IVIS COURSEWARE LIBRARY** |
| DESCRIPTION | Videodiscs designed for Digital's IVIS system on such topics as industrial programmable controllers, introduction to computer systems, laser technology, and mass storage concepts. |

# Videodiscography

| | |
|---|---|
| TYPE | Instructional, Technical |
| FORMAT | Laser CAV |
| LEVEL | 3 |
| SYSTEM DESCRIPTION | IVIS (Interactive Video Information System) |
| CONTACT FOR MORE INFO. | Tim Walsh, (617) 276-4280 |
| PRODUCER | Digital Equipment Corporation |
| DISTRIBUTOR | Digital Equipment Corporation |
| AVAILABILITY | Sale |
| SUGGESTED PRICE | Price of individual discs varies, but average is $2500 |

| | |
|---|---|
| TITLE | **JAILHOUSE ROCK** |
| TYPE | Motion Picture |
| FORMAT | CED |
| RETAIL OUTLET | Movies Unlimited Inc., A2-1002, $24.95 |

| | |
|---|---|
| TITLE | **JAMES TAYLOR IN CONCERT** |
| TYPE | Music |
| FORMAT | Laser CLV, CED |
| FEATURES | Stereo, CX Encoded |
| RUNNING TIME | 90 minutes |
| PRODUCER | CBS Fox Video |
| DISTRIBUTOR | Pioneer Video, Inc., 7023-80, $29.98 |
| RETAIL OUTLET | (CED) Movies Unlimited Inc., A1-7023, $29.98 |

| | |
|---|---|
| TITLE | **JANE FONDA'S WORKOUT** |
| TYPE | Instructional, Dance |
| FORMAT | CED |
| FEATURES | Dual Track |
| DISTRIBUTOR | RCA Videodiscs, 22095, $29.98 |

| | |
|---|---|
| TITLE | **JANE FONDA'S WORKOUT FOR PREGNANCY, BIRTH AND RECOVERY** |
| TYPE | Instructional, Dance |
| FORMAT | CED |
| FEATURES | Dual Track |
| PRODUCER | RCA/KVC |
| DISTRIBUTOR | RCA Videodiscs, 22187, $19.98 |

| | |
|---|---|
| TITLE | **JASON & THE ARGONAUTS** |
| TYPE | Motion Picture |
| FORMAT | CED |
| DISTRIBUTOR | RCA Videodiscs, 03054, $24.98 |

| | |
|---|---|
| TITLE | **JAWS** |
| TYPE | Motion Picture (PG) |
| FORMAT | Laser CLV, CED |
| RUNNING TIME | 124 minutes |
| PRODUCER | Warner Home Video |
| DISTRIBUTOR | Pioneer Video, Inc., 12-001, $29.95 |
| | RCA Videodiscs, 03301, $29.98 |

| | |
|---:|:---|
| TITLE | **JAWS 2** |
| TYPE | Motion Picture |
| FORMAT | Laser CLV, CED |
| RETAIL OUTLET | (CLV) UltraVideo, 16-023, $29.95 |
| | (CED) Movies Unlimited Inc., A5-2010, $29.98 |

| | |
|---:|:---|
| TITLE | **JAZZ IN AMERICA** |
| TYPE | Music |
| FORMAT | CED |
| FEATURES | Stereo |
| RUNNING TIME | 90 minutes |
| PRODUCER | Jazz America |
| DISTRIBUTOR | RCA Videodiscs, 12138, $24.98 |

| | |
|---:|:---|
| TITLE | **JAZZ SINGER, THE** |
| TYPE | Motion Picture, Music (PG) |
| FORMAT | Laser CLV, CED |
| FEATURES | Stereo, CX Encoded |
| PRODUCER | Warner Home Video |
| DISTRIBUTOR | Pioneer Video, Inc., LV2305, $29.95 |
| | RCA Videodiscs, 10668, $29.98 |

| | |
|---:|:---|
| TITLE | **JAZZ SINGER-AL JOLSON** |
| TYPE | Music |
| FORMAT | CED |
| RETAIL OUTLET | Movies Unlimited Inc., A1-4501, $29.98 |

| | |
|---:|:---|
| TITLE | **JAZZERCISE** |
| TYPE | Dance |
| FORMAT | Laser CAV |
| LEVEL | 2 |
| PRODUCER | Optical Programming Associates |
| DISTRIBUTOR | Pioneer Video, Inc., 32-608, $24.95 |

| | |
|---:|:---|
| TITLE | **JEFFERSON STARSHIP** |
| TYPE | Music |
| FORMAT | CED |
| FEATURES | Stereo |
| DISTRIBUTOR | RCA Videodiscs, 12167, $19.98 |

| | |
|---:|:---|
| TITLE | **JEREMIAH JOHNSON** |
| TYPE | Motion Picture |
| FORMAT | CED |
| DISTRIBUTOR | RCA Videodiscs, 03152, $24.98 |

| | |
|---:|:---|
| TITLE | **JERK, THE** |
| TYPE | Motion Picture, Comedy (R) |
| FORMAT | Laser CLV, CED |
| DISTRIBUTOR | RCA Videodiscs, 03306, $29.98 |
| RETAIL OUTLET | (CLV) UltraVideo, 16-015, $29.95 |

# Videodiscography

|   |   |
|---|---|
| TITLE | **JESUS CHRIST SUPERSTAR** |
| TYPE | Motion Picture, Music |
| FORMAT | Laser CLV |
| FEATURES | Stereo |
| RETAIL OUTLET | UltraVideo, 17-002, $29.95 |

|   |   |
|---|---|
| TITLE | **JESUS OF NAZARETH** |
| TYPE | Motion Picture |
| FORMAT | CED (4 Discs) |
| DISTRIBUTOR | RCA Videodiscs, 00510, $99.98 |

|   |   |
|---|---|
| TITLE | **JEZEBEL** |
| TYPE | Motion Picture |
| FORMAT | CED |
| DISTRIBUTOR | RCA Videodiscs, 03411, $24.98 |

|   |   |
|---|---|
| TITLE | **JIM FIXX ON RUNNING** |
| TYPE | Instructional |
| FORMAT | Laser CLV |
| PRODUCER | MCA Videodisc, Inc. |
| DISTRIBUTOR | Pioneer Video, Inc., 32-007, $24.95 |

|   |   |
|---|---|
| TITLE | **JINXED** |
| TYPE | Motion Picture |
| FORMAT | CED |
| RETAIL OUTLET | Movies Unlimited Inc., A2-0216, $29.95 |

|   |   |
|---|---|
| TITLE | **JOAN OF ARC** |
| TYPE | Motion Picture |
| FORMAT | CED |
| DISTRIBUTOR | Movies Unlimited |
| RETAIL OUTLET | Movies Unlimited Inc., A2-0182, $29.95 |

|   |   |
|---|---|
| TITLE | **JOB PERFORMANCE PACKAGE (JPP)** |
| DESCRIPTION | A demonstration disc for the Army designed to demonstrate how technical manual maintenance information and on-the-job training can be effectively accessed using an interactive videodisc. |
| TYPE | Technical, Instructional |
| FORMAT | Laser CAV |
| LEVEL | 3 |
| SYSTEM DESCRIPTION | Designed for a Pioneer player and an Apple computer. |
| DEVELOPER | Computer Sciences Corporation |
| CONTACT FOR MORE INFO. | Dr. William A. Stembler |
| PRODUCER | Dr. William A. Stembler |
| AVAILABILITY | Property of U.S. Army. |

| | |
|---:|:---|
| TITLE | **JOB READINESS MASTERY TEST** |
| DESCRIPTION | Along with 5 floppy diskettes, this videodisc presents a comprehensive 15-to-20 hour examination designed to determine whether or not welfare worker trainees are competent to assume the job of a welfare worker in Florida. A large part of the test is a job simulation, which requires the learner to set up appointments, answer the telephone, interview clients, and complete necessary paperwork. |
| TYPE | Instructional, In-house Training |
| FORMAT | Laser CAV |
| LEVEL | 3 |
| SYSTEM DESCRIPTION | Disc is part of 160-hour curriculum delivered on an Apple II/Pioneer 7820-3/single disc drive system along with ancillary printed reference books. |
| DEVELOPER | University of West Florida/Office for Interactive Technology and Training |
| CONTACT FOR MORE INFO. | Richard Smith; Elizabeth Wright, (413) 256-0444 |
| FEATURES | Dual Audio |
| PRODUCER | University of Nebraska/Nebraska Videodisc Design/Production Group |
| AVAILABILITY | Training program produced under contract with the Florida Department of Health and Rehabilitative Services, Tallahassee, Florida, and is not available for general distribution. |

---

| | |
|---:|:---|
| TITLE | **JOE** |
| TYPE | Motion Picture |
| FORMAT | Laser CLV, CED |
| RUNNING TIME | 107 minutes |
| PRODUCER | Vestron Video |
| DISTRIBUTOR | Pioneer Video, Inc., VL6005, $34.95 |
| RETAIL OUTLET | (CED) Movies Unlimited Inc., A3-6005, $29.95 |

---

| | |
|---:|:---|
| TITLE | **JOHN CURRY'S ICE DANCIN'** |
| TYPE | Dance |
| FORMAT | CED |
| DISTRIBUTOR | RCA Videodiscs, 01510, $19.98 |

---

| | |
|---:|:---|
| TITLE | **JOY OF FAMILY** |
| TYPE | Instructional |
| FORMAT | Laser CLV |
| RETAIL OUTLET | UltraVideo, I220-001, $29.95 |

---

| | |
|---:|:---|
| TITLE | **JOY OF RELAXATION, THE** |
| TYPE | Instructional |
| FORMAT | Laser CAV |
| LEVEL | 1 |
| PRODUCER | Optical Programming Associates |
| DISTRIBUTOR | Pioneer Video, Inc., 37-604, $29.95 |

| | |
|---:|:---|
| TITLE | **JUDY GARLAND** |
| TYPE | Music |
| FORMAT | Laser CLV, CED (Stereo, Banded) |
| FEATURES | Stereo, CX Encoded, Black and white |
| RUNNING TIME | 85 minutes |
| PRODUCER | Pioneer Artists |
| DISTRIBUTOR | Pioneer Video, Inc., PA-83-040, $24.95 RCA Videodiscs, 12051, $24.98 |

| | |
|---:|:---|
| TITLE | **JULIA** |
| TYPE | Motion Picture |
| FORMAT | CED |
| RETAIL OUTLET | Movies Unlimited Inc., A1-1091, $29.98 |

| | |
|---:|:---|
| TITLE | **JULIA CHILD - THE FRENCH CHEF, VOL. 1** |
| TYPE | Instructional |
| FORMAT | CED |
| DISTRIBUTOR | RCA Videodiscs, 02017, $19.98 |

| | |
|---:|:---|
| TITLE | **JUNIOR BONNER** |
| TYPE | Motion Picture |
| FORMAT | CED |
| RETAIL OUTLET | Movies Unlimited Inc., A1-8019, $29.98 |

| | |
|---:|:---|
| TITLE | **KANAKO** |
| TYPE | Music, Adult |
| FORMAT | Laser CAV |
| LEVEL | 1 |
| FEATURES | Stereo, CX Encoded |
| PRODUCER | Pioneer Video Imports |
| DISTRIBUTOR | Pioneer Video, Inc., ME-064US, $27.95 |

| | |
|---:|:---|
| TITLE | **KARATE** |
| DESCRIPTION | Interactive movie videodisc on Karate |
| TYPE | Instructional |
| FORMAT | Laser CAV |
| LEVEL | 3 |
| DEVELOPER | Interactive Television Company |
| CONTACT FOR MORE INFO. | Dr. Steven Levin, Interactive Television Company |
| RUNNING TIME | 30 minutes |
| AVAILABILITY | Contact Interactive Television Company |

| | |
|---:|:---|
| TITLE | **KELLY'S HEROES** |
| TYPE | Motion Picture |
| FORMAT | CED |
| RETAIL OUTLET | Movies Unlimited Inc., A2-0168, $39.95 |

| | |
|---:|:---|
| TITLE | **KENNEDY DOSSIER, THE** |
| RELEASE DATE | 1983 |
| PRODUCER | Sheldon Renan |

| | |
|---:|:---|
| TITLE | **KENNEDY SPACE CENTER - PAD 39A** |
| DESCRIPTION | Surrogate travel. |
| TYPE | In-house Training |
| FORMAT | Laser CLV |
| LEVEL | 3 |
| DEVELOPER | Interactive Television Company |
| CONTACT FOR MORE INFO. | Dr. Steven Levin, Interactive Television Company |
| RUNNING TIME | 30 minutes |
| AVAILABILITY | Contact Interactive Television Company. |

| | |
|---:|:---|
| TITLE | **KENNY LOGGINS ALIVE** |
| TYPE | Music |
| FORMAT | Laser CLV, CED |
| FEATURES | Stereo, CX Encoded |
| RUNNING TIME | 59 minutes |
| PRODUCER | Pioneer Artists |
| DISTRIBUTOR | Pioneer Video, Inc., PA-82-019, $24.95 |
| | RCA Videodiscs, 12120, $24.98 |

| | |
|---:|:---|
| TITLE | **KENTUCKIAN, THE** |
| TYPE | Motion Picture |
| FORMAT | CED |
| RETAIL OUTLET | Movies Unlimited Inc., A1-4649, $29.98 |

| | |
|---:|:---|
| TITLE | **KEY LARGO** |
| TYPE | Motion Picture |
| FORMAT | Laser CLV, CED |
| FEATURES | Black and white |
| RUNNING TIME | 101 minutes |
| PRODUCER | CBS Fox Video |
| DISTRIBUTOR | Pioneer Video, Inc., 4594-80, $34.9 |
| | 8 RCA Videodiscs, 01456, $24.98 |

| | |
|---:|:---|
| TITLE | **KIDNAPPED** |
| TYPE | Motion Picture (G) |
| FORMAT | CED |
| PRODUCER | Walt Disney |
| DISTRIBUTOR | RCA Videodiscs, 00706, $24.98 |

| | |
|---:|:---|
| TITLE | **KIDS ARE ALRIGHT, THE** |
| TYPE | Motion Picture, Music |
| FORMAT | CED |
| FEATURES | Stereo, CX Encoded |
| DISTRIBUTOR | RCA Videodiscs, 12097, $24.98 |

| | |
|---:|:---|
| TITLE | **KIDS FROM FAME, THE - LIVE** |
| TYPE | Motion Picture, Music, Dance |
| FORMAT | Laser CLV |
| RETAIL OUTLET | UltraVideo, ML100205, $34.95 |

# Videodiscography

| | |
|---:|:---|
| TITLE | **KILLER FORCE** |
| TYPE | Motion Picture |
| FORMAT | CED |
| RETAIL OUTLET | Movies Unlimited Inc., A3-4041, $29.98 |

| | |
|---:|:---|
| TITLE | **KING AND I, THE** |
| TYPE | Motion Picture, Music |
| FORMAT | Laser CLV, CED (2 Discs) |
| FEATURES | Stereo |
| RUNNING TIME | 133 minutes |
| PRODUCER | Warner Home Video |
| DISTRIBUTOR | Pioneer Video, Inc., 1004-80, $34.98 |
| RETAIL OUTLET | (CED) Movies Unlimited Inc., A1-1004T, $39.98 |

| | |
|---:|:---|
| TITLE | **KING CREOLE** |
| TYPE | Motion Picture, Music |
| FORMAT | CED |
| DISTRIBUTOR | RCA Videodiscs, 01504, $24.98 |

| | |
|---:|:---|
| TITLE | **KING KONG** |
| TYPE | Motion Picture (PG) |
| FORMAT | Laser CLV, CED (2 Discs) |
| RUNNING TIME | 135 minutes |
| PRODUCER | Paramount Home Video |
| DISTRIBUTOR | Pioneer Video, Inc., LV8872-2, $35.95 |
| | RCA Videodiscs, 00647, $34.98 |

| | |
|---:|:---|
| TITLE | **KING KONG - ORIGINAL VERSION** |
| TYPE | Motion Picture |
| FORMAT | CED |
| DISTRIBUTOR | RCA Videodiscs, 00406, $24.98 |

| | |
|---:|:---|
| TITLE | **KING OF COMEDY, THE** |
| TYPE | Motion Picture (PG) |
| FORMAT | Laser CLV, CED |
| PRODUCER | Arnon Milchan |
| DISTRIBUTOR | RCA Videodiscs, 02170, $29.98 |
| RETAIL OUTLET | (CLV) UltraVideo, VLD3192, $29.95 |

| | |
|---:|:---|
| TITLE | **KING OF HEARTS, THE** |
| TYPE | Motion Picture |
| FORMAT | CED |
| RETAIL OUTLET | Movies Unlimited Inc., A1-4529, $29.98 |

| | |
|---:|:---|
| TITLE | **KINGSTON TRIO/ETC.** |
| TYPE | Music |
| FORMAT | Laser CLV |
| FEATURES | Stereo |
| RETAIL OUTLET | UltraVideo, 81-004, $29.95 |

| | |
|---:|:---|
| TITLE | **KISS ME GOODBYE** |
| TYPE | Motion Picture |
| FORMAT | CED |
| RETAIL OUTLET | Movies Unlimited Inc., A1-1217, $29.98 |

| | |
|---:|:---|
| TITLE | **KLAVIER IM HAUS** |
| DESCRIPTION | This film short features the present life-styles and culture of urban Germany. It is an annotated version completely under conputer control and used to teach German culture and language. |
| TYPE | Motion Picture, Adolescents, Research and Development, Instructional |
| FORMAT | Laser CAV |
| LEVEL | 3 |
| SYSTEM DESCRIPTION | Presently functions on a Sony SMC 70 with player and superimpose unit. |
| DEVELOPER | David O. McKay Institute |
| CONTACT FOR MORE INFO. | Larrie E. Gale |
| FEATURES | In German only with English text files. |
| RUNNING TIME | 13 minutes |
| PRODUCER | Defense Language Institute, Goethe Institute, Brigham Young University |

| | |
|---:|:---|
| TITLE | **KLUTE** |
| TYPE | Motion Picture |
| FORMAT | CED |
| DISTRIBUTOR | RCA Videodiscs, 03129, $24.98 |

| | |
|---:|:---|
| TITLE | **KNACK, THE - LIVE AT CARNEGIE HALL** |
| TYPE | Music |
| FORMAT | Laser CLV |
| FEATURES | Stereo, CX Encoded |
| RUNNING TIME | 50 minutes |
| PRODUCER | Pioneer Artists |
| DISTRIBUTOR | Pioneer Video, Inc., PA-82-016, $24.95 |

| | |
|---:|:---|
| TITLE | **KNOCK ON ANY DOOR** |
| TYPE | Motion Picture |
| FORMAT | CED |
| DISTRIBUTOR | RCA Videodiscs, 03043, $24.98 |

| | |
|---:|:---|
| TITLE | **KOTCH** |
| TYPE | Motion Picture |
| FORMAT | CED |
| RETAIL OUTLET | Movies Unlimited Inc., A1-8008, $29.98 |

| | |
|---:|:---|
| TITLE | **KPR/WINTHROP: UNDERSTANDING CONTRAST MEDIA: THE PROMISE OF NON-IONIC AGENTS** |
| DESCRIPTION | This program uses references and tests to introduce the features and benefits of non-ionic contrast media to physicians. |
| TYPE | Medical, Instructional |
| FORMAT | Laser CAV |
| LEVEL | 3 |

| | |
|---|---|
| SYSTEM DESCRIPTION | System is interfaced with the client's custom-built computer, which allows for switching between computer screens and videodisc images. Total control from computer interfaced to Pioneer 8210. |
| DEVELOPER | Showsound, Inc. |
| FEATURES | Picture Stops |
| PRODUCER | KPR Informedia Corporation |
| DISTRIBUTOR | Winthrop Laboratories |

| | |
|---|---|
| TITLE | **KRAFT PROCESS OVERVIEW** |
| DESCRIPTION | The program gives a brief overview of the process by which wood is converted to pulp. The disc is a prototype intended to demonstrate feasibility and to illustrate the functionality of videodisc. |
| TYPE | Research and Development, Technical, Instructional, In-house Training |
| FORMAT | Laser CAV |
| LEVEL | 2 |
| DEVELOPER | Weyerhaeuser Company/Advanced Computing Unit |
| CONTACT FOR MORE INFO. | Sylvia D. Grislis |
| FEATURES | Digital Dump-Pioneer |
| PRODUCER | Richard Crawford, Mediatek, Tacoma |

| | |
|---|---|
| TITLE | **KRAMER VS. KRAMER** |
| TYPE | Motion Picture (PG) |
| FORMAT | Laser CLV, CED |
| PRODUCER | Warner Home Video |
| DISTRIBUTOR | Pioneer Video, Inc., VLD3205, $29.95 |
| | RCA Videodiscs, 03001, $29.98 |

LA BOHEME. *See* BOHEME, LA

LA FILLE MAL GARDEE. *See* FILLE MAL GARDEE, LA

| | |
|---|---|
| TITLE | **LADY CHATTERLEY'S LOVER** |
| TYPE | Motion Picture (R) |
| FORMAT | Laser CLV, CED |
| RUNNING TIME | 102 minutes |
| PRODUCER | MGM/UA Home Video |
| DISTRIBUTOR | Pioneer Video, Inc., ML100184, $34.95 |
| RETAIL OUTLET | (CED) Movies Unlimited Inc., A2-0184, $29.95 |

| | |
|---|---|
| TITLE | **LADY SINGS THE BLUES** |
| TYPE | Motion Picture, Music (R) |
| FORMAT | Laser CLV, CED (2 Discs) |
| RUNNING TIME | 144 minutes |
| PRODUCER | Warner Home Video |
| DISTRIBUTOR | Pioneer Video, Inc., LV8374, $35.95 |
| | RCA Videodiscs, 00616, $34.98 |

| | |
|---|---|
| TITLE | **LANDISC I** |
| DESCRIPTION | Randomly accessible massive visual inventory of Guilderland, NY. User can find a particular neighborhood, street, or house, and move freely from one to the other. |
| TYPE | Surrogate Travel |
| FORMAT | Laser CAV |
| LEVEL | 1,3 |
| DEVELOPER | John Ciampa |
| CONTACT FOR MORE INFO. | John Ciampa or Lamont Seckman |
| PRODUCER | John Ciampa, American Video Institute/Videodisc and Electronic Publishing Laboratory |
| AVAILABILITY | For government client only. |

| | |
|---|---|
| TITLE | **LANDISC II** |
| DESCRIPTION | Randomly accessible massive visual inventory of Rochester, NY. User can find a particular neighborhood, street, or house, and move freely from one to the other. |
| TYPE | Surrogate Travel |
| FORMAT | Laser CAV |
| LEVEL | 1,3 |
| DEVELOPER | John Ciampa |
| CONTACT FOR MORE INFO. | John Ciampa or Lamont Seckman |
| PRODUCER | John Ciampa, American Video Institute/Videodisc and Electronic Publishing Laboratory |
| AVAILABILITY | For government client only. |

| | |
|---|---|
| TITLE | **LANGUAGE AND LEARNING** |
| DESCRIPTION | Parent/child language skills training for parents hearing-impaired children. |
| TYPE | Instructional |
| FORMAT | Laser CAV |
| LEVEL | 1 |
| CONTACT FOR MORE INFO. | Bureau of Education for the Handicapped/University of Nebraska - Barkley Center |
| PRODUCER | University of Nebraska/Nebraska Videodisc Design/Production Group |
| AVAILABILITY | Not available for release or distribution. |

| | |
|---|---|
| TITLE | **LAST AMERICAN VIRGIN, THE** |
| TYPE | Motion Picture |
| FORMAT | CED |
| FEATURES | Stereo |
| RETAIL OUTLET | Movies Unlimited Inc., A2-0190, $29.95 |

| | |
|---|---|
| TITLE | **LAST CHASE, THE** |
| TYPE | Motion Picture |
| FORMAT | CED |
| RETAIL OUTLET | Movies Unlimited Inc., A3-5004, $29.95 |

| | |
|---:|:---|
| TITLE | **LAST OF THE RED HOT LOVERS, THE** |
| TYPE | Motion Picture, Comedy (PG) |
| FORMAT | Laser CLV |
| RUNNING TIME | 96 minutes |
| PRODUCER | Paramount Home Video |
| DISTRIBUTOR | Pioneer Video, Inc., LV8094, $29.95 |

| | |
|---:|:---|
| TITLE | **LAST TANGO IN PARIS** |
| TYPE | Motion Picture, Adult (X) |
| FORMAT | Laser CLV, CED (2 Discs) |
| RUNNING TIME | 129 minutes |
| PRODUCER | Warner Home Video |
| DISTRIBUTOR | Pioneer Video, Inc., 4507-80, $39.98 |
| | RCA Videodiscs, 01447, $39.98 |

| | |
|---:|:---|
| TITLE | **LAST UNICORN, THE** |
| TYPE | Motion Picture |
| FORMAT | CED |
| FEATURES | Stereo |
| RETAIL OUTLET | Movies Unlimited Inc., A1-9054, $29.98 |

| | |
|---:|:---|
| TITLE | **LAST WALTZ, THE** |
| TYPE | Motion Picture, Music (PG) |
| FORMAT | CED |
| DISTRIBUTOR | RCA Videodiscs, 11436 (Stereo, Banded), 01407, $29.98 |

| | |
|---:|:---|
| TITLE | **LAURA** |
| TYPE | Motion Picture |
| FORMAT | CED |
| RETAIL OUTLET | Movies Unlimited Inc., A1-1094, $29.98 |

| | |
|---:|:---|
| TITLE | **LAWRENCE OF ARABIA** |
| TYPE | Motion Picture |
| FORMAT | CED (2 Discs) |
| FEATURES | Stereo |
| DISTRIBUTOR | RCA Videodiscs, 13053, $39.98 |

| | |
|---:|:---|
| TITLE | **LCA PRESENTS FAMILY ENTERTAINMENT PLAYHOUSE** |
| TYPE | Motion Picture (G) |
| FORMAT | CED |
| DISTRIBUTOR | RCA Videodiscs, 02021, $14.98 |

| | |
|---:|:---|
| TITLE | **LEGEND OF THE LONE RANGER, THE** |
| TYPE | Motion Picture |
| FORMAT | CED |
| RETAIL OUTLET | Movies Unlimited Inc., A1-9034, $29.98 |

| | |
|---|---|
| TITLE | **LENNY** |
| TYPE | Motion Picture (R) |
| FORMAT | Laser CLV, CED |
| FEATURES | Black and white |
| RUNNING TIME | 112 minutes |
| PRODUCER | CBS Fox Video |
| DISTRIBUTOR | Pioneer Video, Inc., 4563-80, $29.98 |
| RETAIL OUTLET | (CED) Movies Unlimited Inc., A1-4563, $29.98 |

| | |
|---|---|
| TITLE | **LENNY BRUCE PERFORMANCE** |
| TYPE | Comedy |
| FORMAT | CED |
| RETAIL OUTLET | Movies Unlimited Inc., A3-3000, $29.95 |

| | |
|---|---|
| TITLE | **LET IT BE** |
| TYPE | Motion Picture, Music (G) |
| FORMAT | Laser CLV, CED |
| RUNNING TIME | 80 minutes |
| PRODUCER | Warner Home Video |
| DISTRIBUTOR | Pioneer Video, Inc., 4508-80, $29.98 |
| | RCA Videodiscs, 01411, $24.98 |

| | |
|---|---|
| TITLE | **LET'S BOWL** |
| TYPE | Motion Picture |
| FORMAT | Laser CLV |
| RETAIL OUTLET | UltraVideo, I100-001, $29.95 |

| | |
|---|---|
| TITLE | **LET'S SPEND THE NIGHT TOGETHER** |
| TYPE | Motion Picture, Music (PG) |
| FORMAT | Laser CLV, CED |
| FEATURES | Stereo, CX Encoded |
| RUNNING TIME | 94 minutes |
| PRODUCER | Embassy Home Entertainment |
| DISTRIBUTOR | Pioneer Video, Inc., 12315, $34.95 |
| RETAIL OUTLET | (CED) Movies Unlimited Inc., A4-2316, $29.95 |

| | |
|---|---|
| TITLE | **LIAR'S MOON** |
| TYPE | Motion Picture |
| FORMAT | CED |
| RETAIL OUTLET | Movies Unlimited Inc., A3-4020, $29.98 |

| | |
|---|---|
| TITLE | **LIFE AND WORKS OF MICHELANGELO, THE** |
| DESCRIPTION | Random access visual tour through the life and works of Michelangelo, concentrating on thirty of his major works. |
| TYPE | Art, Children, Adolescents, Instructional, Archival |
| FORMAT | CED |
| LEVEL | 1 |
| DEVELOPER | Video Properties, Inc. |
| CONTACT FOR MORE INFO. | Diane Smook, (212) 930-4767 |

| | |
|---:|:---|
| FEATURES | Chapters, Dual Audio, Programming Band References, Access to Stills by Page Number |
| PRODUCER | Video Properties, Inc. |
| DISTRIBUTOR | RCA Videodiscs, 25003, $29.98 |

| | |
|---:|:---|
| TITLE | **LION IN WINTER, THE** |
| TYPE | Motion Picture |
| FORMAT | CED (2 Discs) |
| DISTRIBUTOR | RCA Videodiscs, 00802, $34.98 |

| | |
|---:|:---|
| TITLE | **LIPSTICK** |
| TYPE | Motion Picture (R) |
| FORMAT | Laser CLV |
| RUNNING TIME | 90 minutes |
| PRODUCER | Paramount Home Video |
| DISTRIBUTOR | Pioneer Video, Inc., LV8882, $29.95 |

| | |
|---:|:---|
| TITLE | **LITTLE CAESAR** |
| TYPE | Motion Picture |
| FORMAT | CED |
| DISTRIBUTOR | RCA Videodiscs, 03412, $24.98 |

| | |
|---:|:---|
| TITLE | **LITTLE DARLINGS** |
| TYPE | Motion Picture, Comedy (R) |
| FORMAT | Laser CLV |
| RUNNING TIME | 95 minutes |
| PRODUCER | Warner Home Video |
| DISTRIBUTOR | Pioneer Video, Inc., LV1301, $29.95 |

| | |
|---:|:---|
| TITLE | **LITTLE HOUSE ON THE PRAIRIE** |
| TYPE | Motion Picture (G) |
| FORMAT | CED |
| DISTRIBUTOR | RCA Videodiscs, 01206, $19.98 |

| | |
|---:|:---|
| TITLE | **LITTLE PRINCE, THE** |
| TYPE | Motion Picture, Music (G) |
| FORMAT | Laser CLV |
| FEATURES | Stereo, CX Encoded |
| RUNNING TIME | 88 minutes |
| PRODUCER | Paramount Home Video |
| DISTRIBUTOR | Pioneer Video, Inc., LV8017, $29.95 |

| | |
|---:|:---|
| TITLE | **LITTLE RIVER BAND LIVE EXPOSURE** |
| TYPE | Music |
| FORMAT | Laser CLV, CED (Stereo, Banded) |
| FEATURES | Stereo, CX Encoded |
| RUNNING TIME | 58 minutes |
| PRODUCER | Pioneer Artists |
| DISTRIBUTOR | Pioneer Video, Inc., PA-83-038, $24.95 |
| | RCA Videodiscs, 12153, $24.98 |

| | |
|---:|:---|
| TITLE | **LIVE AND LET DIE** |
| TYPE | Motion Picture (PG) |
| FORMAT | CED |
| DISTRIBUTOR | RCA Videodiscs, 01424, $29.98 |

| | |
|---:|:---|
| TITLE | **LIVE FROM THE SANTA BARBARA BOWL - BOB MARLEY AND THE WAILERS** |
| TYPE | Music |
| FORMAT | Laser CLV |
| FEATURES | Stereo, CX Encoded |
| RUNNING TIME | 59 minutes |
| PRODUCER | Pioneer Artists |
| DISTRIBUTOR | Pioneer Video, Inc., PA-82-020, $24.95 |

| | |
|---:|:---|
| TITLE | **LIVE INFIDELITY: REO SPEEDWAGON IN CONCERT** |
| TYPE | Music |
| FORMAT | Laser CLV, CED |
| FEATURES | Stereo, CX Encoded |
| RUNNING TIME | 87 minutes |
| PRODUCER | CBS Fox Video |
| DISTRIBUTOR | Pioneer Video, Inc., 7061-80, $29.98 |
| RETAIL OUTLET | (CED) Movies Unlimited Inc., A2-1022, $24.95 |

| | |
|---:|:---|
| TITLE | **LIZA** |
| TYPE | Music |
| FORMAT | Laser CAV |
| LEVEL | 1 |
| FEATURES | Stereo |
| PRODUCER | Pioneer Artists |
| DISTRIBUTOR | Pioneer Video, Inc., PA-81-002, $24.95 |

| | |
|---:|:---|
| TITLE | **LOCIS INTERACTIVE VIDEODISC, THE** |
| DESCRIPTION | This one-sided disc is intended for the Library's clientele who wish to conduct computer searches with the Scorpio and the Mums computer systems. |
| TYPE | Instructional |
| FORMAT | Laser CAV |
| LEVEL | 3 |
| DEVELOPER | Thomas Held, President, Metamedia Systems, Inc. |
| CONTACT FOR MORE INFO. | Same as developer |
| PRODUCER | Same as developer |

| | |
|---:|:---|
| TITLE | **LOGAN'S RUN** |
| TYPE | Motion Picture |
| FORMAT | CED |
| RETAIL OUTLET | Movies Unlimited Inc., A2-1035, $29.95 |

| | |
|---:|:---|
| TITLE | **LOLITA** |
| TYPE | Motion Picture |
| FORMAT | CED (2 Discs) |
| RETAIL OUTLET | Movies Unlimited Inc., A2-0068, $39.95 |

| | |
|---:|:---|
| TITLE | **LONE WOLF MCQUADE** |
| TYPE | Motion Picture |
| FORMAT | CED |
| DISTRIBUTOR | RCA Videodiscs, 01914, $24.98 |

| | |
|---:|:---|
| TITLE | **LONG RIDERS, THE** |
| TYPE | Motion Picture |
| FORMAT | CED |
| DISTRIBUTOR | RCA Videodiscs, 03422, $24.98 |

| | |
|---:|:---|
| TITLE | **LONGEST DAY, THE** |
| TYPE | Motion Picture |
| FORMAT | CED |
| RETAIL OUTLET | Movies Unlimited Inc., A1-0107, $39.98 |

| | |
|---:|:---|
| TITLE | **LONGEST YARD, THE** |
| TYPE | Motion Picture (R) |
| FORMAT | Laser CLV, CED |
| RUNNING TIME | 120 minutes |
| PRODUCER | Warner Home Video |
| DISTRIBUTOR | Pioneer Video, Inc., LV8708, $29.95 |
| | RCA Videodiscs, 00606, $29.98 |

| | |
|---:|:---|
| TITLE | **LOOKING FOR MR. GOODBAR** |
| TYPE | Motion Picture (R) |
| FORMAT | Laser CLV, CED (2 Discs) |
| RUNNING TIME | 136 minutes |
| PRODUCER | Paramount Home Video |
| DISTRIBUTOR | Pioneer Video, Inc., LV8874, $39.95 |
| | RCA Videodiscs, 00617, $34.98 |

| | |
|---:|:---|
| TITLE | **LOONEY, LOONEY, LOONEY BUGS BUNNY MOVIE** |
| TYPE | Motion Picture |
| FORMAT | CED |
| DISTRIBUTOR | RCA Videodiscs, 03154, $19.98 |

| | |
|---:|:---|
| TITLE | **LORD OF THE RINGS, THE** |
| TYPE | Motion Picture (PG) |
| FORMAT | CED (2 Discs) |
| DISTRIBUTOR | RCA Videodiscs, 02009, $39.98 |

| | |
|---:|:---|
| TITLE | **LORETTA** |
| TYPE | Music |
| FORMAT | Laser CAV |
| LEVEL | 1 |
| FEATURES | Stereo |
| PRODUCER | Warner Home Video |
| DISTRIBUTOR | Pioneer Video, Inc., 74-004, $24.95 |

|                        |                                                |
|-----------------------:|:-----------------------------------------------|
| TITLE                  | **LOSIN' IT**                                  |
| TYPE                   | Motion Picture                                 |
| FORMAT                 | CED                                            |
| RETAIL OUTLET          | Movies Unlimited Inc., A4-0616, $29.95         |

|                        |                                                |
|-----------------------:|:-----------------------------------------------|
| TITLE                  | **LOVE AND DEATH**                             |
| TYPE                   | Motion Picture, Comedy (PG)                    |
| FORMAT                 | Laser CLV, CED                                 |
| RUNNING TIME           | 85 minutes                                     |
| PRODUCER               | CBS Fox Video                                  |
| DISTRIBUTOR            | Pioneer Video, Inc., 4585-80, $34.98           |
|                        | RCA Videodiscs, 01482, $24.98                  |

|                        |                                                |
|-----------------------:|:-----------------------------------------------|
| TITLE                  | **LOVE AT FIRST BITE**                         |
| TYPE                   | Motion Picture, Comedy (PG)                    |
| FORMAT                 | Laser CLV, CED                                 |
| DISTRIBUTOR            | RCA Videodiscs, 01904, $24.98                  |
| RETAIL OUTLET          | (CLV) UltraVideo, VL4052, $34.95               |

|                        |                                                |
|-----------------------:|:-----------------------------------------------|
| TITLE                  | **LOVE BUG, THE**                              |
| TYPE                   | Motion Picture (G)                             |
| FORMAT                 | Laser CLV, CED                                 |
| RUNNING TIME           | 108 minutes                                    |
| PRODUCER               | Walt Disney, Warner Home Video                 |
| DISTRIBUTOR            | Pioneer Video, Inc., 12AS, $34.95              |
|                        | RCA Videodiscs, 00703, $24.98                  |

|                        |                                                |
|-----------------------:|:-----------------------------------------------|
| TITLE                  | **LOVE ME TENDER**                             |
| TYPE                   | Motion Picture                                 |
| FORMAT                 | CED                                            |
| RETAIL OUTLET          | Movies Unlimited Inc., A1-1172, $29.98         |

|                        |                                                |
|-----------------------:|:-----------------------------------------------|
| TITLE                  | **LOVE STORY**                                 |
| TYPE                   | Motion Picture (PG)                            |
| FORMAT                 | Laser CLV, CED                                 |
| RUNNING TIME           | 100 minutes                                    |
| PRODUCER               | Paramount Home Video                           |
| DISTRIBUTOR            | Pioneer Video, Inc., LV8006, $29.95            |
|                        | RCA Videodiscs, 00607, $29.98                  |

|                        |                                                |
|-----------------------:|:-----------------------------------------------|
| TITLE                  | **LOVING COUPLES**                             |
| TYPE                   | Motion Picture                                 |
| FORMAT                 | CED                                            |
| RETAIL OUTLET          | Movies Unlimited Inc., A3-6004, $29.95         |

|                        |                                                |
|-----------------------:|:-----------------------------------------------|
| TITLE                  | **LRN COLLEGE DISC**                           |
| DESCRIPTION            | Recruiting materials for high school graduates.|
| TYPE                   | Instructional, Catalog, Promotional            |
| FORMAT                 | Laser CAV                                      |
| LEVEL                  | 2                                              |
| CONTACT FOR MORE INFO. | Aileen McKenna, (919) 493-3479                 |
| FEATURES               | Chapters, Picture Stops                        |

# Videodiscography

|  |  |
|---|---|
| RUNNING TIME | 30 minutes |
| AVAILABILITY | PREF has laser disc equipment in college placement offices across the country. PREF owns the equipment and distributes discs describing a company's recruiting programs into the network. The college receives the service at no charge, and the corporation pays a distribution fee. The Learning Resources Network is PREF's network for high school guidance offices with information on colleges. |

---

|  |  |
|---|---|
| TITLE | **LUCIANO PAVAROTTI IN CONCERT** |
| TYPE | Music |
| FORMAT | CED |
| FEATURES | Stereo, Banded |
| DISTRIBUTOR | RCA Videodiscs, 12165, $24.98 |

---

|  |  |
|---|---|
| TITLE | **LVT-7A1 DRIVER TRAINER** |
| DESCRIPTION | Pictures of the driving course at Camp Pendleton. Includes land driving, jetty sequences, and surf operations. |
| TYPE | Technical, Instructional |
| FORMAT | Laser CAV |
| LEVEL | 3 |
| SYSTEM DESCRIPTION | Driver training system for Marine Corps. Five driver stations and one computer. Full vehicle operation simulation. |
| DEVELOPER | General Dynamics Electronics |
| CONTACT FOR MORE INFO. | John W. Abraham, (619) 573-7704 |
| AVAILABILITY | Available only from Naval Training and Equipment Center, Orlando, Florida. |

---

|  |  |
|---|---|
| TITLE | **M*A*S*H** |
| TYPE | Motion Picture (PG) |
| FORMAT | Laser CLV, CED |
| FEATURES | Stereo, CX Encoded |
| RUNNING TIME | 116 minutes |
| PRODUCER | Warner Home Video |
| DISTRIBUTOR | Pioneer Video, Inc., 1038-80, $39.98 |
| RETAIL OUTLET | (CED) Movies Unlimited Inc., A1-0108, $29.98 |

---

|  |  |
|---|---|
| TITLE | **M*A*S*H, FAREWELL, GOODBYE, AMEN** |
| TYPE | Motion Picture |
| FORMAT | Laser CLV, CED |
| RUNNING TIME | 119 minutes |
| PRODUCER | CBS Fox Video |
| DISTRIBUTOR | Pioneer Video, Inc., 1215-80, $34.98 |
| RETAIL OUTLET | (CED) Movies Unlimited Inc., A1-1215, $29.98 |

---

|  |  |
|---|---|
| TITLE | **M.J. ZINK PRODUCTIONS, INC. - (NOT GIVEN)** |
| DESCRIPTION | Demonstration of a new product. |
| TYPE | Music, Medical, Technical, Archival, Point-of-Purchase, Advertising |
| LEVEL | 1,2,3 |
| DEVELOPER | Mark Bowman, Tom Sales |
| CONTACT FOR MORE INFO. | Jan Morgan, M.J. Zink, (212) 563-5777 |
| PRODUCER | Mark Bowman |
| DISTRIBUTOR | J.P. Stevens/Compucard |

| | |
|---:|:---|
| TITLE | **MA VLAST (MY FATHERLAND)** |
| TYPE | Music |
| FORMAT | Laser CLV |
| FEATURES | Stereo, CX Encoded |
| PRODUCER | Pioneer Video Imports |
| DISTRIBUTOR | Pioneer Video, Inc., MC-034, $29.95 |

| | |
|---:|:---|
| TITLE | **MACARIO** |
| DESCRIPTION | An annotated version of a classic Mexican film used to teach linguistic and cultural facts about the Spanish language and Mexico. |
| TYPE | Motion Picture, Adolescents, Research and Development, Instructional |
| FORMAT | Laser CAV |
| LEVEL | 3 |
| DEVELOPER | David O. McKay Institute |
| CONTACT FOR MORE INFO. | Larrie E. Gale |
| FEATURES | Dual Audio |
| RUNNING TIME | 92 minutes |
| DISTRIBUTOR | Azteca Films |
| AVAILABILITY | Presently negotiating the distribution of this program with a major publishing house |

| | |
|---:|:---|
| TITLE | **MAD MAX** |
| TYPE | Motion Picture |
| FORMAT | Laser CLV |
| RETAIL OUTLET | UltraVideo, VL3040, $34.95 |

| | |
|---:|:---|
| TITLE | **MAGIC** |
| TYPE | Motion Picture (R) |
| FORMAT | Laser CLV |
| RUNNING TIME | 106 minutes |
| PRODUCER | Warner Home Video |
| DISTRIBUTOR | Pioneer Video, Inc., 150150, $29.95 |
| RETAIL OUTLET | (CED) Movies Unlimited Inc., A4-5016, $29.95 |

| | |
|---:|:---|
| TITLE | **MAGIC PONY, THE** |
| TYPE | Motion Picture |
| FORMAT | Laser CLV, CED |
| RUNNING TIME | 80 minutes |
| PRODUCER | Vestron Video |
| DISTRIBUTOR | Pioneer Video, Inc., VL3011, $34.95 |
| RETAIL OUTLET | (CED) Movies Unlimited Inc., A3-3011, $29.95 |

| | |
|---:|:---|
| TITLE | **MAGNIFICENT SEVEN, THE** |
| TYPE | Motion Picture |
| FORMAT | Laser CLV, CED (2 Discs) |
| RUNNING TIME | 127 minutes |
| PRODUCER | CBS Fox Video |
| DISTRIBUTOR | Pioneer Video, Inc., 4553-80, $39.98 |
| | RCA Videodiscs, 01408, $39.98 |

| | |
|---:|:---|
| TITLE | **MAGNUM FORCE** |
| TYPE | Motion Picture, Adult (R) |
| FORMAT | CED |
| DISTRIBUTOR | RCA Videodiscs, 03130, $29.98 |

| | |
|---:|:---|
| TITLE | **MAHOGANY** |
| TYPE | Motion Picture (PG) |
| FORMAT | Laser CLV, CED |
| RUNNING TIME | 109 minutes |
| PRODUCER | Paramount Home Video |
| DISTRIBUTOR | Pioneer Video, Inc., LV8835, $29.95 |
| | RCA Videodiscs, 00690, $24.98 |

| | |
|---:|:---|
| TITLE | **MAIN EVENT, THE** |
| TYPE | Motion Picture (PG) |
| FORMAT | CED |
| PRODUCER | Warner Home Video |
| DISTRIBUTOR | RCA Videodiscs, 03131, $24.98 |

| | |
|---:|:---|
| TITLE | **MAKING LOVE** |
| TYPE | Motion Picture |
| FORMAT | CED |
| RETAIL OUTLET | Movies Unlimited Inc., A1-1146, $29.98 |

| | |
|---:|:---|
| TITLE | **MAKING OF RAIDERS OF THE LOST ARK, THE** |
| TYPE | Motion Picture |
| FORMAT | Laser CLV, Laser CAV, CED |
| RUNNING TIME | 58 minutes |
| PRODUCER | Paramount Home Video |
| DISTRIBUTOR | Pioneer Video, Inc., LV83049, $29.95 |
| | RCA Videodiscs, 00676, $24.98 |

| | |
|---:|:---|
| TITLE | **MAKING OF STAR WARS AND THE EMPIRE STRIKES BACK, THE** |
| TYPE | Motion Picture |
| FORMAT | Laser CLV, CED |
| FEATURES | Stereo |
| RUNNING TIME | 100 minutes |
| PRODUCER | Warner Home Video |
| DISTRIBUTOR | Pioneer Video, Inc., 1052-80, $29.98 |
| RETAIL OUTLET | (CED) Movies Unlimited Inc., A1-1112, $29.98 |

| | |
|---:|:---|
| TITLE | **MALTESE FALCON, THE** |
| TYPE | Motion Picture |
| FORMAT | Laser CLV, CED |
| FEATURES | Black and white |
| RUNNING TIME | 101 minutes |
| PRODUCER | CBS Fox Video |
| DISTRIBUTOR | Pioneer Video, Inc., 4530-80, $34.98 |
| | RCA Videodiscs, 01439, $24.98 |

| | |
|---:|:---|
| TITLE | **MAN FOR ALL SEASONS, A** |
| TYPE | Motion Picture (G) |
| FORMAT | Laser CLV, CED |
| DISTRIBUTOR | RCA Videodiscs, 03016, $24.98 |
| RETAIL OUTLET | (CLV) UltraVideo, VLD3332, $29.95 |

| | |
|---:|:---|
| TITLE | **MAN OF LA MANCHA** |
| TYPE | Motion Picture |
| FORMAT | CED (2 Discs) |
| FEATURES | Stereo |
| RETAIL OUTLET | Movies Unlimited Inc., A1-4575, $39.98 |

| | |
|---:|:---|
| TITLE | **MAN ON THE MOON** |
| TYPE | Motion Picture |
| FORMAT | CED |
| RETAIL OUTLET | Movies Unlimited Inc., A2-1041, $24.95 |

| | |
|---:|:---|
| TITLE | **MAN WHO CAME TO DINNER** |
| TYPE | Motion Picture |
| FORMAT | CED |
| DISTRIBUTOR | RCA Videodiscs, 03413, $24.98 |

| | |
|---:|:---|
| TITLE | **MAN WHO SHOT LIBERTY VALANCE, THE** |
| TYPE | Motion Picture |
| FORMAT | Laser CLV, CED |
| FEATURES | Black and white |
| RUNNING TIME | 118 minutes |
| PRODUCER | Paramount Home Video |
| DISTRIBUTOR | Pioneer Video, Inc., LV6114, $29.95 |
| | RCA Videodiscs, 00654, $24.98 |

| | |
|---:|:---|
| TITLE | **MAN WITH THE GOLDEN ARM, THE** |
| TYPE | Motion Picture |
| FORMAT | CED |
| RETAIL OUTLET | Movies Unlimited Inc., A1-6101, $29.98 |

| | |
|---:|:---|
| TITLE | **MAN WITH THE GOLDEN GUN, THE** |
| TYPE | Motion Picture (PG) |
| FORMAT | Laser CLV, CED |
| RUNNING TIME | 119 minutes |
| PRODUCER | CBS Fox Video |
| DISTRIBUTOR | Pioneer Video, Inc., 4606-80, $34.98 |
| | RCA Videodiscs, 01452, $29.98 |

| | |
|---:|:---|
| TITLE | **MAN WITH TWO BRAINS, THE** |
| TYPE | Motion Picture |
| FORMAT | Laser CLV |
| RETAIL OUTLET | UltraVideo, W11319, $34.95 |

Videodiscography

| | |
|---|---|
| TITLE | **MANDINGO** |
| TYPE | Motion Picture (R) |
| FORMAT | Laser CLV, CED |
| RUNNING TIME | 127 minutes |
| PRODUCER | Paramount Home Video |
| DISTRIBUTOR | Pioneer Video, Inc., LV8771-2, $39.95 |
| | RCA Videodiscs, 03620, $24.98 |

| | |
|---|---|
| TITLE | **MANHATTAN TRANSFER** |
| TYPE | Music |
| FORMAT | Laser CLV |
| FEATURES | Stereo, CX Encoded |
| RUNNING TIME | 58 minutes |
| PRODUCER | Pioneer Artists |
| DISTRIBUTOR | Pioneer Video, Inc., PA-83-029, $24.95 |

| | |
|---|---|
| TITLE | **MANON** |
| TYPE | Music, Dance |
| FORMAT | Laser CLV |
| FEATURES | Stereo, CX Encoded |
| RUNNING TIME | 112 minutes |
| PRODUCER | Pioneer Artists |
| DISTRIBUTOR | Pioneer Video, Inc., PA-83-047, $34.95 |

| | |
|---|---|
| TITLE | **MANUAL SEAT/MAN SEPARATION FROM A MARTIN-BAKER EJECTION SEAT** |
| DESCRIPTION | Teaches student how to get out of the Martin- Baker Ejection Seat and deploy his parachute when the automatic system fails. Tests student for 100% understanding. |
| TYPE | Medical, Technical, Instructional |
| FORMAT | Laser CAV |
| LEVEL | 3 |
| SYSTEM DESCRIPTION | Pioneer 7820 Model 2 player, Apple IIe with 2 drives, touch panel (infrared) monitor, all in one cabinet. |
| DEVELOPER | Naval Health Sciences Education & Training Command, Camis System for Naval Physiology Training Program (NAPTP) |
| CONTACT FOR MORE INFO. | Carl B. Black, AV Program Mgr, CODE 0042, Naval Medical Center, Bethesda, MD 20814 (301) 295-5593 |
| FEATURES | Stereo, Dual Audio, Touch Panel Monitor |
| PRODUCER | Creativision, Inc. as subcontractor to IIAT |
| DISTRIBUTOR | IIAT for the NAPTP |

| | |
|---|---|
| TITLE | **MANY ADVENTURES OF WINNIE THE POOH, THE** |
| TYPE | Motion Picture (G) |
| FORMAT | Laser CLV, CED |
| RUNNING TIME | 74 minutes |
| PRODUCER | Warner Home Video |
| DISTRIBUTOR | Pioneer Video, Inc., 25AS, $34.95 |
| | RCA Videodiscs, 00725, $24.98 |

| | |
|---:|:---|
| TITLE | **MANY ROADS TO MURDER** |
| DESCRIPTION | Branched mystery game with 16 possible story lines. Object is to guess who committed a murder, motive, and method. |
| TYPE | Game, Adolescents |
| FORMAT | CED |
| LEVEL | 1 |
| DEVELOPER | VIDMAX |
| CONTACT FOR MORE INFO. | Diane Smook, RCA, (212) 930-4767 |
| FEATURES | Chapters, Picture Stops, Programming Band Sequences |
| PRODUCER | VIDMAX |
| DISTRIBUTOR | RCA Videodiscs, 25002, $29.95 |

| | |
|---:|:---|
| TITLE | **MARATHON MAN** |
| TYPE | Motion Picture (R |
| FORMAT | Laser CLV, CED |
| RUNNING TIME | 125 minutes |
| PRODUCER | Warner Home Video |
| DISTRIBUTOR | Pioneer Video, Inc., LV8789, $35.95 |
| | RCA Videodiscs, 00648, $29.98 |

| | |
|---:|:---|
| TITLE | **MARITZ COMMUNICATIONS CO. - (NUMEROUS FOR BUSINESS AND INDUSTRY)** |
| DESCRIPTION | Sales training, service training, product training, customer information. |
| TYPE | Game, Research and Development, Technical, Instructional, In-house Training, Archival, Catalog, Point-of-Purchase, Advertising, Promotional |
| FORMAT | Laser CAV |
| LEVEL | 2, 3 |
| SYSTEM DESCRIPTION | Proprietary |
| FEATURES | All |

| | |
|---:|:---|
| TITLE | **MARTY** |
| TYPE | Motion Picture |
| FORMAT | CED |
| RETAIL OUTLET | Movies Unlimited Inc., A1-4634, $29.98 |

| | |
|---:|:---|
| TITLE | **MARY POPPINS** |
| TYPE | Motion Picture (G) |
| FORMAT | Laser CLV, CED (2 Discs) |
| FEATURES | Stereo |
| RUNNING TIME | 139 minutes |
| PRODUCER | Walt Disney |
| DISTRIBUTOR | Pioneer Video, Inc., 42023AS, $44.95 |
| | RCA Videodiscs, 00714, $39.98 |

| | |
|---:|:---|
| TITLE | **MARY TYLER MOORE SHOW, VOL. 1, THE** |
| TYPE | Comedy |
| FORMAT | CED |
| DISTRIBUTOR | RCA Videodiscs, 02038, $19.98 |

| | |
|---:|:---|
| TITLE | **MARYLAND INSTRUCTIONAL TELEVISION** |
| DESCRIPTION | Provides a 10-minute instructional television program and samplers of MITV's K - 12 and inservice programs. |
| TYPE | Children, Adolescents, Instructional, Catalog |
| FORMAT | Laser CAV |
| LEVEL | 1 |
| DEVELOPER | Maryland Instructional Television |
| CONTACT FOR MORE INFO. | Dr. Martha Cammarata |
| FEATURES | Chapters |
| RUNNING TIME | 29 minutes 35 seconds |
| PRODUCER | Info-Disc Corporation |

| | |
|---:|:---|
| TITLE | **MASTER COOKING COURSE, THE** |
| TYPE | Instructional |
| FORMAT | Laser CAV |
| LEVEL | 1 |
| PRODUCER | Optical Programming Associates |
| DISTRIBUTOR | Pioneer Video, Inc., 31-615, $29.95 |

| | |
|---:|:---|
| TITLE | **MASTERS OF THE DRAGON** |
| DESCRIPTION | Arcade game using live footage and animation of karate. |
| TYPE | Game |
| RELEASE DATE | March 1984 |
| FORMAT | Laser CAV |
| LEVEL | 3 |
| SYSTEM DESCRIPTION | Proprietary |
| DEVELOPER | C.R.T. Company |
| CONTACT FOR MORE INFO. | Bob Newman, (303) 455-1122 |
| FEATURES | Real-time Interaction |
| PRODUCER | Bob Newman |
| DISTRIBUTOR | C.R.T. Company, (303) 431-4274 |
| AVAILABILITY | Rent |
| SUGGESTED PRICE | $7,000 |

| | |
|---:|:---|
| TITLE | **MATERIAL SCIENCE VIDEODISC** |
| DESCRIPTION | To explain, via dramatized extracts of a forensic court case, why failures occur in components, what tests are required to diagnose these failures, and the basics of material science. |
| TYPE | Technical, Instructional, In-house Training |
| RELEASE DATE | 3/84 |
| FORMAT | Laser CAV |
| LEVEL | 3 |
| SYSTEM DESCRIPTION | Philips industrial laser vision player driven by a microcomputer. |
| DEVELOPER | BBC Open University Productions |
| CONTACT FOR MORE INFO. | David Nelson |
| FEATURES | Picture Stops, Dual Audio |
| PRODUCER | Martin Wright |
| DISTRIBUTOR | Open University Education Enterprises |

| | |
|---|---|
| TITLE | **MATH IN BIOLOGY** |
| DESCRIPTION | This program combines a videodisc and a microcomputer to provide interactive basic math instruction. The target groups for the instruction are women and minorities who seek a career in the life and health sciences, but have inadequate math skills. It is applicable to others as well. Biology examples are used throughout. |
| TYPE | Adolescents, Instructional |
| RELEASE DATE | 3/84 |
| FORMAT | Laser CAV |
| LEVEL | 3 |
| DEVELOPER | University of Washington Educational Assessment Center and VideoActive |
| CONTACT FOR MORE INFO. | Dr. Gerald M. Gillmore, PB-30, University of Washington |
| FEATURES | Dual Audio, Sound Over Still |
| PRODUCER | Educational Assessment Center |
| DISTRIBUTOR | Same as producer |
| AVAILABILITY | Sale, Rent |
| SUGGESTED PRICE | $450 |

| | |
|---|---|
| TITLE | **MATILDA** |
| TYPE | Motion Picture |
| FORMAT | CED |
| RETAIL OUTLET | Movies Unlimited Inc., A3-4033, $29.95 |

| | |
|---|---|
| TITLE | **MAUSOLEUM** |
| TYPE | Motion Picture |
| FORMAT | CED |
| RETAIL OUTLET | Movies Unlimited Inc., A4-0876, $29.95 |

| | |
|---|---|
| TITLE | **MAZE FEATURING FRANKIE BEVERLY. HAPPY FEELIN'S- LIVE IN NEW ORLEANS** |
| TYPE | Music |
| FORMAT | Laser CLV |
| FEATURES | Stereo, CX Encoded |
| RUNNING TIME | 57 minutes |
| PRODUCER | Pioneer Artists |
| DISTRIBUTOR | Pioneer Video, Inc., PA-82-023, $24.95 |

| | |
|---|---|
| TITLE | **MAZE MANIA: THE AMAZING MAZE GAME** |
| TYPE | Game |
| FORMAT | Laser CAV |
| LEVEL | 1 |
| PRODUCER | Optical Programming Associates |
| DISTRIBUTOR | Pioneer Video, Inc., 37-603, $29.95 |

| | |
|---|---|
| TITLE | **MEATBALLS** |
| TYPE | Motion Picture, Comedy |
| FORMAT | Laser CLV, CED |
| RUNNING TIME | 92 minutes |
| PRODUCER | Warner Home Video |
| DISTRIBUTOR | Pioneer Video, Inc., VL6009, $34.95 |
| RETAIL OUTLET | (CED) Movies Unlimited Inc., A3-6009, $29.95 |

| | |
|---|---|
| TITLE | **MEDICAL MICROSCOPY** |
| DESCRIPTION | Hematology slide material. |
| TYPE | Medical, Archival, Other |
| FORMAT | Laser CAV |
| LEVEL | 1, 2, 3 |
| SYSTEM DESCRIPTION | Computer text and graphics overlay using Apple II and Microkeyer. Disc can be used without this system, but it is necessary for the highest level of performance. |
| CONTACT FOR MORE INFO. | Dr. James Fine or Dr. Ed Ashwood, (206) 543-9181 |
| PRODUCER | University of Washington/Health Sciences Learning Resources Center |
| DISTRIBUTOR | John Bolles, University of Washington/Center for Educational Resources |

| | |
|---|---|
| TITLE | **MEET ME IN ST. LOUIS** |
| TYPE | Motion Picture |
| FORMAT | CED |
| RETAIL OUTLET | Movies Unlimited Inc., A2-0005, $29.95 |

| | |
|---|---|
| TITLE | **MEET MR. WASHINGTON/MEET MR. LINCOLN** |
| TYPE | Instructional |
| FORMAT | CED |
| DISTRIBUTOR | RCA Videodiscs, 01205, $14.98 |

| | |
|---|---|
| TITLE | **MEGAFORCE** |
| TYPE | Motion Picture (PG) |
| FORMAT | CED |
| FEATURES | Dolby Stereo, Introvision |
| PRODUCER | Twentieth Century Fox |
| RETAIL OUTLET | Movies Unlimited Inc., A1-1182, $29.98 |

| | |
|---|---|
| TITLE | **MEJORE SU PRONUNCIACION (IMPROVE YOUR PRONOUNCIATION)** |
| DESCRIPTION | An "intelligent" disc |
| TYPE | Adolescents, Instructional |
| LEVEL | 2 |
| CONTACT FOR MORE INFO. | Corporation for Public Broadcasting - University of Nebraska KUON-TV |
| PRODUCER | University of Nebraska/Nebraska Videodisc Design/Production Group |
| AVAILABILITY | Other |
| SPECIAL CONDITIONS | Yes |

| | |
|---|---|
| TITLE | **MEL TORME/DELLA REESE IN CONCERT** |
| TYPE | Music |
| FORMAT | Laser CLV |
| FEATURES | Stereo |
| PRODUCER | Warner Home Video |
| DISTRIBUTOR | Pioneer Video, Inc., 74-009, $24.95 |

| | |
|---|---|
| TITLE | **MELVIN AND HOWARD** |
| TYPE | Motion Picture (R) |
| FORMAT | Laser CLV |
| RUNNING TIME | 95 minutes |
| PRODUCER | Warner Home Video |
| DISTRIBUTOR | Pioneer Video, Inc., 10-031, $29.95 |

| | |
|---:|:---|
| TITLE | **MEMORIES OF SUPER BOWL** |
| TYPE | Motion Picture |
| FORMAT | Laser CLV |
| RETAIL OUTLET | UltraVideo, NF-005, $24.95 |

| | |
|---:|:---|
| TITLE | **MERRILL LYNCH** |
| DESCRIPTION | Recruiting materials for college graduates. |
| TYPE | Instructional, Catalog, Promotional |
| FORMAT | Laser CAV |
| LEVEL | 2 |
| CONTACT FOR MORE INFO. | Aileen McKenna, (919) 493-3479 |
| FEATURES | Chapters, Picture Stops |
| RUNNING TIME | 30 minutes |
| AVAILABILITY | PREF has laser disc equipment in college placement offices across the country. PREF owns the equipment and distributes discs describing a company's recruiting programs into the network. The college receives the service at no charge, and the corporation pays a distribution fee. |

| | |
|---:|:---|
| TITLE | **MICK FLEETWOOD - THE VISITOR** |
| TYPE | Music |
| FORMAT | CED |
| DISTRIBUTOR | RCA Videodiscs, 02125, $19.98 |

| | |
|---:|:---|
| TITLE | **MICKEY MOUSE AND DONALD DUCK CARTOONS** |
| TYPE | Motion Picture (G) |
| FORMAT | Laser CLV |
| RUNNING TIME | 82 minutes |
| PRODUCER | Warner Home Video |
| DISTRIBUTOR | Pioneer Video, Inc., 69AS, $34.95 |

| | |
|---:|:---|
| TITLE | **MICKEY MOUSE AND DONALD DUCK CARTOONS** |
| TYPE | Motion Picture (G) |
| FORMAT | Laser CLV |
| RUNNING TIME | 44 minutes |
| PRODUCER | Walt Disney |
| DISTRIBUTOR | Pioneer Video, Inc., 42034AS, $24.95 |

| | |
|---:|:---|
| TITLE | **MICKEY MOUSE AND DONALD DUCK COLLECTION 1** |
| TYPE | Motion Picture |
| FORMAT | Laser CLV |
| RETAIL OUTLET | UltraVideo, 069AS, $34.95 |

| | |
|---:|:---|
| TITLE | **MICKEY MOUSE AND DONALD DUCK COLLECTION 2** |
| TYPE | Motion Picture |
| FORMAT | Laser CLV |
| RETAIL OUTLET | UltraVideo, 070AS, $34.95 |

| | |
|---:|:---|
| TITLE | **MICKEY MOUSE AND DONALD DUCK VOLUME 3** |
| TYPE | Motion Picture |
| FORMAT | Laser CLV |
| RETAIL OUTLET | UltraVideo, 034AS, $24.95 |

| | |
|---:|:---|
| TITLE | **MIDNIGHT COWBOY** |
| TYPE | Motion Picture (R) |
| FORMAT | CED |
| DISTRIBUTOR | RCA Videodiscs, 01485, $29.98 |

| | |
|---:|:---|
| TITLE | **MIDNIGHT EXPRESS** |
| TYPE | Motion Picture (R) |
| FORMAT | Laser CLV, CED |
| PRODUCER | Warner Home Video |
| DISTRIBUTOR | Pioneer Video, Inc., VLD3275, $34.95 |
| | RCA Videodiscs, 03009, $29.98 |

| | |
|---:|:---|
| TITLE | **MILDRED PIERCE** |
| TYPE | Motion Picture |
| FORMAT | CED |
| DISTRIBUTOR | RCA Videodiscs, 03414, $24.98 |

| | |
|---:|:---|
| TITLE | **MILES LEARNING CENTER: APPENDECTOMY** |
| DESCRIPTION | Disc shows a doctor performing an appendectomy on a patient. There is operating footage with an audio voice-over explaining the procedures being done. |
| TYPE | Medical, Instructional |
| RELEASE DATE | 4/81 |
| FORMAT | Laser CAV |
| LEVEL | 2 |
| CONTACT FOR MORE INFO. | Dr. Campbell Moses, Medicus Intercon, (212) 909-9332 |
| FEATURES | Digital Dump-Pioneer |
| PRODUCER | Jeff Silverstein, Fusion Media, Inc. |
| DISTRIBUTOR | Robert Mehan, Videodisc Coordinator, Miles Pharmaceuticals |

| | |
|---:|:---|
| TITLE | **MILES LEARNING CENTER ILEOCOLIC RESECTION FOR CHRON'S DISEASE** |
| DESCRIPTION | Disc shows a doctor performing surgery with a voice-over describing techniques and procedures. The disc includes questions and answers for testing. If a question is answered incorrectly, the disc searches back to show the correct sequence and moves on to the next question. |
| TYPE | Medical, Instructional |
| RELEASE DATE | 4/81 |
| FORMAT | Laser CAV |
| LEVEL | 2 |
| FEATURES | Digital Dump-Pioneer |
| PRODUCER | Jeff Silverstein, Fusion Media, Inc. |
| DISTRIBUTOR | Mr. Robert Mehan, Videodisc Coordinator, Miles Pharmaceuticals |

| | |
|---:|:---|
| TITLE | **MILES LEARNING CENTER: MEDICAL APPLICATIONS/HEMATOLOGY** |
| DESCRIPTION | This disc contains five different parts. The first three parts are slide banks containing 6,000 frames of information on: a) the American Society of Hematology, b) the World Health Organization's International Histologic Classification of Tumours, and c) Western Universities Physical Diagnosis. Part 4 is a film on "Red Cell Shapes in Disease," and Part 5 is a film on "Venipuncture: the Vacutainer System." |
| TYPE | Medical, Instructional |
| RELEASE DATE | 4/81 |

| | |
|---|---|
| FORMAT | Laser CAV |
| LEVEL | 2 |
| CONTACT FOR MORE INFO. | Dr. Campbell Moses, Medicus Intercon, (212) 909-9332 |
| FEATURES | Digital Dump-Pioneer |
| PRODUCER | Jeff Silverstein, Fusion Media Inc. |
| DISTRIBUTOR | Robert Mehan, Videodisc Coordinator, Miles Pharmaceuticals |

| | |
|---|---|
| TITLE | **MILES LEARNING CENTER: ORIENTATION AND OPERATING INSTRUCTIONS** |
| DESCRIPTION | This disc introduces videodisc technology and the Miles Learning Center to residents, doctors, nurses, and interns. Includes operating instructions for the disc player and practice sessions on the disc. |
| TYPE | Medical, Technical, Instructional |
| RELEASE DATE | 4/81 |
| FORMAT | Laser CAV |
| LEVEL | 2 |
| FEATURES | Digital Dump-Pioneer |
| PRODUCER | Jeff Silverstein, Fusion Media, Inc. |
| DISTRIBUTOR | Mr. Robert Mehan, Videodisc Coordinator, Miles Pharmaceuticals |

| | |
|---|---|
| TITLE | **MILTON KEYNES INFORMATION TECHNOLOGY VIDEODISC** |
| DESCRIPTION | This disc was used in a six-month exhibition to simulate all the uses to which a television screen might be put by 1987. Included were multiple TV channels, interactive shopping, voting, education and surrogate journeys. |
| TYPE | Technical, Instructional, In-house Training |
| RELEASE DATE | 11/82 |
| FORMAT | Laser CAV |
| LEVEL | 3 |
| SYSTEM DESCRIPTION | Philips industrial laser vision player, keypad or microcomputer keyboard controlled. |
| DEVELOPER | BBC Open University Production Centre |
| CONTACT FOR MORE INFO. | David Nelson |
| FEATURES | Dual Audio |
| PRODUCER | David Nelson/Martin Wright |
| DISTRIBUTOR | Open University Educational Enterprises |

| | |
|---|---|
| TITLE | **MIRACLE OF LAKE PLACID - WINTER OLYMPICS 1980, THE** |
| TYPE | Motion Picture |
| FORMAT | CED |
| DISTRIBUTOR | RCA Videodiscs, 01702, $19.98 |

| | |
|---|---|
| TITLE | **MISFITS, THE** |
| TYPE | Motion Picture |
| FORMAT | CED |
| DISTRIBUTOR | RCA Videodiscs, 01487, $24.98 |

| | |
|---|---|
| TITLE | **MISS PEACH & KELLY SCHOOL GIRLS** |
| TYPE | Motion Picture |
| FORMAT | CED |
| RETAIL OUTLET | Movies Unlimited Inc., A1-7124, $29.98 |

| | |
|---:|:---|
| TITLE | **MISSING** |
| TYPE | Motion Picture (PG) |
| FORMAT | Laser CLV, CED |
| RUNNING TIME | 122 minutes |
| PRODUCER | Warner Home Video |
| DISTRIBUTOR | Pioneer Video, Inc., 10-034, $29.95 |
| RETAIL OUTLET | (CED) Movies Unlimited Inc., A5-0034, $34.98 |

| | |
|---:|:---|
| TITLE | **MISSION GALACTICA: THE CYLON ATTACK** |
| TYPE | Motion Picture (G) |
| FORMAT | Laser CLV |
| RUNNING TIME | 108 minutes |
| PRODUCER | Warner Home Video |
| DISTRIBUTOR | Pioneer Video, Inc., 13-003, $29.95 |

| | |
|---:|:---|
| TITLE | **MISSIONARY, THE** |
| TYPE | Motion Picture, Comedy |
| FORMAT | CED |
| DISTRIBUTOR | RCA Videodiscs, 02178, $24.98 |

| | |
|---:|:---|
| TITLE | **MISSOURI BREAKS** |
| TYPE | Motion Picture |
| FORMAT | CED |
| RETAIL OUTLET | Movies Unlimited Inc., A1-4578, $29.98 |

| | |
|---:|:---|
| TITLE | **MODERN PROBLEMS** |
| TYPE | Motion Picture, Comedy |
| FORMAT | CED |
| RETAIL OUTLET | Movies Unlimited Inc., A1-1129, $29.98 |

| | |
|---:|:---|
| TITLE | **MODERN TIMES** |
| TYPE | Motion Picture, Comedy |
| FORMAT | CED |
| DISTRIBUTOR | RCA Videodiscs, 02033, $19.98 |

| | |
|---:|:---|
| TITLE | **MOMMIE DEAREST** |
| TYPE | Motion Picture (PG) |
| FORMAT | Laser CLV, CED (2 Discs) |
| RUNNING TIME | 129 minutes |
| PRODUCER | Warner Home Video |
| DISTRIBUTOR | Pioneer Video, Inc., LV1263, $35.95 |
| | RCA Videodiscs, 00677, $34.98 |

| | |
|---:|:---|
| TITLE | **MONSIGNOR** |
| TYPE | Motion Picture |
| FORMAT | CED |
| RETAIL OUTLET | Movies Unlimited Inc., A1-1108, $29.98 |

| | |
|---:|:---|
| TITLE | **MONTEVIDISCO (MALE VERSION)** |
| TYPE | Adolescents, Research and Development, Instructional |
| DESCRIPTION | Through the capabilities of interactive videodisc, the viewers of "Montevidisco" experience a simulated visit to and simulated conversations with the inhabitants of a town in Northern Mexico. The discs was filmed on location in Hermosillo. Due to the gender- intensive nature of Spanish, there are two versions, male and female. |
| FORMAT | Laser CAV |
| LEVEL | 3 |
| DEVELOPER | David O. McKay Institute |
| CONTACT FOR MORE INFO. | Larrie E. Gale |
| RUNNING TIME | 120 minutes |
| PRODUCER | Brigham Young University and FIPSE |
| DISTRIBUTOR | Two publishers have expressed an interest in marketing the program. |

| | |
|---:|:---|
| TITLE | **MONTY PYTHON - LIVE AT THE HOLLYWOOD BOWL** |
| TYPE | Motion Picture, Comedy |
| FORMAT | CED |
| FEATURES | Banded |
| DISTRIBUTOR | RCA Videodiscs, 02144, $24.98 |

| | |
|---:|:---|
| TITLE | **MONTY PYTHON AND THE HOLY GRAIL** |
| TYPE | Motion Picture, Comedy (PG) |
| FORMAT | Laser CLV, CED |
| PRODUCER | RCA/Columbia Pictures Home Video |
| DISTRIBUTOR | Pioneer Video, Inc., VLD3337, $29.95<br>RCA Videodiscs, 03040, $24.98 |

| | |
|---:|:---|
| TITLE | **MONTY PYTHON'S LIFE OF BRIAN** |
| TYPE | Motion Picture, Comedy (R) |
| FORMAT | CED |
| FEATURES | Stereo, CX Encoded |
| DISTRIBUTOR | RCA Videodiscs, 13132, $24.98 |

| | |
|---:|:---|
| TITLE | **MOON IS BLUE, THE** |
| TYPE | Motion Picture, Comedy |
| FORMAT | Laser CLV, CED |
| FEATURES | Black and white |
| RUNNING TIME | 100 minutes |
| PRODUCER | Warner Home Video |
| DISTRIBUTOR | Pioneer Video, Inc., 6102-80, $29.98 |
| RETAIL OUTLET | (CED) Movies Unlimited Inc., A1-6102, $29.98 |

| | |
|---:|:---|
| TITLE | **MOONLIGHTING** |
| TYPE | Motion Picture (PG) |
| FORMAT | Laser CLV |
| RUNNING TIME | 97 minutes |
| PRODUCER | MCA Videodisc, Inc. |
| DISTRIBUTOR | Pioneer Video, Inc., 10-035, $29.95 |

| | |
|---:|:---|
| TITLE | **MOON RAKER** |
| TYPE | Motion Picture (PG) |
| FORMAT | Laser CLV, CED (2 Discs, Stereo, CX Encoded) |
| FEATURES | Stereo |
| RUNNING TIME | 126 minutes |
| PRODUCER | CBS Fox Video |
| DISTRIBUTOR | Pioneer Video, Inc., 4636-80, $39.98 |
| | RCA Videodiscs, 11434, $39.98 |

| | |
|---:|:---|
| TITLE | **MORE ADVENTURES WITH DUSTY** |
| DESCRIPTION | Career education for hearing-impaired students. |
| TYPE | Adolescents, Instructional |
| FORMAT | Laser CAV |
| LEVEL | 3 |
| CONTACT FOR MORE INFO. | California School for the Deaf |
| PRODUCER | University of Nebraska/Nebraska Videodisc Design/Production Group |
| AVAILABILITY | Not available for release or distribution. |

| | |
|---:|:---|
| TITLE | **MOSES** |
| TYPE | Motion Picture |
| FORMAT | CED (2 Discs) |
| RETAIL OUTLET | Movies Unlimited Inc., A1-9047, $39.98 |

| | |
|---:|:---|
| TITLE | **MOUNTED LAND NAVIGATION** |
| DESCRIPTION | Instruction/training videodisc for system on mounted land navigation. |
| TYPE | In-house Training |
| FORMAT | Laser CAV |
| LEVEL | 3 |
| DEVELOPER | Interactive Television Company |
| CONTACT FOR MORE INFO. | Dr. Steven Levin, Interactive Television Company |
| RUNNING TIME | 30 minutes |
| AVAILABILITY | Contact Interactive Television Company. |

| | |
|---:|:---|
| TITLE | **MOVIE MOVIE** |
| TYPE | Motion Picture, Comedy |
| FORMAT | CED |
| DISTRIBUTOR | RCA Videodiscs, 00506, $24.98 |

| | |
|---:|:---|
| TITLE | **MR. MAGOO IN SHERWOOD FOREST** |
| TYPE | Motion Picture (G) |
| FORMAT | Laser CLV, CED |
| RUNNING TIME | 83 minutes |
| PRODUCER | Paramount Home Video |
| DISTRIBUTOR | Pioneer Video, Inc., LV2320A, $29.95 |
| | RCA Videodiscs, 03619, $19.98 |

| | |
|---:|:---|
| TITLE | **MR. MAGOO, VOL. 1** |
| TYPE | Motion Picture |
| FORMAT | CED |
| RETAIL OUTLET | Movies Unlimited Inc., A1-3039, $19.98 |

| | |
|---|---|
| TITLE | **MR. MAGOO'S CHRISTMAS** |
| TYPE | Motion Picture |
| FORMAT | Laser CLV |
| RETAIL OUTLET | UltraVideo, LV2320B, $19.95 |

| | |
|---|---|
| TITLE | **MR. ROBERTS** |
| TYPE | Motion Picture |
| FORMAT | CED |
| DISTRIBUTOR | RCA Videodiscs, 03113, $24.98 |

| | |
|---|---|
| TITLE | **MR. ROGERS - HELPING CHILDREN UNDERSTAND** |
| TYPE | Children |
| FORMAT | CED |
| DISTRIBUTOR | RCA Videodiscs, 02108, $19.98 |

| | |
|---|---|
| TITLE | **MR. ROGERS GOES TO SCHOOL** |
| TYPE | Motion Picture |
| DISTRIBUTOR | RCA Videodiscs, 02101, $19.98 |

| | |
|---|---|
| TITLE | **MUPPET MOVIE, THE** |
| TYPE | Motion Picture, Comedy (G) |
| FORMAT | Laser CLV, CED |
| FEATURES | Stereo |
| RUNNING TIME | 94 minutes |
| PRODUCER | Warner Home Video |
| DISTRIBUTOR | Pioneer Video, Inc., 9001-80, $34.98 |
| | RCA Videodiscs, 00516, $24.98 |

| | |
|---|---|
| TITLE | **MUPPET MUSICIANS OF BREMEN** |
| TYPE | Motion Picture (G) |
| FORMAT | Laser CLV |
| RUNNING TIME | 50 minutes |
| PRODUCER | Walt Disney |
| DISTRIBUTOR | Pioneer Video, Inc., 42807AS, $24.95 |

| | |
|---|---|
| TITLE | **MURDER BY DEATH** |
| TYPE | Motion Picture, Comedy (PG) |
| FORMAT | CED |
| PRODUCER | Columbia Pictures |
| DISTRIBUTOR | RCA Videodiscs, 03047, $24.98 |

| | |
|---|---|
| TITLE | **MURDER BY DECREE** |
| TYPE | Motion Picture (PG) |
| FORMAT | Laser CLV |
| RUNNING TIME | 120 minutes |
| PRODUCER | Warner Home Video |
| DISTRIBUTOR | Pioneer Video, Inc., 4059-80, $34.98 |

**Videodiscography**

| | |
|---:|:---|
| TITLE | **MURDER ON THE ORIENT EXPRESS** |
| TYPE | Motion Picture (G) |
| FORMAT | Laser CLV, CED (2 Discs) |
| RUNNING TIME | 128 minutes |
| PRODUCER | Warner Home Video |
| DISTRIBUTOR | Pioneer Video, Inc., LV1239, $35.95 |
| | RCA Videodiscs, 00649, $34.98 |

| | |
|---:|:---|
| TITLE | **MUSIC OF MELISSA MANCHESTER, THE** |
| TYPE | Music |
| FORMAT | Laser CLV, CED |
| FEATURES | Stereo, CX Encoded |
| RUNNING TIME | 59 minutes |
| PRODUCER | Pioneer Artists |
| DISTRIBUTOR | Pioneer Video, Inc., PA-82-015, $24.95 |
| | RCA Videodiscs, 12100, $24.98 |

| | |
|---:|:---|
| TITLE | **MY BLOODY VALENTINE** |
| TYPE | Motion Picture (R) |
| FORMAT | Laser CLV |
| RUNNING TIME | 91 minutes |
| PRODUCER | Warner Home Video |
| DISTRIBUTOR | Pioneer Video, Inc., LV5431, $29.95 |

| | |
|---:|:---|
| TITLE | **MY BODYGUARD** |
| TYPE | Motion Picture |
| FORMAT | CED |
| RETAIL OUTLET | Movies Unlimited Inc., A1-1111, $29.98 |

| | |
|---:|:---|
| TITLE | **MY FAIR LADY** |
| TYPE | Motion Picture |
| FORMAT | Laser CLV, CED (2 Discs) |
| FEATURES | Stereo |
| RETAIL OUTLET | (CLV) UltraVideo, 80-7038, $39.95 |
| | (CED) Movies Unlimited Inc., A2-1019, $39.95 |

| | |
|---:|:---|
| TITLE | **MY FAVORITE YEAR** |
| TYPE | Motion Picture |
| FORMAT | Laser CLV, CED |
| RETAIL OUTLET | (CLV) UltraVideo, ML100188, $34.95 |
| | (CED) Movies Unlimited Inc., A2-0188, $29.95 |

| | |
|---:|:---|
| TITLE | **MY LITTLE CHICKADEE** |
| TYPE | Motion Picture, Comedy |
| FORMAT | CED |
| DISTRIBUTOR | RCA Videodiscs, 03320, $19.98 |

| | |
|---:|:---|
| TITLE | **MY TUTOR** |
| TYPE | Motion Picture (R) |
| FORMAT | Laser CLV |
| RETAIL OUTLET | UltraVideo, 40-027, $32.95 |

| | |
|---:|:---|
| TITLE | **MYSTERYDISC - MURDER, ANYONE?** |
| DESCRIPTION | Branched mystery game with 16 possible story lines. The object is to guess who comitted a murder, their motive, and method. |
| TYPE | Game, Adolescents |
| FORMAT | CED |
| LEVEL | 1 |
| DEVELOPER | VIDMAX |
| CONTACT FOR MORE INFO. | Diane Smook, RCA, (212) 930-4767 |
| FEATURES | Chapters, Picture Stops, Programming Band Sequences |
| PRODUCER | VIDMAX |
| DISTRIBUTOR | RCA Videodiscs, 25004, $29.98 |

| | |
|---:|:---|
| TITLE | **MYSTERYDISC - MURDER, ANYONE?** |
| TYPE | Game |
| FORMAT | Laser CAV |
| LEVEL | 1 |
| PRODUCER | Warner Home Video |
| DISTRIBUTOR | Pioneer Video, Inc., VMD-82-001, $29.95 |

| | |
|---:|:---|
| TITLE | **MYSTERYDISC 2 - MANY ROADS TO MURDER** |
| TYPE | Game |
| FORMAT | Laser CAV |
| LEVEL | 2 |
| RETAIL OUTLET | UltraVideo, VM83001, $39.95 |

| | |
|---:|:---|
| TITLE | **NASHVILLE** |
| TYPE | Motion Picture (R) |
| FORMAT | Laser CLV, CED (2 Discs) |
| FEATURES | Stereo |
| RUNNING TIME | 160 minutes |
| PRODUCER | Paramount Home Video |
| DISTRIBUTOR | Pioneer Video, Inc., LV8821-2, $35.95 |
| | RCA Videodiscs, 10691, $39.98 |

| | |
|---:|:---|
| TITLE | **NATIONAL GEOGRAPHIC GREAT WHALES/SHARKS** |
| TYPE | Instructional |
| FORMAT | CED |
| DISTRIBUTOR | RCA Videodiscs, 03205, $19.98 |

| | |
|---:|:---|
| TITLE | **NATIONAL GEOGRAPHIC INVISIBLE WORLD/DIVE TO THE EDGE OF CREATION** |
| FORMAT | CED |
| DISTRIBUTOR | RCA Videodiscs, 03204, $19.98 |

| | |
|---:|:---|
| TITLE | **NATIONAL GEOGRAPHIC THE INCREDIBLE MACHINE/MYSTERIES OF THE MIND** |
| TYPE | Instructional |
| FORMAT | CED |
| DISTRIBUTOR | RCA Videodiscs, 03201, $19.98 |

| | |
|---:|:---|
| TITLE | **NATIONAL LAMPOON'S ANIMAL HOUSE** |
| TYPE | Motion Picture, Comedy (R) |
| FORMAT | CED |
| DISTRIBUTOR | RCA Videodiscs, 03302, $29.98 |

| | |
|---:|:---|
| TITLE | **NATIONAL LAMPOON'S CLASS REUNION** |
| TYPE | Motion Picture (R) |
| FORMAT | Laser CLV, CED |
| FEATURES | Stereo |
| PRODUCER | ABC Video Enterprises |
| RETAIL OUTLET | (CLV) UltraVideo, VL5021, $34.95 |
| | (CED) Movies Unlimited Inc., A3-5021, $29.95 |

| | |
|---:|:---|
| TITLE | **NAVY MINE NEUTRALIZATION SYSTEM TRAINER** |
| DESCRIPTION | Disc simulates the monitors watched by the remote control operator on a Navy underwater vehicle. A computer determines the simulated position of the vehicle in the ocean, and selects the appropriate frames on the videodisc for the user to view. |
| TYPE | Research and Development, Technical, Instructional, In-house Training |
| FORMAT | Laser CAV |
| CONTACT FOR MORE INFO. | William A. Johnson, Code 5142, Naval Research Lab |
| FEATURES | Chapters |
| AVAILABILITY | For Navy use only. |

| | |
|---:|:---|
| TITLE | **NEIGHBORS** |
| TYPE | Motion Picture, Comedy (R) |
| FORMAT | Laser CLV, CED |
| PRODUCER | RCA/Columbia Pictures Home Video |
| DISTRIBUTOR | Pioneer Video, Inc., VLD3352, $29.95 |
| | RCA Videodiscs, 03023, $29.98 |

| | |
|---:|:---|
| TITLE | **NEIL SEDAKA IN CONCERT** |
| TYPE | Music |
| FORMAT | Laser CAV |
| LEVEL | 1 |
| FEATURES | Stereo |
| PRODUCER | Warner Home Video |
| DISTRIBUTOR | Pioneer Video, Inc., 74-007, $24.95 |

| | |
|---:|:---|
| TITLE | **NETWORK** |
| TYPE | Motion Picture (R) |
| FORMAT | Laser CLV, CED |
| RUNNING TIME | 116 minutes |
| PRODUCER | MGM/UA Home Video |
| DISTRIBUTOR | Pioneer Video, Inc., ML100012, $34.95 |
| RETAIL OUTLET | (CED) Movies Unlimited Inc., A2-1003, $29.95 |

| | |
|---:|:---|
| TITLE | **NEVER GIVE A SUCKER AN EVEN BREAK** |
| TYPE | Motion Picture, Comedy |
| FORMAT | CED |
| DISTRIBUTOR | RCA Videodiscs, 03319, $19.98 |

| | |
|---:|:---|
| TITLE | **NEW LOOK** |
| TYPE | Adult |
| FORMAT | Laser CLV |
| RETAIL OUTLET | UltraVideo, E100150, $34.95 |

| | |
|---:|:---|
| TITLE | **NEW MEDIA BIBLE I: THE STORY OF JOSEPH** |
| FORMAT | CED |
| FEATURES | Banded |
| DISTRIBUTOR | RCA Videodiscs, 02146, $19.98 |

| | |
|---:|:---|
| TITLE | **NEW MEDIA BIBLE: THE GOSPEL OF LUKE I** |
| FORMAT | CED |
| FEATURES | Banded |
| DISTRIBUTOR | RCA Videodiscs, 02147, $19.98 |

| | |
|---:|:---|
| TITLE | **NEW MEDIA BIBLE THE GOSPEL OF LUKE II** |
| FORMAT | CED |
| FEATURES | Banded |
| DISTRIBUTOR | RCA Videodiscs, 02148, $19.98 |

| | |
|---:|:---|
| TITLE | **NEW YORK YANKEES MIRACLE YEAR - 1978, THE** |
| TYPE | Motion Picture |
| FORMAT | CED |
| DISTRIBUTOR | RCA Videodiscs, 02012, $14.98 |

| | |
|---:|:---|
| TITLE | **NEW YORK, NEW YORK** |
| TYPE | Motion Picture, Music |
| FORMAT | CED (2 Discs) |
| DISTRIBUTOR | RCA Videodiscs, 01489, $39.98 |

| | |
|---:|:---|
| TITLE | **NFL '81 OFFICIAL SEASON YEARBOOK** |
| TYPE | Motion Picture |
| FORMAT | CED |
| DISTRIBUTOR | RCA Videodiscs, 01605, $19.98 |

| | |
|---:|:---|
| TITLE | **NFL SYMFUNNY AND LEGENDS OF THE FALL, THE** |
| TYPE | Motion Picture |
| FORMAT | Laser CAV |
| LEVEL | 1 |
| RUNNING TIME | 46 minutes |
| PRODUCER | Warner Home Video |
| DISTRIBUTOR | Pioneer Video, Inc., LV003, $24.95 |

| | |
|---:|:---|
| TITLE | **NICE DREAMS** |
| TYPE | Motion Picture, Comedy (R) |
| FORMAT | Laser CLV, CED |
| RETAIL OUTLET | (CLV) UltraVideo, VLD3355, $29.95 |
| | (CED) Movies Unlimited Inc., A1-3035, $24.98 |

| | |
|---:|:---|
| TITLE | **NIGHT GAMES** |
| TYPE | Motion Picture |
| FORMAT | CED |
| RETAIL OUTLET | Movies Unlimited Inc., A4-0096, $29.95 |

| | |
|---:|:---|
| TITLE | **NIGHT PORTER, THE** |
| TYPE | Motion Picture (R) |
| FORMAT | Laser CLV, CED |
| RUNNING TIME | 117 minutes |
| PRODUCER | Warner Home Video |
| DISTRIBUTOR | Pioneer Video, Inc., 150250, $29.95 |
| RETAIL OUTLET | (CED) Movies Unlimited Inc., A4-5026, $29.95 |

| | |
|---:|:---|
| TITLE | **NIGHT SHIFT** |
| TYPE | Motion Picture, Comedy (R) |
| FORMAT | Laser CLV, CED |
| FEATURES | Stereo |
| RUNNING TIME | 106 minutes |
| PRODUCER | Warner Home Video |
| DISTRIBUTOR | Pioneer Video, Inc., 20006LV, $29.98 |
| RETAIL OUTLET | (CED) Movies Unlimited Inc., A1-0006, $29.98 |

| | |
|---:|:---|
| TITLE | **NIGHT WITH LOU REED, A** |
| TYPE | Music |
| FORMAT | CED |
| FEATURES | Stereo |
| PRODUCER | RCA |
| DISTRIBUTOR | RCA Videodiscs, 12173, $19.98 |

| | |
|---:|:---|
| TITLE | **NIGHTHAWKS** |
| TYPE | Motion Picture (R) |
| FORMAT | Laser CLV |
| RUNNING TIME | 99 minutes |
| PRODUCER | Warner Home Video |
| DISTRIBUTOR | Pioneer Video, Inc., 12-021, $29.95 |

| | |
|---:|:---|
| TITLE | **NINE TO FIVE** |
| TYPE | Motion Picture, Comedy (PG) |
| FORMAT | Laser CLV, CED |
| FEATURES | Stereo |
| RUNNING TIME | 109 minutes |
| PRODUCER | Warner Home Video |
| DISTRIBUTOR | Pioneer Video, Inc., 1099-80, $29.98 |
| RETAIL OUTLET | (CED) Movies Unlimited Inc., A1-1099, $29.98 |

| | |
|---:|:---|
| TITLE | **1941** |
| TYPE | Motion Picture, Comedy (PG) |
| FORMAT | Laser CLV |
| RUNNING TIME | 118 minutes |
| PRODUCER | Warner Home Video |
| DISTRIBUTOR | Pioneer Video, Inc., 16-014, $29.95 |

| | |
|---:|:---|
| TITLE | **NINETY-SIX: A CATTLE RANCH IN NORTHERN NEVADA** |
| DESCRIPTION | Disc describes the operation of a beef cattle ranch in Humboldt County, Nevada. It is interactive on side A and linear on side B. Includes footage from 1979-81 with some footage from 1945. |
| TYPE | Technical, Instructional |
| RELEASE DATE | Mid 1984 |
| FORMAT | CAV (side A), CLV (side B) |
| LEVEL | 2, 0 |
| SYSTEM DESCRIPTION | Interactive side A works on a Pioneer 7820-3 player. |
| DEVELOPER | Carl Fleischhauer |
| CONTACT FOR MORE INFO. | Carl Fleischhauer, (202) 287-6590 |
| FEATURES | Chapters, Digital Dump-Pioneer, Dual Audio |
| RUNNING TIME | 90 minutes |
| PRODUCER | Library of Congress |
| DISTRIBUTOR | Library of Congress/American Folklife Center |
| AVAILABILITY | Sale |

| | |
|---:|:---|
| TITLE | **NO NUKES: THE MUSE CONCERT** |
| TYPE | Music |
| FORMAT | Laser CLV, CED |
| FEATURES | Stereo |
| RETAIL OUTLET | (CLV) UltraVideo, 80-7065, $29.95 |
| | (CED) Movies Unlimited Inc., A2-1037, $24.95 |

| | |
|---:|:---|
| TITLE | **NORMA RAE** |
| TYPE | Motion Picture |
| FORMAT | CED |
| RETAIL OUTLET | Movies Unlimited Inc., A1-1082, $29.98 |

| | |
|---:|:---|
| TITLE | **NORSEMAN, THE** |
| TYPE | Motion Picture |
| FORMAT | CED |
| RETAIL OUTLET | Movies Unlimited Inc., A3-4036, $29.95 |

| | |
|---:|:---|
| TITLE | **NORTH BY NORTHWEST** |
| TYPE | Motion Picture |
| FORMAT | CED (2 Discs) |
| RETAIL OUTLET | Movies Unlimited Inc., A2-0104, $39.95 |

| | |
|---:|:---|
| TITLE | **NORTH DALLAS FORTY** |
| TYPE | Motion Picture (R) |
| FORMAT | Laser CLV, CED |
| RUNNING TIME | 117 minutes |

| | |
|---:|:---|
| PRODUCER | Warner Home Video |
| DISTRIBUTOR | Pioneer Video, Inc., LV8773, $29.95 |
| | RCA Videodiscs, 00634, $29.98 |

| | |
|---:|:---|
| TITLE | **NOTORIOUS** |
| TYPE | Motion Picture |
| FORMAT | Laser CLV, CED |
| FEATURES | Black and white |
| RUNNING TIME | 103 minutes |
| PRODUCER | Warner Home Video |
| DISTRIBUTOR | Pioneer Video, Inc., 8011-80, $29.98 |
| RETAIL OUTLET | (CED) Movies Unlimited Inc., A1-8011, $29.98 |

| | |
|---:|:---|
| TITLE | **NOW AND FOREVER** |
| TYPE | Motion Picture (R) |
| FORMAT | Laser CLV |
| RETAIL OUTLET | UltraVideo, 11-021, $29.95 |

| | |
|---:|:---|
| TITLE | **NOW VOYAGER** |
| TYPE | Motion Picture |
| FORMAT | CED |
| DISTRIBUTOR | RCA Videodiscs, 03415, $19.98 |

| | |
|---:|:---|
| TITLE | **NUTCRACKER SUITE** |
| TYPE | Music, Dance |
| FORMAT | CED |
| RETAIL OUTLET | Movies Unlimited Inc., A2-1007, $24.95 |

| | |
|---:|:---|
| TITLE | **NUTCRACKER, THE** |
| TYPE | Music, Dance |
| FORMAT | Laser CLV |
| FEATURES | Stereo, CX Encoded |
| RUNNING TIME | 79 minutes |
| PRODUCER | Pioneer Artists |
| DISTRIBUTOR | Pioneer Video, Inc., PA-82-032, $34.95 |

| | |
|---:|:---|
| TITLE | **NUTCRACKER/BARYSHNIKOV** |
| TYPE | Music, Dance |
| FORMAT | CED |
| FEATURES | Stereo |
| RETAIL OUTLET | Movies Unlimited Inc., A2-1065, $29.95 |

| | |
|---:|:---|
| TITLE | **NUTTY PROFESSOR, THE** |
| TYPE | Motion Picture, Comedy |
| FORMAT | CED |
| DISTRIBUTOR | RCA Videodiscs, 03621, $19.98 |

| | |
|---:|:---|
| TITLE | **O.I.S.E. SINGLE POINT THREAD CUTTING** |
| DESCRIPTION | An interactive videodisc training course aimed at teaching high school students how to cut a screw thread, using a machine lathe. Also on the disc is a 12-minute video about a computer-assisted math program. |
| TYPE | Research and Development, Technical, Instructional, Promotional |
| FORMAT | Laser CAV |
| LEVEL | 3 |
| SYSTEM DESCRIPTION | Disc must be used with a Honeywell Level 6 Series 60 minicomputer, plus the CAN-8 system. |
| DEVELOPER | Ontario Institute for Studies in Education |
| CONTACT FOR MORE INFO. | Mary Pat Carroll |
| PRODUCER | Same as developer |
| DISTRIBUTOR | Same as developer |
| AVAILABILITY | Sale |
| ORDER NO. | VSPD-82 |
| SUGGESTED PRICE | $40 |

| | |
|---:|:---|
| TITLE | **OBSERVATION SCENES FOR TRAINING IN NON-VERBAL COMMUNICATION** |
| DESCRIPTION | This disc is used with an interactive computer program to train students in identifying acts specified in the Leary model of nonverbal communication. The scenes show dialogues between two persons. |
| TYPE | Instructional, In-house Training |
| RELEASE DATE | 7/81 |
| FORMAT | Philips for NTSC |
| LEVEL | 3 |
| SYSTEM DESCRIPTION | PDP 11 computer |
| DEVELOPER | Catholic University, the Netherlands/CAI Group and the Psychology Department |
| CONTACT FOR MORE INFO. | Same as developer |
| FEATURES | Chapters, Picture Stops |
| PRODUCER | Same as developer |
| AVAILABILITY | Not commercially available. |

| | |
|---:|:---|
| TITLE | **OCTOPUSSY** |
| TYPE | Motion Picture |
| FORMAT | Laser CLV |
| RETAIL OUTLET | UltraVideo, 80-4715, $39.95 |

| | |
|---:|:---|
| TITLE | **ODD COUPLE, THE** |
| TYPE | Motion Picture, Comedy (G) |
| FORMAT | Laser CLV, CED |
| RUNNING TIME | 106 minutes |
| PRODUCER | Warner Home Video |
| DISTRIBUTOR | Pioneer Video, Inc., LV8026, $29.95 |
| | RCA Videodiscs, 00618, $29.98 |

| | |
|---:|:---|
| TITLE | **OFFICER AND A GENTLEMAN, AN** |
| TYPE | Motion Picture (R) |
| FORMAT | Laser CLV, CED (2 Discs) |
| RUNNING TIME | 126 minutes |
| PRODUCER | Paramount Home Video |
| DISTRIBUTOR | Pioneer Video, Inc., LV1467-2, $39.95 |
| | RCA Videodiscs, 03607, $29.98 |

|             |                                    |
|------------:|------------------------------------|
| TITLE       | **OH, GOD!**                       |
| TYPE        | Motion Picture, Comedy (PG)        |
| FORMAT      | CED                                |
| DISTRIBUTOR | RCA Videodiscs, 03133, $29.98      |

|               |                                          |
|--------------:|------------------------------------------|
| TITLE         | **OKLAHOMA**                             |
| TYPE          | Motion Picture, Music                    |
| FORMAT        | CED (2 Discs)                            |
| RETAIL OUTLET | Movies Unlimited Inc., A2-1005, $39.95   |

|               |                                     |
|--------------:|-------------------------------------|
| TITLE         | **OLD YELLER**                      |
| TYPE          | Motion Picture (G)                  |
| FORMAT        | Laser CLV, CED                      |
| PRODUCER      | Walt Disney                         |
| DISTRIBUTOR   | RCA Videodiscs, 00708, $24.98       |
| RETAIL OUTLET | (CLV) UltraVideo, 037AS, $34.95     |

|             |                                    |
|------------:|------------------------------------|
| TITLE       | **OLIVIA**                         |
| TYPE        | Music                              |
| FORMAT      | Laser CAV                          |
| LEVEL       | 1                                  |
| PRODUCER    | Warner Home Video                  |
| DISTRIBUTOR | Pioneer Video, Inc., 74-005, $24.95 |

|               |                                               |
|--------------:|-----------------------------------------------|
| TITLE         | **OLIVIA IN CONCERT**                         |
| TYPE          | Music                                         |
| FORMAT        | Laser CAV, CED                                |
| LEVEL         | 1                                             |
| FEATURES      | Stereo, CX Encoded                            |
| PRODUCER      | MCA Videodisc, Inc.                           |
| DISTRIBUTOR   | Pioneer Video, Inc., 74-021, $24.95           |
| RETAIL OUTLET | (CED) Movies Unlimited Inc., A5-4021, $24.98  |

|             |                                     |
|------------:|-------------------------------------|
| TITLE       | **OLIVIA: PHYSICAL**                |
| TYPE        | Music                               |
| FORMAT      | Laser CAV                           |
| LEVEL       | 1                                   |
| FEATURES    | Stereo, CX Encoded                  |
| PRODUCER    | Warner Home Video                   |
| DISTRIBUTOR | Pioneer Video, Inc., 74-017, $24.95 |

|               |                                             |
|--------------:|---------------------------------------------|
| TITLE         | **OMEN, THE**                               |
| TYPE          | Motion Picture (R)                          |
| FORMAT        | Laser CLV, CED                              |
| FEATURES      | Stereo                                      |
| RUNNING TIME  | 111 minutes                                 |
| PRODUCER      | Warner Home Video                           |
| DISTRIBUTOR   | Pioneer Video, Inc., 1079-80, $29.98        |
| RETAIL OUTLET | (CED) Movies Unlimited Inc., A1-1079, $29.98 |

| | |
|---:|:---|
| TITLE | **ON GOLDEN POND** |
| TYPE | Motion Picture (PG) |
| FORMAT | Laser CLV, CED |
| RUNNING TIME | 109 minutes |
| PRODUCER | Warner Home Video |
| DISTRIBUTOR | Pioneer Video, Inc., 9037-80, $39.98 |
| | RCA Videodiscs, 00527, $29.98 |

| | |
|---:|:---|
| TITLE | **ON HER MAJESTY'S SECRET SERVICE** |
| TYPE | Motion Picture |
| FORMAT | CED (2 Discs) |
| DISTRIBUTOR | RCA Videodiscs, 01443, $39.98 |

| | |
|---:|:---|
| TITLE | **ON THE BEACH** |
| TYPE | Motion Picture |
| FORMAT | CED (2 Discs) |
| RETAIL OUTLET | Movies Unlimited Inc., A1-4521, $39.98 |

| | |
|---:|:---|
| TITLE | **ON THE TOWN** |
| TYPE | Motion Picture |
| FORMAT | CED |
| RETAIL OUTLET | Movies Unlimited Inc., A2-0057, $29.95 |

| | |
|---:|:---|
| TITLE | **ONE & ONLY, GENUINE ORIGINAL FAMILY BAND, THE** |
| TYPE | Motion Picture, Music |
| FORMAT | Laser CLV |
| RUNNING TIME | 110 minutes |
| PRODUCER | Warner Home Video |
| DISTRIBUTOR | Pioneer Video, Inc., 42017AS, $34.95 |

| | |
|---:|:---|
| TITLE | **ONE FLEW OVER THE CUCKOO'S NEST** |
| TYPE | Motion Picture (R) |
| FORMAT | CED (2 Discs) |
| DISTRIBUTOR | RCA Videodiscs, 02008, $34.98 |

| | |
|---:|:---|
| TITLE | **ONE FROM THE HEART** |
| TYPE | Motion Picture, Music |
| FORMAT | Laser CLV, CED |
| FEATURES | Stereo |
| DISTRIBUTOR | RCA Videodiscs, 13050, $29.98 |
| RETAIL OUTLET | (CLV) UltraVideo, 80-4715, $39.95 |

| | |
|---:|:---|
| TITLE | **ONE NIGHT STAND: A KEYBOARD EVENT** |
| TYPE | Music |
| FORMAT | Laser CLV, CED |
| FEATURES | Stereo, CX Encoded |
| RUNNING TIME | 98 minutes |
| PRODUCER | CBS Fox Video |
| DISTRIBUTOR | Pioneer Video, Inc., 7044-80, $29.98 |
| RETAIL OUTLET | (CED) Movies Unlimited Inc., A1-7044, $29.98 |

| | |
|---|---|
| TITLE | **ONE THOUSAND EIGHT HUNDRED AND FIFTY TWO HUMPBACK WHALES** |
| DESCRIPTION | Includes single frame photographs of individually identified humpback whales for use with computer sorting programs to assist researchers in recognizing individuals repeatedly. |
| TYPE | Research and Development, Technical, Archival, Catalog |
| FORMAT | Laser CAV |
| LEVEL | 3 |
| SYSTEM DESCRIPTION | Apple computer sorts data base for coded data, which matches (with variable acceptance criteria) a candidate image, then directs videodisc player to display image for visual comparison. |
| DEVELOPER | Ken Balcomb |
| CONTACT FOR MORE INFO. | Ken Balcomb |
| DISTRIBUTOR | Ocean Research and Education Society |
| AVAILABILITY | Sale |
| SPECIAL CONDITIONS | To cooperating researchers |
| SUGGESTED PRICE | $20 |

---

| | |
|---|---|
| TITLE | **ONLY WHEN I LAUGH** |
| TYPE | Motion Picture |
| FORMAT | Laser CLV |
| RETAIL OUTLET | UltraVideo, VLD3362, $29.95 |

---

| | |
|---|---|
| TITLE | **ORCA, THE KILLER WHALE** |
| TYPE | Motion Picture (PG) |
| FORMAT | Laser CLV |
| RUNNING TIME | 92 minutes |
| PRODUCER | Warner Home Video |
| DISTRIBUTOR | Pioneer Video, Inc., LV8935, $29.95 |

---

| | |
|---|---|
| TITLE | **ORDINARY PEOPLE** |
| TYPE | Motion Picture (R) |
| FORMAT | Laser CLV, CED (2 Discs) |
| RUNNING TIME | 124 minutes |
| PRODUCER | Warner Home Video |
| DISTRIBUTOR | Pioneer Video, Inc., LV8964, $35.95<br>RCA Videodiscs, 00641, $34.98 |

---

| | |
|---|---|
| TITLE | **ORIENTAL DREAMS** |
| TYPE | Music, Adult |
| FORMAT | Laser CAV |
| LEVEL | 1 |
| FEATURES | Stereo, CX Encoded |
| PRODUCER | Pioneer Video Imports |
| DISTRIBUTOR | Pioneer Video, Inc., ME-003, $27.95 |

---

| | |
|---|---|
| TITLE | **OSCAR, THE** |
| TYPE | Motion Picture |
| FORMAT | CED |
| RETAIL OUTLET | Movies Unlimited Inc., A4-0606, $29.95 |

| | |
|---:|:---|
| TITLE | **OUR TOWN** |
| TYPE | Motion Picture |
| FORMAT | CED |
| DISTRIBUTOR | RCA Videodiscs, 02002, $19.98 |

| | |
|---:|:---|
| TITLE | **OUTLAND** |
| TYPE | Motion Picture |
| FORMAT | CED |
| FEATURES | Stereo |
| DISTRIBUTOR | RCA Videodiscs, 13146, $29.98 |

| | |
|---:|:---|
| TITLE | **OWL AND THE PUSSYCAT, THE** |
| TYPE | Motion Picture, Music |
| FORMAT | CED |
| DISTRIBUTOR | RCA Videodiscs, 03026, $29.98 |

| | |
|---:|:---|
| TITLE | **PADDINGTON BEAR, VOL. 1** |
| TYPE | Children |
| FORMAT | CED |
| FEATURES | Banded |
| DISTRIBUTOR | RCA Videodiscs, 02179, $19.98 |

| | |
|---:|:---|
| TITLE | **PADDINGTON BEAR, VOL. 2** |
| TYPE | Children |
| FORMAT | CED |
| DISTRIBUTOR | RCA Videodiscs, 02201, $19.98 |

| | |
|---:|:---|
| TITLE | **PAINT YOUR WAGON** |
| TYPE | Motion Picture, Music (PG) |
| FORMAT | Laser CLV |
| FEATURES | Stereo, CX Encoded |
| RUNNING TIME | 164 minutes |
| PRODUCER | Paramount Home Video |
| DISTRIBUTOR | Pioneer Video, Inc., LV6933-2, $35.95 |

| | |
|---:|:---|
| TITLE | **PAPER CHASE, THE** |
| TYPE | Motion Picture |
| FORMAT | CED |
| RETAIL OUTLET | Movies Unlimited Inc., A1-1046, $29.98 |

| | |
|---:|:---|
| TITLE | **PAPER MOON** |
| TYPE | Motion Picture, Comedy (PG) |
| FORMAT | Laser CLV, CED |
| FEATURES | Black and white |
| RUNNING TIME | 102 minutes |
| PRODUCER | Warner Home Video |
| DISTRIBUTOR | Pioneer Video, Inc., LV8465, $29.95 |
| | RCA Videodiscs, 00626, $24.98 |

# Videodiscography

| | |
|---|---|
| TITLE | **PAPILLON** |
| TYPE | Motion Picture |
| FORMAT | CED (2 Discs) |
| RETAIL OUTLET | Movies Unlimited Inc., A1-7090, $39.98 |

| | |
|---|---|
| TITLE | **PARADISE** |
| TYPE | Motion Picture |
| FORMAT | CED |
| RETAIL OUTLET | Movies Unlimited Inc., A4-6036, $29.95 |

| | |
|---|---|
| TITLE | **PARASITE** |
| TYPE | Motion Picture |
| FORMAT | CED |
| RETAIL OUTLET | Movies Unlimited Inc., A4-0546, $29.95 |

| | |
|---|---|
| TITLE | **PARIS-PARIS** |
| DESCRIPTION | Three videodisc sides run on six players to inform and entertain the customers of a large drug store with information about Paris cultural events, special effects, and advertising. The program is shown in split-screen images on a big monitor wall. |
| TYPE | Comedy, Music, Dance, Art, Advertising, Promotional |
| RELEASE DATE | January 1984 |
| FORMAT | Laser CAV |
| LEVEL | 3 |
| SYSTEM DESCRIPTION | Fifty-four monitors with fiber optic correction to make vanish frames (VidiWall, a CAT-Philips development) fed by six VP-835s, one U-Matic machine, a tv camera for live shots, and one graphic computer. All are computer-controlled. |
| DEVELOPER | Patric Martin, CAT Paris |
| FEATURES | Stereo, Dual Audio, Sound Over Still |
| PRODUCER | CAT Paris- Computer Assisted Televideo S.a.r.l. |
| DISTRIBUTOR | Video Banque, Versailles |
| AVAILABILITY | Video Banque will place systems in various locations, such as railroad stations. |

| | |
|---|---|
| TITLE | **PASSION OF LOVE** |
| TYPE | Motion Picture |
| FORMAT | CED |
| RETAIL OUTLET | Movies Unlimited Inc., A3-4026, $29.95 |

| | |
|---|---|
| TITLE | **PATERNITY** |
| TYPE | Motion Picture (PG) |
| FORMAT | Laser CLV, CED |
| RUNNING TIME | 94 minutes |
| PRODUCER | Warner Home Video |
| DISTRIBUTOR | Pioneer Video, Inc., LV1401, $29.95 |
| | RCA Videodiscs, 00673, $24.98 |

| | |
|---:|:---|
| TITLE | **PATHS OF GLORY** |
| TYPE | Motion Picture |
| FORMAT | CED |
| RETAIL OUTLET | Movies Unlimited Inc., A1-4692, $29.98 |

| | |
|---:|:---|
| TITLE | **PATRIOT CLS OJT PHASE 1** |
| DESCRIPTION | Training of selected radar maintenance tasks. |
| TYPE | Technical, Instructional, Promotional |
| FORMAT | Laser CAV |
| LEVEL | 3 |
| DEVELOPER | T. Stroud |
| CONTACT FOR MORE INFO. | E. Dextraze, (617) 272-9300 |
| FEATURES | Sound Over Still |
| PRODUCER | Raytheon Service Company |
| AVAILABILITY | Not for sale. |

| | |
|---:|:---|
| TITLE | **PATRIOT MISSILE SYSTEM PROTOTYPE MAINTENANCE DISC** |
| DESCRIPTION | A prototype Army maintenance manual on videodisc. Multiple manuals are accessible on one disc. |
| TYPE | Research and Development, Technical |
| FORMAT | Laser CAV |
| LEVEL | 3 |
| SYSTEM DESCRIPTION | Disc works on a Sony player and computer; system has a custom keypad designed to accomplish many special functions. |
| DEVELOPER | Computer Sciences Corporation |
| CONTACT FOR MORE INFO. | Dr. William A. Stembler |
| PRODUCER | Same as contact |
| AVAILABILITY | Property of U.S. Army. |

| | |
|---:|:---|
| TITLE | **PATTON** |
| TYPE | Motion Picture (PG) |
| FORMAT | Laser CLV, CED (2 Discs) |
| FEATURES | Stereo, CX Encoded |
| RUNNING TIME | 171 minutes |
| PRODUCER | Warner Home Video |
| DISTRIBUTOR | Pioneer Video, Inc., 1005-80, $49.98 |
| RETAIL OUTLET | (CED) Movies Unlimited Inc., A1-0110, $39.98 |

| | |
|---:|:---|
| TITLE | **PAUL MCCARTNEY AND WINGS - ROCKSHOW** |
| TYPE | Music |
| FORMAT | Laser CLV, CED |
| FEATURES | Stereo, CX Encoded |
| RUNNING TIME | 102 minutes |
| PRODUCER | Pioneer Artists |
| DISTRIBUTOR | Pioneer Video, Inc., PA-82-027, $29.95<br>RCA Videodiscs, 12098, $24.98 |

| | |
|---:|:---|
| TITLE | **PAUL SIMON** |
| TYPE | Music |
| FORMAT | Laser CAV |
| LEVEL | 1 |

| | |
|---:|:---|
| FEATURES | Stereo |
| RUNNING TIME | 53 minutes |
| PRODUCER | Pioneer Artists |
| DISTRIBUTOR | Pioneer Video, Inc., PA-81-001, $24.95 |

| | |
|---:|:---|
| TITLE | **PAUL SIMON IN CONCERT** |
| TYPE | Music |
| FORMAT | CED |
| FEATURES | Stereo, CX Encoded |
| DISTRIBUTOR | RCA Videodiscs, 12075, $24.98 |

| | |
|---:|:---|
| TITLE | **PAVAROTTI IN LONDON** |
| TYPE | Music |
| FORMAT | Laser CLV, CED |
| FEATURES | Stereo, CX Encoded |
| RUNNING TIME | 51 minutes |
| PRODUCER | Pioneer Artists |
| DISTRIBUTOR | Pioneer Video, Inc., PA-83-043, $24.95 |
| | RCA Videodiscs, 12165, $24.98 |

| | |
|---:|:---|
| TITLE | **PBS CLOSED CAPTIONED DISC** |
| DESCRIPTION | The first experiment in encoding line 21 closed captions on videodisc. |
| TYPE | Research and Development |
| LEVEL | 1,2 |
| CONTACT FOR MORE INFO. | Bureau of Education for the Handicapped/University of Nebraska - Barkley Center |
| PRODUCER | University of Nebraska/Nebraska Videodisc Design/Production Group |
| AVAILABILITY | Not available for release or distribution. |

| | |
|---:|:---|
| TITLE | **PEDIATRIC CARDIOVASCULAR DEFECTS** |
| DESCRIPTION | Audiovisual, textual data base reference of 23 pediatric cardiovascular defects. |
| TYPE | Medical, Instructional |
| FORMAT | Laser CAV |
| LEVEL | 3 |
| SYSTEM DESCRIPTION | Coloney Learning Carrel System |
| DEVELOPER | Interactive Video Concepts, Inc. (IVC) |
| CONTACT FOR MORE INFO. | Peg Callahan, IVC/Jayne Parker, Florida Department of Health and Rehabilitative Services, Children's Medical Services. |
| RUNNING TIME | 8 hours |
| PRODUCER | Interactive Video Concepts, Inc. |
| AVAILABILITY | Used for continuing nursing education by Children's Medical Services, Florida Department of Health and Rehabilitative Services. |

| | |
|---:|:---|
| TITLE | **PEDIATRIC HEMATOLOGICAL DISEASES** |
| DESCRIPTION | Developed to teach counseling skills in sickle cell anemia based on case management skills used with leukemia patients. Also provides a data base reference of hematological diseases. |
| TYPE | Medical, Instructional, In-house Training |
| FORMAT | Laser CAV |
| LEVEL | 3 |

| | |
|---|---|
| SYSTEM DESCRIPTION | Coloney Learning Carrel System |
| DEVELOPER | Interactive Video Concepts, Inc. (IVC) |
| CONTACT FOR MORE INFO. | Peg Callahan, IVC or Jayne Parker, Florida Department of Health and Rehabilitative Services, Children's Medical Services. |
| RUNNING TIME | 8 hours |
| PRODUCER | Interactive Video Concepts, Inc. |
| AVAILABILITY | Used for continuing nursing education by Children's Medical Services, Florida Department of Health and Rehabilitative Services. |

| | |
|---|---|
| TITLE | **PENNIES FROM HEAVEN** |
| TYPE | Motion Picture, Music |
| FORMAT | CED |
| RETAIL OUTLET | Movies Unlimited Inc., A2-1062, $29.95 |

| | |
|---|---|
| TITLE | **PERCEPTRONICS, INC. - (NOT GIVEN)** |
| DESCRIPTION | Video map discs. |
| TYPE | Technical, Instructional, Archival, Catalog |
| FORMAT | Laser CAV |
| LEVEL | 3 |
| CONTACT FOR MORE INFO. | Steve Johnston, Perceptronics, Inc. |
| DISTRIBUTOR | Perceptronics, Inc. |
| AVAILABILITY | A price catalog will be available soon. |

| | |
|---|---|
| TITLE | **PERMS AND HAIRSTYLES** |
| DESCRIPTION | L'Oreal program about permanents and styles. Side A is information for customers and side B is for hairdressers. |
| TYPE | Instructional, Point-of-Purchase, Promotional |
| RELEASE DATE | October 1983 |
| FORMAT | Laser CAV |
| LEVEL | 1 |
| DEVELOPER | CAT Stuttgart- Computer Assisted Televideo AG |
| CONTACT FOR MORE INFO. | Maria Schlipt |
| FEATURES | Chapters, Picture Stops |
| PRODUCER | L'Oreal, Gmbh, Karlsruhe, Germany |
| DISTRIBUTOR | L'Oreal |
| AVAILABILITY | Not for sale. Disc has been placed in 200 hair salons in Germany on a test basis. |

| | |
|---|---|
| TITLE | **PERSONAL COMPUTER DIAGNOSTICS** |
| DESCRIPTION | Highly interactive two-sided videodisc designed to train computer repair technicians in the diagnosis and repair of common personal computer faults. |
| TYPE | Technical, Instructional |
| FORMAT | Laser CAV |
| LEVEL | 3 |
| SYSTEM DESCRIPTION | Player, IBM Personal Computer, and two touch- sensitive monitors |
| DEVELOPER | Creativision, Inc. |
| CONTACT FOR MORE INFO. | Mr. Rick Glasby |
| PRODUCER | Creativision, Inc. |
| AVAILABILITY | Custom project. Disc not available. |

| | |
|---:|:---|
| TITLE | **PERSONAL COMPUTER EXHIBIT** |
| DESCRIPTION | This videodisc is an exhibit at the Ontario Science Centre and Ontario Place in Toronto. It was sponsored by IBM Canada. Simple programming and basic computer functions are explained. |
| TYPE | Instructional, Promotional |
| FORMAT | Laser CAV |
| LEVEL | 3 |
| SYSTEM DESCRIPTION | The IBM PC controls a Pioneer model 3 player via SIA interface. Graphics from the PC are used as well as graphics from the videodisc. |
| DEVELOPER | Interactive Image Technology and Exhibit Technology, Inc. |
| CONTACT FOR MORE INFO. | Patrick Lee |
| PRODUCER | same as contact |
| DISTRIBUTOR | IBM Canada |

| | |
|---:|:---|
| TITLE | **PETE'S DRAGON** |
| TYPE | Motion Picture, Music |
| FORMAT | Laser CLV |
| FEATURES | Stereo, CX Encoded |
| RUNNING TIME | 128 minutes |
| PRODUCER | Warner Home Video |
| DISTRIBUTOR | Pioneer Video, Inc., 42010AS, $44.95 |

| | |
|---:|:---|
| TITLE | **PETER ALLEN AND THE ROCKETTES AT RADIO CITY MUSIC HALL** |
| TYPE | Music, Dance |
| FORMAT | Laser CLV, CED |
| FEATURES | Stereo |
| RUNNING TIME | 57 minutes |
| PRODUCER | Warner Home Video |
| DISTRIBUTOR | Pioneer Video, Inc., 74-016, $24.95 |
| RETAIL OUTLET | (CED) Movies Unlimited Inc., A1-6130, $29.98 |

| | |
|---:|:---|
| TITLE | **PETER GRIMES** |
| TYPE | Music |
| FEATURES | Stereo, CX Encoded |
| RUNNING TIME | 150 minutes |
| PRODUCER | Pioneer Artists |
| DISTRIBUTOR | Pioneer Video, Inc., PA-82-008, $59.95 |

| | |
|---:|:---|
| TITLE | **PETRIFIED FOREST, THE** |
| TYPE | Motion Picture |
| FORMAT | CED |
| DISTRIBUTOR | RCA Videodiscs, 03416, $19.98 |

| | |
|---:|:---|
| TITLE | **PHANTASM** |
| TYPE | Motion Picture (R) |
| FORMAT | Laser CLV, CED |
| FEATURES | Stereo |
| RUNNING TIME | 90 minutes |
| PRODUCER | Warner Home Video |
| DISTRIBUTOR | Pioneer Video, Inc., 4066-80, $34.98 |
| | RCA Videodiscs, 00806, $19.98 |

| | |
|---:|:---|
| TITLE | **PHANTOM TOLLBOOTH** |
| TYPE | Motion Picture |
| FORMAT | CED |
| RETAIL OUTLET | Movies Unlimited Inc., A2-0155, $29.95 |

| | |
|---:|:---|
| TITLE | **PHILIPS VP-835 DEMONSTRATION DISC** |
| DESCRIPTION | Enables Philips salesmen to demonstrate the features of the new modular industrial videodisc player VP- 835. |
| TYPE | Music, Dance, Art, Technical, Instructional, Catalog, Promotional |
| FORMAT | Laser CAV |
| LEVEL | 2,3 |
| SYSTEM DESCRIPTION | May be used with player RAM, EPROM to be put into player and external computer. |
| DEVELOPER | CAT London- Computer Assisted Televideo Ltd. |
| CONTACT FOR MORE INFO. | Judith C. Elliott |
| FEATURES | Stereo, Dual Audio, EPROM |
| PRODUCER | Philips Gloeilampenfabrieken, Eindhoven, Holland |
| DISTRIBUTOR | Philips National Organizations |
| AVAILABILITY | Loan |

| | |
|---:|:---|
| TITLE | **PHYSICS I & II** |
| DESCRIPTION | Laboratory simulation for college level. |
| TYPE | Instructional |
| FORMAT | Laser CAV |
| LEVEL | 3 |
| CONTACT FOR MORE INFO. | Corporation for Public Broadcasting/Annenberg |
| PRODUCER | University of Nebraska/Nebraska Videodisc Design/Production Group |
| AVAILABILITY | Not available for release or distribution. |

| | |
|---:|:---|
| TITLE | **PIAF** |
| TYPE | Music |
| FORMAT | CED |
| RETAIL OUTLET | Movies Unlimited Inc., A2-1044, $24.95 |

| | |
|---:|:---|
| TITLE | **PINK FLOYD - THE WALL** |
| TYPE | Motion Picture, Music |
| FORMAT | Laser CLV |
| FEATURES | Stereo |
| RETAIL OUTLET | UltraVideo, ML100268, $34.95 |

| | |
|---:|:---|
| TITLE | **PINK FLOYD AT POMPEII** |
| TYPE | Music |
| FORMAT | CED |
| FEATURES | Stereo, CX Encoded |
| DISTRIBUTOR | RCA Videodiscs, 12061, $24.98 |

| | |
|---:|:---|
| TITLE | **PINK PANTHER, THE** |
| TYPE | Motion Picture, Comedy (PG) |
| FORMAT | Laser CLV, CED |
| RUNNING TIME | 113 minutes |
| PRODUCER | Warner Home Video |
| DISTRIBUTOR | Pioneer Video, Inc., 4509-80, $34.98 |
| | RCA Videodiscs, 01405, $29.98 |

| | |
|---|---|
| TITLE | **PINK PANTHER STRIKES AGAIN, THE** |
| TYPE | Motion Picture, Comedy (PG) |
| FORMAT | Laser CLV, CED |
| RUNNING TIME | 110 minutes |
| PRODUCER | Warner Home Video |
| DISTRIBUTOR | Pioneer Video, Inc., 4564-80, $34.98 |
| | RCA Videodiscs, 01427, $24.98 |

| | |
|---|---|
| TITLE | **PIPPIN** |
| TYPE | Music, Comedy |
| FORMAT | Laser CLV, CED |
| FEATURES | Stereo, CX Encoded |
| RUNNING TIME | 110 minutes |
| PRODUCER | Pioneer Artists |
| DISTRIBUTOR | Pioneer Video, Inc., PA-82-009, $34.95 |
| | RCA Videodiscs, 12064, $24.98 |

| | |
|---|---|
| TITLE | **PIRATE, THE** |
| TYPE | Motion Picture |
| FORMAT | CED |
| RETAIL OUTLET | Movies Unlimited Inc., A2-1045, $29.95 |

| | |
|---|---|
| TITLE | **PIRATE MOVIE, THE** |
| TYPE | Motion Picture |
| FORMAT | CED |
| FEATURES | Stereo |
| RETAIL OUTLET | Movies Unlimited Inc., A1-1185, $29.98 |

| | |
|---|---|
| TITLE | **PIRATES OF PENZANCE, THE** |
| TYPE | Motion Picture, Music (G) |
| FORMAT | Laser CLV, CED |
| FEATURES | Stereo |
| RUNNING TIME | 112 minutes |
| PRODUCER | MCA Videodisc, Inc. |
| DISTRIBUTOR | Pioneer Video, Inc., 17-011, $29.95 |
| RETAIL OUTLET | (CED) Movies Unlimited Inc., A5-7011, $24.98 |

| | |
|---|---|
| TITLE | **PLACE IN THE SUN, A** |
| TYPE | Motion Picture |
| FORMAT | Laser CLV, CED |
| RUNNING TIME | 123 minutes |
| PRODUCER | Warner Home Video |
| DISTRIBUTOR | Pioneer Video, Inc., LV5815, $35.95 |
| | RCA Videodiscs, 00696, $24.98 |

| | |
|---:|:---|
| TITLE | **PLANET OF THE APES** |
| TYPE | Motion Picture |
| FORMAT | CED |
| RETAIL OUTLET | Movies Unlimited Inc., A1-0109, $29.98 |

| | |
|---:|:---|
| TITLE | **PLAY IT AGAIN, SAM** |
| TYPE | Motion Picture, Comedy (PG) |
| FORMAT | Laser CLV, CED |
| RUNNING TIME | 102 minutes |
| PRODUCER | Warner Home Video |
| DISTRIBUTOR | Pioneer Video, Inc., LV8112, $29.95 |
| | RCA Videodiscs, 00619, $24.98 |

| | |
|---:|:---|
| TITLE | **PLAY MISTY FOR ME** |
| TYPE | Motion Picture (R) |
| FORMAT | Laser CLV |
| RUNNING TIME | 102 minutes |
| PRODUCER | Warner Home Video |
| DISTRIBUTOR | Pioneer Video, Inc., 12-024, $29.95 |

| | |
|---:|:---|
| TITLE | **PLAYBOY MAGAZINE, VOLUME 1** |
| TYPE | Motion Picture, Adult |
| FORMAT | CED |
| FEATURES | Stereo |
| PRODUCER | CBS Fox Video |
| RETAIL OUTLET | Movies Unlimited Inc., A1-6201, $29.98 |

| | |
|---:|:---|
| TITLE | **PLAYBOY MAGAZINE, VOLUME 2** |
| TYPE | Motion Picture, Adult |
| FORMAT | CED |
| PRODUCER | CBS Fox Video |
| RETAIL OUTLET | Movies Unlimited Inc., A1-6202, $29.98 |

| | |
|---:|:---|
| TITLE | **PLAYBOY VIDEO, VOLUME 1** |
| TYPE | Adult |
| FORMAT | Laser CLV |
| RUNNING TIME | 79 minutes |
| PRODUCER | Warner Home Video |
| DISTRIBUTOR | Pioneer Video, Inc., 6201-80, $29.98 |

| | |
|---:|:---|
| TITLE | **PLAYBOY VIDEO, VOLUME 2** |
| TYPE | Adult |
| FORMAT | Laser CLV |
| FEATURES | Stereo, CX Encoded |
| RUNNING TIME | 81 minutes |
| PRODUCER | CBS Fox Video |
| DISTRIBUTOR | Pioneer Video, Inc., 6202-80, $29.98 |

| | |
|---|---|
| TITLE | **PLAYBOY COLLECTOR'S EDITION VOLUME 3** |
| TYPE | Adult |
| FORMAT | Laser CLV |
| FEATURES | Stereo, CX Encoded |
| RETAIL OUTLET | UltraVideo, 80-6203, $29.95 |

| | |
|---|---|
| TITLE | **PLAYBOY'S PLAYMATE REVIEW (1ST ANNUAL)** |
| TYPE | Motion Picture, Adult |
| FORMAT | Laser CLV, CED |
| FEATURES | Stereo |
| PRODUCER | CBS Fox Video |
| RETAIL OUTLET | (CLV) UltraVideo, 80-6255, $29.95 |
| | (CED) Movies Unlimited Inc., A1-6255, $29.98 |

| | |
|---|---|
| TITLE | **POINT OF SALE DEMONSTRATION** |
| DESCRIPTION | Arc welding, artificial insemination, and pesticide safety demonstrations. |
| FORMAT | Laser CAV |
| TYPE | Research and Development |
| LEVEL | 3 |
| CONTACT FOR MORE INFO. | Utah State University/Control Data Corporation |
| PRODUCER | University of Nebraska/Nebraska Videodisc Design/Production Group |
| AVAILABILITY | Not available for release or distribution. |

| | |
|---|---|
| TITLE | **POLLYANNA** |
| TYPE | Motion Picture |
| FORMAT | Laser CLV |
| RETAIL OUTLET | UltraVideo, 045AS, $34.95 |

| | |
|---|---|
| TITLE | **POLTERGEIST** |
| TYPE | Motion Picture (PG) |
| FORMAT | Laser CLV, CED |
| FEATURES | Stereo, CX Encoded |
| RUNNING TIME | 115 minutes |
| PRODUCER | MGM/UA Home Video |
| DISTRIBUTOR | Pioneer Video, Inc., ML100165, $34.95 |
| RETAIL OUTLET | (CED) Movies Unlimited Inc., A2-0165, $29.95 |

| | |
|---|---|
| TITLE | **POPEYE** |
| TYPE | Motion Picture, Music, Comedy (PG) |
| FORMAT | Laser CLV, CED (Stereo, CX Encoded) |
| FEATURES | Stereo |
| RUNNING TIME | 114 minutes |
| PRODUCER | Warner Home Video |
| DISTRIBUTOR | Pioneer Video, Inc., LV1171, $29.95 |
| | RCA Videodiscs, 10665, $29.98 |

| | |
|---:|:---|
| TITLE | **PORKY'S** |
| TYPE | Motion Picture, Comedy (R) |
| FORMAT | Laser CLV, CED |
| RUNNING TIME | 94 minutes |
| PRODUCER | CBS Fox Video |
| DISTRIBUTOR | Pioneer Video, Inc., 1149-80, $39.98 |
| RETAIL OUTLET | (CED) Movies Unlimited Inc., A1-1149, $29.98 |

| | |
|---:|:---|
| TITLE | **POSEIDON ADVENTURE** |
| TYPE | Motion Picture |
| FORMAT | CED |
| RETAIL OUTLET | Movies Unlimited Inc., A1-1058, $29.98 |

| | |
|---:|:---|
| TITLE | **POSTMAN ALWAYS RINGS TWICE, THE** |
| TYPE | Motion Picture (R) |
| FORMAT | Laser CLV, CED |
| RUNNING TIME | 120 minutes |
| PRODUCER | CBS Fox Video, Inc. |
| DISTRIBUTOR | Pioneer Video, Inc., 7077-80, $34.98 |
| RETAIL OUTLET | (CED) Movies Unlimited Inc., A2-1033, $24.95 |

| | |
|---:|:---|
| TITLE | **POTPOURRI** |
| DESCRIPTION | A series of "how-to's" and a state-of-the-art stereo audio experiment designed for the consumer player. |
| TYPE | Promotional |
| LEVEL | 1,2,3 |
| CONTACT FOR MORE INFO. | Corporation for Public Broadcasting - University of Nebraska KUON-TV/ Sony |
| PRODUCER | University of Nebraska/Nebraska Videodisc Design/Production Group |
| AVAILABILITY | Not available for release or distribution. |

| | |
|---:|:---|
| TITLE | **POWER PLANT TOUR** |
| DESCRIPTION | Power plant tour covered on a disc for engineers. |
| TYPE | Instructional |
| FORMAT | Laser CAV |
| LEVEL | 3 |
| CONTACT FOR MORE INFO. | Combustion Engineering, Inc. |
| PRODUCER | University of Nebraska/Nebraska Videodisc Design/Production Group |
| AVAILABILITY | Not available for release or distribution. |

| | |
|---:|:---|
| TITLE | **PRESLEY - KING OF ROCK 'N' ROLL** |
| TYPE | Music |
| FORMAT | CED |
| DISTRIBUTOR | RCA Videodiscs, 02056, $24.98 |

| | |
|---:|:---|
| TITLE | **PRETTY BABY** |
| TYPE | Motion Picture, Adult (R) |
| FORMAT | Laser CLV, CED |
| RUNNING TIME | 109 minutes |
| PRODUCER | Warner Home Video |
| DISTRIBUTOR | Pioneer Video, Inc., LV8940, $29.95 |
| | RCA Videodiscs, 00671, $29.98 |

**Videodiscography**

| | |
|---:|:---|
| TITLE | **PRIDE OF THE YANKEES, THE** |
| TYPE | Motion Picture |
| FORMAT | Laser CLV |
| FEATURES | Black and white |
| RETAIL OUTLET | (CLV) UltraVideo, VLD3375, $29.95 |

| | |
|---:|:---|
| TITLE | **PRINCE AND THE PAUPER** |
| TYPE | Motion Picture |
| FORMAT | CED |
| RETAIL OUTLET | Movies Unlimited Inc., A1-4637, $29.98 |

| | |
|---:|:---|
| TITLE | **PRINCE OF THE CITY** |
| TYPE | Motion Picture (R) |
| FORMAT | CED (2 Discs) |
| PRODUCER | Warner Home Video |
| DISTRIBUTOR | RCA Videodiscs, 03151, $39.98 |

| | |
|---:|:---|
| TITLE | **PRINCIPLES OF OPERANT CONDITIONING** |
| DESCRIPTION | A computer-controlled disc using a graphic overlay designed to teach psychology at the college level. |
| TYPE | Instructional |
| FORMAT | Laser CAV |
| LEVEL | 3 |
| CONTACT FOR MORE INFO. | Corporation for Public Broadcasting - University of Nebraska KUON-TV |
| PRODUCER | University of Nebraska/Nebraska Videodisc Design/Production Group |
| AVAILABILITY | Not available for release or distribution. |

| | |
|---:|:---|
| TITLE | **PRIVATE BENJAMIN** |
| TYPE | Motion Picture (R) |
| FORMAT | Laser CLV, CED |
| RUNNING TIME | 110 minutes |
| PRODUCER | Warner Home Video |
| DISTRIBUTOR | Pioneer Video, Inc., 11075LV, $29.98 |
| | RCA Videodiscs, 03140, $29.98 |

| | |
|---:|:---|
| TITLE | **PRIVATE EYES, THE** |
| TYPE | Motion Picture, Comedy (PG) |
| FORMAT | Laser CLV, CED |
| RUNNING TIME | 97 minutes |
| PRODUCER | Vestron Video |
| DISTRIBUTOR | Pioneer Video, Inc., VL5001, $34.95 |
| RETAIL OUTLET | (CED) Movies Unlimited Inc., A3-5001, $29.95 |

| | |
|---:|:---|
| TITLE | **PRIVATE LESSONS** |
| TYPE | Motion Picture, Comedy (R) |
| FORMAT | Laser CLV, CED |
| RUNNING TIME | 87 minutes |
| PRODUCER | Warner Home Video |
| DISTRIBUTOR | Pioneer Video, Inc., 16-027, $29.95 |
| RETAIL OUTLET | (CED) Movies Unlimited Inc., A5-6027, $29.98 |

| | |
|---:|:---|
| TITLE | **PRIVATE POPSICLE** |
| TYPE | Motion Picture |
| FORMAT | CED |
| RETAIL OUTLET | Movies Unlimited Inc., A2-0228, $29.95 |

| | |
|---:|:---|
| TITLE | **PRODUCERS, THE** |
| TYPE | Motion Picture, Comedy |
| FORMAT | Laser CLV, CED |
| RUNNING TIME | 88 minutes |
| PRODUCER | Warner Home Video |
| DISTRIBUTOR | Pioneer Video, Inc., 4058-80, $29.98 |
| | RCA Videodiscs, 00804, $19.98 |

| | |
|---:|:---|
| TITLE | **PRODUCING INTERACTIVE VIDEODISCS** |
| DESCRIPTION | A two-disc set covering the basic design and production parameters of producing interactive videodiscs. |
| TYPE | Instructional |
| FORMAT | Laser CAV |
| LEVEL | 1,2 |
| CONTACT FOR MORE INFO. | Optical Recording Project/3M |
| PRODUCER | University of Nebraska/Nebraska Videodisc Design/Production Group |
| AVAILABILITY | Available from the Optical Recording Project/3M. See their entry in the Mastering and Replication Resources section. |

| | |
|---:|:---|
| TITLE | **PRODUCTION ASSOCIATES, INC. - (NOT GIVEN)** |
| TYPE | Instructional |
| LEVEL | 3 |
| DEVELOPER | Sperry Gyroscope |
| FEATURES | Chapters, Picture Stops, Dual Audio |
| RUNNING TIME | 60 minutes |
| PRODUCER | Production Associates, Inc./Sperry Gyroscope |

| | |
|---:|:---|
| TITLE | **PROFESSIONAL SELLING SKILLS SYSTEM III** |
| DESCRIPTION | A research-based, 13-sided videodisc program to teach salespeople skills that make for selling success. Four of the 13 videodisc sides instruct sales managers how to coach their salespeople to reinforce the skills. Professional Selling Skills System III is designed for group instruction (3-4 days) and includes non-videodisc components, such as trainee workbooks and audio tapes. |
| TYPE | Instructional, In-house Training |
| FORMAT | Laser CAV |
| LEVEL | 2 |
| DEVELOPER | Judith H. Steele |
| CONTACT FOR MORE INFO. | Elizabeth Moser |
| FEATURES | Digital Dump-Sony, Dual Audio |
| PRODUCER | Xerox Learning Systems |
| DISTRIBUTOR | Same as producer |
| AVAILABILITY | Sale. Must purchase discs along with total training system. |
| SUGGESTED PRICE | Varies according to volume. Contact Xerox Learning Systems for more information. |

| | |
|---:|:---|
| TITLE | **PROM NIGHT** |
| TYPE | Motion Picture (R) |
| FORMAT | Laser CLV |
| RETAIL OUTLET | UltraVideo, 12-022, $29.95 |

| | |
|---:|:---|
| TITLE | **PSYCHO** |
| TYPE | Motion Picture (M) |
| FORMAT | Laser CLV, CED |
| FEATURES | Black and white |
| RUNNING TIME | 108 minutes |
| PRODUCER | Warner Home Video |
| DISTRIBUTOR | Pioneer Video, Inc., 11-003, $29.95 |
| | RCA Videodiscs, 03315, $24.98 |

| | |
|---:|:---|
| TITLE | **PSYCHO 2** |
| TYPE | Motion Picture (M) |
| FORMAT | Laser CLV |
| RETAIL OUTLET | UltraVideo, 40-008, $32.95 |

| | |
|---:|:---|
| TITLE | **PUBLIC ENEMY** |
| TYPE | Motion Picture |
| FORMAT | CED |
| DISTRIBUTOR | RCA Videodiscs, 01458, $24.98 |

| | |
|---:|:---|
| TITLE | **PUMPING IRON** |
| TYPE | Motion Picture (PG) |
| FORMAT | CED |
| DISTRIBUTOR | RCA Videodiscs, 01805, $19.98 |

| | |
|---:|:---|
| TITLE | **PUZZLE OF THE TACOMA NARROWS BRIDGE COLLAPSE, THE** |
| DESCRIPTION | A low cost and innovative approach to videodisc education designed for college level physics. |
| TYPE | Instructional |
| FORMAT | Laser CAV |
| LEVEL | 1,2,3 |
| CONTACT FOR MORE INFO. | National Science Foundation/University of Nebraska Physics Department |

| | |
|---:|:---|
| TITLE | **QUADROPHENIA** |
| TYPE | Music, Motion Picture (R) |
| FORMAT | CED |
| FEATURES | Stereo |
| PRODUCER | World Northal |
| DISTRIBUTOR | RCA Videodiscs, 12169, $24.98 |

| | |
|---:|:---|
| TITLE | **QUARTERHORSE** |
| DESCRIPTION | Sixty quarterhorse races with two independent audio tracks, multiple freeze frames, and an overlay of alphanumerics for the finish of each race. |
| TYPE | Game |

| | |
|---:|:---|
| FORMAT | Laser CAV |
| DEVELOPER | Rodesch & Associates, Inc./Escondido, CA 92025 |
| CONTACT FOR MORE INFO. | Dale F. Rodesch, (619) 743-7681 |
| FEATURES | Dual Audio |
| RUNNING TIME | 26 minutes |
| PRODUCER | Rodesch & Associates, Inc. |
| AVAILABILITY | Sale |
| SPECIAL CONDITIONS | Sold as part of finished game. |

| | |
|---:|:---|
| TITLE | **QUEEN - GREATEST FLIX** |
| TYPE | Music |
| FORMAT | Laser CLV, CED |
| FEATURES | Stereo, CX Encoded |
| RUNNING TIME | 58 minutes |
| PRODUCER | Pioneer Artists |
| DISTRIBUTOR | Pioneer Video, Inc., PA-82-026, $24.95 |
| | RCA Videodiscs, 12136, $24.98 |

| | |
|---:|:---|
| TITLE | **QUEST FOR FIRE** |
| TYPE | Motion Picture (R) |
| FORMAT | Laser CLV, CED |
| FEATURES | Stereo, CX Encoded |
| RUNNING TIME | 100 minutes |
| PRODUCER | Warner Home Video |
| DISTRIBUTOR | Pioneer Video, Inc., 1148-80, $34.98 |
| RETAIL OUTLET | (CED) Movies Unlimited Inc., A1-1148, $29.98 |

| | |
|---:|:---|
| TITLE | **QUICK DOG TRAINING** |
| TYPE | Motion Picture |
| FORMAT | CED |
| RETAIL OUTLET | Movies Unlimited Inc., A4-1086, $29.95 |

| | |
|---:|:---|
| TITLE | **QUIET MAN, THE** |
| TYPE | Motion Picture |
| FORMAT | CED (2 Discs) |
| DISTRIBUTOR | RCA Videodiscs, 00304, $34.98 |

| | |
|---:|:---|
| TITLE | **RACE FOR YOUR LIFE, CHARLIE BROWN** |
| TYPE | Motion Picture (G) |
| FORMAT | Laser CLV, CED |
| RUNNING TIME | 76 minutes |
| PRODUCER | Warner Home Video |
| DISTRIBUTOR | Pioneer Video, Inc., LV8850, $29.95 |
| | RCA Videodiscs, 00630, $24.98 |

| | |
|---:|:---|
| TITLE | **RADAR, MULTIPLE BEAM ACQUISITION AND WIDE APERTURE SURVEILLANCE** |
| DESCRIPTION | Introduction and familiarization explanation of various types of radars and their functions. |
| TYPE | Technical, Instructional, Promotional |

|   |   |
|---|---|
| FORMAT | Laser CAV |
| LEVEL | 3 |
| DEVELOPER | Ed Dextraze |
| CONTACT FOR MORE INFO. | Ed Dextraze, (617) 272-9300 |
| FEATURES | Sound Over Still |
| PRODUCER | Raytheon Service Company |
| AVAILABILITY | Not for sale. |

|   |   |
|---|---|
| TITLE | **RAGGEDY ANN & ANDY** |
| TYPE | Motion Picture |
| FORMAT | CED |
| RETAIL OUTLET | Movies Unlimited Inc., A1-7089, $29.98 |

|   |   |
|---|---|
| TITLE | **RAGGEDY MAN** |
| TYPE | Motion Picture (PG) |
| FORMAT | Laser CLV |
| RUNNING TIME | 93 minutes |
| PRODUCER | Warner Home Video |
| DISTRIBUTOR | Pioneer Video, Inc., 10-032, $29.95 |

|   |   |
|---|---|
| TITLE | **RAGING BULL** |
| TYPE | Motion Picture (R) |
| FORMAT | Laser CLV, CED (2 Discs) |
| FEATURES | Stereo, CX Encoded, Black and white |
| RUNNING TIME | 129 minutes |
| PRODUCER | Warner Home Video |
| DISTRIBUTOR | Pioneer Video, Inc., 4523-80, $34.98 |
|   | RCA Videodiscs, 01422, $39.98 |

|   |   |
|---|---|
| TITLE | **RAGTIME** |
| TYPE | Motion Picture (PG) |
| FORMAT | Laser CLV, CED (2 Discs) |
| RUNNING TIME | 156 minutes |
| PRODUCER | Warner Home Video |
| DISTRIBUTOR | Pioneer Video, Inc., LV1486, $39.95 |
|   | RCA Videodiscs, 00678, $34.98 |

|   |   |
|---|---|
| TITLE | **RAIDERS OF THE LOST ARK** |
| TYPE | Motion Picture |
| FORMAT | Laser CLV, Laser CAV |
| LEVEL | (CLV) 0, (CAV) 1 |
| RETAIL OUTLET | UltraVideo, (CLV) LV1376, $29.95; (CAV) LV1376C, $49.95 |

|   |   |
|---|---|
| TITLE | **RAINBOW GOBLINS STORY** |
| TYPE | Music |
| FORMAT | Laser CAV |
| LEVEL | 1 |

|  |  |
|---|---|
| FEATURES | Stereo |
| PRODUCER | Pioneer Video Imports |
| DISTRIBUTOR | Pioneer Video, Inc., MP023, $27.95 |

|  |  |
|---|---|
| TITLE | **RAINBOW: LIVE BETWEEN THE EYES** |
| TYPE | Music |
| FORMAT | Laser CAV |
| LEVEL | 1 |
| FEATURES | Stereo |
| RETAIL OUTLET | UltraVideo, 83-052, $24.95 |

|  |  |
|---|---|
| TITLE | **RAISE THE TITANIC** |
| TYPE | Motion Picture (PG) |
| FORMAT | Laser CLV, CED |
| RUNNING TIME | 115 minutes |
| PRODUCER | Warner Home Video |
| DISTRIBUTOR | Pioneer Video, Inc., 9023-80, $34.98 |
| RETAIL OUTLET | (CED) Movies Unlimited Inc., A1-9023, $29.98 |

|  |  |
|---|---|
| TITLE | **RALSTON-PURINA** |
| DESCRIPTION | A product information program for use at trade shows and conventions. |
| TYPE | Instructional, Promotional |
| FORMAT | Laser CAV |
| LEVEL | 2,3 |
| DEVELOPER | Jack Lieb Productions, Inc. |
| FEATURES | Chapters, Picture Stops, Digital Dump-Sony |
| AVAILABILITY | Not for sale. |

|  |  |
|---|---|
| TITLE | **RAPUNZEL** |
| TYPE | Motion Picture |
| FORMAT | Laser CLV |
| FEATURES | Closed captioned |
| RETAIL OUTLET | UltraVideo, 80-6370, $29.95 |

|  |  |
|---|---|
| TITLE | **RAVEN, THE** |
| TYPE | Motion Picture |
| FORMAT | Laser CLV, CED |
| RUNNING TIME | 86 minutes |
| PRODUCER | Vestron Video |
| DISTRIBUTOR | Pioneer Video, Inc., VL-4021, $34.95 |
| RETAIL OUTLET | (CED) Movies Unlimited Inc., A3-4021, $29.95 |

|  |  |
|---|---|
| TITLE | **RAY CHARLES CONCERT** |
| TYPE | Music |
| FORMAT | Laser CLV |
| FEATURES | Stereo |
| RETAIL OUTLET | UltraVideo, 74-612, $24.95 |

# Videodiscography

| | |
|---:|:---|
| TITLE | **RCA ENGINEERING DEMO DISC** |
| DESCRIPTION | Disc includes a self-paced tutorial on the interactive functions of videodiscs, a demonstration of features cross-referenced to a glossary, and a video clip data base of tennis professionals accessible through menu frames. The disc uses graphics, menu choice frames and animation. |
| TYPE | Research and Development, Technical, Instructional, In-house Training, Archival, Catalog |
| FORMAT | CED |
| LEVEL | 2 |
| DEVELOPER | John Vaughan and W.B. Porter of The Communication Studio |
| CONTACT FOR MORE INFO. | John Vaughan or W.B. Porter, (212) 924-3729 |
| FEATURES | Chapters, Picture Stops, Dual Audio |
| RUNNING TIME | 60 minutes |
| PRODUCER | Interactive Systems Organization (defunct) |
| DISTRIBUTOR | RCA Engineering Labs (Richard Allen) |
| AVAILABILITY | Used for in-house research and development. Not available outside. |

---

| | |
|---:|:---|
| TITLE | **RCA'S ALL-STAR COUNTRY MUSIC FAIR** |
| TYPE | Music |
| FORMAT | CED |
| FEATURES | Stereo, CX Encoded |
| DISTRIBUTOR | RCA Videodiscs, 12124, $19.98 |

---

| | |
|---:|:---|
| TITLE | **READING DEVELOPMENT VIDEODISC** |
| DESCRIPTION | Disc to help parents teach their child to read. |
| TYPE | Instructional |
| FORMAT | Laser CAV |
| LEVEL | 2 |
| DEVELOPER | BBC Open University Production Centre |
| CONTACT FOR MORE INFO. | Vic Lockwood |
| FEATURES | Picture Stops, Dual Audio |
| PRODUCER | Vic Lockwood |
| RETAIL OUTLET | W.H. Smith (Britain's largest newspaper and book retailer.) |

---

| | |
|---:|:---|
| TITLE | **REBEL WITHOUT A CAUSE** |
| TYPE | Motion Picture |
| FORMAT | CED |
| DISTRIBUTOR | RCA Videodiscs, 03109, $24.98 |

---

| | |
|---:|:---|
| TITLE | **RED ON ROUNDBALL** |
| TYPE | Instructional |
| FORMAT | CED |
| DISTRIBUTOR | RCA Videodiscs, 02122, $19.98 |

---

| | |
|---:|:---|
| TITLE | **RED RIVER** |
| TYPE | Motion Picture |
| FORMAT | CED (2 Discs) |
| DISTRIBUTOR | RCA Videodiscs, 01454, $34.98 |

| | |
|---:|:---|
| TITLE | **RED SHOES, THE** |
| TYPE | Motion Picture, Dance |
| FORMAT | CED (2 Discs) |
| DISTRIBUTOR | RCA Videodiscs, 01001, $34.98 |

| | |
|---:|:---|
| TITLE | **REDD FOXX - VIDEO IN A PLAIN BROWN WRAPPER** |
| TYPE | Adult |
| FORMAT | Laser CLV |
| RETAIL OUTLET | UltraVideo, VL2008, $34.95 |

| | |
|---:|:---|
| TITLE | **REDS** |
| TYPE | Motion Picture (PG) |
| FORMAT | Laser CLV, CED (2 Discs) |
| RUNNING TIME | 195 minutes |
| PRODUCER | Paramount Home Video |
| DISTRIBUTOR | Pioneer Video, Inc., LV1331, $39.95 |
| | RCA Videodiscs, 03610, $39.98 |

| | |
|---:|:---|
| TITLE | **RENAL ANALYSIS** |
| DESCRIPTION | Teaches renal analysis screening skills to CMS nurses. |
| TYPE | Medical, Instructional |
| FORMAT | Laser CAV |
| LEVEL | 3 |
| SYSTEM DESCRIPTION | Discovision 7820-1 player, Apple II Plus microcomputer, and a VAI-1 interface card. |
| DEVELOPER | MGT of America, Inc., Tallahassee, FL |
| CONTACT FOR MORE INFO. | Jayne Parker |
| RUNNING TIME | Varies, but approximately 6 hours |
| AVAILABILITY | Not for sale, but program may become available in the future. |

| | |
|---:|:---|
| TITLE | **REPAIR PARTS AND SPECIAL TOOLS LIST - AN UGC 74 (RPSTL)** |
| DESCRIPTION | A visual catalog of repair parts and special tools for an Army teletypewriter. Developed to demonstrate the cost-effectiveness and functional value of having an RPSTL on videodisc. |
| TYPE | Research and Development, Technical, Catalog |
| FORMAT | Laser CAV |
| LEVEL | 2,3 |
| SYSTEM DESCRIPTION | Pioneer player, Apple computer |
| DEVELOPER | Computer Sciences Corporation |
| CONTACT FOR MORE INFO. | Dr. William A. Stembler |
| FEATURES | Digital Dump-Pioneer |
| PRODUCER | Dr. William A. Stembler |
| AVAILABILITY | Property of U.S. Army. |

| | |
|---:|:---|
| TITLE | **RESEARCH DISC** |
| DESCRIPTION | An experimental disc designed to test the comparative effectiveness of visual and auditory elements. |
| TYPE | Research and Development |
| FORMAT | Laser CAV |
| LEVEL | 1, 2, 3 |
| CONTACT FOR MORE INFO. | Bureau of Education for the Handicapped/University of Nebraska Barkley Center |

# Videodiscography

| | |
|---|---|
| PRODUCER | University of Nebraska/Nebraska Videodisc Design/Production Group |
| AVAILABILITY | Not available for release or distribution. |

| | |
|---|---|
| TITLE | **RETURN OF A MAN CALLED HORSE** |
| TYPE | Motion Picture |
| FORMAT | CED |
| RETAIL OUTLET | Movies Unlimited Inc., A1-4591, $29.98 |

| | |
|---|---|
| TITLE | **RETURN OF THE KING** |
| TYPE | Motion Picture |
| FORMAT | CED |
| DISTRIBUTOR | RCA Videodiscs, 02089, $19.98 |

| | |
|---|---|
| TITLE | **RETURN OF THE PINK PANTHER, THE** |
| TYPE | Motion Picture, Comedy (G) |
| FORMAT | Laser CLV, CED |
| RUNNING TIME | 115 minutes |
| PRODUCER | Warner Home Video |
| DISTRIBUTOR | Pioneer Video, Inc., 9031-80, $34.98 |
| | RCA Videodiscs, 00521, $29.98 |

| | |
|---|---|
| TITLE | **RETURN OF THE STREET FIGHTER** |
| TYPE | Motion Picture |
| FORMAT | CED |
| RETAIL OUTLET | Movies Unlimited Inc., A1-7099, $29.98 |

| | |
|---|---|
| TITLE | **RETURN TO MACON COUNTY** |
| TYPE | Motion Picture |
| FORMAT | CED |
| RETAIL OUTLET | Movies Unlimited Inc., A3-4039, $29.95 |

| | |
|---|---|
| TITLE | **REVENGE OF THE PINK PANTHER** |
| TYPE | Motion Picture, Comedy (PG) |
| FORMAT | Laser CLV, CED |
| RUNNING TIME | 99 minutes |
| PRODUCER | CBS Fox Video |
| DISTRIBUTOR | Pioneer Video, Inc., 4610-80, $34.98 |
| | RCA Videodiscs, 01419, $29.98 |

| | |
|---|---|
| TITLE | **RICH AND FAMOUS** |
| TYPE | Motion Picture |
| FORMAT | Laser CLV, CED |
| RETAIL OUTLET | (CLV) UltraVideo, ML100111, $34.95 |
| | (CED) Movies Unlimited Inc., A2-1055, $29.95 |

| | |
|---|---|
| TITLE | **RICHARD PRYOR & STEVE MARTIN HOST SATURDAY NIGHT LIVE, VOL. 2** |
| TYPE | Comedy |
| FORMAT | CED |
| DISTRIBUTOR | RCA Videodiscs, 01210, $19.98 |

| | |
|---:|:---|
| TITLE | **RICHARD PRYOR LIVE IN CONCERT** |
| TYPE | Motion Picture, Comedy |
| FORMAT | CED |
| DISTRIBUTOR | RCA Videodiscs, 02001, $24.98 |

| | |
|---:|:---|
| TITLE | **RICHARD PRYOR LIVE ON THE SUNSET STRIP** |
| TYPE | Comedy (R) |
| FORMAT | Laser CLV, CED |
| DISTRIBUTOR | RCA Videodiscs, 03034, $29.98 |
| RETAIL OUTLET | (CLV) UltraVideo, VLD3375, $29.95 |

| | |
|---:|:---|
| TITLE | **RING OF BRIGHT WATER** |
| TYPE | Motion Picture |
| FORMAT | CED |
| ORDER NO. | A1-8022 |
| SUGGESTED PRICE | $29.98 |

| | |
|---:|:---|
| TITLE | **RIO BRAVO** |
| TYPE | Motion Picture |
| FORMAT | CED (2 Discs) |
| DISTRIBUTOR | RCA Videodiscs, 03155, $39.98 |

| | |
|---:|:---|
| TITLE | **RIO LOBO** |
| TYPE | Motion Picture |
| FORMAT | CED |
| RETAIL OUTLET | Movies Unlimited Inc., A2-1027, $24.95 |

| | |
|---:|:---|
| TITLE | **ROAD GAMES** |
| TYPE | Motion Picture |
| FORMAT | CED |
| RETAIL OUTLET | Movies Unlimited Inc., A4-0166, $29.95 |

| | |
|---:|:---|
| TITLE | **ROAD WARRIOR, THE** |
| TYPE | Motion Picture (R) |
| FORMAT | Laser CLV, CED |
| FEATURES | Dolby Stereo |
| RUNNING TIME | 95 minutes |
| PRODUCER | Warner Home Video |
| DISTRIBUTOR | Pioneer Video, Inc., 11181LV, $29.98 |
| RETAIL OUTLET | (CED) Movies Unlimited Inc., A1-1181W, $29.98 |

| | |
|---:|:---|
| TITLE | **ROARING TWENTIES, THE** |
| TYPE | Motion Picture |
| FORMAT | CED |
| RUNNING TIME | 95 minutes |
| PRODUCER | United Artists |
| DISTRIBUTOR | RCA Videodiscs, 03417, $24.98 |

| | |
|---:|:---|
| TITLE | **ROBE, THE** |
| TYPE | Motion Picture |
| FORMAT | CED (2 Discs) |
| RETAIL OUTLET | Movies Unlimited Inc., A1-1022, $39.98 |

| | |
|---:|:---|
| TITLE | **ROCK ADVENTURE** |
| TYPE | Music |
| FORMAT | Laser CAV |
| LEVEL | 1 |
| FEATURES | Stereo, CX Encoded |
| PRODUCER | Pioneer Video Imports |
| DISTRIBUTOR | Pioneer Video, Inc., MS009, $27.95 |

| | |
|---:|:---|
| TITLE | **ROCKY** |
| TYPE | Motion Picture (PG) |
| FORMAT | Laser CLV, CED |
| FEATURES | Stereo |
| RUNNING TIME | 118 minutes |
| PRODUCER | Warner Home Video |
| DISTRIBUTOR | Pioneer Video, Inc., 4546-80, $34.98 |
| | RCA Videodiscs, 01401, $29.98 |

| | |
|---:|:---|
| TITLE | **ROCKY 2** |
| TYPE | Motion Picture (PG) |
| FORMAT | Laser CLV, CED |
| FEATURES | Stereo, CX Encoded |
| RUNNING TIME | 119 minutes |
| PRODUCER | Warner Home Video |
| DISTRIBUTOR | Pioneer Video, Inc., 4565-80, $39.98 |
| | RCA Videodiscs, 01428, $29.98 |

| | |
|---:|:---|
| TITLE | **ROCKY 3** |
| TYPE | Motion Picture (PG) |
| FORMAT | Laser CLV, CED |
| FEATURES | Stereo, CX Encoded |
| RUNNING TIME | 100 minutes |
| PRODUCER | CBS Fox Video |
| DISTRIBUTOR | Pioneer Video, Inc., 4708-80, $39.98 |
| | RCA Videodiscs, 13421, $29.98 |

| | |
|---:|:---|
| TITLE | **ROD STEWART - TONIGHT HE'S YOURS** |
| TYPE | Music |
| FORMAT | Laser CLV, CED |
| FEATURES | Stereo, CX Encoded |
| RUNNING TIME | 90 minutes |
| PRODUCER | Warner Home Video |
| DISTRIBUTOR | Pioneer Video, Inc., 121150, $29.95 |
| RETAIL OUTLET | (CED) Movies Unlimited Inc., A4-2116, $21.95 |

| | |
|---:|:---|
| TITLE | **ROD STEWART LIVE AT THE LOS ANGELES FORUM** |
| TYPE | Music |
| FORMAT | CED |
| FEATURES | Stereo, CX Encoded |
| DISTRIBUTOR | RCA Videodiscs, 12074, $24.98 |

| | |
|---:|:---|
| TITLE | **RODAN** |
| TYPE | Motion Picture |
| FORMAT | Laser CLV, CED |
| RUNNING TIME | 72 minutes |
| PRODUCER | Vestron Video |
| DISTRIBUTOR | Pioneer Video, Inc., VL3013, $34.95 |
| RETAIL OUTLET | (CED) Movies Unlimited Inc., A3-3013, $29.95 |

| | |
|---:|:---|
| TITLE | **ROLLERBALL** |
| TYPE | Motion Picture |
| FORMAT | CED |
| RETAIL OUTLET | Movies Unlimited Inc., A1-4559, $29.98 |

| | |
|---:|:---|
| TITLE | **ROLLING THUNDER** |
| TYPE | Motion Picture |
| FORMAT | CED |
| RUNNING TIME | 119 minutes |
| PRODUCER | Paramount Pictures |
| DISTRIBUTOR | RCA Videodiscs, 00697, $24.98 |

| | |
|---:|:---|
| TITLE | **ROMAN HOLIDAY** |
| TYPE | Motion Picture |
| FORMAT | Laser CLV |
| RETAIL OUTLET | UltraVideo, LV6204, $29.95 |

| | |
|---:|:---|
| TITLE | **ROMEO AND JULIET** |
| TYPE | Motion Picture (PG) |
| FORMAT | Laser CLV, CED (2 Discs) |
| RUNNING TIME | 138 minutes |
| PRODUCER | Warner Home Video |
| DISTRIBUTOR | Pioneer Video, Inc., LV7809, $35.95 |
| | RCA Videodiscs, 00620, $34.98 |

| | |
|---:|:---|
| TITLE | **ROSE, THE** |
| TYPE | Motion Picture, Music (R) |
| FORMAT | Laser CLV, CED (2 Discs) |
| FEATURES | Stereo, CX Encoded |
| RUNNING TIME | 134 minutes |
| PRODUCER | Warner Home Video |
| DISTRIBUTOR | Pioneer Video, Inc., 1092-80, $39.98 |
| RETAIL OUTLET | (CED) Movies Unlimited Inc., A1-1092, $39.98 |

| | |
|---:|:---|
| TITLE | **ROSEMARY'S BABY** |
| TYPE | Motion Picture (R) |
| FORMAT | Laser CLV, CED (2 Discs) |

| | |
|---:|:---|
| RUNNING TIME | 137 minutes |
| PRODUCER | Warner Home Video |
| DISTRIBUTOR | Pioneer Video, Inc., LV6831, $35.95 |
| | RCA Videodiscs, 00621, $34.98 |

| | |
|---:|:---|
| TITLE | **ROSTROPOVICH: DVORAK CELLO CONCERTO/SAINT-SAENS** |
| TYPE | Music |
| FORMAT | Laser CLV |
| FEATURES | Stereo, CX Encoded |
| RUNNING TIME | 65 minutes |
| PRODUCER | Pioneer Artists |
| DISTRIBUTOR | Pioneer Video, Inc., PA-82-024, $24.95 |

| | |
|---:|:---|
| TITLE | **ROUGH CUT** |
| TYPE | Motion Picture (PG) |
| FORMAT | Laser CLV |
| RUNNING TIME | 111 minutes |
| PRODUCER | Warner Home Video |
| DISTRIBUTOR | Pioneer Video, Inc., LV1213, $29.95 |

| | |
|---:|:---|
| TITLE | **ROXY MUSIC** |
| TYPE | Music |
| FORMAT | CED |
| FEATURES | Stereo, Banded |
| DISTRIBUTOR | RCA Videodiscs, 12215, $19.98 |

| | |
|---:|:---|
| TITLE | **ROYAL WEDDING** |
| TYPE | Motion Picture |
| FORMAT | CED |
| RETAIL OUTLET | Movies Unlimited Inc., A2-1036, $29.95 |

| | |
|---:|:---|
| TITLE | **RUN SILENT, RUN DEEP** |
| TYPE | Motion Picture |
| FORMAT | CED |
| RETAIL OUTLET | Movies Unlimited Inc., A1-4657, $29.98 |

| | |
|---:|:---|
| TITLE | **RUSH - EXIT STAGE LEFT** |
| TYPE | Music |
| FORMAT | Laser CLV, CED |
| FEATURES | Stereo, CX Encoded |
| PRODUCER | Pioneer Artists |
| DISTRIBUTOR | Pioneer Video, Inc., PA-83-035, $24.95 |
| | RCA Videodiscs, 12127, $24.98 |

| | |
|---:|:---|
| TITLE | **RUSSIANS ARE COMING, THE RUSSIANS ARE COMING, THE** |
| TYPE | Motion Picture, Comedy (G) |
| FORMAT | CED |
| RUNNING TIME | 120 minutes |
| PRODUCER | United Artists |
| DISTRIBUTOR | RCA Videodiscs, 01492, $29.98 |

| | |
|---:|:---|
| TITLE | **RUST NEVER SLEEPS** |
| TYPE | Motion Picture, Music |
| FORMAT | CED |
| FEATURES | Stereo, CX Encoded |
| DISTRIBUTOR | RCA Videodiscs, 12062, $24.98 |

| | |
|---:|:---|
| TITLE | **S.O.B.** |
| TYPE | Motion Picture (R) |
| FORMAT | CED (2 Discs) |
| PRODUCER | Paramount Pictures |
| RETAIL OUTLET | Movies Unlimited Inc., A1-7110, $39.98 |

| | |
|---:|:---|
| TITLE | **SACRED MUSIC OF DUKE ELLINGTON** |
| TYPE | Music |
| FORMAT | CED |
| FEATURES | Stereo |
| RETAIL OUTLET | Movies Unlimited Inc., A2-1067, $29.95 |

| | |
|---:|:---|
| TITLE | **SAILOR WHO FELL FROM GRACE WITH THE SEA, THE** |
| TYPE | Motion Picture (R) |
| FORMAT | CED |
| DISTRIBUTOR | RCA Videodiscs, 00807, $24.98 |

| | |
|---:|:---|
| TITLE | **SALEM II** |
| DESCRIPTION | Surrogate travel videodisc of the Salem II nuclear power plant. |
| TYPE | In-house Training |
| FORMAT | Laser CAV |
| LEVEL | 3 |
| DEVELOPER | Interactive Television Company |
| CONTACT FOR MORE INFO. | Dr. Steven Levin, Interactive Television Company |
| RUNNING TIME | 30 minutes |
| AVAILABILITY | Contact Interactive Television Company. |

| | |
|---:|:---|
| TITLE | **SAMSON & DELILAH** |
| TYPE | Motion Picture |
| FORMAT | Laser CLV |
| RUNNING TIME | 128 minutes |
| PRODUCER | Paramount Home Video |
| DISTRIBUTOR | Pioneer Video, Inc., LV6726-2, $35.95 |

| | |
|---:|:---|
| TITLE | **SAMSON ET DALILA** |
| TYPE | Music |
| FORMAT | Laser CLV |
| FEATURES | Stereo, CX Encoded |
| RUNNING TIME | 118 minutes |
| PRODUCER | Pioneer Artists |
| DISTRIBUTOR | Pioneer Video, Inc., PA-82-014, $59.95 |

| | |
|---:|:---|
| TITLE | **SAN FRANCISCO** |
| DESCRIPTION | Photography of San Francisco harbor and 800 miles of harbor coastline. Includes corresponding nautical charts. |
| TYPE | Archival |
| FORMAT | Laser CAV |
| LEVEL | 3 |
| DEVELOPER | Interactive Television Company |
| CONTACT FOR MORE INFO. | Dr. Steven Levin, Interactive Television Company |
| RUNNING TIME | 30 minutes |
| AVAILABILITY | Contact Interactive Television Company. |

| | |
|---:|:---|
| TITLE | **SANDS OF IWO JIMA** |
| TYPE | Motion Picture |
| FORMAT | CED |
| DISTRIBUTOR | RCA Videodiscs, 00303, $24.98 |

| | |
|---:|:---|
| TITLE | **SATURDAY NIGHT FEVER** |
| TYPE | Motion Picture, Music, Dance (R) |
| FORMAT | Laser CLV |
| FEATURES | Stereo, CX Encoded |
| RUNNING TIME | 118 minutes |
| PRODUCER | Warner Home Video |
| DISTRIBUTOR | Pioneer Video, Inc., LV1113, $35.95 |
| | RCA Videodiscs, 00602, $29.98 |

| | |
|---:|:---|
| TITLE | **SATURN 3** |
| TYPE | Motion Picture (R) |
| FORMAT | Laser CLV, CED |
| FEATURES | Stereo |
| RUNNING TIME | 88 minutes |
| PRODUCER | Warner Home Video |
| DISTRIBUTOR | Pioneer Video, Inc., 9004-80, $29.98 |
| RETAIL OUTLET | (CED) Movies Unlimited Inc., A1-9004, $29.98 |

| | |
|---:|:---|
| TITLE | **SAVANNAH SMILES** |
| TYPE | Motion Picture |
| FORMAT | Laser CLV, CED |
| RETAIL OUTLET | (CLV) UltraVideo, E205850, $34.95 |
| | (CED) Movies Unlimited Inc., A4-0586, $29.95 |

| | |
|---:|:---|
| TITLE | **SAVE THE TIGER** |
| TYPE | Motion Picture (R) |
| FORMAT | Laser CLV |
| RUNNING TIME | 100 minutes |
| PRODUCER | Warner Home Video |
| DISTRIBUTOR | Pioneer Video, Inc., LV8479, $29.95 |

| | |
|---:|:---|
| TITLE | **SCANNERS** |
| TYPE | Motion Picture |
| FORMAT | Laser CLV |
| RETAIL OUTLET | UltraVideo, E208050, $34.95 |

| | |
|---:|:---|
| TITLE | **SCE: SONGS UNITS 2 & 3 SURROGATE TRAVEL** |
| DESCRIPTION | Surrogate travel through a nuclear power plant. Includes detailed close-ups of areas that are relatively inaccessible once the unit is in operation. |
| TYPE | Technical, Instructional, In-house Training, Archival |
| RELEASE DATE | January 1984 |
| LEVEL | 3 |
| SYSTEM DESCRIPTION | Joystick-controlled surrogate travel with push buttons to select additional information. |
| DEVELOPER | Combustion Engineering, Inc. |
| CONTACT FOR MORE INFO. | J.F. Church, W.B. Renfro |
| PRODUCER | Combustion Engineering and Southern California Edison Co. |
| AVAILABILITY | Not available |

| | |
|---:|:---|
| TITLE | **SCHOLASTIC PRODUCTIONS: AS WE GROW** |
| TYPE | Children, Instructional |
| FORMAT | CED |
| DISTRIBUTOR | RCA Videodiscs, 02031, $14.98 |

| | |
|---:|:---|
| TITLE | **SEA HAWK, THE** |
| TYPE | Motion Picture |
| FORMAT | CED |
| DISTRIBUTOR | RCA Videodiscs, 03418, $24.98 |

| | |
|---:|:---|
| TITLE | **SEA WOLVES** |
| TYPE | Motion Picture |
| FORMAT | CED (2 Discs) |
| RETAIL OUTLET | Movies Unlimited Inc., A1-7072, $39.98 |

| | |
|---:|:---|
| TITLE | **SEAFOOD DEPARTMENT OPERATIONS** |
| DESCRIPTION | Procedures for operating a supermarket seafood department. |
| TYPE | In-house Training |
| FORMAT | Laser CAV |
| LEVEL | 3 |
| DEVELOPER | Interactive Television Company |
| CONTACT FOR MORE INFO. | Dr. Steven Levin |
| RUNNING TIME | 30 minutes |
| AVAILABILITY | Contact Interactive Television Company. |

| | |
|---:|:---|
| TITLE | **SEARCHERS, THE** |
| TYPE | Motion Picture |
| FORMAT | CED |
| DISTRIBUTOR | RCA Videodiscs, 03106, $29.98 |

| | |
|---:|:---|
| TITLE | **SECRET OF N-I-M-H, THE** |
| TYPE | Motion Picture (G) |
| FORMAT | Laser CLV, CED |
| FEATURES | Stereo, CX Encoded |
| RUNNING TIME | 83 minutes |
| PRODUCER | MGM/UA Home Video |
| DISTRIBUTOR | Pioneer Video, Inc., ML100211, $34.95 |
| RETAIL OUTLET | (CED) Movies Unlimited Inc., A2-0211, $29.95 |

| | |
|---:|:---|
| TITLE | **SECRET POLICEMAN'S OTHER BALL** |
| TYPE | Motion Picture |
| FORMAT | CED |
| FEATURES | Stereo |
| RETAIL OUTLET | Movies Unlimited Inc., A2-1064, $29.95 |

| | |
|---:|:---|
| TITLE | **SEDUCTION OF JOE TYNAN, THE** |
| TYPE | Motion Picture (R) |
| FORMAT | Laser CLV |
| RUNNING TIME | 107 minutes |
| PRODUCER | Warner Home Video |
| DISTRIBUTOR | Pioneer Video, Inc., 10-024, $29.95 |

| | |
|---:|:---|
| TITLE | **SEDUCTION, THE** |
| TYPE | Motion Picture |
| FORMAT | CED |
| RETAIL OUTLET | Movies Unlimited Inc., A4-0556, $29.95 |

| | |
|---:|:---|
| TITLE | **SEEMS LIKE OLD TIMES** |
| TYPE | Motion Picture, Comedy (PG) |
| FORMAT | Laser CLV, CED |
| PRODUCER | Warner Home Video |
| DISTRIBUTOR | Pioneer Video, Inc., VLD5545, $29.95 |
| | RCA Videodiscs, 03031, $24.98 |

| | |
|---:|:---|
| TITLE | **SELLING THE PSYCHOLOGICAL** |
| DESCRIPTION | A general course in salesmanship, using an interactive videodisc under the control of a PLATO terminal. |
| TYPE | Instructional |
| RELEASE DATE | 1979 |
| FORMAT | Laser CAV |
| LEVEL | 3 |
| DEVELOPER | Control Data Education Co. - PLATO Advanced Development |
| PRODUCER | Howard Mark |
| DISTRIBUTOR | Control Data Corporation |
| AVAILABILITY | Out of production. |

| | |
|---:|:---|
| TITLE | **SEMI-TOUGH** |
| TYPE | Motion Picture, Comedy (R) |
| FORMAT | Laser CLV, CED |
| RUNNING TIME | 108 minutes |
| PRODUCER | Warner Home Video |
| DISTRIBUTOR | Pioneer Video, Inc., 4517-80, $34.98 |
| | RCA Videodiscs, 01450, $24.98 |

| | |
|---:|:---|
| TITLE | **SENDER, THE** |
| TYPE | Motion Picture |
| FORMAT | Laser CLV |
| RETAIL OUTLET | UltraVideo, LV1537, $29.95 |

| | |
|---:|:---|
| TITLE | **SENIORS** |
| TYPE | Motion Picture |
| FORMAT | CED |
| RETAIL OUTLET | Movies Unlimited Inc., A3-4011, $29.95 |

| | |
|---:|:---|
| TITLE | **SEPARATE WAYS** |
| TYPE | Motion Picture |
| FORMAT | CED |
| RETAIL OUTLET | Movies Unlimited Inc., A3-5010, $29.95 |

| | |
|---:|:---|
| TITLE | **SERGEANT YORK** |
| TYPE | Motion Picture |
| FORMAT | CED (2 Discs) |
| DISTRIBUTOR | RCA Videodiscs, 03419, $34.98 |

| | |
|---:|:---|
| TITLE | **SERIAL** |
| TYPE | Motion Picture, Comedy (R) |
| FORMAT | Laser CLV |
| RUNNING TIME | 90 minutes |
| PRODUCER | Warner Home Video |
| DISTRIBUTOR | Pioneer Video, Inc., LV1191, $29.95 |

| | |
|---:|:---|
| TITLE | **SERPICO** |
| TYPE | Motion Picture (R) |
| FORMAT | Laser CLV, CED (2 Discs) |
| RUNNING TIME | 130 minutes |
| PRODUCER | Warner Home Video |
| DISTRIBUTOR | Pioneer Video, Inc., LV8689, $39.95 |
| | RCA Videodiscs, 00651, $39.98 |

| | |
|---:|:---|
| TITLE | **SEVEN BRIDES FOR SEVEN BROTHERS** |
| TYPE | Motion Picture |
| FORMAT | CED |
| RETAIL OUTLET | Movies Unlimited Inc., A2-1042, $29.95 |

| | |
|---:|:---|
| TITLE | **SEVEN YEAR ITCH, THE** |
| TYPE | Motion Picture |
| FORMAT | CED |
| RETAIL OUTLET | Movies Unlimited Inc., A1-0114, $29.98 |

| | |
|---:|:---|
| TITLE | **SEX ON THE RUN** |
| TYPE | Motion Picture, Adult (R) |
| FORMAT | Laser CLV, CED |
| RUNNING TIME | 88 minutes |
| PRODUCER | Warner Home Video |
| DISTRIBUTOR | Pioneer Video, Inc., WL5501, $29.95 |
| RETAIL OUTLET | (CED) Movies Unlimited Inc., A3-5501, $24.95 |

| | |
|---:|:---|
| TITLE | **SGT. PEPPER'S LONELY HEARTS CLUB BAND** |
| TYPE | Music |
| FORMAT | Laser CLV |
| FEATURES | Stereo |
| RETAIL OUTLET | UltraVideo, 17-004, $29.95 |

| | |
|---:|:---|
| TITLE | **SHADOWS AND LIGHT** |
| TYPE | Music |
| FORMAT | Laser CLV, CED |
| FEATURES | Stereo, (CED) CX Encoded |
| RUNNING TIME | 58 minutes |
| PRODUCER | Pioneer Artists |
| DISTRIBUTOR | Pioneer Video, Inc., PA-81-003, $24.95 |
| | RCA Videodiscs, 12079, $24.98 |

| | |
|---:|:---|
| TITLE | **SHAFT** |
| TYPE | Motion Picture |
| FORMAT | CED |
| RETAIL OUTLET | Movies Unlimited Inc., A2-0191, $29.95 |

| | |
|---:|:---|
| TITLE | **SHAGGY DOG, THE** |
| TYPE | Motion Picture (G) |
| FORMAT | Laser CLV, CED |
| FEATURES | Black and white |
| RUNNING TIME | 101 minutes |
| PRODUCER | Walt Disney |
| DISTRIBUTOR | Pioneer Video, Inc., 42043AS, $34.95 |
| | RCA Videodiscs, 00713, $24.98 |

| | |
|---:|:---|
| TITLE | **SHANE** |
| TYPE | Motion Picture |
| FORMAT | Laser CLV, CED |
| RUNNING TIME | 117 minutes |
| PRODUCER | Warner Home Video |
| DISTRIBUTOR | Pioneer Video, Inc., LV6522, $29.95 |
| | RCA Videodiscs, 00622, $24.98 |

| | |
|---:|:---|
| TITLE | **SHARKEY'S MACHINE** |
| TYPE | Motion Picture |
| FORMAT | CED |
| DISTRIBUTOR | RCA Videodiscs, 03141, $29.98 |

| | |
|---:|:---|
| TITLE | **SHEENA EASTON LIVE AT THE PALACE** |
| TYPE | Music |
| FORMAT | Laser CLV, CED |
| FEATURES | Stereo, CX Encoded |
| RUNNING TIME | 59 minutes |
| PRODUCER | Pioneer Artists |
| DISTRIBUTOR | Pioneer Video, Inc., PA-83-045, $24.95 |
| | RCA Videodiscs, 12177, $19.98 |

| | |
|---:|:---|
| TITLE | **SHINING, THE** |
| TYPE | Motion Picture |
| FORMAT | CED (2 Discs) |
| DISTRIBUTOR | RCA Videodiscs, 03143, $39.98 |

| | |
|---:|:---|
| TITLE | **SHOGUN** |
| TYPE | Motion Picture (PG) |
| FORMAT | Laser CLV, CED |
| RUNNING TIME | 119 minutes |
| PRODUCER | Warner Home Video |
| DISTRIBUTOR | Pioneer Video, Inc., LV1423, $29.95 |
| | RCA Videodiscs, 00658, $29.98 |

| | |
|---:|:---|
| TITLE | **SHOGUN ASSASSIN** |
| TYPE | Motion Picture (R) |
| FORMAT | Laser CLV |
| RETAIL OUTLET | UltraVideo, 12-023, $29.95 |

| | |
|---:|:---|
| TITLE | **SHOOT THE MOON** |
| TYPE | Motion Picture |
| FORMAT | CED |
| RETAIL OUTLET | Movies Unlimited Inc., A2-1066, $29.95 |

| | |
|---:|:---|
| TITLE | **SHOOTIST, THE** |
| TYPE | Motion Picture (PG) |
| FORMAT | Laser CLV, CED |
| RUNNING TIME | 100 minutes |
| PRODUCER | Warner Home Video |
| DISTRIBUTOR | Pioneer Video, Inc., LV8904, $29.95 |
| | RCA Videodiscs, 00640, $24.98 |

| | |
|---:|:---|
| TITLE | **SHOT IN THE DARK, A** |
| TYPE | Motion Picture, Comedy |
| FORMAT | Laser CLV, CED |
| RUNNING TIME | 101 minutes |
| PRODUCER | CBS Fox Video |
| DISTRIBUTOR | Pioneer Video, Inc., 4528-80, $34.98 |
| | RCA Videodiscs, 01437, $24.98 |

| | |
|---:|:---|
| TITLE | **SHOWDOWN: SUGAR RAY LEONARD VS. THOMAS HEARNS** |
| TYPE | Motion Picture |
| FORMAT | CED |
| DISTRIBUTOR | RCA Videodiscs, 02085, $19.98 |

| | |
|---:|:---|
| TITLE | **SHUTTLE** |
| DESCRIPTION | Disc includes approximately 3,400 NASA photos of STS development and on-board photography for STS 1, 2, 3, and 4; development movies; engineering movies; staging; views from orbit, cargo bay, manipulator arm experiments, and cabin interior; and landing video from chase planes. |

|  |  |
|---|---|
| TYPE | Technical, Instructional, Archival |
| FORMAT | Laser CAV |
| LEVEL | 1 |
| DEVELOPER | Drew University/Center for Aerospace Education in cooperation with NASA |
| CONTACT FOR MORE INFO. | Patrick Binnes, (201) 377-0302 |
| FEATURES | Chapters, Picture Stops, Stereo |
| PRODUCER | Video Vision Associates, Ltd. |
| DISTRIBUTOR | Same as producer |
| AVAILABILITY | Sale |
| ORDER NO. | SPACE DISC 3 |
| SUGGESTED PRICE | $320 |

|  |  |
|---|---|
| TITLE | **SIGHT THROUGH SOUND, AN INTERACTIVE INTRODUCTION TO MEDICAL DIAGNOSTIC ULTRASOUND** |
| DESCRIPTION | A highly interactive introduction to ultrasound, covering the physics of ultrasound, ultrasound equipment operation, image interpretation, scanning techniques, and clinical applications. Includes case studies, games, and simulations of equipment operation. |
| TYPE | Medical, Technical, Instructional, In-house Training |
| RELEASE DATE | 1/84 |
| FORMAT | Laser CAV |
| LEVEL | 1,2 |
| DEVELOPER | Interactive Video Concepts, Inc. (IVC) |
| CONTACT FOR MORE INFO. | Peg Callahan |
| FEATURES | Chapters, Picture Stops, Digital Dump-Pioneer |
| PRODUCER | Interactive Video Concepts, Inc. |
| DISTRIBUTOR | W.B. Saunders (CBS Educational and Professional Publishing) |
| AVAILABILITY | Sale |
| ORDER NO. | 0721611036 |
| SUGGESTED PRICE | $495 |

|  |  |
|---|---|
| TITLE | **SILENT PARTNER, THE** |
| TYPE | Motion Picture, Comedy (R) |
| FORMAT | Laser CLV, CED |
| RUNNING TIME | 103 minutes |
| PRODUCER | Vestron Video |
| DISTRIBUTOR | Pioneer Video, Inc., VL5007, $34.95 |
| RETAIL OUTLET | (CED) Movies Unlimited Inc., A3-5007, $29.95 |

|  |  |
|---|---|
| TITLE | **SILENT RAGE** |
| FORMAT | CED |
| DISTRIBUTOR | RCA Videodiscs, 03076, $29.95 |

|  |  |
|---|---|
| TITLE | **SILVER STREAK** |
| TYPE | Motion Picture |
| FORMAT | CED |
| RETAIL OUTLET | Movies Unlimited Inc., A1-1080, $29.98 |

| | |
|---:|:---|
| TITLE | **SIMON & GARFUNKEL IN CONCERT** |
| TYPE | Music |
| FORMAT | CED |
| FEATURES | Stereo |
| RETAIL OUTLET | Movies Unlimited Inc., A2-1060, $24.95 |

| | |
|---:|:---|
| TITLE | **SIMON & GARFUNKEL: THE CONCERT IN CENTRAL PARK** |
| TYPE | Music |
| FORMAT | Laser CLV |
| FEATURES | Stereo, CX Encoded |
| RUNNING TIME | 87 minutes |
| PRODUCER | CBS Fox Video |
| DISTRIBUTOR | Pioneer Video, Inc., 7133-80, $34.98 |

| | |
|---:|:---|
| TITLE | **SINBAD AND THE EYE OF THE TIGER** |
| TYPE | Motion Picture (G) |
| FORMAT | Laser CLV, CED |
| PRODUCER | Columbia Pictures |
| DISTRIBUTOR | RCA Videodiscs, 03019, $19.98 |
| RETAIL OUTLET | (CLV) UltraVideo, VLD5550, $29.95 |

| | |
|---:|:---|
| TITLE | **SINGIN' IN THE RAIN** |
| TYPE | Motion Picture |
| FORMAT | Laser CLV, CED |
| RETAIL OUTLET | (CLV) UltraVideo, ML100185, $34.95 |
| | (CED) Movies Unlimited Inc., A2-0185, $29.95 |

| | |
|---:|:---|
| TITLE | **SIX PACK** |
| TYPE | Motion Picture |
| FORMAT | CED |
| RETAIL OUTLET | Movies Unlimited Inc., A1-1183, $29.98 |

| | |
|---:|:---|
| TITLE | **SIX WEEKS** |
| TYPE | Motion Picture (PG) |
| FORMAT | Laser CLV, CED |
| PRODUCER | Columbia Pictures |
| DISTRIBUTOR | RCA Videodiscs, 03055, $24.98 |
| RETAIL OUTLET | UltraVideo, VLD3377, $29.95 |

| | |
|---:|:---|
| TITLE | **SKF: DIAGNOSTIC CHALLENGES IN GASTROENTEROLOGY** |
| DESCRIPTION | Patient simulation that allows a doctor to diagnose and treat a patient who has ulcers. |
| TYPE | Medical, Instructional |
| RELEASE DATE | 11/82 |
| FORMAT | Laser CAV |
| LEVEL | 2 |
| DEVELOPER | Fusion Media, Inc. |
| FEATURES | Digital Dump-Pioneer |
| PRODUCER | EPIC, Ltd. London, England |
| DISTRIBUTOR | Smith, Kline & French Pharmaceuticals (SKF), United Kingdom |

# Videodiscography

| | |
|---:|:---|
| TITLE | **SKF: THE HIDDEN LANGUAGE OF ARTHRITIS** |
| DESCRIPTION | Patient simulation that allows a doctor to learn the symptoms described by a patient that lead to a diagnosis of arthritis. |
| TYPE | Medical, Instructional |
| RELEASE DATE | 10/83 |
| FORMAT | Laser CAV |
| LEVEL | 2 |
| FEATURES | Digital Dump-Pioneer |
| PRODUCER | Fusion Media, Inc. |
| DISTRIBUTOR | Smith, Kline & French Pharmaceuticals (SKF), United Kingdom |

| | |
|---:|:---|
| TITLE | **SLAP SHOT** |
| TYPE | Motion Picture (R) |
| FORMAT | Laser CLV |
| RUNNING TIME | 123 minutes |
| PRODUCER | Warner Home Video |
| DISTRIBUTOR | Pioneer Video, Inc., 16-004, $29.95 |

| | |
|---:|:---|
| TITLE | **SLAVE OF THE CANNIBAL GODS** |
| TYPE | Motion Picture |
| FORMAT | CED |
| RETAIL OUTLET | Movies Unlimited Inc., A3-C035, $29.98 |

| | |
|---:|:---|
| TITLE | **SLEEPER** |
| TYPE | Motion Picture, Comedy (PG) |
| FORMAT | Laser CLV, CED |
| RUNNING TIME | 88 minutes |
| PRODUCER | CBS Fox Video |
| DISTRIBUTOR | Pioneer Video, Inc., 4522-80, $34.98 |
| | RCA Videodiscs, 01417, $24.98 |

| | |
|---:|:---|
| TITLE | **SLEEPING BEAUTY** |
| TYPE | Motion Picture |
| FORMAT | Laser CLV |
| FEATURES | Closed captioned |
| RETAIL OUTLET | UltraVideo, 80-6171, $34.95 |

| | |
|---:|:---|
| TITLE | **SLEEPWALKER** |
| DESCRIPTION | An odyssey into the twilight between waking and dreaming. Presents a subjective narrative cued by an icongraphic audio montage. |
| TYPE | Art, Adolescents |
| RELEASE DATE | March 1984 |
| FORMAT | CED, Laser CAV |
| LEVEL | 1 |
| DEVELOPER | Leviathan Studios |
| CONTACT FOR MORE INFO. | Michael Harvey, (415) 931-1378 |
| FEATURES | Chapters, Stereo, CX Encoded |
| RUNNING TIME | 50 minutes |
| PRODUCER | Paul Radomski |
| DISTRIBUTOR | Leviathan Studios, $35 |

| | |
|---:|:---|
| TITLE | **SLIPSTREAM - STARRING JETHRO TUL** |
| TYPE | Music |
| FORMAT | CED |
| FEATURES | Stereo, CX Encoded, 12116, $24.98 |

| | |
|---:|:---|
| TITLE | **SLUMBER PARTY '57** |
| TYPE | Motion Picture, Comedy, Adults |
| FORMAT | Laser CLV, CED |
| RUNNING TIME | 89 minutes |
| PRODUCER | Vestron Video |
| DISTRIBUTOR | Pioneer Video, Inc., VL4007, $29.95 |
| RETAIL OUTLET | (CED) Movies Unlimited Inc., A3-4007, $24.95 |

| | |
|---:|:---|
| TITLE | **SMALL TOWN IN TEXAS, A** |
| TYPE | Motion Picture |
| FORMAT | CED |
| RETAIL OUTLET | Movies Unlimited Inc., A3-4034, $29.95 |

| | |
|---:|:---|
| TITLE | **SMOKEY AND THE BANDIT** |
| TYPE | Motion Picture, Comedy (PG) |
| FORMAT | Laser CLV, CED |
| PRODUCER | Warner Home Video |
| DISTRIBUTOR | Pioneer Video, Inc., 12-004, $29.95 |
| | RCA Videodiscs, 03305, $29.98 |

| | |
|---:|:---|
| TITLE | **SMOKEY AND THE BANDIT II** |
| TYPE | Motion Picture, Comedy (PG) |
| FORMAT | Laser CLV |
| RUNNING TIME | 101 minutes |
| PRODUCER | Warner Home Video |
| DISTRIBUTOR | Pioneer Video, Inc., 12-012, $29.95 |
| RETAIL OUTLET | (CLV) UltraVideo, VLD5951, $29.95 |

| | |
|---:|:---|
| TITLE | **SMOKING: HOW TO STOP** |
| TYPE | Instructional |
| FORMAT | Laser CLV |
| RETAIL OUTLET | UltraVideo, 52-003, $9.95 |

| | |
|---:|:---|
| TITLE | **SOCIAL STYLE SALES STRATEGIES** |
| DESCRIPTION | Teaches the salesperson how to observe and identify different styles of behavior in the prospective customer; and how to tailor sales strategies to those individual differences. |
| TYPE | Instructional |
| RELEASE DATE | Spring 1984 |
| FORMAT | Laser CAV |
| LEVEL | 3 |
| DEVELOPER | IT, Inc. |
| CONTACT FOR MORE INFO. | Gary Quinlin, President, or Constance Buffalo, Vice President for Production, (800) 328-7937 |
| RUNNING TIME | 8-10 hours |
| DISTRIBUTOR | Wilson Learning Corporation |

| | |
|---:|:---|
| TITLE | **SOLDIER, THE** |
| TYPE | Motion Picture |
| FORMAT | CED |
| RETAIL OUTLET | Movies Unlimited Inc., A4-0016, $29.95 |

| | |
|---:|:---|
| TITLE | **SOME KIND OF HERO** |
| TYPE | Motion Picture, Comedy (R) |
| FORMAT | Laser CLV, CED |
| RUNNING TIME | 97 minutes |
| PRODUCER | Warner Home Video |
| DISTRIBUTOR | Pioneer Video, Inc., LV1118, $29.95 |
| | RCA Videodiscs, 03601, $24.98 |

| | |
|---:|:---|
| TITLE | **SOME LIKE IT HOT** |
| TYPE | Motion Picture, Comedy |
| FORMAT | Laser CLV, CED |
| FEATURES | Black and white |
| RUNNING TIME | 121 minutes |
| PRODUCER | Warner Home Video |
| DISTRIBUTOR | Pioneer Video, Inc., 4577-80, $34.98 |
| | RCA Videodiscs, 01441, $24.98 |

| | |
|---:|:---|
| TITLE | **SOMETHING WICKED THIS WAY COMES** |
| TYPE | Motion Picture |
| FORMAT | Laser CLV |
| RETAIL OUTLET | UltraVideo, 0166AS, $34.95 |

| | |
|---:|:---|
| TITLE | **SOMEWHERE IN TIME** |
| TYPE | Motion Picture (PG) |
| FORMAT | Laser CLV |
| RUNNING TIME | 103 minutes |
| PRODUCER | MCA Videodisc, Inc. |
| DISTRIBUTOR | Pioneer Video, Inc., 13-005, $29.95 |

| | |
|---:|:---|
| TITLE | **SON OF FOOTBALL FOLLIES AND BIG GAME AMERICA, THE** |
| TYPE | Motion Picture |
| FORMAT | Laser CAV |
| LEVEL | 1 |
| RUNNING TIME | 45 minutes |
| PRODUCER | Warner Home Video |
| DISTRIBUTOR | Pioneer Video, Inc., LV002, $24.95 |

| | |
|---:|:---|
| TITLE | **SONS OF KATIE ELDER, THE** |
| TYPE | Motion Picture |
| FORMAT | CED |
| DISTRIBUTOR | RCA Videodiscs, 03606, $24.98 |

| | |
|---:|:---|
| TITLE | **SONY OVERVIEW** |
| DESCRIPTION | An overview and guide to videodisc production. |
| TYPE | Technical, Instructional, In-house Training, Promotional |
| FORMAT | Laser CAV |

| | |
|---|---|
| LEVEL | 2,3 |
| DEVELOPER | Sony Corporation of America |
| CONTACT FOR MORE INFO. | Mary Ann Axt, (201) 930-1000 |
| FEATURES | Digital Dump-Pioneer, Dual Audio, CX Encoded |
| RUNNING TIME | Under 30 minutes |
| PRODUCER | Sony Corporation of America |
| DISTRIBUTOR | Same as producer |
| AVAILABILITY | Sale |
| ORDER NO. | V8739 |
| SUGGESTED PRICE | $50 |

---

| | |
|---|---|
| TITLE | **SOPHIE'S CHOICE** |
| TYPE | Motion Picture (R) |
| FORMAT | Laser CLV, CED (2 Discs) |
| RUNNING TIME | 150 minutes |
| PRODUCER | CBS Fox Video |
| DISTRIBUTOR | Pioneer Video, Inc., 9076-80, $49.98 |
| | RCA Videodiscs, 00530, $39.98 |

---

| | |
|---|---|
| TITLE | **SOUND OF MUSIC, THE** |
| TYPE | Motion Picture, Music (G) |
| FORMAT | Laser CLV, CED |
| FEATURES | Stereo, CX Encoded |
| RUNNING TIME | 174 minutes |
| PRODUCER | Warner Home Video |
| DISTRIBUTOR | Pioneer Video, Inc., 1051-80, $49.98 |
| RETAIL OUTLET | (CED) Movies Unlimited Inc., A1-1051, $39.98 |

---

| | |
|---|---|
| TITLE | **SOUTH PACIFIC** |
| TYPE | Motion Picture |
| FORMAT | CED (2 Discs) |
| RETAIL OUTLET | Movies Unlimited Inc., A1-7045, $39.98 |

---

| | |
|---|---|
| TITLE | **SOUTHERN COMFORT** |
| TYPE | Motion Picture |
| FORMAT | Laser CLV |
| RETAIL OUTLET | UltraVideo, E302650, $34.95 |

---

| | |
|---|---|
| TITLE | **SOYLENT GREEN** |
| TYPE | Motion Picture |
| FORMAT | CED |
| RETAIL OUTLET | Movies Unlimited Inc., A2-1050, $29.95 |

---

| | |
|---|---|
| TITLE | **SPACE AGE, THE** |
| TYPE | Motion Picture, Technical, Instructional, Archival |
| FORMAT | Laser CAV |
| LEVEL | 1 |
| DEVELOPER | Drew University/Center for Aerospace Education in cooperation with NASA |
| CONTACT FOR MORE INFO. | Patrick Binnes (201) 377-0302 |

| | |
|---|---|
| FEATURES | Chapters, Picture Stops, Stereo |
| PRODUCER | Video Vision Associates, Ltd. |
| DISTRIBUTOR | Video Vision Associates, Ltd. |
| ORDER NO. | SPACE DISC 5 |
| SUGGESTED PRICE | $320 |

| | |
|---|---|
| TITLE | **SPACE SHUTTLE MISSION REPORTS: S.T.S. 5, 6 & 7** |
| DESCRIPTION | Flight of shuttle missions 5, 6, and 7. Includes on-board orbital activity and about 900 still frames. |
| TYPE | Children, Adolescents, Technical, Instructional, Archival |
| FORMAT | Laser CAV |
| LEVEL | 1, 2 |
| DEVELOPER | Video Vision Associates, Ltd. |
| CONTACT FOR MORE INFO. | Patrick Binnes, (201) 377-0302 |
| FEATURES | Chapters, Picture Stops, Dual Audio |
| PRODUCER | Video Vision Associates, Ltd. |
| DISTRIBUTOR | Same as producer |
| SUGGESTED PRICE | $39.95 |

| | |
|---|---|
| TITLE | **SPACE SHUTTLE: MISSION CONTROL** |
| DESCRIPTION | Documentary footage and stills from Space Shuttle missions 5, 6, and 7 (7 is with Sally Ride). There are 300 stills per mission. |
| TYPE | Children, Adolescents, Instructional, Archival |
| RELEASE DATE | 3/84 |
| FORMAT | CED |
| LEVEL | 1 |
| DEVELOPER | Video Vision Associates, Ltd. |
| CONTACT FOR MORE INFO. | Diane Smook (212) 930-4767 |
| FEATURES | Chapters, Picture Stops, Dual Audio |
| PRODUCER | Video Vision Associates, Ltd. |
| DISTRIBUTOR | RCA Videodiscs, 25006, $29.95 |

| | |
|---|---|
| TITLE | **SPACEHUNTER: ADVENTURES IN THE FORBIDDEN ZONE** |
| TYPE | Motion Picture (PG) |
| FORMAT | CED |
| FEATURES | Stereo |
| DISTRIBUTOR | RCA Videodiscs, 13056, $24.98 |

| | |
|---|---|
| TITLE | **SPANISH MAP READING CONVERSION PROJECT** |
| DESCRIPTION | Films on map reading skills, available with English and Spanish narration, were converted into three level 2 interactive discs. |
| TYPE | Research and Development, Technical, Instructional |
| FORMAT | Laser CAV |
| LEVEL | 2 |
| DEVELOPER | Computer Sciences Corporation |
| CONTACT FOR MORE INFO. | Dr. William A. Stembler |
| FEATURES | Dual Audio |
| PRODUCER | Dr. William A. Stembler |
| AVAILABILITY | Property of U.S. Army. |

| | |
|---:|:---|
| TITLE | **SPARROW MISSILE TEST SET TRAINING VIDEODISC** |
| DESCRIPTION | Calibration of Raytheon's Sparrow AIU-7F missile test equipment. The disc focuses on the calibration on ranging frequency and modulation of the signal being generated within the test set's electronic unit. It is designed primarily as a demonstration disc. |
| TYPE | Adult, Technical, Instructional |
| RELEASE DATE | 6/83 |
| FORMAT | Laser CAV |
| LEVEL | 3 |
| SYSTEM DESCRIPTION | The disc can be used with the Raytrain system and the Sony SMC-70 microcomputer system. |
| DEVELOPER | Lou Hattman, J.J. Wall, or E. Dextraze |
| CONTACT FOR MORE INFO. | Howard Lavin or E. Dextraze, (617) 272-9300 |
| FEATURES | Chapters, Picture Stops, Digital Dump-Sony, Dual Audio |
| RUNNING TIME | 30 minutes |
| PRODUCER | J.J. Wall |
| AVAILABILITY | Not for sale. |

| | |
|---:|:---|
| TITLE | **SPECTRUM OF INTEGRATED LOGISTICS SUPPORT** |
| DESCRIPTION | An interactive videodisc program for trade show exhibitions. Some of the topics covered are concept of ILS, support equipment engineering, electronics readiness center, technical data and training operations. |
| TYPE | Research and Development, Instructional, In-house Training |
| FORMAT | Laser CAV |
| LEVEL | 3 |
| SYSTEM DESCRIPTION | Two-sided videodisc that is completely menu-driven with user inputs via the touch-sensitive color monitor. |
| DEVELOPER | Westinghouse Technical Training Operations |
| CONTACT FOR MORE INFO. | Steve G. Kukla, (301) 995-5482 |
| FEATURES | Text/graphics Overlays, Touch Panel |
| PRODUCER | Westinghouse Multi-Media Production Center |
| AVAILABILITY | Not available. |

| | |
|---:|:---|
| TITLE | **SPELLBOUND** |
| TYPE | Motion Picture |
| FORMAT | CED |
| RETAIL OUTLET | Movies Unlimited Inc., A1-8035, $29.98 |

| | |
|---:|:---|
| TITLE | **SPERRY DEMO** |
| DESCRIPTION | A two-disc set demonstrating various applications of videodisc for interactive training. |
| TYPE | Promotional |
| FORMAT | Laser CAV |
| LEVEL | 3 |
| CONTACT FOR MORE INFO. | Sperry Gyroscope |
| PRODUCER | University of Nebraska/Nebraska Videodisc Design/Production Group |
| AVAILABILITY | Not available for release or distribution. |

| | |
|---:|:---|
| TITLE | **SPINA BIFIDA** |
| DESCRIPTION | Basic training for nurses in the care of spina bifida clients. |
| TYPE | Medical, Instructional |
| FORMAT | Laser CAV |
| LEVEL | 3 |

| | |
|---|---|
| SYSTEM DESCRIPTION | PR 7820-1 player, Apple II Plus computer, Coloney interface. |
| DEVELOPER | Florida State University/Multimedia Laboratories |
| CONTACT FOR MORE INFO. | Dr. Brent Hewlett |
| FEATURES | Dual Audio |
| PRODUCER | MGT of America, Tallahassee, FL |
| DISTRIBUTOR | Children's Medical Services/Department of HRS, Jayne Parker |
| AVAILABILITY | Not available at this time. |

| | |
|---|---|
| TITLE | **SPIRAL STAIRCASE** |
| TYPE | Motion Picture |
| FORMAT | CED |
| RETAIL OUTLET | Movies Unlimited Inc., A1-8030, $29.98 |

| | |
|---|---|
| TITLE | **SPLIT IMAGE** |
| TYPE | Motion Picture |
| FORMAT | CED |
| RETAIL OUTLET | Movies Unlimited Inc., A4-3226, $29.95 |

| | |
|---|---|
| TITLE | **SPORTS AND FITNESS IN ONTARIO** |
| DESCRIPTION | Provides a rundown on sports activities and their availability throughout Ontario, as well as offering an informative sports quiz. The viewer can select the activities about which he or she wishes to see information. |
| TYPE | Promotional |
| FORMAT | Laser CAV |
| LEVEL | 2 |
| PRODUCER | Patrick Lee |
| DEVELOPER | Interactive Image Technologies Inc. |
| CONTACT FOR MORE INFO. | Patrick Lee/Joseph Koenig |
| FEATURES | Other (English/French) |
| DISTRIBUTOR | Ontario Ministry of Tourism and Recreation |

| | |
|---|---|
| TITLE | **SPRING BREAK** |
| FORMAT | CED |
| DISTRIBUTOR | RCA Videodiscs, 03076, $29.95 |

| | |
|---|---|
| TITLE | **SPS-49 RADAR** |
| DESCRIPTION | Training on radar equipment. |
| TYPE | Technical, Instructional, Promotional |
| FORMAT | Laser CAV |
| LEVEL | 3 |
| DEVELOPER | W. Bell/E. Dextraze |
| CONTACT FOR MORE INFO. | Ed Dextraze, (617) 272-9300 |
| FEATURES | Sound Over Still |
| PRODUCER | Raytheon Service Company |
| AVAILABILITY | Not for sale. |

| | |
|---|---|
| TITLE | **SPY DISC, THE** |
| PRODUCER | Sheldon Renan, Peter Bloch |
| DISTRIBUTOR | Optical Programming Associates |

| | |
|---:|:---|
| TITLE | **SPY WHO LOVED ME, THE** |
| TYPE | Motion Picture (PG) |
| FORMAT | Laser CLV, CED |
| RUNNING TIME | 119 minutes |
| PRODUCER | CBS Fox Video |
| DISTRIBUTOR | Pioneer Video, Inc., 4638-80, $34.98 |
| | RCA Videodiscs, 01451, $29.98 |

| | |
|---:|:---|
| TITLE | **SQUIRM** |
| TYPE | Motion Picture |
| FORMAT | CED |
| RETAIL OUTLET | Movies Unlimited Inc., A3-3019, $29.95 |

| | |
|---:|:---|
| TITLE | **STAGECOACH** |
| TYPE | Motion Picture |
| FORMAT | CED |
| DISTRIBUTOR | RCA Videodiscs, 01101, $24.98 |

| | |
|---:|:---|
| TITLE | **STALAG 17** |
| TYPE | Motion Picture |
| FORMAT | Laser CLV, CED |
| FEATURES | Black and white |
| RUNNING TIME | 120 minutes |
| PRODUCER | Paramount Home Video |
| DISTRIBUTOR | Pioneer Video, Inc., LV5816, $29.95 |
| | RCA Videodiscs, 00627, $24.98 |

| | |
|---:|:---|
| TITLE | **STANDARD HEATS OF SOLUTION** |
| DESCRIPTION | An interactive general chemistry pre-laboratory experiment at the college freshman level. Program contains self-evaluation quizzes, and remedial segments. |
| TYPE | Research and Development, Instructional, In-house Training |
| FORMAT | Laser CAV |
| LEVEL | 2 |
| DEVELOPER | UCLA Office of Instructional Development and UCLA Department of Chemistry |
| CONTACT FOR MORE INFO. | Bernie L. Mitchell, (213) 825-7771 |
| FEATURES | Digital Dump-Pioneer |
| PRODUCER | Bernie L. Mitchell |
| DISTRIBUTOR | UCLA Instructional Media Library |
| SUGGESTED PRICE | $150 |

| | |
|---:|:---|
| TITLE | **STAR IS BORN, A** |
| TYPE | Motion Picture, Music (PG) |
| FORMAT | CED (2 Discs) |
| FEATURES | Stereo, CX Encoded |
| DISTRIBUTOR | RCA Videodiscs, 13134, $39.98 |

# Videodiscography

| | |
|---|---|
| TITLE | **STAR TALK** |
| DESCRIPTION | This disc is designed to allow Bowling Museum visitors to interview their favorite professional bowling stars. Chyron graphic menu pages form the interactive nature of the disc. |
| TYPE | Adolescents, Instructional, Archival, Museum Exhibit |
| RELEASE DATE | Spring 1984 |
| FORMAT | Laser CAV |
| LEVEL | 2 |
| DEVELOPER | National Bowling Hall of Fame (NBHOF)/Exhibit Technology, Inc. (ETI) |
| CONTACT FOR MORE INFO. | Bruce Pluckhahn (NBHOF) or Don Cochrane, President, Marketing ETI |
| FEATURES | Chapters, Picture Stops, Dual Audio |
| PRODUCER | Patrick Dillon, Exhibit Technology, Inc. |
| RETAIL OUTLET | NBHOF & Museum, St. Louis, MO 53129 |
| AVAILABILITY | Available for viewing as part of ETI's product line sampling. |

---

| | |
|---|---|
| TITLE | **STAR TREK - THE MOTION PICTURE** |
| TYPE | Motion Picture (G) |
| FORMAT | Laser CLV, CED (2 Discs) |
| FEATURES | Stereo |
| RUNNING TIME | 132 minutes |
| PRODUCER | Warner Home Video |
| DISTRIBUTOR | Pioneer Video, Inc., LV1180, $35.95 |
| | RCA Videodiscs, 00636, $39.98 |

---

| | |
|---|---|
| TITLE | **STAR TREK - THE MOTION PICTURE** **(Special longer version)** |
| TYPE | Motion Picture (G) |
| FORMAT | Laser CLV |
| FEATURES | Stereo, CX Encoded |
| RUNNING TIME | 143 minutes |
| PRODUCER | Paramount Home Video |
| DISTRIBUTOR | Pioneer Video, Inc., LV8858-2A, $39.95 |

---

| | |
|---|---|
| TITLE | **STAR TREK 2 - THE WRATH OF KHAN** |
| TYPE | Motion Picture (PG) |
| FORMAT | Laser CLV, CED |
| FEATURES | Stereo, CX Encoded |
| RUNNING TIME | 113 minutes |
| PRODUCER | Warner Home Video |
| DISTRIBUTOR | Pioneer Video, Inc., LV1180, $29.95 |
| | RCA Videodiscs, 13605, $29.98 |

---

| | |
|---|---|
| TITLE | **STAR TREK 6 AMOK TIME/JOURNEY TO BABEL** |
| TYPE | Motion Picture |
| FORMAT | CED |
| RUNNING TIME | 102 minutes |
| PRODUCER | Paramount Pictures |
| DISTRIBUTOR | RCA Videodiscs, 03609, $19.98 |

| | |
|---:|:---|
| TITLE | **STAR TREK, VOL. 1** |
| TYPE | Motion Picture |
| FORMAT | CED |
| DISTRIBUTOR | RCA Videodiscs, 00631, $19.98 |

| | |
|---:|:---|
| TITLE | **STAR TREK, VOL. 2** |
| TYPE | Motion Picture |
| FORMAT | CED |
| DISTRIBUTOR | RCA Videodiscs, 00632, $19.98 |

| | |
|---:|:---|
| TITLE | **STAR TREK, VOL. 3** |
| TYPE | Motion Picture |
| FORMAT | CED |
| DISTRIBUTOR | RCA Videodiscs, 00664, $19.98 |

| | |
|---:|:---|
| TITLE | **STAR TREK, VOL. 4** |
| TYPE | Motion Picture |
| FORMAT | CED |
| DISTRIBUTOR | RCA Videodiscs, 03602, $19.98 |

| | |
|---:|:---|
| TITLE | **STAR TREK, VOL. 5** |
| TYPE | Motion Picture |
| FORMAT | CED |
| DISTRIBUTOR | RCA Videodiscs, 00672, $19.98 |

| | |
|---:|:---|
| TITLE | **STAR WARS** |
| TYPE | Motion Picture (PG) |
| FORMAT | Laser CLV, CED |
| FEATURES | Stereo, CX Encoded |
| RUNNING TIME | 118 minutes |
| PRODUCER | Warner Home Video |
| DISTRIBUTOR | Pioneer Video, Inc., 1130-80, $34.98 |
| RETAIL OUTLET | (CED) Movies Unlimited Inc., A1-1130, $34.98 |

| | |
|---:|:---|
| TITLE | **STARDUST MEMORIES** |
| TYPE | Motion Picture |
| FEATURES | Black and white |
| FORMAT | Laser CLV, CED |
| RETAIL OUTLET | (CLV) UltraVideo, 80-4554, $34.95 |
| | (CED) Movies Unlimited Inc., A1-4554, $29.98 |

| | |
|---:|:---|
| TITLE | **START TO FINISH - THE GRAND PRIX** |
| TYPE | Motion Picture |
| FORMAT | CED |
| RETAIL OUTLET | Movies Unlimited Inc., A2-0232, $29.95 |

# Videodiscography

| | |
|---:|:---|
| TITLE | **STARTING OVER** |
| TYPE | Motion Picture, Comedy (R) |
| FORMAT | Laser CLV |
| RUNNING TIME | 105 minutes |
| PRODUCER | Warner Home Video |
| DISTRIBUTOR | Pioneer Video, Inc., LV1239, $29.95 |
| | RCA Videodiscs, 00633, $29.98 |

| | |
|---:|:---|
| TITLE | **STEVE MILLER BAND LIVE** |
| TYPE | Music |
| FORMAT | CED |
| FEATURES | Stereo, Banded |
| DISTRIBUTOR | RCA Videodiscs, 12182, $19.98 |

| | |
|---:|:---|
| TITLE | **STEVIE NICKS IN CONCERT** |
| TYPE | Music |
| FORMAT | Laser CLV, CED |
| FEATURES | Stereo, CX Encoded |
| RUNNING TIME | 55 minutes |
| PRODUCER | Pioneer Artists |
| DISTRIBUTOR | Pioneer Video, Inc., PA-83-033, $24.95 |
| RETAIL OUTLET | (CED) Movies Unlimited Inc., A1-7136, $29.98 |

| | |
|---:|:---|
| TITLE | **STILL OF THE NIGHT** |
| TYPE | Motion Picture (PG) |
| FORMAT | CED |
| PRODUCER | United Artists Corporation |
| DISTRIBUTOR | MGM/UA Entertainment Co. |
| RETAIL OUTLET | Movies Unlimited Inc., A1-4711, $29.98 |

| | |
|---:|:---|
| TITLE | **STING, THE** |
| TYPE | Motion Picture, Comedy (PG) |
| FORMAT | Laser CLV, CED (2 Discs) |
| RUNNING TIME | 129 minutes |
| PRODUCER | Warner Home Video |
| DISTRIBUTOR | Pioneer Video, Inc., 11-011, $29.95 |
| | RCA Videodiscs, 03003, $39.98 |

| | |
|---:|:---|
| TITLE | **STING 2, THE** |
| TYPE | Motion Picture (PG) |
| FORMAT | Laser CLV, CED |
| PRODUCER | Universal City Studios, Inc. |
| RETAIL OUTLET | (CLV) UltraVideo, 11-017, $32.95 |
| | (CED) Movies Unlimited Inc., A5-1017, $29.98 |

| | |
|---:|:---|
| TITLE | **STIR CRAZY** |
| TYPE | Motion Picture, Comedy (R) |
| FORMAT | Laser CLV, CED |
| PRODUCER | Warner Home Video |
| DISTRIBUTOR | Pioneer Video, Inc., VLD3380, $29.95 |
| | RCA Videodiscs, 03002, $29.98 |

| | |
|---:|:---|
| TITLE | **STORY OF O, THE** |
| TYPE | Motion Picture, Adult (X) |
| FORMAT | Laser CLV, CED |
| RETAIL OUTLET | (CLV) UltraVideo, ML100202, $34.95 |
| | (CED) Movies Unlimited Inc., A2-0202, $29.95 |

| | |
|---:|:---|
| TITLE | **STRANGE BEHAVIOR** |
| TYPE | Motion Picture |
| FORMAT | CED |
| DISTRIBUTOR | RCA Videodiscs, 02190, $24.98 |

| | |
|---:|:---|
| TITLE | **STRANGER IS WATCHING, A** |
| TYPE | Motion Picture |
| FORMAT | CED |
| RETAIL OUTLET | Movies Unlimited Inc., A2-1057, $29.95 |

| | |
|---:|:---|
| TITLE | **STRANGERS ON A TRAIN** |
| TYPE | Motion Picture |
| FORMAT | CED |
| DISTRIBUTOR | RCA Videodiscs, 03156, $24.98 |

| | |
|---:|:---|
| TITLE | **STRAW DOGS** |
| TYPE | Motion Picture (R) |
| FORMAT | Laser CLV, CED |
| FEATURES | Stereo |
| RUNNING TIME | 117 minutes |
| PRODUCER | Warner Home Video |
| DISTRIBUTOR | Pioneer Video, Inc., 8005-80, $34.98 |
| RETAIL OUTLET | (CED) Movies Unlimited Inc., A1-8005, $29.98 |

| | |
|---:|:---|
| TITLE | **STREET FIGHTER, THE** |
| TYPE | Motion Picture |
| FORMAT | CED |
| RETAIL OUTLET | Movies Unlimited Inc., A2-1052, $24.95 |

| | |
|---:|:---|
| TITLE | **STREETCAR NAMED DESIRE, A** |
| TYPE | Motion Picture |
| FEATURES | Black and white |
| FORMAT | Laser CLV, CED |
| DISTRIBUTOR | RCA Videodiscs, 01457, $24.98 |
| RETAIL OUTLET | (CLV) UltraVideo, 80-4571, $34.95 |

| | |
|---:|:---|
| TITLE | **STRIPES** |
| TYPE | Motion Picture, Comedy (R) |
| FORMAT | Laser CLV, CED |
| DISTRIBUTOR | RCA Videodiscs, 03022, $29.98 |
| RETAIL OUTLET | (CLV) UltraVideo, VLD5557, $29.95 |

| | |
|---:|:---|
| TITLE | **STUDENT BODIES** |
| TYPE | Motion Picture, Comedy (R) |
| FORMAT | Laser CLV |
| RUNNING TIME | 86 minutes |
| PRODUCER | Warner Home Video |
| DISTRIBUTOR | Pioneer Video, Inc., LV1476, $29.95 |

| | |
|---:|:---|
| TITLE | **STUNT DOG** |
| TYPE | Game |
| RELEASE DATE | 1983 |
| PRODUCER | Sheldon Renan |

| | |
|---:|:---|
| TITLE | **STUNT MAN, THE** |
| TYPE | Motion Picture |
| FORMAT | CED (2 Discs) |
| RETAIL OUTLET | Movies Unlimited Inc., A1-1110, $39.98 |

| | |
|---:|:---|
| TITLE | **SUMMER LOVERS** |
| TYPE | Motion Picture (R) |
| FORMAT | Laser CLV, CED |
| FEATURES | Stereo |
| RUNNING TIME | 98 minutes |
| PRODUCER | Embassy Home Entertainment |
| DISTRIBUTOR | Pioneer Video, Inc., 17045, $29.95 |
| RETAIL OUTLET | (CED) Movies Unlimited Inc., A4-7046, $29.95 |

| | |
|---:|:---|
| TITLE | **SUMMER OF '42** |
| TYPE | Motion Picture (PG) |
| FORMAT | CED |
| PRODUCER | Warner Home Video |
| DISTRIBUTOR | RCA Videodiscs, 03153, $24.98 |

| | |
|---:|:---|
| TITLE | **SUNSET BOULEVARD** |
| TYPE | Motion Picture |
| FORMAT | CED |
| DISTRIBUTOR | RCA Videodiscs, 00628, $24.98 |

| | |
|---:|:---|
| TITLE | **SUNSHINE BOYS, THE** |
| TYPE | Motion Picture |
| FORMAT | CED |
| RETAIL OUTLET | Movies Unlimited Inc., A2-1028, $29.95 |

| | |
|---:|:---|
| TITLE | **SUPER BOWL XIV: STEELERS VS. RAMS PLUS SEASON'S HIGHLIGHTS** |
| TYPE | Motion Picture |
| FORMAT | CED |
| DISTRIBUTOR | RCA Videodiscs, 01602, $19.98 |

| | |
|---:|:---|
| TITLE | **SUPER BOWL XV: RAIDERS VS. EAGLES PLUS SEASON'S HIGHLIGHTS** |
| TYPE | Motion Picture |
| FORMAT | CED |
| DISTRIBUTOR | RCA Videodiscs, 01603, $19.98 |

| | |
|---:|:---|
| TITLE | **SUPER MEMORIES OF THE SUPER BOWLS** |
| TYPE | Motion Picture |
| FORMAT | Laser CAV |
| LEVEL | 1 |
| RUNNING TIME | 47 minutes |
| PRODUCER | Warner Home Video |
| DISTRIBUTOR | Pioneer Video, Inc., LV005, $24.95 |

| | |
|---:|:---|
| TITLE | **SUPER SEVENTIES, THE** |
| TYPE | Motion Picture |
| FORMAT | Laser CAV |
| LEVEL | 1 |
| RUNNING TIME | 49 minutes |
| PRODUCER | Warner Home Video |
| DISTRIBUTOR | Pioneer Video, Inc., LV004, $24.95 |

| | |
|---:|:---|
| TITLE | **SUPERMAN - THE MOVIE** |
| TYPE | Motion Picture (PG) |
| FORMAT | Laser CLV, CED |
| FEATURES | Stereo, CX Encoded |
| RUNNING TIME | 144 minutes |
| PRODUCER | Warner Home Video |
| DISTRIBUTOR | Pioneer Video, Inc., 1013LV, $34.98 |
| | RCA Videodiscs, 03101, $39.98 |

| | |
|---:|:---|
| TITLE | **SUPERMAN 2** |
| TYPE | Motion Picture (PG) |
| FORMAT | Laser CLV, CED (2 Discs) |
| FEATURES | Stereo, CX Encoded |
| RUNNING TIME | 127 minutes |
| PRODUCER | Warner Home Video |
| DISTRIBUTOR | Pioneer Video, Inc., 11120LV, $34.98 |
| | RCA Videodiscs, 13136, $39.98 |

| | |
|---:|:---|
| TITLE | **SURVIVAL ANGLIA'S WORLD OF WILDLIFE, VOL. 1** |
| TYPE | Motion Picture, Instructional |
| FORMAT | CED |
| DISTRIBUTOR | RCA Videodiscs, 02020, $14.98 |

| | |
|---:|:---|
| TITLE | **SURVIVAL ANGLIA'S WORLD OF WILDLIFE, VOL. 2** |
| TYPE | Instructional |
| FORMAT | CED |
| DISTRIBUTOR | RCA Videodiscs, 02129, $14.98 |

# Videodiscography

| | |
|---:|:---|
| TITLE | **SURVIVORS, THE** |
| TYPE | Motion Picture, Comedy (R) |
| FORMAT | CED |
| PRODUCER | Columbia Pictures |
| DISTRIBUTOR | RCA Videodiscs, 03057, $24.98 |

| | |
|---:|:---|
| TITLE | **SWAMP THING** |
| TYPE | Motion Picture |
| FORMAT | Laser CLV, CED |
| RETAIL OUTLET | (CLV) UltraVideo, E160550, $34.95 |
| | (CED) Movies Unlimited Inc., A4-6056, $29.95 |

| | |
|---:|:---|
| TITLE | **SWAN LAKE** |
| TYPE | Music, Dance |
| FORMAT | Laser CLV, CED (2 Discs) |
| FEATURES | Stereo, CX Encoded |
| RUNNING TIME | 128 minutes |
| PRODUCER | Pioneer Artists |
| DISTRIBUTOR | Pioneer Video, Inc., PA-82-018, $39.95 |
| | RCA Videodiscs, 12212, $39.98 |

| | |
|---:|:---|
| TITLE | **SWEPT AWAY** |
| TYPE | Motion Picture (R) |
| FORMAT | Laser CLV, CED |
| DISTRIBUTOR | RCA Videodiscs, 01803, $19.98 |
| RETAIL OUTLET | (CLV) UltraVideo, VLD5568, $29.95 |

| | |
|---:|:---|
| TITLE | **SWING TIME** |
| TYPE | Motion Picture, Music, Dance |
| FORMAT | CED |
| DISTRIBUTOR | RCA Videodiscs, 00405, $19.98 |

| | |
|---:|:---|
| TITLE | **SWISS FAMILY ROBINSON** |
| TYPE | Motion Picture (G) |
| FORMAT | Laser CLV, CED |
| DISTRIBUTOR | RCA Videodiscs, 00726, $24.98 |
| RETAIL OUTLET | (CLV) UltraVideo, 053AS, $34.95 |

| | |
|---:|:---|
| TITLE | **SWITCH ON** |
| DESCRIPTION | Program to learn English. |
| TYPE | In-house Training |
| FORMAT | Laser CAV |
| LEVEL | 1 |
| CONTACT FOR MORE INFO. | Mr. Walter Stenzel |
| FEATURES | Chapters, Picture Stops, Digital Dump-Sony, Dual Audio |
| PRODUCER | Nelson Filmscan |
| DISTRIBUTOR | Langenscheidt KG |
| AVAILABILITY | Rent |
| SUGGESTED PRICE | DM 128 |

| | |
|---|---|
| TITLE | **SWORD AND THE SORCERER, THE** |
| TYPE | Motion Picture (R) |
| FORMAT | Laser CLV |
| FEATURES | Stereo, CX Encoded |
| RUNNING TIME | 100 minutes |
| PRODUCER | MCA Videodisc, Inc. |
| DISTRIBUTOR | Pioneer Video, Inc., 13-010, $29.95 |

| | |
|---|---|
| TITLE | **SYBIL** |
| TYPE | Motion Picture |
| FORMAT | CED |
| RETAIL OUTLET | Movies Unlimited Inc., A1-7122, $29.98 |

| | |
|---|---|
| TITLE | **SYSCON CORPORATION, INC. - (NOT GIVEN)** |
| DESCRIPTION | Instructional disc for U.S. Navy AEGIS shipbuilding project. Graphics are mixed with live action footage to introduce fleet users to the many elements and complex capabilities of the AEGIS combat system. |
| TYPE | Technical, Instructional |
| FORMAT | Laser CAV |
| LEVEL | 3 |
| DEVELOPER | SYSCON Corporation, Inc. in collaboration with WICAT Systems, Inc. |
| CONTACT FOR MORE INFO. | Mr. Thomas F. Koehl, SYSCON |
| FEATURES | Chapters, Picture Stops |
| PRODUCER | SYSCON/WICAT (Mr. Jack Warden of SYSCON is the Project Coordinator.) |
| DISTRIBUTOR | SYSCON (for Commander, Naval Sea Systems Command) |
| AVAILABILITY | U.S. Navy only |

| | |
|---|---|
| TITLE | **SYSTEM TEST ENHANCEMENT PROJECT (STEP)** |
| DESCRIPTION | Supports system testing of the E-3 AWACS equipment. Disc provides reference data, training information, maintenance and troubleshooting data. Features motion, audio, still frames, computer-graphics, and total user inputs via the touch-sensitive color monitor. |
| TYPE | Research and Development, Instructional, In-house Training |
| FORMAT | Laser CAV |
| LEVEL | 3 |
| SYSTEM DESCRIPTION | Stand-alone training and maintenance-aiding system housed in a carrel and comprised of a player, touch-sensitive color monitor, microcomputer/keyboard, and floppy disk drives. |
| DEVELOPER | Westinghouse Technical Training Operations |
| CONTACT FOR MORE INFO. | Steve G. Kukla, (301) 995-5482 |
| FEATURES | Other |
| PRODUCER | Westinghouse Multi-Media Production Center |
| AVAILABILITY | Not available. |

| | |
|---|---|
| TITLE | **TABLE FOR FIVE** |
| TYPE | Motion Picture |
| FORMAT | Laser CLV |
| RETAIL OUTLET | UltraVideo, 80-7043, $29.95 |

| | |
|---:|:---|
| TITLE | **TAKANAKA WORLD** |
| TYPE | Music |
| FORMAT | Laser CAV |
| LEVEL | 1 |
| FEATURES | Stereo, CX Encoded |
| PRODUCER | Pioneer Video Imports |
| DISTRIBUTOR | Pioneer Video, Inc., MP020, $27.95 |

| | |
|---:|:---|
| TITLE | **TAKE THE MONEY AND RUN** |
| TYPE | Motion Picture, Comedy (R) |
| FORMAT | Laser CLV, CED |
| RUNNING TIME | 85 minutes |
| PRODUCER | Warner Home Video |
| DISTRIBUTOR | Pioneer Video, Inc., 8007-80, $29.98 |
| RETAIL OUTLET | (CED) Movies Unlimited Inc., A1-8007, $29.98 |

| | |
|---:|:---|
| TITLE | **TAKING OF PELHAM 1-2-3, THE** |
| TYPE | Motion Picture |
| FORMAT | CED |
| RETAIL OUTLET | Movies Unlimited Inc., A1-4647, $29.98 |

| | |
|---:|:---|
| TITLE | **TALE OF THE FROG PRINCE** |
| TYPE | Motion Picture |
| FORMAT | Laser CLV |
| RETAIL OUTLET | UltraVideo, 80-6172, $29.95 |

| | |
|---:|:---|
| TITLE | **TALES FROM MUPPETLAND** |
| TYPE | Motion Picture (G) |
| FORMAT | CED |
| DISTRIBUTOR | RCA Videodiscs, 02067, $19.98 |

| | |
|---:|:---|
| TITLE | **TALES FROM MUPPETLAND 2** |
| TYPE | Motion Picture (G) |
| FORMAT | CED |
| DISTRIBUTOR | RCA Videodiscs, 02071, $19.98 |

| | |
|---:|:---|
| TITLE | **TALES OF HOFFMANN, THE** |
| TYPE | Music |
| FORMAT | Laser CLV, CED (Stereo, Banded, 2 Discs) |
| FEATURES | Stereo, CX Encoded |
| PRODUCER | Pioneer Artists |
| DISTRIBUTOR | Pioneer Video, Inc., PA-81-006, $59.95 |
| | RCA Videodiscs, 12154, $39.98 |

| | |
|---:|:---|
| TITLE | **TAMING OF THE SHREW, THE** |
| TYPE | Motion Picture |
| FORMAT | CED |
| DISTRIBUTOR | RCA Videodiscs, 03029, $19.98 |

| | |
|---:|:---|
| TITLE | **TAPS** |
| TYPE | Motion Picture |
| FORMAT | CED |
| RETAIL OUTLET | Movies Unlimited Inc., A1-1128, $29.98 |

| | |
|---:|:---|
| TITLE | **TARZAN, THE APE MAN** |
| TYPE | Motion Picture (R) |
| FORMAT | Laser CLV, CED |
| FEATURES | Stereo |
| RUNNING TIME | 112 minutes |
| PRODUCER | MGM/UA Home Video |
| DISTRIBUTOR | Pioneer Video, Inc., ML100109, $34.95 |
| RETAIL OUTLET | (CED) Movies Unlimited Inc., A2-1040, $29.95 |

| | |
|---:|:---|
| TITLE | **TATTOO** |
| TYPE | Motion Picture |
| FORMAT | CED |
| RETAIL OUTLET | Movies Unlimited Inc., A1-1123, $29.98 |

| | |
|---:|:---|
| TITLE | **TAXI DRIVER** |
| TYPE | Motion Picture (R) |
| FORMAT | Laser CLV, CED |
| DISTRIBUTOR | RCA Videodiscs, 03032, $29.98 |
| RETAIL OUTLET | (CLV) UltraVideo, VLD5920, $29.95 |

| | |
|---:|:---|
| TITLE | **TEDDY PENDERGRASS IN CONCERT** |
| TYPE | Music |
| FORMAT | CED |
| FEATURES | Stereo |
| RETAIL OUTLET | Movies Unlimited Inc., A1-7135, $29.98 |

| | |
|---:|:---|
| TITLE | **TELEMART** |
| DESCRIPTION | Disc is part of a teletext information retrieval system in a Toronto shopping mall. Different goods and services are offered on the disc to complement teletext information on a separate display screen. |
| TYPE | Advertising |
| RELEASE DATE | 11/82 |
| FORMAT | Laser CAV |
| LEVEL | 3 |
| SYSTEM DESCRIPTION | The computer selects images from a visual "data bank" and plays the appropriate section of the videodisc. Subscribers can show single frames or motion picture sequences relating to their goods and services. |
| DEVELOPER | Interactive Image Technologies with Telemart Ltd. |
| CONTACT FOR MORE INFO. | Patrick Lee, Joseph Koenig |
| PRODUCER | Patrick Lee |

| | |
|---:|:---|
| TITLE | **TEMPEST, THE** |
| FORMAT | CED (2 Discs) |
| FEATURES | Stereo |
| DISTRIBUTOR | RCA Videodiscs, 13078, $34.95 |

# Videodiscography

| | |
|---:|:---|
| TITLE | "10" |
| TYPE | Motion Picture, Comedy (R) |
| FORMAT | Laser CLV, CED |
| RUNNING TIME | 120 minutes |
| PRODUCER | Warner Home Video |
| DISTRIBUTOR | Pioneer Video, Inc., 2002LV, $29.98 |
| | RCA Videodiscs, 03102, $29.98 |

| | |
|---:|:---|
| TITLE | **TEN COMMANDMENTS, THE** |
| TYPE | Motion Picture (G) |
| FORMAT | Laser CLV, CED (2 Discs) |
| FEATURES | Stereo, CX Encoded |
| RUNNING TIME | 220 minutes |
| PRODUCER | Warner Home Video |
| DISTRIBUTOR | Pioneer Video, Inc., LV6524, $39.95 |
| | RCA Videodiscs, 00623, $39.98 |

| | |
|---:|:---|
| TITLE | **TERRYTOONS, VOL. 1 FEATURING MIGHTY MOUSE** |
| TYPE | Motion Picture (G) |
| FORMAT | CED |
| DISTRIBUTOR | RCA Videodiscs, 01503, $14.98 |

| | |
|---:|:---|
| TITLE | **TESS** |
| TYPE | Motion Picture (PG) |
| FORMAT | Laser CLV, CED (2 Discs) |
| FEATURES | Stereo, CX Encoded |
| PRODUCER | Warner Home Video |
| DISTRIBUTOR | Pioneer Video, Inc., LV5945, $34.95 |
| | RCA Videodiscs, 03010, $34.98 |

| | |
|---:|:---|
| TITLE | **TEST REACTOR** |
| DESCRIPTION | Surrogate travel of an experimental reactor facility. |
| TYPE | Research and Development |
| FORMAT | Laser CAV |
| LEVEL | 3 |
| DEVELOPER | Interactive Television Company |
| CONTACT FOR MORE INFO. | Dr. Steven Levin, Interactive Television Company |
| RUNNING TIME | 30 minutes |
| AVAILABILITY | Contact Interactive Television Company. |

| | |
|---:|:---|
| TITLE | **TEX** |
| TYPE | Motion Picture (PG) |
| FORMAT | Laser CLV, CED |
| RUNNING TIME | 102 minutes |
| PRODUCER | Walt Disney |
| DISTRIBUTOR | Pioneer Video, Inc., 42142AS, $34.95 |
| RETAIL OUTLET | (CED) Movies Unlimited Inc., A1-D142, $21.98 |

| | |
|---:|:---|
| TITLE | **TEXAS CHAINSAW MASSACRE** |
| TYPE | Motion Picture (R) |
| FORMAT | Laser CLV, CED |
| RETAIL OUTLET | (CLV) UltraVideo, ML100151, $39.95 |
| | (CED) Movies Unlimited Inc., A3-C034, $29.95 |

| | |
|---:|:---|
| TITLE | **TEXAS INSTRUMENTS** |
| DESCRIPTION | Recruiting materials for college graduates. |
| TYPE | Instructional, Catalog, Promotional, Other |
| FORMAT | Laser CAV |
| LEVEL | 2 |
| CONTACT FOR MORE INFO. | Aileen McKenna, (919) 493-3479 |
| FEATURES | Chapters, Picture Stops |
| RUNNING TIME | 30 minutes |
| AVAILABILITY | PREF has laser disc equipment in college placement offices across the country. PREF owns the equipment and distributes discs describing a company's recruiting programs into the network. The college receives the service at no charge, and the corporation pays a distribution fee. |

| | |
|---:|:---|
| TITLE | **THAT CHAMPIONSHIP SEASON** |
| TYPE | Motion Picture |
| FORMAT | CED |
| RETAIL OUTLET | Movies Unlimited Inc., A2-0221, $29.95 |

| | |
|---:|:---|
| TITLE | **THAT'S ENTERTAINMENT** |
| TYPE | Motion Picture, Music (G) |
| FORMAT | Laser CLV, CED |
| FEATURES | Stereo |
| RUNNING TIME | 120 minutes |
| PRODUCER | MGM/UA Home Video |
| DISTRIBUTOR | Pioneer Video, Inc., ML10007, $34.95 |
| RETAIL OUTLET | (CED) Movies Unlimited Inc., A2-1016, $29.98 |

| | |
|---:|:---|
| TITLE | **THE END** |
| TYPE | Motion Picture |
| FORMAT | CED |
| RETAIL OUTLET | Movies Unlimited Inc., A1-4607, $29.98 |

| | |
|---:|:---|
| TITLE | **THERE'S A MEETIN' HERE TONIGHT** |
| TYPE | Music |
| FORMAT | Laser CLV |
| FEATURES | Stereo |
| RUNNING TIME | 107 minutes |
| PRODUCER | Pioneer Artists |
| DISTRIBUTOR | Pioneer Video, Inc., PA-81-004, $29.95 |

| | |
|---:|:---|
| TITLE | **THERE'S NO BUSINESS LIKE SHOW BUSINESS** |
| TYPE | Motion Picture |
| FORMAT | CED |
| RETAIL OUTLET | Movies Unlimited Inc., A1-1086, $29.98 |

# Videodiscography

| | |
|---:|:---|
| TITLE | **THEY ALL LAUGHED** |
| TYPE | Motion Picture, Comedy (PG) |
| FORMAT | Laser CLV, CED |
| RUNNING TIME | 115 minutes |
| PRODUCER | Vestron Video |
| DISTRIBUTOR | Pioneer Video, Inc., VL6008, $34.95 |
| RETAIL OUTLET | (CED) Movies Unlimited Inc., A3-5005, $29.95 |

| | |
|---:|:---|
| TITLE | **THEY CALL ME BRUCE?** |
| TYPE | Motion Picture |
| FORMAT | CED |
| RETAIL OUTLET | Movies Unlimited Inc., A3-5015, $29.95 |

| | |
|---:|:---|
| TITLE | **THEY DRIVE BY NIGHT** |
| TYPE | Motion Picture |
| FORMAT | CED |
| PRODUCER | CBS/FOX VIDEO |
| RETAIL OUTLET | Movies Unlimited Inc., A1-4640, $29.98 |

| | |
|---:|:---|
| TITLE | **THEY SHOOT HORSES, DON'T THEY** |
| FORMAT | Laser CLV, CED |
| RUNNING TIME | 120 minutes |
| PRODUCER | Warner Home Video |
| DISTRIBUTOR | Pioneer Video, Inc., LV8004-80, $29.99 |
| RETAIL OUTLET | (CED) Movies Unlimited Inc., A1-8004, $29.98 |

| | |
|---:|:---|
| TITLE | **THIEF** |
| TYPE | Motion Picture (R) |
| FORMAT | CED |
| PRODUCER | United Artists Corporation |
| DISTRIBUTOR | United Artists |
| RETAIL OUTLET | Movies Unlimited Inc., A1-4550, $29.98 |

| | |
|---:|:---|
| TITLE | **THING, THE** |
| TYPE | Motion Picture (R) |
| FORMAT | Laser CLV |
| FEATURES | Stereo, CX Encoded |
| RUNNING TIME | 108 minutes |
| PRODUCER | MCA Videodisc Inc. |
| DISTRIBUTOR | Pioneer Video, Inc., 11-015, $29.95 |

| | |
|---:|:---|
| TITLE | **THING, THE (ORIGINAL 1951 VERSION)** |
| TYPE | Motion Picture |
| FORMAT | CED |
| DISTRIBUTOR | RCA Videodiscs, 00404, $19.98 |

| | |
|---:|:---|
| TITLE | **THINGS ARE TOUGH ALL OVER** |
| FORMAT | CED |
| DISTRIBUTOR | RCA Videodiscs, 03077, $29.95 |

| | |
|---:|:---|
| TITLE | **THINK IT THROUGH I** |
| DESCRIPTION | "Something's Missing," a computer-controlled "sleuth" game designed to develop the reasoning processes in deaf learners. |
| TYPE | Game, Children, Instructional |
| RELEASE DATE | 1981 |
| FORMAT | Laser CAV |
| LEVEL | 3 |
| CONTACT FOR MORE INFO. | Bureau of Education for the Handicapped/University of Nebraska Barkley Center |
| PRODUCER | University of Nebraska/Nebraska Videodisc Design/Production Group |
| AVAILABILITY | Not available for release or distribution. |

| | |
|---:|:---|
| TITLE | **THINK IT THROUGH II** |
| DESCRIPTION | "Help, I'm Lost," a computer-controlled "sleuth" game designed to develop the reasoning processes in deaf learners. |
| TYPE | Game, Children, Instructional |
| RELEASE DATE | 1981 |
| FORMAT | Laser CAV |
| LEVEL | 3 |
| CONTACT FOR MORE INFO. | Bureau of Education for the Handicapped/University of Nebraska Barkley Center |
| PRODUCER | University of Nebraska/ Nebraska Videodisc Design/Production Group |
| AVAILABILITY | Not available for release or distribution. |

| | |
|---:|:---|
| TITLE | **35MM PHOTOGRAPHY** |
| TYPE | Instructional |
| FORMAT | Laser CAV |
| RETAIL OUTLET | UltraVideo, I700-001, $49.95 |

| | |
|---:|:---|
| TITLE | **THIRTY-NINE STEPS, THE** |
| TYPE | Motion Picture |
| FORMAT | CED |
| DISTRIBUTOR | RCA Videodiscs, 01016, $19.98 |

| | |
|---:|:---|
| TITLE | **THOMAS CROWN AFFAIR, THE** |
| TYPE | Motion Picture |
| FORMAT | CED |
| RETAIL OUTLET | Movies Unlimited Inc., A1-4510, $29.98 |

| | |
|---:|:---|
| TITLE | **THOUSAND CLOWNS, A** |
| TYPE | Motion Picture |
| FORMAT | CED |
| RETAIL OUTLET | Movies Unlimited Inc., A1-4574, $29.98 |

| | |
|---:|:---|
| TITLE | **THREE DAYS OF THE CONDOR** |
| TYPE | Motion Picture (R) |
| FORMAT | Laser CLV, CED |
| RUNNING TIME | 118 minutes |
| PRODUCER | Warner Home Video |
| DISTRIBUTOR | Pioneer Video, Inc., LV8803, $29.95<br>RCA Videodiscs, 00655, $29.98 |

| | |
|---:|:---|
| TITLE | **THREE MUSKETEERS, THE** |
| TYPE | Motion Picture (PG) |
| FORMAT | CED |
| DISTRIBUTOR | RCA Videodiscs, 02102, $24.98 |

| | |
|---:|:---|
| TITLE | **THREE STOOGES VIDEODISC, VOL. 1, THE** |
| TYPE | Motion Picture, Comedy |
| FORMAT | CED |
| DISTRIBUTOR | RCA Videodiscs, 03021, $24.98 |

| | |
|---:|:---|
| TITLE | **THREE STOOGES VIDEODISC, VOL. 2, THE** |
| TYPE | Motion Picture, Comedy |
| FORMAT | CED |
| DISTRIBUTOR | RCA Videodiscs, 03049, $24.98 |

| | |
|---:|:---|
| TITLE | **THUNDERBALL** |
| TYPE | Motion Picture (PG) |
| FORMAT | Laser CLV, CED (2 Discs) |
| DISTRIBUTOR | RCA Videodiscs, 01494, $39.98 |
| RETAIL OUTLET | (CLV) UltraVideo, 80-4611, $39.95 |

| | |
|---:|:---|
| TITLE | **THUNDERBIRDS ARE GO** |
| TYPE | Motion Picture |
| FORMAT | CED |
| RETAIL OUTLET | Movies Unlimited Inc., A2-0231, $29.95 |

| | |
|---:|:---|
| TITLE | **THUNDERBOLT AND LIGHTFOOT** |
| TYPE | Motion Picture |
| FORMAT | CED |
| DISTRIBUTOR | RCA Videodiscs, 01495, $29.98 |

| | |
|---:|:---|
| TITLE | **TICKET TO HEAVEN** |
| TYPE | Motion Picture |
| FORMAT | CED |
| RETAIL OUTLET | Movies Unlimited Inc., A2-0150, $29.95 |

| | |
|---:|:---|
| TITLE | **TILL MARRIAGE DO US PART** |
| TYPE | Motion Picture, Adult (R) |
| FORMAT | Laser CLV, CED |
| RUNNING TIME | 97 minutes |
| PRODUCER | Warner Home Video |
| DISTRIBUTOR | Pioneer Video, Inc., VL3005, $34.95 |
| RETAIL OUTLET | (CED) Movies Unlimited Inc., A3-3005, $29.95 |

| | |
|---:|:---|
| TITLE | **TIME BANDITS** |
| TYPE | Motion Picture, Comedy (PG) |
| FORMAT | Laser CLV, CED |
| FEATURES | Stereo, CX Encoded |
| RUNNING TIME | 116 minutes |

| | |
|---:|:---|
| PRODUCER | Warner Home Video |
| DISTRIBUTOR | Pioneer Video, Inc., LV2310, $29.95 |
| | RCA Videodiscs, 12168 (Stereo, Banded), 02099, $29.98 |

| | |
|---:|:---|
| TITLE | **TIME MACHINE, THE** |
| TYPE | Motion Picture |
| FORMAT | CED |
| RETAIL OUTLET | Movies Unlimited Inc., A2-0152, $29.95 |

| | |
|---:|:---|
| TITLE | **TJS** |
| DESCRIPTION | Instructional content related to maintenance of the Tactical Jamming System (TJS) of the EA6B jet plane. |
| TYPE | Instructional |
| RELEASE DATE | 8/82 |
| FORMAT | Laser CAV |
| LEVEL | 3 |
| DEVELOPER | Grumman Aerospace Corporation |
| CONTACT FOR MORE INFO. | Roger Schaefer |
| PRODUCER | Windsor |
| DISTRIBUTOR | U.S. Navy |
| AVAILABILITY | Availability limited to distribution by the U.S. Navy. |

| | |
|---:|:---|
| TITLE | **TO BE OR NOT TO BE** |
| FORMAT | CED |
| DISTRIBUTOR | RCA Videodiscs, 01104, $24.98 |

| | |
|---:|:---|
| TITLE | **TO CATCH A THIEF** |
| TYPE | Motion Picture |
| FORMAT | Laser CLV |
| RUNNING TIME | 103 minutes |
| PRODUCER | Warner Home Video |
| DISTRIBUTOR | Pioneer Video, Inc., LV6308, $29.95 |

| | |
|---:|:---|
| TITLE | **TO RUSSIA...WITH ELTON JOHN** |
| TYPE | Music |
| FORMAT | CED |
| DISTRIBUTOR | RCA Videodiscs, 00515, $24.98 |

| | |
|---:|:---|
| TITLE | **TOM AND JERRY** |
| TYPE | Motion Picture |
| FORMAT | CED |
| RETAIL OUTLET | Movies Unlimited Inc., A2-1004, $29.95 |

| | |
|---:|:---|
| TITLE | **TOM AND JERRY CARTOON FESTIVAL, VOL. 1** |
| TYPE | Motion Picture (G) |
| FORMAT | Laser CLV |
| RUNNING TIME | 58 minutes |
| PRODUCER | MGM/UA Home Video |
| DISTRIBUTOR | Pioneer Video, Inc., ML100019, $25.95 |

| | |
|---:|:---|
| TITLE | **TOM AND JERRY, VOL. 2** |
| TYPE | Motion Picture |
| FORMAT | CED |
| RETAIL OUTLET | Movies Unlimited Inc., A2-1049, $29.95 |

| | |
|---:|:---|
| TITLE | **TOM JONES** |
| TYPE | Motion Picture, Comedy |
| FORMAT | Laser CLV, CED (2 Discs) |
| RUNNING TIME | 127 minutes |
| PRODUCER | Warner Home Video |
| DISTRIBUTOR | Pioneer Video, Inc., 4511-80, $34.98 |
| | RCA Videodiscs, 01414, $34.98 |

| | |
|---:|:---|
| TITLE | **TOMMY** |
| TYPE | Motion Picture, Music (PG) |
| FORMAT | Laser CLV, CED |
| FEATURES | Stereo, CX Encoded |
| DISTRIBUTOR | RCA Videodiscs, 13025, $29.98 |
| RETAIL OUTLET | (CLV) UltraVideo, VLD5951, $29.95 |

| | |
|---:|:---|
| TITLE | **TOMORROW TODAY** |
| DESCRIPTION | Demonstration videodisc featuring programs from the WNET/13 program library. Chapters include "A Tennis Lesson," "Dancing with a Disc," "Writing Tune-Up," "Looking at Art," and "Public Affairs." |
| TYPE | Dance, Art, Instructional, Promotional |
| RELEASE DATE | 12/81 |
| FORMAT | Laser CAV |
| LEVEL | 2 |
| DEVELOPER | Educational Broadcasting Corporation (WNET/13) |
| CONTACT FOR MORE INFO. | Jeffrey Honeyman, Producer/Programmer, (212) 664-7194 |
| FEATURES | Digital Dump-Pioneer, Stereo, Dual Audio |
| PRODUCER | Jeffrey Honeyman |
| AVAILABILITY | Loan. This is a demonstration-only disc; not intended for sale. |

| | |
|---:|:---|
| TITLE | **TONY BENNETT SONGBOOK - RECORDED LIVE IN NEW YORK** |
| TYPE | Music |
| FORMAT | CED |
| FEATURES | Stereo, CX Encoded |
| DISTRIBUTOR | RCA Videodiscs, 12117, $24.98 |

| | |
|---:|:---|
| TITLE | **TOOTSIE** |
| TYPE | Motion Picture |
| FORMAT | CED |
| DISTRIBUTOR | RCA Videodiscs, 03071, $29.95 |

| | |
|---:|:---|
| TITLE | **TORA! TORA! TORA!** |
| TYPE | Motion Picture (G) |
| FORMAT | Laser CLV, CED (2 Discs) |
| FEATURES | Stereo |
| RUNNING TIME | 144 minutes |
| PRODUCER | Warner Home Video |
| DISTRIBUTOR | Pioneer Video, Inc., 1017-80, $39.98 |
| RETAIL OUTLET | (CED) Movies Unlimited Inc., A1-1017, $39.98 |

| | |
|---:|:---|
| TITLE | **TOTALLY GO-GOS** |
| TYPE | Music |
| FORMAT | CED |
| FEATURES | Stereo, CX Encoded |
| DISTRIBUTOR | RCA Videodiscs, 12126, $24.98 |

| | |
|---:|:---|
| TITLE | **TOUCH OF LOVE...MASSAGE, THE** |
| TYPE | Adult |
| FORMAT | Laser CAV |
| LEVEL | 1 |
| FEATURES | Stereo |
| RUNNING TIME | 27 minutes |
| PRODUCER | Warner Home Video |
| DISTRIBUTOR | Pioneer Video, Inc., 30-001, $24.95 |

| | |
|---:|:---|
| TITLE | **TOWERING INFERNO** |
| TYPE | Motion Picture |
| FORMAT | CED (2 Discs) |
| RETAIL OUTLET | Movies Unlimited Inc., A1-1071, $39.98 |

| | |
|---:|:---|
| TITLE | **TOY, THE** |
| TYPE | Motion Picture, Comedy |
| FORMAT | CED |
| DISTRIBUTOR | RCA Videodiscs, 03044, $29.98 |

| | |
|---:|:---|
| TITLE | **TPS-43 INTERACTIVE "SAMPLER"** |
| DESCRIPTION | Provides a variety of preventive procedures, circuit theory descriptions, illustrated parts breakdowns, parts lists, etc. for the TPS-43 ground-based surveillance radar. |
| TYPE | Research and Development, Instructional, In-house Training |
| FORMAT | Laser CAV |
| LEVEL | 3 |
| DEVELOPER | Westinghouse Technical Training Operations |
| CONTACT FOR MORE INFO. | Steve G. Kukla, (301) 995-5482 |
| FEATURES | Text/graphics Overlays, Touch Panel |
| PRODUCER | Westinghouse Multi-Media Production Center |
| AVAILABILITY | Not available. |

| | |
|---:|:---|
| TITLE | **TRACING A GROUND** |
| DESCRIPTION | Disc shows how to trace a ship's switchboard ground indication to the fault, through power and distribution panels and using ship diagrams. |
| TYPE | Technical, Instructional |
| FORMAT | Laser CAV |
| LEVEL | 3 |
| SYSTEM DESCRIPTION | Pioneer 7820 Model 2 player, Apple IIe with two drives, and a touch panel (infrared monitor), all in one cabinet. |
| DEVELOPER | Creativision, Inc. |
| CONTACT FOR MORE INFO. | Mr. Rick Glasby |
| FEATURES | Stereo, Dual Audio, Touch Panel Monitor |
| PRODUCER | Creativision, Inc. |
| DISTRIBUTOR | Creativision, Inc. |
| AVAILABILITY | Produced with contract funds from the U.S. Navy. |

| | |
|---:|:---|
| TITLE | **TRADITION AND CHANGE** |
| DESCRIPTION | Bi-lingual introduction to Canadian Indian art, and the work of the Dept. of Indian and Northern Affairs, Federal Government of Canada. Used as a travelling exhibit throughout Canada. |
| TYPE | Promotional |
| FORMAT | Laser CAV |
| LEVEL | 2 |
| DEVELOPER | Interactive Image Technologies and Exhibit Technology, Inc. |
| CONTACT FOR MORE INFO. | Patrick Lee |
| FEATURES | Dual Audio (English/French) |
| PRODUCER | Patrick Lee |
| DISTRIBUTOR | Department of Indian and Northern Affairs, Government of Canada |

| | |
|---:|:---|
| TITLE | **TRAIL OF THE PINK PANTHER** |
| TYPE | Motion Picture, Comedy (PG) |
| FORMAT | Laser CLV, CED |
| RUNNING TIME | 97 minutes |
| PRODUCER | CBS Fox Video |
| DISTRIBUTOR | Pioneer Video, Inc., 4710-80, $34.98 |
| RETAIL OUTLET | (CED) Movies Unlimited Inc., A1-4710, $29.98 |

| | |
|---:|:---|
| TITLE | **TRANSFORMING OFFICE PRODUCTIVITY (TOP) SEMINAR** |
| DESCRIPTION | Led by Xerox experts, the TOP Seminar explains the motivations for and nature of the Ethernet network, and the full family of Xerox office products. Using videodisc as the source of the visuals eliminated six slide projectors, two computers, a 1/4" tape recorder, a video tape player, a motion picture projector, and many cables. |
| TYPE | Instructional |
| FORMAT | Laser CAV |
| LEVEL | 2 |
| DEVELOPER | The Glyn Group, Inc. |
| CONTACT FOR MORE INFO. | J.M. McCarthy, (212) 255-5156 |
| FEATURES | Digital Dump-Sony |
| PRODUCER | Xerox Corporation |
| AVAILABILITY | Disc is not in distribution. |

| | |
|---:|:---|
| TITLE | **TRAPEZE** |
| TYPE | Motion Picture |
| FORMAT | CED |
| RETAIL OUTLET | Movies Unlimited Inc., A1-4702, $29.98 |

| | |
|---:|:---|
| TITLE | **TREASURE** |
| PRODUCER | Sheldon Renan, Peter Bloch |
| DISTRIBUTOR | National Video Systems, Inc. |

|  |  |
|---:|:---|
| TITLE | **TREASURE ISLAND** |
| TYPE | Motion Picture (G) |
| FORMAT | Laser CLV, CED |
| RUNNING TIME | 87 minutes |
| PRODUCER | Walt Disney |
| DISTRIBUTOR | Pioneer Video, Inc., 42041AS, $34.95 |
|  | RCA Videodiscs, 00730, $24.98 |

|  |  |
|---:|:---|
| TITLE | **TREASURE ISLAND (WITH WALLACE BEERY)** |
| FORMAT | CED |
| DISTRIBUTOR | RCA Videodiscs, 00726, $24.98 |
| RETAIL OUTLET | Movies Unlimited Inc., A2-1020, $24.95 |

|  |  |
|---:|:---|
| TITLE | **TREASURE OF SIERRA MADRE** |
| TYPE | Motion Picture |
| FORMAT | Laser CLV, CED |
| FEATURES | Black and white |
| RUNNING TIME | 126 minutes |
| PRODUCER | CBS Fox Video |
| DISTRIBUTOR | Pioneer Video, Inc., 4639-80, $34.98 |
|  | RCA Videodiscs, 01430, $24.98 |

|  |  |
|---:|:---|
| TITLE | **TRIBUTE** |
| TYPE | Motion Picture (PG) |
| FORMAT | Laser CLV, CED (2 Discs) |
| RUNNING TIME | 133 minutes |
| PRODUCER | Warner Home Video |
| DISTRIBUTOR | Pioneer Video, Inc., VL6003, $39.95 |
| RETAIL OUTLET | (CED) Movies Unlimited Inc., A3-6003, $39.95 |

|  |  |
|---:|:---|
| TITLE | **TRON** |
| TYPE | Motion Picture (PG) |
| FORMAT | Laser CLV, CED |
| FEATURES | Stereo, CX Encoded |
| RUNNING TIME | 95 minutes |
| PRODUCER | Warner Home Video |
| DISTRIBUTOR | Pioneer Video, Inc., 42122AS, $34.95 |
|  | RCA Videodiscs, 10732, $29.98 |

|  |  |
|---:|:---|
| TITLE | **TROUBLESHOOTING HORSEPOWER LOSS** |
| DESCRIPTION | Diagnostic procedures program designed to train service personnel on the proper techniques to follow when servicing diesel engines used in heavy farm equipment. |
| TYPE | Technical, Instructional |
| FORMAT | Laser CAV |
| LEVEL | 2 |
| DEVELOPER | Signal Communications and Northwest Teleproductions |
| CONTACT FOR MORE INFO. | Jim Kreiser, Signal Communications |
| FEATURES | Chapters, Digital Dump-Sony |
| PRODUCER | Jim Kreiser (Signal), Bob Kerr (Northwest Teleproductions) |
| AVAILABILITY | Loan. Produced for interactive demonstration only. |

| | |
|---|---|
| TITLE | **TRUE CONFESSIONS** |
| TYPE | Motion Picture (R) |
| FORMAT | Laser CLV, CED |
| RUNNING TIME | 107 minutes |
| PRODUCER | MGM/UA Home Video, Inc. |
| DISTRIBUTOR | Pioneer Video, Inc., ML100145, $34.95 |
| RETAIL OUTLET | (CED) Movies Unlimited Inc., A2-1053, $29.95 |

| | |
|---|---|
| TITLE | **TRUE GRIT** |
| TYPE | Motion Picture (G) |
| FORMAT | Laser CLV, CED (2 Discs) |
| RUNNING TIME | 128 minutes |
| PRODUCER | Warner Home Video |
| DISTRIBUTOR | Pioneer Video, Inc., LV6833, $35.95 |
| | RCA Videodiscs, 00608, $34.98 |

| | |
|---|---|
| TITLE | **TUBES VIDEO, THE** |
| TYPE | Music |
| FORMAT | Laser CAV, CED (Stereo, Banded) |
| LEVEL | 1 |
| FEATURES | Stereo, CX Encoded |
| RUNNING TIME | 53 minutes |
| PRODUCER | Pioneer Artists |
| DISTRIBUTOR | Pioneer Video, Inc., PA-82-012, $24.95 |
| | RCA Videodiscs, 12141, $24.98 |

| | |
|---|---|
| TITLE | **TUMBLING (CLOSED CAPTION)** |
| DESCRIPTION | "Basic Tumbling Skills" for hearing-impaired learners. |
| TYPE | Children, Instructional |
| RELEASE DATE | 1980 |
| FORMAT | Laser CAV |
| LEVEL | 2 |
| CONTACT FOR MORE INFO. | Bureau of Education for the Handicapped/ University of Nebraska Barkley Center |
| PRODUCER | University of Nebraska/Nebraska Videodisc Design/Production Group |
| AVAILABILITY | Not available for release or distribution. |

| | |
|---|---|
| TITLE | **TURNING POINT, THE** |
| TYPE | Motion Picture |
| FORMAT | CED |
| RETAIL OUTLET | Movies Unlimited Inc., A1-1089, $29.98 |

| | |
|---|---|
| TITLE | **TUT, THE BOY KING/THE LOUVRE** |
| TYPE | Art |
| FORMAT | CED |
| DISTRIBUTOR | RCA Videodiscs, 01202, $19.98 |

| | |
|---|---|
| TITLE | **TWELVE ANGRY MEN** |
| TYPE | Motion Picture |
| FORMAT | CED |
| DISTRIBUTOR | RCA Videodiscs, 01496, $24.98 |

| | |
|---:|:---|
| TITLE | **TWELVE O'CLOCK HIGH** |
| TYPE | Motion Picture |
| FORMAT | CED (2 Discs) |
| RETAIL OUTLET | Movies Unlimited Inc., A1-1075, $39.98 |

| | |
|---:|:---|
| TITLE | **20,000 LEAGUES UNDER THE SEA** |
| TYPE | Motion Picture (G) |
| FORMAT | Laser CLV, CED |
| RUNNING TIME | 118 minutes |
| PRODUCER | Walt Disney |
| DISTRIBUTOR | Pioneer Video, Inc., 42112AS, $34.95 |
| DISTRIBUTOR | RCA Videodiscs, 00701, $24.98 |

| | |
|---:|:---|
| TITLE | **2001: A SPACE ODYSSEY** |
| TYPE | Motion Picture |
| FORMAT | Laser CLV, CED (2 Discs) |
| FEATURES | Stereo |
| RETAIL OUTLET | (CLV) UltraVideo, ML100002, $39.95 (CED) Movies Unlimited Inc., A2-1018, $39.95 |

| | |
|---:|:---|
| TITLE | **U.S. DEPARTMENT OF EDUCATION WORLD OF WORK** |
| DESCRIPTION | This series focuses on job opportunities in the new technologies, and is available for vocational counseling in a number of videodisc formats. |
| TYPE | Instructional |
| RELEASE DATE | 4/84 |
| FORMAT | Laser CAV |
| LEVEL | 1,3 |
| SYSTEM DESCRIPTION | Level 3 is used in conjunction with the Sony SMC-70 with superimposed graphics and text. The Sony will control the player as well. |
| DEVELOPER | Fusion Media, Inc. |
| FEATURES | Chapters, Picture Stops, Dual Audio |
| PRODUCER | Technivision, Inc. |
| DISTRIBUTOR | U.S. Department of Education |

| | |
|---:|:---|
| TITLE | **UNDERSEA WORLD OF JACQUES COUSTEAU, VOL. 1** |
| TYPE | Instructional |
| FORMAT | CED |
| DISTRIBUTOR | RCA Videodiscs, 02006, $19.98 |

| | |
|---:|:---|
| TITLE | **UNIVERSITY OF LONDON PROGRAMME DEVELOPMENT VIDEODISC** |
| DESCRIPTION | Disc consists of various independent extracts from previous recordings and original material, all used to exploit the properties of, and illustrate the options provided by interactivevideodisc technology. Shows which kinds of source material work well and which work badly to help those considering producing a videodisc. |
| TYPE | Medical, Research and Development, Technical, Instructional, In-house Training, Archival, Catalog |
| RELEASE DATE | 2/84 |
| FORMAT | Laser CAV |
| LEVEL | 3 |
| DEVELOPER | University of London Audio-Visual Centre |
| CONTACT FOR MORE INFO. | David R. Clark |

# Videodiscography

|  |  |
|---|---|
| PRODUCER | University of London Audio-Visual Centre |
| DISTRIBUTOR | Same as producer |
| SUGGESTED PRICE | Price given on application. |

|  |  |
|---|---|
| TITLE | **UNIVERSITY OF NEBRASKA/NEBRASKA VIDEODISC DESIGN/PRODUCTION GROUP - (NOT GIVEN)** |
| DESCRIPTION | MAT Project. Proprietary experimental disc for major U.S. corporation. |
| TYPE | Research and Development |
| FORMAT | Laser CAV |
| LEVEL | 3 |
| PRODUCER | University of Nebraska/Nebraska Videodisc Design/Production Group |
| AVAILABILITY | Not available for release or distribution. |

|  |  |
|---|---|
| TITLE | **UNIVERSITY OF NEBRASKA/NEBRASKA VIDEODISC DESIGN/PRODUCTION GROUP - (NOT GIVEN)** |
| DESCRIPTION | NEM Project. Proprietary experimental disc for major U.S. corporation. |
| TYPE | Research and Development |
| FORMAT | Laser CAV |
| LEVEL | 3 |
| PRODUCER | University of Nebraska/Nebraska Videodisc Design/Production Group |
| AVAILABILITY | Not available for release or distribution. |

|  |  |
|---|---|
| TITLE | **UP IN SMOKE** |
| TYPE | Motion Picture, Comedy (R) |
| FORMAT | Laser CLV, CED |
| RUNNING TIME | 87 minutes |
| PRODUCER | Warner Home Video |
| DISTRIBUTOR | Pioneer Video, Inc., LV8966, $29.95<br>RCA Videodiscs, 00652, $29.98 |

|  |  |
|---|---|
| TITLE | **URBAN COWBOY** |
| TYPE | Motion Picture (PG) |
| FORMAT | Laser CLV, CED (2 Discs) |
| RUNNING TIME | 132 minutes |
| PRODUCER | Warner Home Video |
| DISTRIBUTOR | Pioneer Video, Inc., LV1285, $35.95<br>RCA Videodiscs, 00639, $34.98 |

|  |  |
|---|---|
| TITLE | **URINARY CATHETERIZATION** |
| DESCRIPTION | Demonstrates the correct procedures for urinary catheterization, including how to prepare the patient and tray, and proper catheterization techniques for female and male patients. |
| TYPE | Instructional |
| FORMAT | Laser CAV |
| LEVEL | 2 |
| DEVELOPER | Alberta Vocational Centre, Calgary, Alberta |
| CONTACT FOR MORE INFO. | Russell Sawchuk at ACCESS Alberta |
| FEATURES | Digital Dump-Sony |
| PRODUCER | ACCESS Alberta and Alberta Vocational Centre |
| DISTRIBUTOR | Marketing Department, ACCESS Alberta |
| AVAILABILITY | Sale. Distribution is limited to accredited educational and health care institutions. |
| SUGGESTED PRICE | $225 |

|  |  |
|---|---|
| TITLE | **URODYNAMICS IN CLINICAL PRACTICE** |
| DESCRIPTION | Designed to instruct physicians in the use of combined urological studies, and encourage their use in the diagnosis of disorders of the urinary tract. It premiered at the American Urodynamic Association convention in Las Vegas, Nevada, in 1983. |
| TYPE | Instructional |
| RELEASE DATE | 4/83 |
| FORMAT | Laser CAV |
| LEVEL | 3 |
| DEVELOPER | The Glyn Group, Stephen Weinstein & Associates Advertising |
| CONTACT FOR MORE INFO. | J.M. McCarthy, Glyn Group, Inc., (212) 255-5156 |
| FEATURES | Digital Dump-Pioneer |
| PRODUCER | Roerig - a Division of Pfizer Pharmaceuticals |
| AVAILABILITY | Disc is not in distribution. |

|  |  |
|---|---|
| TITLE | **UROLOGY** |
| DESCRIPTION | Disc consists of 500 color images transferred from 35,000 existing slides. Each image was identified by the catalog number within the Urology Slide Library. They were arranged in four major categories, based on a retrieval scheme developed at the University of Cincinnati. Each image may exist in more than one category. The categories that users may want to view may either be selected outright, or joined or intersected to form new categories such as "X rays of Kidneys." |
| TYPE | Medical, Research and Development, Archival |
| FORMAT | Thomson-CSF transmissive |
| LEVEL | 3 |
| SYSTEM DESCRIPTION | Thomson-CSF TTV 3620 player, Apple II microcomputer with single disk drive, a Sony color monitor for the videodisc, a Sanyo black and white monitor for the computer, and Apple Thomson driver software developed at the CAI Lab for an IEEE 488 board from California Computer Systems. |
| CONTACT FOR MORE INFO. | Joan Sustik Huntley, (319) 353-3170 |
| PRODUCER | University of Iowa/Weeg Computing Center |
| AVAILABILITY | Since the project was for research and development purposes only, the disc is not publicly available. Interested persons may view the videodisc at the Weeg Computer Center. |

|  |  |
|---|---|
| TITLE | **USED CARS** |
| TYPE | Motion Picture, Comedy (R) |
| FORMAT | CED |
| DISTRIBUTOR | RCA Videodiscs, 03027, $24.98 |

|  |  |
|---|---|
| TITLE | **USING THE APR** |
| DESCRIPTION | Along with 11 floppy diskettes, this videodisc focuses on the procedures for completing a complex form that welfare workers in Florida must complete accurately in order for needy families with dependent children to receive the aid to which they are entitled. |
| TYPE | Instructional, In-house Training |
| RELEASE DATE | 9/15/83 |
| FORMAT | Laser CAV |
| LEVEL | 3 |

| | |
|---:|:---|
| SYSTEM DESCRIPTION | Disc is part of a 160-hour curriculum delivered on an Apple II/Pioneer 7820-3/single disk drive system along with ancillary printed reference books. |
| DEVELOPER | University of West Florida/Office for Interactive Technology and Training |
| CONTACT FOR MORE INFO. | Richard Smith; Elizabeth Wright, (413) 256-0444 |
| FEATURES | Dual Audio |
| PRODUCER | University of Nebraska/Nebraska Videodisc Design/Production Group |
| AVAILABILITY | Training program produced under contract with the Florida Department of Health and Rehabilitative Services, Tallahassee, Florida, and is not available for general distribution. |

| | |
|---:|:---|
| TITLE | **VALLEY GIRL** |
| TYPE | Motion Picture |
| FORMAT | Laser CLV |
| RETAIL OUTLET | UltraVideo, VL5016, $34.95 |

| | |
|---:|:---|
| TITLE | **VEHICLE EMISSION CONTROL** |
| DESCRIPTION | Service training for the Ford Video network. |
| TYPE | Technical, Instructional, In-house Training |
| RELEASE DATE | 2/82 |
| LEVEL | 2 |
| CONTACT FOR MORE INFO. | Raymond J. Marx |
| FEATURES | Chapters, Picture Stops, Digital Dump-Sony |
| RUNNING TIME | 24 minutes |
| PRODUCER | Ron Herman |
| DISTRIBUTOR | Ford Video Network |

| | |
|---:|:---|
| TITLE | **VENOM** |
| TYPE | Motion Picture |
| FORMAT | CED |
| RETAIL OUTLET | Movies Unlimited Inc., A3-3025, $29.98 |

| | |
|---:|:---|
| TITLE | **VERDICT, THE** |
| TYPE | Motion Picture (R) |
| FORMAT | Laser CLV, CED |
| RUNNING TIME | 128 minutes |
| PRODUCER | CBS Fox Video |
| DISTRIBUTOR | Pioneer Video, Inc., 1188-80, $39.98 |
| RETAIL OUTLET | (CED) Movies Unlimited Inc., A1-1188, $39.98 |

| | |
|---:|:---|
| TITLE | **VHD INTERACTIVE VIDEODISC DEMONSTRATION** |
| DESCRIPTION | Demonstration of the capabilities of the VHD videodisc system. |
| TYPE | Promotional |
| FORMAT | VHD |
| LEVEL | 3 |
| CONTACT FOR MORE INFO. | Thorn-EMI |
| PRODUCER | University of Nebraska/Nebraska Videodisc Design/Production Group |
| AVAILABILITY | Not available for release or distribution. |

| | |
|---:|:---|
| TITLE | **VICE SQUAD** |
| TYPE | Motion Picture (R) |
| FORMAT | Laser CLV, CED |
| RUNNING TIME | 97 minutes |
| PRODUCER | Embassy Home Entertainment |
| DISTRIBUTOR | Pioneer Video, Inc., 20155, $34.95 |
| RETAIL OUTLET | (CED) Movies Unlimited Inc., A4-0156, $29.95 |

| | |
|---:|:---|
| TITLE | **VICTOR, VICTORIA** |
| TYPE | Motion Picture |
| FORMAT | Laser CLV, CED (2 Discs) |
| FEATURES | Stereo |
| RETAIL OUTLET | (CLV) UltraVideo, ML100151, $39.95 |
| | (CED) Movies Unlimited Inc., A2-0151, $39.95 |

| | |
|---:|:---|
| TITLE | **VICTORY** |
| TYPE | Motion Picture |
| FORMAT | CED |
| RETAIL OUTLET | Movies Unlimited Inc., A2-1047, $24.95 |

| | |
|---:|:---|
| TITLE | **VICTORY AT SEA** |
| TYPE | Motion Picture, Music |
| FORMAT | CED |
| DISTRIBUTOR | RCA Videodiscs, 01201, $14.98 |

| | |
|---:|:---|
| TITLE | **VIDEO BOOKMARK, THE** |
| DESCRIPTION | A demonstration of the random access capabilities of the Sony player. |
| TYPE | Promotional |
| FORMAT | Laser CAV |
| LEVEL | 2 |
| CONTACT FOR MORE INFO. | Sony Corporation of America |
| PRODUCER | University of Nebraska/Nebraska Videodisc Design/Production Group |
| AVAILABILITY | Not available for release or distribution. |

| | |
|---:|:---|
| TITLE | **VIDEO ENHANCED GERMAN GATEWAY** |
| DESCRIPTION | Introduces the German language and culture. Includes interactive videodisc vignettes and practices. Uses video segments originally produced by the BBC. |
| TYPE | Research and Development, Instructional |
| RELEASE DATE | Winter 1984 |
| FORMAT | Laser CAV |
| LEVEL | 3 |
| SYSTEM DESCRIPTION | Sony SMC 70 and player with superimpose unit. |
| DEVELOPER | Defense Language Institute and Brigham Young University |
| CONTACT FOR MORE INFO. | Larrie E. Gale |
| FEATURES | In German only. |

| | |
|---:|:---|
| TITLE | **VIDEO MUSEUM, THE** |
| DESCRIPTION | A demonstration of the still storage capabilities of the videodisc. |
| TYPE | Promotional |
| FORMAT | Laser CAV |
| LEVEL | 1 |
| CONTACT FOR MORE INFO. | Vision Machine |
| PRODUCER | University of Nebraska/Nebraska Videodisc Design/Production Group |
| AVAILABILITY | Not available for release or distribution. |

| | |
|---:|:---|
| TITLE | **VIDEO VERSATILITY** |
| DESCRIPTION | Explains the advantages of discs for hearing-impaired learners. |
| TYPE | Promotional |
| RELEASE DATE | 1980 |
| FORMAT | Laser CAV |
| LEVEL | 2 |
| CONTACT FOR MORE INFO. | Bureau of Education for the Handicapped/University of Nebraska Barkley Center |
| PRODUCER | University of Nebraska/Nebraska Videodisc Design/Production Group |
| AVAILABILITY | Not available for release or distribution. |

| | |
|---:|:---|
| TITLE | **VIDEODISC SERIES/APPRENTICE & JOURNEYMAN TRAINING** |
| DESCRIPTION | Forty 1/2 hour videodiscs dealing with soldering, welding, safety, architectural sheet metal, testing and balancing, solar energy, air pollution, industrial sheet metal, etc. |
| TYPE | Technical, Instructional |
| RELEASE DATE | 7/83 |
| FORMAT | Laser CAV |
| LEVEL | 1,2,3 |
| DEVELOPER | National Training Fund, Sheet Metal and Air Conditioning Industry |
| CONTACT FOR MORE INFO. | Robert Drucker |
| RUNNING TIME | 1200 minutes |
| PRODUCER | Robert Drucker & Company |
| AVAILABILITY | This is proprietary material and is not available for sale, rental, or preview to any organizations or individuals without permission from the National Training Fund of the Sheet Metal and Air Conditioning Industry. |

| | |
|---:|:---|
| TITLE | **VIDEODISC, THE: A TOOL FOR COMMUNICATION** |
| DESCRIPTION | A basic overview of optical videodiscs and their applications. |
| TYPE | Instructional, Promotional |
| FORMAT | Laser CAV |
| LEVEL | 1 |
| CONTACT FOR MORE INFO. | Optical Recording Project/3M |
| PRODUCER | University of Nebraska/Nebraska Videodisc Design/Production Group |
| AVAILABILITY | Not available for release or distribution. |

| | |
|---:|:---|
| TITLE | **VIDEODROME** |
| TYPE | Motion Picture |
| FORMAT | Laser CLV |
| RUNNING TIME | 88 minutes |
| PRODUCER | MCA Videodisc, Inc. |
| DISTRIBUTOR | Pioneer Video, Inc., 11-018, $29.95 |

| | |
|---|---|
| TITLE | **VILLA ALEGRE** |
| DESCRIPTION | A 10-disc set in Spanish language education for elementary school students. |
| TYPE | Children, Instructional |
| FORMAT | Laser CAV |
| LEVEL | 1, 3 |
| CONTACT FOR MORE INFO. | Bilingual Children Television (BCTV) |
| PRODUCER | University of Nebraska/Nebraska Videodisc Design/Production Group |

| | |
|---|---|
| TITLE | **VILLAGE OF THE DAMNED** |
| TYPE | Motion Picture |
| FORMAT | CED |
| RETAIL OUTLET | Movies Unlimited Inc., A2-0174, $29.95 |

| | |
|---|---|
| TITLE | **VINCENT VAN GOGH: A PORTRAIT IN TWO PARTS** |
| DESCRIPTION | Side 1 contains a study of Van Gogh's development as an artist, hosted by Leonard Nimoy. There are two sound tracks, 16 chapters, and 700 still frames. Side 2 contains a 60-minute dramatization, starring Leondard Nimoy. |
| TYPE | Art, Instructional, Catalog |
| RELEASE DATE | March 1982 |
| FORMAT | Laser CAV |
| LEVEL | 1 |
| DEVELOPER | Vivian L. Ebersman, North American Philips Corporation |
| CONTACT FOR MORE INFO. | Same as developer |
| FEATURES | Chapters, Picture Stops, Dual Audio |
| PRODUCER | Same as developer |
| DISTRIBUTOR | NAPCEC; Pioneer Video, Inc. |
| AVAILABILITY | Sale |
| SUGGESTED PRICE | $49.95 |

| | |
|---|---|
| TITLE | **VINSON PROJECT, THE** |
| DESCRIPTION | Interactive training in the operation of aircraft carrier elevators. |
| TYPE | Technical, Instructional |
| FORMAT | Laser CAV |
| LEVEL | 3 |
| CONTACT FOR MORE INFO. | Perceptronics, Inc. |
| PRODUCER | University of Nebraska/Nebraska Videodisc Design/Production Group |
| AVAILABILITY | Not available for release or distribution. |

| | |
|---|---|
| TITLE | **VISION - UNIX INTERACTIVE VIDEODISC TRAINING CURRICULUM** |
| DESCRIPTION | The Vision Curriculum, consisting of 3 courses divided into 37 units, encompasses a complete range of training courses for UNIX. Due to the modular design of Vision, an individual can take only the training needed to accomplish his or her job. |
| TYPE | Technical, Instructional, In-house Training |
| RELEASE DATE | 12/83 |
| FORMAT | Laser CAV |
| LEVEL | 3 |
| SYSTEM DESCRIPTION | Laser videodisc player, IBM Personal Computer, high-resolution color monitor with light pen, and the ITS 2000 Controller. |
| DEVELOPER | Interactive Training Systems, Inc. |

# Videodiscography

| | |
|---|---|
| CONTACT FOR MORE INFO. | David Snow |
| FEATURES | Dual Audio |
| AVAILABILITY | Contact Interactive Training Systems, (800) 227-1127 |

| | |
|---|---|
| TITLE | **VISITING HOURS** |
| TYPE | Motion Picture |
| FORMAT | CED |
| RETAIL OUTLET | Movies Unlimited Inc., A1-1171, $29.98 |

| | |
|---|---|
| TITLE | **VISTA (VIDEODISC INTERPERSONAL SKILLS TRAINING AND ASSESSMENT)** |
| DESCRIPTION | The VISTA project was designed to teach leadership and counseling skills to second lieutenants in the Army. The final products are 11 highly interactive scenarios stored on 9 videodiscs with the following instructional options game paddle or light pen input device, optional record keeping, and pedagogical or experiential instructional designs. |
| TYPE | Research and Development, Instructional, In-house Training |
| RELEASE DATE | May 1983 |
| FORMAT | Laser CAV |
| LEVEL | 3 |
| SYSTEM DESCRIPTION | Original system was an Apple II Plus with two disk drives, Pascal language system, Mountain Hardware clock card, Coloney video interface, Pioneer 7820 Model 3 player, Symtec light pen, and Sony monitor. More recently, software has been written for a large number of alternate hardware configurations. |
| DEVELOPER | U.S. Army Research Institute (ARI) and Mellonics Systems Development Division, Litton Systems, Inc. |
| CONTACT FOR MORE INFO. | Dr. James E. Schroeder of ARI, (404) 545-1278, or Dr. Edward W. Youngling of Mellonics, (404) 561-0064 |

| | |
|---|---|
| TITLE | **VIVA LAS VEGAS** |
| TYPE | Motion Picture |
| FORMAT | CED |
| RETAIL OUTLET | Movies Unlimited Inc., A2-1048, $29.95 |

| | |
|---|---|
| TITLE | **VLADIMIR HOROWITZ IN LONDON** |
| TYPE | Music |
| FORMAT | CED |
| FEATURES | Stereo, CX Encoded |
| DISTRIBUTOR | RCA Videodiscs, 12128, $24.98 |

| | |
|---|---|
| TITLE | **VON HERRING CAPER, THE "A CHOICE MYSTERY"** |
| DESCRIPTION | The player's quest is to figure out who killed Mrs. Von Herring. The player participates in making up the story at numerous points in the game, and must respond within specified time limits in order to kiss the blond, physically catch the killer, etc. |
| TYPE | Game |
| FORMAT | Laser CAV |
| LEVEL | 2 |
| DEVELOPER | Joey Silvian |
| CONTACT FOR MORE INFO. | Joey Silvian, (213) 395-6962 |

| | |
|---:|:---|
| FEATURES | Digital Dump-Pioneer |
| PRODUCER | Mattel Electronics/Light Tracks |
| AVAILABILITY | Personal demonstration is available on request. |

| | |
|---:|:---|
| TITLE | **VOYAGER** |
| DESCRIPTION | Visual results from the exploration of Jupiter and Saturn by Voyagers I and II. Images include computer animations, data animation, and still-frames. |
| TYPE | Technical, Instructional, Archival |
| RELEASE DATE | 3/83 |
| FORMAT | Laser CAV |
| LEVEL | 1 |
| DEVELOPER | Drew University/Center for Aerospace Education in cooperation with NASA. |
| CONTACT FOR MORE INFO. | Patrick Binnes, (201) 377-0302 |
| FEATURES | Chapters, Picture Stops, Stereo |
| PRODUCER | Video Vision Associates, Ltd. |
| DISTRIBUTOR | Same as producer |
| AVAILABILITY | Sale |
| ORDER NO. | Space Disc 1 |
| SUGGESTED PRICE | $320 |

| | |
|---:|:---|
| TITLE | **WALK THROUGH THE UNIVERSE, A** |
| DESCRIPTION | An interactive introduction to astronomy based on CBS "Project Universe" telecourse. |
| TYPE | Research and Development, Instructional, Promotional |
| FORMAT | CED |
| LEVEL | 1, 3 |
| SYSTEM DESCRIPTION | RCA interactive player, Model 400 |
| DEVELOPER | CBS Publishing Group |
| CONTACT FOR MORE INFO. | Mr. Jay Plourde, CBS College Publishing Group, (212) 872-2591 |
| FEATURES | Chapters, Picture Stops, Stereo, Dual Audio, CX Encoded, Sound Over Still |
| PRODUCER | Peter A. Kreisky, CBS Publishing Group |
| DISTRIBUTOR | CBS College Publishing |
| AVAILABILITY | For market research and selective distribution initially. |

| | |
|---:|:---|
| TITLE | **WAR GAMES** |
| TYPE | Motion Picture (PG) |
| FORMAT | Laser CLV, CED |
| FEATURES | Stereo |
| PRODUCER | United Artists |
| DISTRIBUTOR | RCA Videodiscs, 13425, $29.98 |
| RETAIL OUTLET | (CLV) UltraVideo, 80-4714, $34.95 |

| | |
|---:|:---|
| TITLE | **WAR OF THE WORLDS, THE** |
| TYPE | Motion Picture |
| FORMAT | Laser CLV, CED |
| RUNNING TIME | 85 minutes |
| PRODUCER | Warner Home Video |
| DISTRIBUTOR | Pioneer Video, Inc., LV5303, $29.95<br>RCA Videodiscs, 00629, $29.98 |

| | |
|---:|:---|
| TITLE | **WAR PRAYER, THE** |
| DESCRIPTION | An interactive educational disc aimed at secondary school students. |
| TYPE | Motion Picture, Adolescents, Instructional |
| FORMAT | Laser CAV |
| LEVEL | 2 |
| DEVELOPER | Forrester Productions/Corporation for Public Broadcasting |
| CONTACT FOR MORE INFO. | David Kappes, (212) 884-1919 |
| FEATURES | Digital Dump-Pioneer, Dual Audio |
| PRODUCER | Forrester Productions |
| DISTRIBUTOR | Same as producer |
| AVAILABILITY | Not publicly available. |

| | |
|---:|:---|
| TITLE | **WARRIORS, THE** |
| TYPE | Motion Picture (R) |
| FORMAT | Laser CLV |
| RUNNING TIME | 94 minutes |
| PRODUCER | Warner Home Video |
| DISTRIBUTOR | Pioneer Video, Inc., LV1122, $29.95 |

| | |
|---:|:---|
| TITLE | **WASHINGTON AFFAIR, THE** |
| TYPE | Motion Picture |
| FORMAT | CED |
| RETAIL OUTLET | Movies Unlimited Inc., A4-3346, $29.95 |

| | |
|---:|:---|
| TITLE | **WASN'T THAT A TIME?** |
| TYPE | Motion Picture |
| FORMAT | CED |
| RETAIL OUTLET | Movies Unlimited Inc., A2-0218, $29.95 |

| | |
|---:|:---|
| TITLE | **WATERSHIP DOWN** |
| TYPE | Motion Picture (PG) |
| FORMAT | CED |
| DISTRIBUTOR | RCA Videodiscs, 02042, $19.98 |

| | |
|---:|:---|
| TITLE | **WAY WE LIVE, THE** |
| DESCRIPTION | Allows a group of viewers to select one of four characters and follow him/her through a day-in-the-life simulation, making choices about lifestyle that affect the character's healthy longevity. |
| TYPE | Medical |
| RELEASE DATE | 3/83 |
| FORMAT | Laser CAV |
| LEVEL | 3 |
| SYSTEM DESCRIPTION | Disc is controlled by an Atari 800 via a DM1000 interface. Up to 15 people can vote on life-style choices offered by the disc. The system tabulates and displays votes, and also graphs results. Player branches on choice. Offers choice of 4 subjects. |
| CONTACT FOR MORE INFO. | Patrick Lee/Joseph Koenig |
| PRODUCER | Patrick Lee |
| AVAILABILITY | Sale, Rent |
| SUGGESTED PRICE | $4,400 |

| | |
|---:|:---|
| TITLE | **WAY WE WERE, THE** |
| TYPE | Motion Picture (PG) |
| FORMAT | CED |
| DISTRIBUTOR | RCA Videodiscs, 03037, $29.98 |

| | |
|---:|:---|
| TITLE | **WEEK AT THE RACES, A** |
| DESCRIPTION | A variety of toteboards are paired with 48 videotape horse races. The disc is packaged with video "fun money," win and place cards, and game selector chips. |
| TYPE | Game, Children, Adolescents |
| RELEASE DATE | 9/83 |
| FORMAT | CED |
| LEVEL | 1 |
| DEVELOPER | RCA Videodiscs |
| CONTACT FOR MORE INFO. | Diane Smook, (212) 930-4767 |
| FEATURES | Chapters, Picture Stops |
| PRODUCER | Roy Freeman, Video Capture, Inc. for RCA |
| DISTRIBUTOR | RCA Videodiscs |
| AVAILABILITY | Sale |
| ORDER NO. | 25001 |
| SUGGESTED PRICE | $29.95 |

| | |
|---:|:---|
| TITLE | **WE'RE NO ANGELS** |
| TYPE | Motion Picture |
| FORMAT | Laser CLV |
| RETAIL OUTLET | UltraVideo, LV5414, $29.95 |

| | |
|---:|:---|
| TITLE | **WEST SIDE STORY** |
| TYPE | Motion Picture, Music, Dance |
| FORMAT | Laser CLV, CED (2 Discs) |
| FEATURES | Stereo, CX Encoded |
| RUNNING TIME | 150 minutes |
| PRODUCER | Warner Home Video |
| DISTRIBUTOR | Pioneer Video, Inc., 4519-80, $39.98 |
| | RCA Videodiscs, 11432, $34.98 |

| | |
|---:|:---|
| TITLE | **WESTERN ELECTRIC** |
| DESCRIPTION | An interactive disc providing product information for various Western Electric Products. |
| TYPE | Promotional |
| RELEASE DATE | 1983 |
| FORMAT | Laser CAV |
| LEVEL | 3 |
| PRODUCER | Brien Lee & Company |
| DISTRIBUTOR | Western Electric, Morristown, NJ |
| AVAILABILITY | For trade show exhibition in a terminal packaged by Bell Labs. |

| | |
|---:|:---|
| TITLE | **WESTWORLD** |
| TYPE | Motion Picture |
| FORMAT | CED |
| RETAIL OUTLET | Movies Unlimited Inc., A2-1043, $29.95 |

| | |
|---:|:---|
| TITLE | **WEYERHAEUSER** |
| DESCRIPTION | Recruiting material for college graduates. |
| TYPE | Instructional, Catalog, Promotional |
| FORMAT | Laser CAV |
| LEVEL | 2 |
| CONTACT FOR MORE INFO. | Aileen McKenna, (919) 493-3479 |
| FEATURES | Chapters, Picture Stops |
| RUNNING TIME | 30 minutes |
| AVAILABILITY | PREF has laser disc equipment in college placement offices across the country. PREF owns the equipment and distributes discs describing a company's recruiting programs into the network. The college receives the service at no charge, and the corporation pays a distribution fee. |

| | |
|---:|:---|
| TITLE | **WHALES** |
| DESCRIPTION | An interchangeable level 1 and 2 disc designed for grades 5-8. |
| TYPE | Children, Instructional |
| LEVEL | 1, 2 |
| CONTACT FOR MORE INFO. | National Geographic Society |
| PRODUCER | University of Nebraska/Nebraska Videodisc Design/Production Group |
| AVAILABILITY | Not available for release or distribution. |

| | |
|---:|:---|
| TITLE | **WHAT'S NEW, PUSSYCAT?** |
| TYPE | Motion Picture, Comedy |
| FORMAT | CED |
| RETAIL OUTLET | Movies Unlimited Inc., A1-4583, $29.98 |

| | |
|---:|:---|
| TITLE | **WHAT'S UP, DOC?** |
| TYPE | Motion Picture, Comedy (G) |
| FORMAT | CED |
| DISTRIBUTOR | RCA Videodiscs, 03112, $24.98 |

| | |
|---:|:---|
| TITLE | **WHAT'S UP, TIGER LILY?** |
| TYPE | Motion Picture, Comedy |
| FORMAT | Laser CLV, CED |
| RUNNING TIME | 90 minutes |
| PRODUCER | Vestron Video |
| DISTRIBUTOR | Pioneer Video, Inc., VL4018, $34.95 |
| RETAIL OUTLET | (CED) Movies Unlimited Inc., A3-4018, $29.95 |

| | |
|---:|:---|
| TITLE | **WHEN A STRANGER CALLS** |
| TYPE | Motion Picture |
| FORMAT | Laser CLV |
| RETAIL OUTLET | UltraVideo, VLD6050, $29.95 |

| | |
|---:|:---|
| TITLE | **WHEN WORLDS COLLIDE** |
| TYPE | Motion Picture (G) |
| FORMAT | Laser CLV |
| RUNNING TIME | 81 minutes |
| PRODUCER | Warner Home Video |
| DISTRIBUTOR | Pioneer Video, Inc., LV5106, $29.95 |

| | |
|---|---|
| TITLE | **WHITE HEAT** |
| TYPE | Motion Picture |
| FORMAT | CED |
| DISTRIBUTOR | RCA Videodiscs, 01459, $19.98 |

| | |
|---|---|
| TITLE | **WHITE LIGHTNING** |
| TYPE | Motion Picture (PG) |
| FORMAT | CED |
| DISTRIBUTOR | RCA Videodiscs, 01418, $24.98 |

| | |
|---|---|
| TITLE | **WHITE MUSIC** |
| TYPE | Music |
| FORMAT | Laser CAV |
| LEVEL | 1 |
| FEATURES | Stereo, CX Encoded |
| PRODUCER | Pioneer Video Imports |
| DISTRIBUTOR | Pioneer Video, Inc., MS008, $27.95 |

| | |
|---|---|
| TITLE | **WHO - LAST AMERICAN TOUR, THE** |
| TYPE | Music |
| FORMAT | Laser CLV, CED |
| FEATURES | Stereo |
| RETAIL OUTLET | (CLV) UltraVideo, 80-6234, $29.95 |
| | (CED) Movies Unlimited Inc., A1-6234, $29.98 |

| | |
|---|---|
| TITLE | **WHO ROCKS AMERICA, THE** |
| TYPE | Music |
| FORMAT | Laser CLV |
| FEATURES | Stereo, CX Encoded |
| RUNNING TIME | 118 minutes |
| PRODUCER | CBS Fox Video |
| DISTRIBUTOR | Pioneer Video, Inc., 6234-80, $34.98 |

| | |
|---|---|
| TITLE | **WHO'S AFRAID OF OPERA? VOL. 1** |
| TYPE | Music |
| FORMAT | CED |
| RETAIL OUTLET | Movies Unlimited Inc., A2-0187, $29.95 |

| | |
|---|---|
| TITLE | **WHO'S AFRAID OF OPERA? VOL. 2** |
| TYPE | Music |
| FORMAT | CED |
| RETAIL OUTLET | Movies Unlimited Inc., A2-0212, $29.95 |

| | |
|---|---|
| TITLE | **WHO'S AFRAID OF OPERA? VOL. 3** |
| TYPE | Music |
| FORMAT | CED |
| RETAIL OUTLET | Movies Unlimited Inc., A2-0213, $29.95 |

| | |
|---:|:---|
| TITLE | **WHOLLY MOSES!** |
| TYPE | Motion Picture, Comedy (PG) |
| FORMAT | Laser CLV, CED |
| PRODUCER | RCA/Columbia Pictures Home Video |
| DISTRIBUTOR | Pioneer Video, Inc., VLD6060, $29.95 |
| | RCA Videodiscs, 03024, $24.98 |

| | |
|---:|:---|
| TITLE | **WHOSE LIFE IS IT ANYWAY?** |
| TYPE | Motion Picture (R) |
| FORMAT | CED |
| PRODUCER | Metro-Goldwyn-Mayer Film Co. |
| DISTRIBUTOR | MGM/United Artists |
| RETAIL OUTLET | Movies Unlimited Inc., A2-0140, $29.95 |

| | |
|---:|:---|
| TITLE | **WIFEMISTRESS** |
| TYPE | Motion Picture |
| FORMAT | CED |
| RETAIL OUTLET | Movies Unlimited Inc., A2-1034, $24.95 |

| | |
|---:|:---|
| TITLE | **WILD BUNCH, THE** |
| TYPE | Motion Picture |
| FORMAT | CED (2 Discs) |
| DISTRIBUTOR | RCA Videodiscs, 03135, $34.98 |

| | |
|---:|:---|
| TITLE | **WILD IN THE COUNTRY** |
| TYPE | Motion Picture |
| FORMAT | CED |
| FEATURES | Stereo |
| RETAIL OUTLET | Movies Unlimited Inc., A1-1174, $29.98 |

| | |
|---:|:---|
| TITLE | **WILS - WANG INTERACTIVE LEARNING SYSTEM - WORD PROCESSING PROGRAM** |
| DESCRIPTION | Videodisc-based self-paced computer-assisted instruction in Wang Word Processing on OIS and VS/IIS systems. WILS enables the user to acquire the full range of basic word processing skills from an introduction to the screen and keyboard through merge printing. |
| TYPE | Technical, Instructional, In-house Training, |
| RELEASE DATE | 7/1/83 |
| FORMAT | Laser CAV |
| LEVEL | 3 |
| SYSTEM DESCRIPTION | Wang OIS and VS/IIs systems |
| DEVELOPER | Wang Laboratories, Inc. |
| CONTACT FOR MORE INFO. | Local Wang Office |
| FEATURES | Dual Audio |
| PRODUCER | Jam Productions, Rochester, NY |
| DISTRIBUTOR | Wang Laboratories, Inc. |
| AVAILABILITY | Sale. Requires Wang equipment. |
| ORDER NO. | PN200-1041&2 |
| SUGGESTED PRICE | $5000 each |

| | |
|---:|:---|
| TITLE | **WIMBLEDON 1981/WIMBLEDON A CENTURY OF GREATNESS** |
| TYPE | Motion Picture |
| FORMAT | CED |
| DISTRIBUTOR | RCA Videodiscs, 02078, $14.98 |

| | |
|---:|:---|
| TITLE | **WIMBLEDON 1979-1980 FEATURING BORG AND MCENROE** |
| TYPE | Motion Picture |
| FORMAT | CED |
| DISTRIBUTOR | RCA Videodiscs, 02048, $14.98 |

| | |
|---:|:---|
| TITLE | **WINNIE THE POOH** |
| TYPE | Motion Picture |
| FORMAT | Laser CLV |
| RETAIL OUTLET | UltraVideo, 025AS, $34.95 |

| | |
|---:|:---|
| TITLE | **WINTER KILLS** |
| TYPE | Motion Picture |
| FORMAT | CED |
| RETAIL OUTLET | Movies Unlimited Inc., A4-0566, $29.95 |

| | |
|---:|:---|
| TITLE | **WIZ, THE** |
| TYPE | Motion Picture, Music, Dance (G) |
| FORMAT | Laser CLV |
| FEATURES | Stereo |
| RUNNING TIME | 134 minutes |
| PRODUCER | Warner Home Video |
| DISTRIBUTOR | Pioneer Video, Inc., 17-005, $29.95 |

| | |
|---:|:---|
| TITLE | **WIZARD OF OZ, THE** |
| TYPE | Motion Picture, Music |
| FORMAT | Laser CLV, CED |
| RUNNING TIME | 101 minutes |
| PRODUCER | MGM/UA Home Video |
| DISTRIBUTOR | Pioneer Video, Inc., ML100001, $34.95 |
| RETAIL OUTLET | (CED) Movies Unlimited Inc., A2-1015, $29.95 |

| | |
|---:|:---|
| TITLE | **WOMEN IN LOVE** |
| TYPE | Motion Picture |
| FORMAT | CED (2 Discs) |
| RETAIL OUTLET | Movies Unlimited Inc., A1-4576, $39.98 |

| | |
|---:|:---|
| TITLE | **WOODSTOCK** |
| TYPE | Motion Picture, Music |
| FORMAT | CED (2 Discs) |
| FEATURES | Stereo, Banded |
| DISTRIBUTOR | RCA Videodiscs, 13110, $39.98 |

# Videodiscography

| | |
|---|---|
| TITLE | **WOODY WOODPECKER AND HIS FRIENDS** |
| TYPE | Motion Picture (G) |
| FORMAT | Laser CLV |
| RUNNING TIME | 80 minutes |
| PRODUCER | MCA Videodisc, Inc. |
| DISTRIBUTOR | Pioneer Video, Inc., 61-008, $29.95 |

| | |
|---|---|
| TITLE | **WORDS IN MOTION I** |
| DESCRIPTION | Fingerspelling for elementary hearing-impaired learners. |
| TYPE | Children, Instructional |
| RELEASE DATE | 1980 |
| FORMAT | Laser CAV |
| LEVEL | 1, 2 |
| CONTACT FOR MORE INFO. | Bureau of Education for the Handicapped/University of Nebraska Barkley Center |
| PRODUCER | University of Nebraska/Nebraska Videodisc Design/Production Group |
| AVAILABILITY | Not available for release or distribution. |

| | |
|---|---|
| TITLE | **WORDS IN MOTION II** |
| DESCRIPTION | Fingerspelling. |
| TYPE | Children, Instructional |
| RELEASE DATE | 1980 |
| FORMAT | Laser CAV |
| LEVEL | 1, 2 |
| CONTACT FOR MORE INFO. | Bureau of Education for the Handicapped/University of Nebraska Barkley Center |
| AVAILABILITY | Not available for release or distribution. |

| | |
|---|---|
| TITLE | **WORLD ACCORDING TO GARP, THE** |
| TYPE | Motion Picture |
| FORMAT | CED |
| RETAIL OUTLET | Movies Unlimited Inc., A1-1261, $39.98 |

| | |
|---|---|
| TITLE | **WORLD OF MARTIAL ARTS, THE** |
| TYPE | Instructional |
| FORMAT | Laser CAV |
| LEVEL | 1 |
| PRODUCER | Optical Programming Associates |
| DISTRIBUTOR | Pioneer Video, Inc., 37-605, $29.95 |

| | |
|---|---|
| TITLE | **WORLD OF WILDLIFE, VOL. 1** |
| FORMAT | CED |
| DISTRIBUTOR | RCA Videodiscs, 02020, $14.98 |

| | |
|---|---|
| TITLE | **WORLD ON A SILVER PLATTER, THE** |
| DESCRIPTION | Program designed to support sale of the Sylvania videodisc player. It introduces the player, allows an interactive demonstration of its features, and presents the best selling method. |
| TYPE | Instructional, In-house Training, |
| RELEASE DATE | 6/82 |

| | |
|---:|:---|
| FORMAT | Laser CAV |
| LEVEL | 1 |
| DEVELOPER | The Glyn Group, Inc. |
| CONTACT FOR MORE INFO. | J.M. McCarthy, (212) 255-5156 |
| FEATURES | Chapters, Picture Stops, Stereo, Dual Audio |
| PRODUCER | Scott Dobbins/The Glyn Group/North American Philips, Consumer Electronics Corporation |
| AVAILABILITY | Not available. 3M demonstration program. Contact Scott Dobbins, (212) 741-1383 for demonstration. |

| | |
|---:|:---|
| TITLE | **WORLD SERIES 1980 - PHILADELPHIA PHILLIES VS. KANSAS CITY ROYALS** |
| TYPE | Motion Picture |
| FORMAT | CED |
| DISTRIBUTOR | RCA Videodiscs, 01209, $14.98 |

| | |
|---:|:---|
| TITLE | **WORLD'S GREATEST LOVER, THE** |
| TYPE | Motion Picture |
| FORMAT | CED |
| RETAIL OUTLET | Movies Unlimited Inc., A1-1105, $29.98 |

| | |
|---:|:---|
| TITLE | **XANADU** |
| TYPE | Motion Picture, Music, (PG) |
| FORMAT | Laser CLV |
| FEATURES | Stereo |
| RUNNING TIME | 97 minutes |
| PRODUCER | Warner Home Video |
| DISTRIBUTOR | Pioneer Video, Inc., 17-006, $29.95 |

| | |
|---:|:---|
| TITLE | **YANKEE DOODLE DANDY** |
| TYPE | Motion Picture, Music, Dance |
| FORMAT | Laser CLV, CED |
| FEATURES | Black and white |
| RUNNING TIME | 126 minutes |
| PRODUCER | CBS Fox Video |
| DISTRIBUTOR | Pioneer Video, Inc., 4513-80, $39.98<br>RCA Videodiscs, 01433, $24.98 |

| | |
|---:|:---|
| TITLE | **YEAR OF LIVING DANGEROUSLY, THE** |
| TYPE | Motion Picture |
| FORMAT | Laser CLV |
| RETAIL OUTLET | UltraVideo, ML100243, $34.95 |

| | |
|---:|:---|
| TITLE | **YELLOWBEARD** |
| TYPE | Motion Picture, Comedy |
| FORMAT | CED |
| DISTRIBUTOR | RCA Videodiscs, 01915, $24.98 |

| | |
|---:|:---|
| TITLE | **YES, GEORGIO** |
| TYPE | Motion Picture |
| FORMAT | Laser CLV, CED |
| FEATURES | Stereo |
| RETAIL OUTLET | (CLV) UltraVideo, ML100208, $34.95 |
| | (CED) Movies Unlimited Inc., A2-0208, $29.95 |

| | |
|---:|:---|
| TITLE | **YOR: THE HUNTER FROM THE FUTURE** |
| TYPE | Motion Picture |
| FORMAT | CED |
| DISTRIBUTOR | RCA Videodiscs, 03072, $29.95 |

| | |
|---:|:---|
| TITLE | **YOU AND AFDC** |
| DESCRIPTION | A high interest, low-level interactive program designed to introduce social workers to the job of a welfare worker in Florida and to a 160-hour interactive job skills training program. The disc introduces the trainee to the computer-based instruction system and program; and sets the tone for a comfortable familiarity with the materials in as short a time as possible. It is a single, continuous motion sequence with only three pauses designed and positioned so that the trainee will have time to read pertinent materials at the appropriate times. |
| TYPE | Instructional, In-house Training |
| RELEASE DATE | 9/15/83 |
| FORMAT | Laser CAV |
| LEVEL | 3 |
| SYSTEM DESCRIPTION | Disc is part of a 160-hour curriculum delivered on an Apple II/Pioneer 7820-3/single disk drive system along with ancillary printed reference books. |
| DEVELOPER | University of West Florida/Office for Interactive Technology and Training |
| CONTACT FOR MORE INFO. | Richard Smith; Elizabeth Wright, (413) 256-0444 |
| FEATURES | Dual Audio |
| PRODUCER | University of Nebraska/Nebraska Videodisc Design/Production Group |
| AVAILABILITY | Training program produced under contract with the Florida Department of Health and Rehabilitative Services, Tallahassee, Florida, and is not available for general distribution. |

| | |
|---:|:---|
| TITLE | **YOU CAN BUY A COMPUTER!** |
| DESCRIPTION | An interactive videodisc to help salespeople save time teaching potential customers about computer concepts and terms. Computer uses, how they operate, equipment needed, and buying tips are covered in a short, fast, audiovisual presentation. A "Customer Needs" questionnaire uses a voice responder and produces a hard copy. There is a glossary and a game-formatted computer IQ test. |
| TYPE | Instructional, In-house Training, Promotional |
| RELEASE DATE | 11/28/83 |
| FORMAT | Laser CAV |
| LEVEL | 3 |
| SYSTEM DESCRIPTION | Pioneer LP 7820-3 player, Apple IIe with two disk drives, Amdek color monitor, and an Allen interface card. |

| | |
|---:|:---|
| DEVELOPER | Sharon Brown |
| CONTACT FOR MORE INFO. | Same as developer |
| FEATURES | Dual Audio |
| PRODUCER | Dr. David L. Winn |
| AVAILABILITY | Sale |
| SUGGESTED PRICE | Linear version, $39 |
| | Interactive version, $395 |

| | |
|---:|:---|
| TITLE | **YOU ONLY LIVE TWICE** |
| TYPE | Motion Picture |
| FORMAT | Laser CLV, CED |
| DISTRIBUTOR | RCA Videodiscs, 01425, $29.98 |
| RETAIL OUTLET | (CLV) UltraVideo, 80-4526, $34.95 |

| | |
|---:|:---|
| TITLE | **YOUNG DOCTORS IN LOVE** |
| TYPE | Motion Picture (R) |
| FORMAT | Laser CLV, CED |
| RUNNING TIME | 97 minutes |
| PRODUCER | Vestron Video |
| DISTRIBUTOR | Pioneer Video, Inc., LV5012, $34.95 |
| RETAIL OUTLET | (CED) Movies Unlimited Inc., A3-5012, $29.95 |

| | |
|---:|:---|
| TITLE | **YOUNG FRANKENSTEIN** |
| TYPE | Motion Picture, Comedy (PG) |
| FORMAT | Laser CLV, CED |
| FEATURES | (CLV) Stereo, Black and white |
| RUNNING TIME | 106 minutes |
| PRODUCER | Warner Home Video |
| DISTRIBUTOR | Pioneer Video, Inc., 1103-80, $34.98 |
| RETAIL OUTLET | (CED) Movies Unlimited Inc., A1-1103, $29.98 |

| | |
|---:|:---|
| TITLE | **YUM-YUM GIRLS** |
| TYPE | Motion Picture, Adult (R) |
| FORMAT | Laser CLV |
| RUNNING TIME | 93 minutes |
| PRODUCER | Warner Home Video |
| DISTRIBUTOR | Pioneer Video, Inc., 28-010, $29.95 |

| | |
|---:|:---|
| TITLE | **Z** |
| TYPE | Motion Picture |
| FORMAT | Laser CLV, CED |
| RETAIL OUTLET | (CLV) UltraVideo, VLD6100, $29.95 |
| | (CED) Movies Unlimited Inc., A1-1804, $24.98 |

| | |
|---:|:---|
| TITLE | **ZAPPED!** |
| TYPE | Motion Picture (R) |
| FORMAT | Laser CLV, CED |
| RUNNING TIME | 98 minutes |
| PRODUCER | Embassy Home Entertainment |
| DISTRIBUTOR | Pioneer Video, Inc., 16045, $29.95 |
| RETAIL OUTLET | (CED) Movies Unlimited Inc., A4-6046, $29.95 |

# Videodiscography

| | |
|---:|:---|
| TITLE | **ZIEGFELD FOLLIES** |
| TYPE | Motion Picture |
| FORMAT | CED |
| RETAIL OUTLET | Movies Unlimited Inc., A2-0173, $29.95 |

| | |
|---:|:---|
| TITLE | **ZORRO, THE GAY BLADE** |
| TYPE | Motion Picture |
| FORMAT | CED |

# INDEXES

## INDIVIDUAL NAME AFFILIATION

| | | |
|---|---|---|
| Abbott, Anthony | MECKLER PUBLISHING | P&D, 153 |
| Abell, R.A. (Bob) | ALPHATEL SYSTEMS LTD. | P&D, 121; H/S/IS, 180 |
| Adkins, Allen Lee | INTERACTIVE ARTS INTERNATIONAL | P&D, 144; H/S/IS, 192 |
| Allan, Colin | DIGITAL EQUIPMENT CORPORATION | P&D, 132 |
| Allen, David P. | BOSTON MEDIA CONSULTANTS | P&D, 123; H/S/IS, 181 |
| Allen, Steven | ALLEN COMMUNICATION | H/S/IS, 180 |
| Allen, Rex | ALLEN COMMUNICATION | H/S/IS, 180 |
| Anderson, Frances | ILLINOIS STATE UNIV. COLLEGE OF FINE ARTS | P&D, 140 |
| Armstrong, Jim | SOUTHERN ALBERTA INST. OF TECH./ LEARNING RES. CEN. | P&D, 165 |
| Axt, Mary Ann | SONY CORPORATION OF AMERICA | P&D, 164 |
| Baer, Ralph | SANDERS ASSOCIATES, INC. | P&D, 161 |
| Barnes, Stan | INTERMEDIA | P&D, 146; H/S/IS, 193 |
| Barreto, Virginia | TRILOGIC, INC. | P&D, 168; H/S/IS, 208 |
| Barry, Leslie | FUSION MEDIA, INC. | P&D, 137 |
| Bashista, Michael | BASHISTA, MICHAEL | P&D, 122 |
| Batcho, Joseph J. | COOPERLASER SONICS | H/S/IS, 184 |
| Becker, Jurgen | INNOVISION | P&D, 143 |
| Becker, Jurgen | MAZ & MOVIE | H/S/IS, 196 |
| Behrend, Sam | PRESIDIO VIDEO | P&D, 158 |
| Behrend, Sam | UNIV. OF ARIZONA/ BIOMEDICAL COMMUNICATIONS | P&D, 169 |
| Beiser, Leo | LEO BEISER INC. | P&D, 150 |
| Belzile, Jan | INFOTECH CONSULTANTS | P&D, 142 |
| Bergsma-Owen, Judy | CUTTING EDGE PRODUCTIONS, INC. | P&D, 130 |
| Berry, Jack | MULTI IMAGE INTERNATIONAL | H/S/IS, 197 |
| Bertram, Henry A. | UNIV. OF CALIFORNIA AT SAN DIEGO | PT. 1, 101 |
| Bloch, Peter | RENAN PRODUCTIONS | P&D, 161 |
| Bouley, Raymond J. | INTERACTIVE AUTHORING, INC. | P&D, 144 |
| Brancaccio, Dan | PRIMARIUS | H/S/IS, 201 |
| Branson, Robert | FLORIDA STATE UNIV./ CENTER FOR EDUC. TECHNOLOGY | P&D, 135 |
| Brown, Sharon | INTERACTIVE VIDEO CORPORATION | P&D, 146 |
| Bruce, Robert | CORNELL UNIVERSITY/ DEPARTMENT OF EDUCATION | P&D, 129 |
| Buckley, David | MULTI IMAGE INTERNATIONAL | H/S/IS, 197 |
| Buffalo, Constance | IT, INC. | P&D, 147; H/S/IS, 194 |
| Burley, Jim | ADVANCED VIDEO, INC. | P&D, 120 |
| Butler, Lauralee | INTERACTIVE TECHNOLOGIES | PT. 1, 27 |
| Butler, Madeleine L. | BUTLER, MADELEINE L. | P&D, 124 |
| Butler, Michael | HEARTLAND COMMUNICATIONS | P&D, 138 |
| Butler, Ron | PIONEER VIDEO, INC. | H/S/IS, 200; M&R, 213 |
| Callahan, Peg | INTERACTIVE VIDEO CONCEPTS, INC. | P&D, 145; H/S/IS, 193 |
| Cammarata, Martha | MARYLAND INSTRUCTIONAL TELEVISION | P&D, 153 |
| Carlisle, Joan | OPTICAL SOCIETY OF AMERICA | P&D, 157 |
| Carlquist, William | HORIZONTAL EDITING STUDIOS | P&D, 139 |

PT. 1 = Part One
P&D = Production and Development Resources Section
H/S/IS = Hardware/Software/Integrated Systems Section
M&R = Mastering and Replication Resources Section

| | | |
|---|---|---|
| Carlson, Gary | JOHN WILEY & SONS | PT. 1, 107 |
| Carrington, David | WDC MARKETING ASSOCIATES, INC. | P&D, 175; H/S/IS, 209 |
| Carroll, Mary Pat | ONTARIO INSTITUTE FOR STUDIES IN EDUCATION | P&D, 156 |
| Cavagnol, Richard M. | WICAT SYSTEMS, INC. | P&D, 176; H/S/IS, 210 |
| Cella, Al | GRUMMAN AEROSPACE CORPORATION | P&D, 138; H/S/IS, 190 |
| Cheffins, P. | ENERGY, MINES & RESOURCES | H/S/IS, 188 |
| Chick, Leon | TECHNIDISC | M&R, 216 |
| Christie, Ken | KEN S. CHRISTIE AND ASSOCIATES | PT. 1, 100; P&D, 149 |
| Church, James F. | COMBUSTION ENGINEERING, INC. | P&D, 127 |
| Ciampa, John | AMERICAN VIDEO INST./ VIDEODISC & ELEC. PUBL. LAB. | P&D, 121; H/S/IS, 180 |
| Clark, David R. | CENTRE FOR REMOTE SENSING | H/S/IS, 182 |
| Clark, David R. | CLARK, DAVID R. | P&D, 126 |
| Clark, David R. | UNIV. OF LONDON AUDIO-VISUAL CENTRE | P&D, 171 |
| Clark, Bill | VIDEO VISION ASSOCIATES, LTD. | P&D, 174 |
| Clark, D. Joseph | VIDEODISCOVERY | P&D, 175 |
| Claxton, William | SCRIPTECH | P&D, 162; H/S/IS, 203 |
| Cochrane, Don | EXHIBIT TECHNOLOGY, INC. | P&D, 135; H/S/IS, 188 |
| Cohen, Joel | SOUND CONCEPTS | H/S/IS, 205 |
| Cooper, Jack | INTERACTIVE ARTS & SCIENCE/IAS, INC. | P&D, 143 |
| Crosby, Walter R. | PROPERTY TECHNOLOGY, INC. | P&D, 159; H/S/IS, 201 |
| Dargan, Tom | SONY VIDEO UTILIZATION SERVICES | P&D, 121 |
| Dark, Paul | VIDEO MASTERS, INC. | P&D, 173; H/S/IS, 209 |
| Davidson, Stuart | DAVIDSON, STUART P. | P&D, 131 |
| Davis, Barbara | DAVIS, BARBARA GROSS | P&D, 131 |
| Daynes, Rod | INTERACTIVE TECHNOLOGIES CORPORATION | PT. 1, 7; P&D, 144 |
| Devlin, Sandra | DEVLIN PRODUCTIONS, INC. | P&D, 132 |
| Dextraze, Ed | RAYTHEON SERVICE COMPANY | P&D, 159; H/S/IS, 201 |
| Dietz, Lawrence | ALEC GROUP | P&D, 121 |
| Dobbins, Scott | SCOTT DOBBINS PRODUCTIONS | P&D, 162 |
| Dobyns, Frank | EDITEL-NEW YORK | P&D, 134; H/S/IS, 188 |
| Donne, Eric | INDIS INTERNATIONAL | P&D, 141 |
| Downes, Judith B. | CENTURY III TELEPRODUCTIONS | P&D, 125 |
| Doyle, Brian | I.S.D. #279/PUBLIC SCHOOL SYSTEM | P&D, 140 |
| Drucker, Robert | ROBERT DRUCKER & COMPANY | P&D, 161 |
| Dsabbarzadeh, Rudy | PARALLEL COMMUNICATIONS | P&D, 157 |
| Dugan, Allen | DBS FILMS, INC. | P&D, 132 |
| Dupuy, Alain | LOGIVISION | P&D, 150; H/S/IS, 195 |
| Durix, Christian | LOGIVISION | P&D, 150; H/S/IS, 195 |
| Dyer, Mike | VIDEO ASSOCIATES LABS | H/S/IS, 209 |
| Faller, Karl | WREN ASSOCIATES INC. | P&D, 177 |
| Farrow, Jason | SONY CORPORATION OF AMERICA | P&D, 164; H/S/IS, 204 |
| Feeney, Ray | SPECTRA-IMAGE, INC. | PT. 1, 67 |
| Fenhagen, Don | INFORMATION DELIVERY SYSTEMS, INC. | P&D, 142 |
| Ferguson, Bruce | DEPT. OF NATIONAL DEFENCE | P&D, 132; H/S/IS, 185 |
| Fertner, Ken | FRANKLIN RESEARCH CENTER | P&D, 136; H/S/IS, 189 |
| Firestone, Susan | WREN ASSOCIATES INC. | P&D, 177 |
| Fisher, Rick | ADVANCED VIDEO, INC. | P&D, 120 |
| Flanigan, Pomela | COURSEWARE, INC. | P&D, 130; H/S/IS, 184 |
| Fleischmann, Jonathan | DIGITAL TECHNIQUES INC. | H/S/IS, 186 |
| Franke, Arthur | NEW MEDIA GRAPHICS CORPORATION | H/S/IS, 198 |
| Frauenhoffer, Paul | IBM | H/S/IS, 191 |
| Fukuda, Dave M. | HITACHI SALES CORPORATION OF AMERICA | H/S/IS, 191 |
| Fuller, Robert G. | UNIV. OF NEBRASKA - LINCOLN | PT. 1, 87 |
| Gale, Larrie | DAVID O. MCKAY INSTITUTE | P&D, 130; H/S/IS, 185 |
| Garant, George | TERAK CORPORATION | H/S/IS, 207 |
| Garhart, Casey | GARHART, CASEY | PT.1, 45 |
| Garripoli, Garri | CITIDISC PRODUCTIONS | P&D, 126; H/S/IS, 182 |
| Gay, Geri | CORNELL UNIVERSITY/ DEPARTMENT OF EDUCATION | P&D, 129 |
| Gayeski, Diane | OMNICOM ASSOCIATES | P&D, 156; H/S/IS, 198 |
| Gaynor, Joseph | INNOVATIVE TECHNOLOGY ASSOCIATES | P&D, 142 |
| Gerstenmaier, William | GERSTENMAIER, WILLIAM | P&D, 137 |
| Ghislandi, Patrizia | CENTRO TELEVISIVO UNIVERSITARIO (CTU) | P&D, 125 |
| Gillmore, Gerald | VIDEOACTIVE, INC. | P&D, 174 |
| Glasby, Rick | CREATIVISION, INC. | P&D, 130; H/S/IS, 184 |
| Goddard, Larry D. | WINDMILL PRODUCTIONS, INC. | P&D, 177 |
| Gordon, Cindy | INFOTECH CONSULTANTS | P&D, 142 |

# Individual Name Affiliation

| | | |
|---|---|---|
| Gordon, Robert | PRODUCTION ASSOCIATES, INC. | P&D, 158 |
| Gordon, Ronald | BELL TELEPHONE LABORATORIES | H/S/IS, 181 |
| Grefe, Richard | CORPORATION FOR PUBLIC BROADCASTING | P&D, 129 |
| Gregg, D.P. | GREGG, D.P. | H/S/IS, 190 |
| Gregory, Claire | INDIANA UNIVERSITY RADIO AND TELEVISION SERVICE | P&D, 141 |
| Guggenheim, Rolph | LUCASFILM, LTD. | P&D, 151 |
| Hallas, Joyce | ACTRONICS, INC. | P&D, 119; H/S/IS, 179 |
| Halpern, Charles | LUMIERE PRODUCTIONS | P&D, 151 |
| Hansen, Mark | MARK IV DESIGNS | P&D, 152; H/S/IS, 195 |
| Hargan, Carol | HUMAN RESOURCES RESEARCH ORGANIZATION (HUMRRO) | P&D, 139 |
| Harris, Colin | ICS-INTEXT, A DIVISION OF NATIONAL EDUC. CORP. | H/S/IS, 191 |
| Hartigan, John | SONY COMMUNICATIONS PRODUCTS COMPANY | P&D, 164; H/S/IS, 204; M&R, 215 |
| Haupt, Judy | MEDIA CONCEPTS, INC. | P&D, 154 |
| Hayes, B. | HAYES PRODUCTIONS, INC. | P&D, 138 |
| Hedges, Thomas | VCM SYSTEMS, INC. | P&D, 172; H/S/IS, 208 |
| Hegg, Vivian L. | PERFORMANCE TECHNOLOGY, INC. | H/S/IS, 200 |
| Heigl, Ralph | VIDEO VISION ASSOCIATES, LTD. | P&D, 174; H/S/IS, 209 |
| Helgerson, Linda W. | HELGERSON ASSOCIATES | P&D, 139 |
| Hewlett, Brent | FLORIDA STATE UNIV./ CENTER FOR EDUC. TECHNOLOGY | P&D, 135 |
| Hewlett, Brent | HEWLETT, BRENT | P&D, 139 |
| Heyer, Mark | BRADEN GROUP | PT. 1, 102 |
| Hildred, Christopher | SIMON FRASER UNIVERSITY/ INSTRUCTIONAL MEDIA CENTRE | P&D, 163 |
| Hoekema, Jim | WICAT SYSTEMS, INC. | PT. 1, 35 |
| Hoffman, Susan | INTERMEDIA | P&D, 146; H/S/IS, 193 |
| Homme, Evie | OPTICAL RECORDING PROJECT/3M | H/S/IS, 199; M&R, 211 |
| Honeyman, Jeffrey | EDUCATIONAL BROADCASTING CORPORATION (WNET/13) | P&D, 134 |
| Hopwood, David | VIDEO SOFTWARE ASSOCIATES, INC. | P&D, 173 |
| Howard, Diane | BCD ASSOCIATES, INC. | H/S/IS, 180 |
| Howell, Jeffrey A. | ACTRONICS, INC. | H/S/IS, 179 |
| Huffman, David | SOUTHWEST TEXAS STATE UNIVERSITY | H/S/IS, 205 |
| Hughes, Lyric | TRANS-LINGUAL COMMUNICATIONS, INC. | P&D, 168; H/S/IS, 207 |
| Humphrey, Darrell F. | MARTIN MARIETTA DATA SYSTEMS/ TECH. SYSTEMS GROUP | P&D, 152; H/S/IS, 195 |
| Huntley, Joan Sustik | UNIV. OF IOWA/ COMPUTER ASSISTED INSTRUCTION LAB | P&D, 170 |
| Isbouts, Jean-Pierre | ADVANCED IMAGE TECHNOLOGY, INC. | P&D, 120; H/S/IS, 179 |
| Ittelson, John C. | ITTELSON, JOHN C. | P&D, 148 |
| Jasionowski, Anthony | MATSUSHITA TECHNOLOGY CENTER | H/S/IS, 196 |
| Johnston, Steven | PERCEPTRONICS, INC. | P&D, 157; H/S/IS, 199 |
| Kappes, David | FORRESTER PRODUCTIONS | P&D, 136 |
| Kar, Saroj K. | TCT TECHNICAL TRAINING, INC. ("T-CUBED") | P&D, 166; H/S/IS, 206 |
| Kelbaugh, Larry | KELBAUGH, LARRY E. | P&D, 149 |
| Kelly, Susan Brooks | VIDEO SOFTWARE ASSOCIATES, INC. | P&D, 173 |
| Kemph, Jeff | DIGITAL CONTROLS | PT. 1, 61 |
| Kessinger, Phil | EUGENE PUBLIC SCHOOLS | PT. 1, 93; P&D, 134 |
| Keyerleber, Joe | JOSEPH KEYERLEBER PRODUCTIONS | P&D, 148 |
| Kindleberger, Charles | APPLIED VIDEO TECHNOLOGY, INC. | P&D, 121 |
| King, Bert | INTEGRATED AUTOMATION | H/S/IS, 192 |
| King, Mike | OPTICAL RECORDING PROJECT/3M | H/S/IS, 192 |
| Kirchner, Glenn | SIMON FRASER UNIVERSITY/ INSTRUCTIONAL MEDIA CENTRE | P&D, 163 |
| Kite, Charles R. | JACK LIEB PRODUCTIONS, INC. | P&D, 148; H/S/IS, 194 |
| Klotz, J. Taylor | FUSION MEDIA, INC. | P&D, 137 |
| Koehl, Thomas | SYSCON CORPORATION, INC. | P&D, 166 |
| Krasney, Stuart | INNOTECH | P&D, 142; H/S/IS, 192 |
| Kreiser, Jim | SIGNAL COMMUNICATIONS | P&D, 163 |
| Kribs, H. Dewey | INSTRUCTIONAL SCIENCE AND DEVELOPMENT, INC. | P&D, 143 |
| Kukla, Steve G. | WESTINGHOUSE ELECTRIC CORP. | P&D, 176; H/S/IS, 210 |
| Lacoursiere, William | SOFI INC. (INTERACTIVE TRAINING & INFO. CORP.) | P&D, 164; H/S/IS, 204 |

PT. 1 = Part One
P&D = Production and Development Resources Section
H/S/IS = Hardware/Software/Integrated Systems Section
M&R = Mastering and Replication Resources Section

| | | |
|---|---|---|
| Laffey, James | VIDEOACTIVE, INC. | P&D, 174 |
| Lahey, George | NAVAL PERSONNEL RESEARCH & DEVELOPMENT CENTER | P&D, 155; H/S/IS, 198 |
| Lake, Peter A. | LAKE, PETER A. | P&D, 149 |
| L'Allier, James J. | WILSON LEARNING CORPORATION | PT. 1, 106 |
| Laurier, Nicole | SOFI INC. (INTERACTIVE TRAINING & INFO. CORP.) | P&D, 164; H/S/IS, 204 |
| Lavin, Howard | RAYTHEON SERVICE COMPANY | P&D, 159 |
| Leader, Donna | BALL COMMUNICATIONS, INC. | P&D, 122 |
| Lee, Brien | BRIEN LEE & COMPANY | P&D, 123 |
| Lee, Patrick | INTERACTIVE IMAGE TECHNOLOGIES INC. | P&D, 144 |
| Lee, Neville | MEMOREX/BURROUGHS - OPTICAL MEMORY TECHNOLOGY | P&D, 154 |
| Lenfest, David | LENFEST & ASSOCIATES | P&D, 150 |
| Lesnick, Dolores E. | WANG LABORATORIES, INC. | H/S/IS, 209 |
| Levin, Steven L. | INTERACTIVE TELEVISION COMPANY | P&D, 145; H/S/IS, 193 |
| Lieb, Warren | JACK LIEB PRODUCTIONS, INC. | P&D, 148; H/S/IS, 194 |
| Lind, Daniel G. | UNIV. OF IOWA/UNIVERSITY VIDEO CENTER | P&D, 170 |
| Lippincott, Rob | WGBH EDUCATIONAL FOUNDATION | P&D, 176 |
| Lisowski, James A. | FUTUREDAY | P&D, 137 |
| Liu, Chelcie | SPIE-THE INTERNATIONAL OPTICAL ENGINEERING SOCIETY | P&D, 165 |
| Loney, Charles | CONTROL DATA CORPORATION | P&D, 129; H/S/IS, 184 |
| Lotecka, L. | INTERACTOR COMPUTER-MEDIA SYSTEMS, INC./ APAL FOUN. | P&D, 146 |
| Lowe, Larry | VIDEODISCDESIGN | P&D, 175 |
| Lynett, Patricia | UNIV. OF WEST FLA./ OFF. FOR INTERACT. TECH. & TRNG | P&D, 172 |
| Mahy, Isabelle | CEDI INFORMATIQUE | P&D, 125 |
| Marcus, Aaron | AARON MARCUS AND ASSOCIATES | P&D, 119 |
| Marsh, Chris | RECENT PRODUCTIONS LIMITED | P&D, 160; H/S/IS, 202 |
| Matheny, Stephen | DIGITAL VIDEO CORPORATION | P&D, 133; H/S/IS, 187 |
| Mathieson, Michael | SUN RESEARCH, INC. | H/S/IS, 205 |
| McCabe, George S. | CABLESHARE INC. | P&D, 124; H/S/IS, 182 |
| McCann, Frank | RCA VIDEODISCS | P&D, 160; H/S/IS, 202 |
| McCarthy, Joseph | GLYN GROUP, INC. | P&D, 138 |
| McElhatten, David | D'VIDEO AND ASSOCIATES | P&D, 133 |
| McHugh, Joe | VHD CORPORATION OF AMERICA | H/S/IS, 208 |
| McKenna, Aileen | PREF, THE PLACEMENT REFERENCE NETWORK | P&D, 158 |
| Meckler, Alan | MECKLER PUBLISHING | P&D, 153 |
| Mengel, Craig A. | FUSION MEDIA, INC. | P&D, 137 |
| Messerschmitt, John | NORTH AMERICAN PHILIPS CORPORATION | H/S/IS, 198 |
| Michaud, B. | SOFI INC. (INTERACTIVE TRAINING & INFO. CORP.) | P&D, 164; H/S/IS, 204 |
| Michels, Paul | COASTAL VIDEO COMMUNICATIONS | P&D, 127 |
| Miller, Rockley | FUTURE SYSTEMS, INC. | P&D, 137 |
| Mishkin, Andrew H. | ORENDA, INCORPORATED | P&D, 157; H/S/IS, 199 |
| Moes, Robert | NORTH AMERICAN PHILIPS CORPORATION | H/S/IS, 198 |
| Monaco, Don | CLINICAL INTERACTIONS, INC. | P&D, 126 |
| Monaco, Don | REACTIVE SYSTEMS, INC. | H/S/IS, 202 |
| Morin, Chris | MATROX ELECTRONIC SYSTEMS LTD. | P&D, 153; H/S/IS, 196 |
| Myer, Tom | COMPUTER HARDWARE, INC. | H/S/IS, 183 |
| Nelson, David | BBC OPEN UNIVERSITY PRODUCTION CENTRE | P&D, 123 |
| Nelson, G. | HAYES PRODUCTIONS, INC. | P&D, 138 |
| Netta, Franz | TELEMEDIA GmbH | P&D, 168; H/S/IS, 206 |
| Nevin, Darius G. | FIRST CAPITAL CORPORATION OF CHICAGO | P&D, 135 |
| Newman, Robert A., III | DAVIS AUDIO-VISUAL, INC. | P&D, 131; H/S/IS, 185 |
| Nicholson, Thomas J. | CARLOS RAMIREZ & ALBERT H. WOODS INC. | P&D, 124 |
| Niedzwiecki, Joann K. | INTERNATIONAL RESOURCE DEVELOPMENT INC. | P&D, 147 |
| Niemeyer, John J. | NIEMEYER, JOHN J. | P&D, 155 |
| Noelting, Bill | GENERAL TECHNICAL CORPORATION (GENTECH) | H/S/IS, 189 |
| North, Michael | MATRIX INFORMATION SYSTEMS | P&D, 153; H/S/IS, 196 |
| Nugent, Ron | UNIV. OF NEB./ NEB. VIDEODISC DESIGN/PROD. GROUP | PT. 1, 77; P&D, 171 |
| O'Connor, Richard | U.S. PUBLIC HEALTH SERVICE | P&D, 169 |
| Oakley, Brent | XENON | P&D, 177 |

# Individual Name Affiliation

| | | |
|---|---|---|
| Ockerse, A.B. | UNIV. OF ILL. AT CHICAGO/ HEALTH SERVICES CENTER | P&D, 169 |
| Paddock, Sandie | INTERACTIVE RESEARCH CORPORATION | P&D, 144; H/S/IS, 192 |
| Palmer, J.G. | COMPUTER SCIENCES CORPORATION | P&D, 128; H/S/IS, 183 |
| Pannoni, Robert | EASYWAY COMPUTER SOLUTIONS | P&D, 133 |
| Paris, Judith | MECKLER PUBLISHING | P&D, 153 |
| Payne, Jay | TRILOGIC, INC. | H/S/IS, 208 |
| Pearlman, Lenard | EDITEL-CHICAGO | P&D, 133 |
| Pellegrini, Umberto | TECNETRA-TECHNOLOGY TRANSFER COMPANY | P&D, 167; H/S/IS, 206 |
| Petro, Michael J. | MICHAEL J. PETRO LTD. | P&D, 154; H/S/IS, 197 |
| Pogue, Richard E. | MED. COLL. OF GA./ RESEARCH & EDUC. SUPPORT | H/S/IS, 196 |
| Porter, Art | INTERACT, INC. | P&D, 143 |
| Price, Frank | OPTICAL RECORDING PROJECT/3M | H/S/IS, 199 |
| Quinlin, Gary | IT, INC. | P&D, 147; H/S/IS, 194 |
| Raab, Fredric | TELEMATIC SYSTEMS | P&D, 167; H/S/IS, 206 |
| Ramsay, Gene | IMAGE PREMASTERING SERVICES | P&D, 141 |
| Rapp, Philip | LASER 7 GROUP | P&D, 149 |
| Reeves, Perry | SONY | PT. 1, 105 |
| Reeves, Thomas C. | REEVES, THOMAS C. | P&D, 160 |
| Reilly, Harry | SOCIETY FOR VISUAL EDUCATION, INC. | P&D, 164; H/S/IS, 204 |
| Renfro, W.B. | COMBUSTION ENGINEERING, INC. | P&D, 127 |
| Ring, William | TELEMEDIA CORPORATION | P&D, 167 |
| Ritchie, Wes | SIGNAL COMMUNICATIONS | P&D, 163 |
| Roach, D.K. | UNIV. COLLEGE/ CENTRE FOR EDUCATIONAL TECHNOLOGY | P&D, 169 |
| Roberts, Linda G. | U.S. DEPARTMENT OF EDUCATION | PT. 1, 97 |
| Rodesch, Dale | RODESCH DEVELOPMENT CORP. | H/S/IS, 203 |
| Rosensweet, Bob | LEARNINGWARE CORP. | P&D, 150 |
| Sala, Bonnie D. | DISCOVISION ASSOCIATES | H/S/IS, 187 |
| Salyer, Stephen | EDUCATIONAL BROADCASTING CORPORATION (WNET/13) | P&D, 134 |
| Sammon, Christine | SOUTHERN ALBERTA INST. OF TECH./ LEARNING RES. CEN. | P&D, 165 |
| Sanzo, Robert A. | FROST & SULLIVAN, INC. | P&D, 136 |
| Sawchuk, Russell | ALBERTA EDUCATIONAL COMMUNICATIONS CORP. (ACCESS) | P&D, 120 |
| Sayers, John | FUTURE SYSTEMS, INC. | P&D, 137 |
| Schecter, Ellen | SCHECTER, ELLEN | P&D, 162 |
| Schroeder, James E. | U.S. ARMY RESEARCH INSTITUTE | P&D, 168 |
| Seckman, Lamont | AMERICAN VIDEO INST./ VIDEODISC & ELEC. PUBL. LAB. | H/S/IS, 180 |
| Seibold, David | COMPUTER CENTER/WARNER AMEX QUBE | H/S/IS, 183 |
| Selz, Nina | ICS-INTEXT, A DIVISION OF NATIONAL EDUC. CORP. | P&D, 140 |
| Seward, Park | VIDEO PARK, INC. | P&D, 173 |
| Short, Larry | CREATIVE UNIVERSAL, INC. | P&D, 130 |
| Siecko, Nick | EDUCATIONAL COMPUTER CORPORATION | P&D, 134; H/S/IS, 188 |
| Silber, Jay | DBS FILMS, INC. | P&D, 132 |
| Silverstein, Jeffrey | FUSION MEDIA, INC. | P&D, 137 |
| Silvian, Joey | LIGHT TRACKS | P&D, 150; H/S/IS, 195 |
| Simmons, Truxton | VIDEO IMAGE CONSULTANTS | P&D, 173 |
| Sit, David | EDUCATIONAL BROADCASTING CORPORATION (WNET/13) | P&D, 134 |
| Smeloff, Jim | UNIV. OF NEB./ NEB. VIDEODISC DESIGN/PROD. GROUP | PT. 1, 59 |
| Smeloff, Nick | SMELOFF TELEPRODUCTIONS (SMELOFF, INC.) | P&D, 163 |
| Smith, Charles T. | DIGITAL EQUIPMENT CO. | H/S/IS, 186 |
| Smith, Ed | MARITZ INC. (MAIN OFFICE) | P&D, 151 |
| Smith, V. Elliott | CRANBROOK INSTITUTE OF SCIENCE | H/S/IS, 184 |
| Snow, David | INTERACTIVE TRAINING SYSTEMS, INC. | P&D, 145; H/S/IS, 193 |
| Spence, Sharon | JACK LIEB PRODUCTIONS, INC. | P&D, 148; H/S/IS, 194 |
| Staff, Todd | BALL TELEVISION GROUP | P&D, 122 |
| Stark, Donald | CLINICAL INTERACTIONS, INC. | P&D, 126 |
| Stembler, William A. | COMPUTER SCIENCES CORPORATION | P&D, 128; H/S/IS, 183 |
| Stever, Kent | PRODUCTIVE LEARNING, INC. | P&D, 159 |
| Storey, Meg | STOREY, MEG | P&D, 166 |
| Storey, Meg | WECHSLER, JUDITH | P&D, 175 |

PT. 1 = Part One
P&D = Production and Development Resources Section
H/S/IS = Hardware/Software/Integrated Systems Section
M&R = Mastering and Replication Resources Section

| | | |
|---|---|---|
| Streibel, Michael J. | UNIV. OF WISCONSIN/ EDUCATIONAL TECHNOLOGY PROGRAM | P&D, 172 |
| Styer, Michael | MARYLAND CENTER FOR PUBLIC BROADCASTING | P&D, 152 |
| Sun, Chen | SUN INTERACTIVE VIDEO | P&D, 166; H/S/IS, 205 |
| Sundt, Jerry | MARITZ VIDEO CENTER | P&D, 152 |
| Sutliff, Dean N. | COMPUTER VIDEO PRODUCTIONS, INC. | P&D, 128; H/S/IS, 183 |
| Szabo, Mike | SAPAN | P&D, 162 |
| Temple, Nick | TERAK CORPORATION | H/S/IS, 207 |
| Tombs, Don | VIDEODISC TECHNOLOGY LIMITED | P&D, 174 |
| Treneer, George | EECO INCORPORATED | H/S/IS, 188 |
| Trugman, Ronald | COAST COMMUNICATIONS | P&D, 127 |
| Underhill, William | HUMAN RESOURCES RESEARCH ORGANIZATION (HUMBRO) | H/S/IS, 191 |
| Vaughan, John | COMMUNICATION STUDIO | P&D, 127; H/S/IS, 183 |
| Vincelette, Robert H., Jr. | TELEVISION AUDIO SUPPORT ACTIVITY | H/S/IS, 207 |
| Walker, Arnold | UNIV. OF MINNESOTA/ UNIV. MEDIA RES. TELEVISION | P&D, 171 |
| Walker, Richard | INTERACTIVE TECHNOLOGIES CORPORATION | PT. 1, 27 |
| Wechsler, Judith | WECHSLER, JUDITH | P&D, 175 |
| Weltman, Gershon | PERCEPTRONICS, INC. | PT. 1, 95; H/S/IS, 199 |
| Wescott, Glenn L. | SPERRY MEDIA ENGINEERING CLEARWATER | P&D, 165 |
| White, D. | HAYES PRODUCTIONS, INC. | P&D, 138 |
| Wickre, Terry | WICKERWORKS VIDEO PRODUCTIONS, INC. | P&D, 176 |
| Williams, David | OMNICOM ASSOCIATES | P&D, 156; H/S/IS, 198 |
| Williams, James | CREATIVISION, INC. | H/S/IS, 184 |
| Williams, Mark B. | WILLIAMS, MARK B., INTERACTIVE VIDEO DESIGN | P&D, 177 |
| Williams, Stan | FULL CIRCLE COMMUNICATIONS, INC. | P&D, 136; H/S/IS, 189 |
| Wilson, J. | CUBIC CORP. DEFENSE SYS. DIV. | H/S/IS, 185 |
| Wilson, Roger | INTERACTIVE VIDEO PROGRAMMES (DRAGONFLAIR LTD) | P&D, 146 |
| Windsor, Tom | SONY OF CANADA LTD. | H/S/IS, 205 |
| Winslow, John | OPTICAL DISC CORPORATION (ODC) | PT. 1, 83 |
| Wise, Billy | CONSULTANT'S CHOICE, INC. | P&D, 128 |
| Withrow, Frank B. | U.S. DEPARTMENT OF EDUCATION | PT. 1, 97 |
| Witt, Cliff | BNA COMMUNICATIONS INC. | P&D, 123 |
| Worcester, Chad | UNIV. OF NEB./ NEB. VIDEODISC DESIGN/PROD. GROUP | PT. 1, 51 |
| Wright, Elizabeth | NATIONAL EVALUATION SYSTEMS, INC. | P&D, 155 |
| Wright, Barbara | PRODUCTION CONSULTANTS | P&D, 159 |
| Yampolsky, Michael | JAM, INC. | P&D, 148; H/S/IS, 194 |
| Yaworski, Dorothy | SIGNATURE PRODUCTIONS | P&D, 163 |
| Yeazel, L.A. | PIONEER VIDEO, INC. | P&D, 158 |
| Zickel, J.H. | RAYTHEON SERVICE COMPANY | H/S/IS, 201 |
| Zimmerman, Kurt | ITM SYSTEMS | P&D, 147; H/S/IS, 194 |
| Zingg, L. | HAZELTINE CORPORATION/ TRAINING SYSTEMS CENTER | H/S/IS, 190 |
| Zink, Jan | M. J. ZINK PRODUCTIONS, INC. | P&D, 151 |
| Zollman, Dean | KANSAS STATE UNIVERSITY | PT. 1, 73 |

## HARDWARE/SOFTWARE/INTEGRATED SYSTEMS COMPANIES BY TYPE OF PRODUCT

### HARDWARE

#### DIGITAL AUDIO

BELL TELEPHONE LABORATORIES, 180
INTERACTIVE RESEARCH CORPORATION, 192
MATROX ELECTRONIC SYSTEMS LTD., 196
MATSUSHITA TECHNOLOGY CENTER, 196
NORTH AMERICAN PHILIPS CORPORATION, 198
SONY COMMUNICATIONS PRODUCTS COMPANY, 204
SONY CORPORATION OF AMERICA, 204
SONY OF CANADA LTD., 205
VHD CORPORATION OF AMERICA, 208

#### INPUT DEVICES

ADVANCED IMAGE TECHNOLOGY, INC., 179
BELL TELEPHONE LABORATORIES, 180
CAVRI SYSTEMS, INC., 182
CENTRE FOR REMOTE SENSING, 182
CUBIC CORP. DEFENSE SYS. DIV., 185
DIGITAL EQUIPMENT CO., 186
DIGITAL VIDEO CORPORATION, 187
INTEGRATED AUTOMATION, 192
INTERACTIVE TELEVISION COMPANY, 193
INTERMEDIA, 193
NEW MEDIA GRAPHICS CORPORATION, 198
NORTH AMERICAN PHILIPS CORPORATION, 198
ORENDA, INCORPORATED, 199
PERCEPTRONICS, INC. (WASH. DC OPERATIONS), 199
PERFORMANCE TECHNOLOGY, INC., 200
PIONEER VIDEO, INC., 200
REACTIVE SYSTEMS, INC., 202
SONY CORPORATION OF AMERICA, 204
SONY OF CANADA LTD., 205
VIDEO MASTERS, INC., 209
WICAT SYSTEMS, INC., 210

#### INTERFACES

ADVANCED IMAGE TECHNOLOGY, INC., 179
ALLEN COMMUNICATIONS, 179
AMERICAN VIDEO INST./VIDEODISC & ELEC. PUBL. LAB., 180
ANTHRO-DIGITAL, INC., 180
BCD ASSOCIATES, INC., 180
BLUE LAKES COMPUTING, 181
CAVRI SYSTEMS, INC., 182
CENTRE FOR REMOTE SENSING, 182
CUBIC CORP. DEFENSE SYS. DIV., 185
DAVID O. MCKAY INSTITUTE, 185
DIGITAL CONTROLS INC., 185
DIGITAL EQUIPMENT CO., 186
DIGITAL TECHNIQUES INC., 187
DIGITAL VIDEO CORPORATION, 187
EDUCATIONAL COMPUTER CORPORATION, 188
EXHIBIT TECHNOLOGY, INC., 188
GENERAL TECHNICAL CORPORATION (GENTECH), 189
HAZELTINE CORPORATION/TRAINING SYSTEMS CENTER, 190
IBM, 191
INTEGRATED AUTOMATION, 192
INTERACTIVE RESEARCH CORPORATION, 192
INTERACTIVE TELEVISION COMPANY, 193
INTERACTIVE TRAINING SYSTEMS, INC., 193
INTERMEDIA, 193
JAM, INC., 194
LOGIVISION, 195
MARK IV DESIGNS, 195
MATROX ELECTRONIC SYSTEMS LTD., 196
NEW MEDIA GRAPHICS CORPORATION, 198
NUVATEC, INC., 198
ONLINE COMPUTER SYSTEMS, 198
ORENDA, INCORPORATED, 199
OWL MICRO-COMMUNICATIONS, 199
PERGAMON INTERNATIONAL INFORMATION CORPORATION, 200
PHILIPS INTERNATIONAL/AUDIO DIV./LASERVISION DEPT., 200
PIONEER VIDEO, INC., 200
POSITRON, 200
PRIMARIUS, 201
SCRIPTECH, 203
SONY COMMUNICATIONS PRODUCTS COMPANY, 204
SONY CORPORATION OF AMERICA, 204
SONY OF CANADA LTD., 205
SYMTEC, INC., 206
TECNETRA-TECHNOLOGY TRANSFER COMPANY, 206

TEXAS INSTRUMENTS, 207
VIDEO VISION ASSOCIATES, LTD., 209
WDC MARKETING ASSOCIATES, INC., 209
WHITNEY EDUCATIONAL SYSTEMS, 210
WICAT SYSTEMS, INC., 210
WINDELOTT CO., 210

### OVERLAYS/SUPERIMPOSE UNITS

ADVANCED IMAGE TECHNOLOGY, INC., 179
ALLEN COMMUNICATIONS, 179
BELL TELEPHONE LABORATORIES, 180
CABLESHARE INC., 181
CUBIC CORP. DEFENSE SYS. DIV., 185
DEPARTMENT OF NATIONAL DEFENCE, 185
DIGITAL EQUIPMENT CO., 186
DIGITAL TECHNIQUES INC., 186
HAZELTINE CORPORATION/TRAINING SYSTEMS CENTER, 190
IBM, 191
INTERACTIVE TRAINING SYSTEMS, INC., 193
MATROX ELECTRONIC SYSTEMS LTD., 196
NEW MEDIA GRAPHICS CORPORATION, 198
ORENDA, INCORPORATED, 199
PERCEPTRONICS, INC. (WASH. DC OPERATIONS), 199
PHILIPS INTERNATIONAL/AUDIO DIV./LASERVISION DEPT., 200
RAYTHEON SERVICE COMPANY, 201
SONY COMMUNICATIONS PRODUCTS COMPANY, 204
SONY CORPORATION OF AMERICA, 204
SONY OF CANADA LTD., 205
TERAK CORPORATION, 207
VHD CORPORATION OF AMERICA, 208
VIDEO ASSOCIATES LABS, 209
WICAT SYSTEMS, INC., 210

### STILL-FRAME AUDIO

ADVANCED IMAGE TECHNOLOGY, INC., 179
ALLEN COMMUNICATIONS, 179
BELL TELEPHONE LABORATORIES, 180
DIGITAL EQUIPMENT CO., 186
EECO INCORPORATED, 188
MATROX ELECTRONIC SYSTEMS LTD., 196
MATSUSHITA TECHNOLOGY CENTER, 196
NORTH AMERICAN PHILIPS CORPORATION, 198
PIONEER VIDEO, INC., 200
PRIMARIUS, 201
SONY COMMUNICATIONS PRODUCTS COMPANY, 204
SONY CORPORATION OF AMERICA, 204
SONY OF CANADA LTD., 205

### VIDEODISC PLAYERS

HITACHI SALES CORPORATION OF AMERICA, 191
MATSUSHITA TECHNOLOGY CENTER, 196
NORTH AMERICAN PHILIPS CORPORATION, 198
PHILIPS INTERNATIONAL/AUDIO DIV./LASERVISION DEPT., 200
PIONEER VIDEO, INC., 200
RCA VIDEODISCS, 202
SONY COMMUNICATIONS PRODUCTS COMPANY, 204
SONY CORPORATION OF AMERICA, 204
SONY OF CANADA LTD., 205
VHD CORPORATION OF AMERICA, 208

## SOFTWARE

### AUTHORING SYSTEMS

ACTRONICS, INC., 179
ADVANCED IMAGE TECHNOLOGY, INC., 179
ALLEN COMMUNICATIONS, 179
ALPHATEL SYSTEMS LTD., 180
BCD ASSOCIATES, INC., 180
BELL TELEPHONE LABORATORIES, 180
CABLESHARE INC., 181
CAT ZURICH- COMPUTER ASSISTED TELEVIDEO AG, 182
CITIDISC PRODUCTIONS, 182
COMMUNICATION STUDIO, 183
COMPUTER SCIENCES CORPORATION, 183
CONTROL DATA CORPORATION, 184
COURSEWARE, INC., 184
CUBIC CORP. DEFENSE SYS. DIV., 185
DAVID O. MCKAY INSTITUTE, 185
DIGITAL EQUIPMENT CO., 186
DIGITAL TECHNIQUES INC., 186
DIGITAL VIDEO CORPORATION, 187
EDITEL-NEW YORK, 188
EXHIBIT TECHNOLOGY, INC., 188
FRANKLIN RESEARCH CENTER, 189
FULL CIRCLE COMMUNICATIONS, INC., 189
GENERAL TECHNICAL CORPORATION (GENTECH), 189
GRUMMAN AEROSPACE CORPORATION, 190
HAZELTINE CORPORATION/TRAINING SYSTEMS CENTER, 190
HUMAN RESOURCES RESEARCH ORGANIZATION (HUMRRO), 191
IBM, 191
ICS- INTEXT, A DIVISION OF NATIONAL EDUC. CORP., 191
INTERACTIVE RESEARCH CORPORATION, 192
INTERACTIVE TELEVISION COMPANY, 193
INTERACTIVE TRAINING SYSTEMS, INC., 193
INTERACTIVE VIDEO CONCEPTS, INC., 193
JAM, INC., 194
MARK IV DESIGNS, 195
MATRIX INFORMATION SYSTEMS, 196
MED. COLL. OF GA./RESEARCH & EDUC. SUPPORT, 196
MICHAEL J. PETRO LTD., 197
NEW MEDIA GRAPHICS CORPORATION, 198
OMNICOM ASSOCIATES, 198
PERCEPTRONICS, INC., 199
PERCEPTRONICS, INC. (WASH. DC OPERATIONS), 199
PHILIPS INTERNATIONAL/AUDIO DIV./LASERVISION DEPT., 200
PIONEER VIDEO, INC., 200
PRIMARIUS, 201
RAYTHEON SERVICE COMPANY, 201
SCRIPTECH, 203
SOFI INC. (INTERACTIVE TRAINING & INFO. CORP.), 204
SONY OF CANADA LTD., 205
SUN INTERACTIVE VIDEO, 205
TCT TECHNICAL TRAINING, INC. ("T-CUBED"), 206
TELEMATIC SYSTEMS, 206
TERAK CORPORATION, 207
TRANS-LINGUAL COMMUNICATIONS, INC., 207
VIDEO VISION ASSOCIATES, LTD., 209
WDC MARKETING ASSOCIATES, INC., 209
WICAT SYSTEMS, INC., 210

### DISC SIMULATORS

BELL TELEPHONE LABORATORIES, 180
CAT ZURICH- COMPUTER ASSISTED TELEVIDEO AG, 182

# Hardware/Software/Integrated Systems Companies By Type Of Product

COMPUTER SCIENCES CORPORATION, 183
DAVIS AUDIO-VISUAL, INC., 185
DIGITAL EQUIPMENT CO., 186
EDITEL-NEW YORK, 188
INNOTECH, 192
INTERACTIVE RESEARCH CORPORATION, 192
MICHAEL J. PETRO LTD., 197
NEW MEDIA GRAPHICS CORPORATION, 198
OMNICOM ASSOCIATES, 198
PHILIPS INTERNATIONAL/AUDIO DIV./LASERVISION DEPT., 200
PIONEER VIDEO, INC., 200
RECENT PRODUCTIONS LIMITED, 202
SCRIPTECH, 203
SONY COMMUNICATIONS PRODUCTS COMPANY, 204
SONY CORPORATION OF AMERICA, 204
SONY OF CANADA LTD., 205
TECNETRA-TECHNOLOGY TRANSFER COMPANY, 206
TELEMATIC SYSTEMS, 206
TRANS-LINGUAL COMMUNICATIONS, INC., 207

### DRIVERS/VIDEO UTILITIES

ACTRONICS, INC., 179
ADVANCED IMAGE TECHNOLOGY, INC., 179
ALLEN COMMUNICATIONS, 179
ALPHATEL SYSTEMS LTD., 180
BELL TELEPHONE LABORATORIES, 180
CABLESHARE INC., 181
CAT ZURICH- COMPUTER ASSISTED TELEVIDEO AG, 182
COMMUNICATION STUDIO, 183
COMPUTER SCIENCES CORPORATION, 183
CONTROL DATA CORPORATION, 184
COURSEWARE, INC., 184
CUBIC CORP. DEFENSE SYS. DIV., 185
DAVIS AUDIO-VISUAL, INC., 185
DIGITAL TECHNIQUES INC., 186
DIGITAL VIDEO CORPORATION, 187
ENERGY, MINES & RESOURCES, 188
EXHIBIT TECHNOLOGY, INC., 188
FRANKLIN RESEARCH CENTER, 189
FULL CIRCLE COMMUNICATIONS, INC., 189
IBM, 191
INTERACTIVE RESEARCH CORPORATION, 192
INTERACTIVE TELEVISION COMPANY, 193
MARK IV DESIGNS, 195
MARTIN MARIETTA DATA SYSTEMS/TECH. SYSTEMS GROUP, 195
MATRIX INFORMATION SYSTEMS, 196
MICHAEL J. PETRO LTD., 197
NEW MEDIA GRAPHICS CORPORATION, 198
ORENDA, INCORPORATED, 199
PERCEPTRONICS, INC. (WASH. DC OPERATIONS), 199
PHILIPS INTERNATIONAL/AUDIO DIV./LASERVISION DEPT., 200
PRIMARIUS, 201
SOFI INC. (INTERACTIVE TRAINING & INFO. CORP.), 204
SONY COMMUNICATIONS PRODUCTS COMPANY, 204
SONY CORPORATION OF AMERICA, 204
SONY OF CANADA LTD., 205
TCT TECHNICAL TRAINING, INC. ("T-CUBED"), 206
TELEMATIC SYSTEMS, 206
VIDEO ASSOCIATES LABS, 209

### GRAPHICS EDITORS/TITLERS

ADVANCED IMAGE TECHNOLOGY, INC., 179
BELL TELEPHONE LABORATORIES, 180
BOSTON MEDIA CONSULTANTS, 181
CABLESHARE INC., 181
CAT ZURICH- COMPUTER ASSISTED TELEVIDEO AG, 182
CITIDISC PRODUCTIONS, 182
COMMUNICATION STUDIO, 183
COMPUTER SCIENCES CORPORATION, 183
CONTROL DATA CORPORATION, 184
COURSEWARE, INC., 184
CUBIC CORP. DEFENSE SYS. DIV., 185
DIGITAL EQUIPMENT CO., 186
DIGITAL TECHNIQUES INC., 187
EDITEL-NEW YORK, 188
ENERGY, MINES & RESOURCES, 188
INTERACTIVE RESEARCH CORPORATION, 192
MED. COLL. OF GA./RESEARCH & EDUC. SUPPORT, 196
NEW MEDIA GRAPHICS CORPORATION, 198
PERCEPTRONICS, INC. (WASH. DC OPERATIONS), 199
PRIMARIUS, 201
SONY COMMUNICATIONS PRODUCTS COMPANY, 204
SONY CORPORATION OF AMERICA, 204
SONY OF CANADA LTD., 205
TELEMATIC SYSTEMS, 206
TERAK CORPORATION, 207
VHD CORPORATION OF AMERICA, 208

## INTEGRATED SYSTEMS

### INFORMATION DISPLAYS

ADVANCED IMAGE TECHNOLOGY, INC., 179
ALLEN COMMUNICATIONS, 179
AMERICAN VIDEO INST./VIDEODISC & ELEC. PUBL. LAB., 180
BELL TELEPHONE LABORATORIES, 180
CABLESHARE INC., 181
CAT ZURICH- COMPUTER ASSISTED TELEVIDEO AG, 182
CITIDISC PRODUCTIONS, 182
COMPUTER CENTER/WARNER AMEX QUBE, 183
COMPUTER VIDEO PRODUCTIONS, INC., 183
CREATIVISION, INC., 184
CUBIC CORP. DEFENSE SYS. DIV., 185
DAVIS AUDIO-VISUAL, INC., 185
DIGITAL EQUIPMENT CO., 186
DIGITAL TECHNIQUES INC., 186
DIGITAL VIDEO CORPORATION, 187
ENERGY, MINES & RESOURCES, 188
EXHIBIT TECHNOLOGY, INC., 188
FULL CIRCLE COMMUNICATIONS, INC., 189
ICS- INTEXT, A DIVISION OF NATIONAL EDUC. CORP., 191
INTEGRATED AUTOMATION, 192
INTERACTIVE TELEVISION COMPANY, 193
INTERACTIVE TRAINING SYSTEMS, INC., 193
INTERMEDIA, 193
JACK LIEB PRODUCTIONS, INC., 194
LOGIVISION, 195
MICHAEL J. PETRO LTD., 197
NEW MEDIA GRAPHICS CORPORATION, 198
ORENDA, INCORPORATED, 199
PERCEPTRONICS, INC., 199
PERCEPTRONICS, INC. (WASH. DC OPERATIONS), 199
PHILIPS INTERNATIONAL/AUDIO DIV./LASERVISION DEPT., 200
PRIMARIUS, 201
PROPERTY TECHNOLOGY, INC., 201
SOFI INC. (INTERACTIVE TRAINING & INFO. CORP.), 204
SUN INTERACTIVE VIDEO, 205
TELEMEDIA GmbH, 206

TERAK CORPORATION, 207
TRILOGIC, INC., 208
VCM SYSTEMS, INC., 208
VHD CORPORATION OF AMERICA, 208
VIDEO MASTERS, INC., 209
WDC MARKETING ASSOCIATES, INC., 209
WICAT SYSTEMS, INC., 210

### POINT-OF-PURCHASE SYSTEMS

ADVANCED IMAGE TECHNOLOGY, INC., 179
ALLEN COMMUNICATIONS, 179
BELL TELEPHONE LABORATORIES, 180
CABLESHARE INC., 181
CAT ZURICH- COMPUTER ASSISTED TELEVIDEO AG, 182
COMPUTER VIDEO PRODUCTIONS, INC., 183
CREATIVISION, INC., 184
DAVIS AUDIO-VISUAL, INC., 185
DIGITAL EQUIPMENT CO., 186
DIGITAL TECHNIQUES INC., 186
DIGITAL VIDEO CORPORATION, 187
EDITEL-NEW YORK, 188
EXHIBIT TECHNOLOGY, INC., 188
FULL CIRCLE COMMUNICATIONS, INC., 189
IBM, 191
ICS- INTEXT, A DIVISION OF NATIONAL EDUC. CORP., 191
INNOTECH, 192
INTERACTIVE TRAINING SYSTEMS, INC., 193
INTERMEDIA, 193
IT, INC., 194
JACK LIEB PRODUCTIONS, INC., 194
LOGIVISION, 195
MAZ & MOVIE, 196
MICHAEL J. PETRO LTD., 197
NEW MEDIA GRAPHICS CORPORATION, 198
OMNICOM ASSOCIATES, 198
ORENDA, INCORPORATED, 199
PERCEPTRONICS, INC. (WASH. DC OPERATIONS), 199
PHILIPS INTERNATIONAL/AUDIO DIV./LASERVISION DEPT., 200
SOFI INC. (INTERACTIVE TRAINING & INFO. CORP.), 104
SUN INTERACTIVE VIDEO, 205
TELEMATIC SYSTEMS, 206
TELEMEDIA GmbH, 206
TRILOGIC, INC., 208
VCM SYSTEMS, INC., 208
VIDEO MASTERS, INC., 209
WDC MARKETING ASSOCIATES, INC., 209

### TRAINING WORK STATIONS/CARRELS

ACTRONICS, INC., 179
ADVANCED IMAGE TECHNOLOGY, INC., 179
ALLEN COMMUNICATIONS, 179
ALPHATEL SYSTEMS LTD., 180
BELL TELEPHONE LABORATORIES, 180
CAT ZURICH- COMPUTER ASSISTED TELEVIDEO AG, 182
COMPUTER SCIENCES CORPORATION, 183
COMPUTER VIDEO PRODUCTIONS, INC., 183
CONTROL DATA CORPORATION, 184
CREATIVISION, INC., 184
CUBIC CORP. DEFENSE SYS. DIV., 185

DAVIS AUDIO-VISUAL, INC., 185
DIGITAL EQUIPMENT CO., 186
DIGITAL TECHNIQUES INC., 186
DIGITAL VIDEO CORPORATION, 187
EDUCATIONAL COMPUTER CORPORATION, 188
EXHIBIT TECHNOLOGY, INC., 188
FRANKLIN RESEARCH CENTER, 189
FULL CIRCLE COMMUNICATIONS, INC., 189
GENERAL TECHNICAL CORPORATION (GENTECH), 190
GRUMMAN AEROSPACE CORPORATION, 190
HAZELTINE CORPORATION/TRAINING SYSTEMS CENTER, 190
IBM, 191
ICS- INTEXT, A DIVISION OF NATIONAL EDUC. CORP., 191
INTERACTIVE TELEVISION COMPANY, 193
INTERACTIVE TRAINING SYSTEMS, INC., 193
IT, INC., 194
JAM, INC., 194
LOGIVISION, 195
MATRIX INFORMATION SYSTEMS, 196
MATROX ELECTRONIC SYSTEMS LTD., 196
MAZ & MOVIE, 197
MICHAEL J. PETRO LTD., 198
NEW MEDIA GRAPHICS CORPORATION, 198
OMNICOM ASSOCIATES, 199
ORENDA, INCORPORATED, 199
PERCEPTRONICS, INC., 199
PERCEPTRONICS, INC. (WASH. DC OPERATIONS), 200
PHILIPS INTERNATIONAL/AUDIO DIV./LASERVISION DEPT., 201
PRIMARIUS, 102
PROPERTY TECHNOLOGY, INC., 201
RAYTHEON SERVICE COMPANY, 205
SUN INTERACTIVE VIDEO, 206
SYNSOR CORPORATION, 206
TCT TECHNICAL TRAINING, INC. ("T-CUBED"), 206
TECNETRA-TECHNOLOGY TRANSFER COMPANY, 206
TELEMATIC SYSTEMS, 207
TERAK CORPORATION, 207
TRILOGIC, INC., 208
VIDEO MASTERS, INC., 209
WANG LABORATORIES, INC., 209
WICAT SYSTEMS, INC., 210

### VIDEO GAMES

BELL TELEPHONE LABORATORIES, 180
CITIDISC PRODUCTIONS, 182
DAVIS AUDIO-VISUAL, INC., 185
EDITEL-NEW YORK, 188
ENERGY, MINES & RESOURCES, 188
EXHIBIT TECHNOLOGY, INC., 188
INTERACTIVE ARTS INTERNATIONAL, 192
ITM SYSTEMS, 194
LOGIVISION, 195
NEW MEDIA GRAPHICS CORPORATION, 198
ORENDA, INCORPORATED, 199
PERCEPTRONICS, INC., 199
PERCEPTRONICS, INC. (WASH. DC OPERATIONS), 199
PHILIPS INTERNATIONAL/AUDIO DIV./LASERVISION DEPT., 200
RODESCH DEVELOPMENT CORP., 203

## ORGANIZATIONS BY GEOGRAPHIC LOCATION

| UNITED STATES | PAGE |
|---|---|
| **ALABAMA** | |
| WDC MARKETING ASSOCIATES, INC. | P&D, 175; H/S/IS, 209 |
| **ARIZONA** | |
| PRESIDIO VIDEO | P&D, 158 |
| TERAK CORPORATION | H/S/IS, 207 |
| UNIV. OF ARIZONA/BIOMEDICAL COMMUNICATIONS | P&D, 169 |
| **CALIFORNIA** | |
| AARON MARCUS AND ASSOCIATES | P&D, 119 |
| ADVANCED VIDEO, INC. | P&D, 120 |
| ALEC GROUP | P&D, 121 |
| BASHISTA, MICHAEL | P&D, 122 |
| CAT CORPORATION—LOS ANGELES | P&D, 125 |
| COAST COMMUNICATIONS | P&D, 126 |
| COOPERLASER SONICS | H/S/IS, 184 |
| COURSEWARE, INC. | P&D, 129; H/S/IS, 184 |
| CUBIC CORP. DEFENSE SYS. DIV. | H/S/IS, 185 |
| DAVIS, BARBARA GROSS | P&D, 131 |
| DISCOVISION ASSOCIATES | H/S/IS, 187 |
| D'VIDEO AND ASSOCIATES | P&D, 133 |
| EASYWAY COMPUTER SOLUTIONS | P&D, 133 |
| EECO INCORPORATED | H/S/IS, 188 |
| GREGG, D.P. | H/S/IS, 190 |
| HORIZONTAL EDITING STUDIOS | P&D, 139 |
| INNOTECH | P&D, 142; H/S/IS, 192 |
| INNOVATIVE TECHNOLOGY ASSOCIATES | P&D, 142 |
| INSTRUCTIONAL SCIENCE AND DEVELOPMENT, INC. | P&D, 143 |
| INTEGRATED AUTOMATION | H/S/IS, 192 |
| INTERACT, INC. | P&D, 143 |
| INTERACTIVE ARTS & SCIENCE/IAS, INC. | P&D, 143 |
| INTERACTIVE ARTS INTERNATIONAL | P&D, 143; H/S/IS, 192 |
| INTERACTIVE RESEARCH CORPORATION | P&D, 144; H/S/IS, 192 |
| INTERACTIVE TECHNOLOGIES CORPORATION | P&D, 144 |
| INTERACTOR COMPUTER-MEDIA SYSTEMS, INC./APAL FOUN. | P&D, 146 |
| ITM SYSTEMS | P&D, 147; H/S/IS, 194 |
| ITTELSON, JOHN C. | P&D, 147 |
| KEN S. CHRISTIE AND ASSOCIATES | P&D, 149 |
| LAKE, PETER A. | P&D, 149 |
| LASER 7 GROUP | P&D, 149 |
| LIGHT TRACKS | P&D, 150; H/S/IS, 195 |
| LUCASFILM, LTD. | P&D, 151 |
| LUMIERE PRODUCTIONS | P&D, 151 |
| MATRIX INFORMATION SYSTEMS | P&D, 153; H/S/IS, 196 |
| MEMOREX/BURROUGHS—OPTICAL MEMORY TECHNOLOGY | P&D, 154 |
| NAVAL PERSONNEL RESEARCH & DEVELOPMENT CENTER (NPRDC) | P&D, 155; H/S/IS, 198 |
| ONE PASS FILM AND VIDEO | P&D, 156 |
| ORENDA, INCORPORATED | P&D, 157; H/S/IS, 199 |
| PERCEPTRONICS, INC. | H/S/IS, 199 |
| PIONEER VIDEO, INC. | P&D, 158; H/S/IS, 200; M&R, 213 |
| PRIMARIUS | H/S/IS, 201 |
| RENAN PRODUCTIONS | P&D, 161 |
| ROBERT DRUCKER & COMPANY | P&D, 161 |
| RODESCH DEVELOPMENT CORP. | H/S/IS, 203 |
| SCRIPTECH | P&D, 162; H/S/IS, 203 |
| SONY VIDEO UTILIZATION SERVICES | P&D, 165 |
| SUN INTERACTIVE VIDEO | P&D, 166; H/S/IS, 205 |
| TCT TECHNICAL TRAINING, INC. ("T-CUBED") | P&D, 166; H/S/IS, 206 |
| TELEVISION AUDIO SUPPORT ACTIVITY | H/S/IS, 208 |
| TRILOGIC, INC. | P&D, 168; H/S/IS, 208 |
| VHD CORPORATION OF AMERICA | H/S/IS, 208 |
| VIDEODISCDESIGN | P&D, 175 |
| WHITNEY EDUCATIONAL SYSTEMS | H/S/IS, 210 |
| WILLIAMS, MARK B., INTERACTIVE VIDEO DESIGN | P&D, 177 |
| XENON | P&D, 177 |

P&D = Production and Development Resources Section
H/S/IS = Hardware/Software/Integrated Systems Section
M&R = Mastering and Replication Resources Section

## COLORADO

| | |
|---|---|
| DAVIS AUDIO-VISUAL, INC. | P&D, 131; H/S/IS, 185 |
| LENFEST & ASSOCIATES | P&D, 150 |
| MARTIN MARIETTA DATA SYSTEMS/TECH. SYSTEMS GROUP | P&D, 152; H/S/IS, 195 |
| SMELOFF TELEPRODUCTIONS (SMELOFF, INC.) | P&D, 163 |
| VIDEO IMAGE CONSULTANTS | P&D, 173 |
| WICKERWORKS VIDEO PRODUCTIONS, INC. | P&D, 176 |

## CONNECTICUT

| | |
|---|---|
| CAVRI SYSTEMS, INC. | H/S/IS, 182 |
| COMBUSTION ENGINEERING, INC. | P&D, 127 |
| ICS-INTEXT, A DIVISION OF NATIONAL EDUC. CORP. | P&D, 140; H/S/IS, 191 |
| INTERNATIONAL RESOURCE DEVELOPMENT INC. | P&D, 147 |
| MECKLER PUBLISHING | P&D, 153 |
| OPTICAL RECORDING PROJECT/ 3M (EAST COAST OFFICE) | H/S/IS, 199 |
| PRODUCTION CONSULTANTS | P&D, 159 |

## FLORIDA

| | |
|---|---|
| CREATIVISION, INC. | P&D, 130; H/S/IS, 184 |
| DIGITAL VIDEO CORPORATION | P&D, 133; H/S/IS, 187 |
| FLORIDA STATE UNIV./CENTER FOR EDUC. TECHNOLOGY | P&D, 135 |
| HEWLETT, BRENT | P&D, 139 |
| IBM | H/S/IS, 191 |
| PRODUCTION ASSOCIATES, INC. | P&D, 158 |
| SPERRY MEDIA ENGINEERING CLEARWATER | P&D, 165 |
| UNIV. OF WEST FLA./OFF. FOR INTERACT. TECH. & TRNG | P&D, 172 |

## GEORGIA

| | |
|---|---|
| CONSULTANT'S CHOICE, INC. | P&D, 128 |
| DIGITAL CONTROLS INC. | H/S/IS, 185 |
| MED. COLL. OF GA./RESEARCH & EDUC. SUPPORT | H/S/IS, 196 |
| REEVES, THOMAS C. | P&D, 160 |
| U.S. ARMY RESEARCH INSTITUTE | P&D, 168 |
| WINDMILL PRODUCTIONS, INC. | P&D, 177 |

## IOWA

| | |
|---|---|
| SIGNAL COMMUNICATIONS | P&D, 163 |
| UNIV. OF IOWA/COMPUTER ASSISTED INSTRUCTION LAB | P&D, 170 |
| UNIV. OF IOWA/UNIVERSITY VIDEO CENTER | P&D, 170 |
| VCM SYSTEMS, INC. | P&D, 172; H/S/IS, 208 |

## ILLINOIS

| | |
|---|---|
| EDITEL-CHICAGO | P&D, 133 |
| ILLINOIS STATE UNIV./COLLEGE OF FINE ARTS | P&D, 140 |
| INFORMATION DELIVERY SYSTEMS, INC. | P&D, 142 |
| JACK LIEB PRODUCTIONS, INC. | P&D, 148; H/S/IS, 194 |
| NUVATEC, INC. | H/S/IS, 198 |
| SOCIETY FOR VISUAL EDUCATION, INC. | P&D, 164; H/S/IS, 204 |
| TRANS-LINGUAL COMMUNICATIONS, INC. | P&D, 168; H/S/IS, 207 |
| UNIV. OF ILL. AT CHICAGO/ HEALTH SCIENCES CENTER | P&D, 169 |

## INDIANA

| | |
|---|---|
| BALL COMMUNICATIONS, INC. | P&D, 122 |
| GENERAL TECHNICAL CORPORATION (GENTECH) | H/S/IS, 189 |
| INDIANA UNIVERSITY RADIO AND TELEVISION SERVICE | P&D, 141 |
| RCA VIDEODISC OPERATIONS | M&R, 215 |

## KANSAS

| | |
|---|---|
| SIGNATURE PRODUCTIONS | P&D, 163 |

## LOUISIANA

| | |
|---|---|
| U.S. PUBLIC HEALTH SERVICE | P&D, 169 |
| VIDEO PARK, INC. | P&D, 173 |

## MASSACHUSETTS

| | |
|---|---|
| ANTHRO-DIGITAL, INC. | H/S/IS, 180 |
| BOSTON MEDIA CONSULTANTS | P&D, 123; H/S/IS, 181 |
| BUTLER, MADELEINE L. | P&D, 124 |
| CENTURY III TELEPRODUCTIONS | P&D, 125 |
| DAVIDSON, STUART P. | P&D, 131 |
| DIGITAL EQUIPMENT CORPORATION | P&D, 132; H/S/IS, 186 |
| DIGITAL TECHNIQUES INC. | H/S/IS, 186 |
| FIRST CAPITAL CORPORATION OF CHICAGO | P&D, 135 |
| INTERACTIVE TRAINING SYSTEMS, INC. | P&D, 145; H/S/IS, 193 |
| NATIONAL EVALUATION SYSTEMS, INC. | P&D, 155 |
| NEW MEDIA GRAPHICS CORPORATION | H/S/IS, 198 |
| PROPERTY TECHNOLOGY, INC. | P&D, 159; H/S/IS, 201 |
| RAYTHEON SERVICE COMPANY | P&D, 159; H/S/IS, 201 |
| SOUND CONCEPTS | H/S/IS, 205 |
| STOREY, MEG | P&D, 166 |
| TELEMATIC SYSTEMS | P&D, 167; H/S/IS, 206 |
| WANG LABORATORIES, INC. | H/S/IS, 209 |
| WECHSLER, JUDITH | P&D, 175 |
| WGBH EDUCATIONAL FOUNDATION | P&D, 176 |

## MARYLAND

| | |
|---|---|
| BNA COMMUNICATIONS INC. | P&D, 123 |
| MARYLAND CENTER FOR PUBLIC BROADCASTING | P&D, 152 |
| MARYLAND INSTRUCTIONAL TELEVISION | P&D, 153 |
| ONLINE COMPUTER SYSTEMS | H/S/IS, 198 |
| WESTINGHOUSE ELECTRIC CORP. | P&D, 176; H/S/IS, 210 |

# Organizations By Geographic Location

## MICHIGAN

| | |
|---|---|
| CAT CORPORATION—DETROIT | P&D, 124 |
| CRANBROOK INSTITUTE OF SCIENCE | H/S/IS, 184 |
| CREATIVE UNIVERSAL, INC. | P&D, 130 |
| FULL CIRCLE COMMUNICATIONS, INC. | P&D, 136; H/S/IS, 189 |
| MARITZ VIDEO CENTER | P&D, 152 |
| SYMTEC, INC. | H/S/IS, 206 |
| TECHNIDISC | M&R, 216 |

## MINNESOTA

| | |
|---|---|
| COMPUTER VIDEO PRODUCTIONS, INC. | P&D, 128; H/S/IS, 183 |
| CONTROL DATA CORPORATION | P&D, 129; H/S/IS, 184 |
| I.S.D. #279/PUBLIC SCHOOL SYSTEM | P&D, 140 |
| IMAGE PREMASTERING SERVICES | P&D, 141 |
| OPTICAL RECORDING PROJECT/3M | H/S/IS, 199; M&R, 211 |
| PRODUCTIVE LEARNING, INC. | P&D, 159 |
| UNIV. OF MINNESOTA/UNIV. MEDIA RES. TELEVISION | P&D, 171 |

## MISSOURI

| | |
|---|---|
| APPLIED VIDEO TECHNOLOGY, INC. | P&D, 121 |
| MARITZ INC. (MAIN OFFICE) | P&D, 151 |
| NIEMEYER, JOHN J. | P&D, 155 |
| VIDEO MASTERS, INC. | P&D, 173; H/S/IS, 209 |

## NEBRASKA

| | |
|---|---|
| COMPUTER HARDWARE, INC. | H/S/IS, 183 |
| MARK IV DESIGNS | P&D, 152; H/S/IS, 195 |
| UNIV. OF NEB./NEB. VIDEODISC DESIGN/PROD. GROUP | P&D, 171 |

## NEW HAMPSHIRE

| | |
|---|---|
| SANDERS ASSOCIATES, INC. | P&D, 161 |
| SUN RESEARCH, INC. | H/S/IS, 205 |

## NEW JERSEY

| | |
|---|---|
| BELL TELEPHONE LABORATORIES | H/S/IS, 180 |
| CLINICAL INTERACTIONS, INC. | P&D, 126 |
| HITACHI SALES CORPORATION OF AMERICA | H/S/IS, 191 |
| MATSUSHITA TECHNOLOGY CENTER | H/S/IS, 196 |
| PIONEER VIDEO, INC. | P&D, 158; H/S/IS, 200; M&R, 213 |
| REACTIVE SYSTEMS, INC. | H/S/IS, 202 |
| SONY COMMUNICATIONS PRODUCTS COMPANY | P&D, 164; H/S/IS, 204; M&R, 215 |
| SONY CORPORATION OF AMERICA | P&D, 164; H/S/IS, 204 |
| SYSCON CORPORATION, INC. | P&D, 166 |
| TELEMEDIA CORPORATION | P&D, 167 |
| VIDEO VISION ASSOCIATES, LTD. | P&D, 174; H/S/IS, 209 |
| WREN ASSOCIATES INC. | P&D, 177 |

## NEW MEXICO

| | |
|---|---|
| IT, INC. | P&D, 147; H/S/IS, 194 |

## NEW YORK

| | |
|---|---|
| ADVANCED IMAGE TECHNOLOGY, INC. | P&D, 120; H/S/IS, 179 |
| AMERICAN VIDEO INST./VIDEO-DISC & ELEC. PUBL. LAB. | P&D, 121; H/S/IS, 180 |
| CARLOS RAMIREZ & ALBERT H. WOODS INC. | P&D, 124 |
| CAT CORPORATION—NEW YORK | P&D, 125 |
| CBS VIDEO ENTERPRISES | M&R, 211 |
| COMMUNICATION STUDIO | P&D, 127; H/S/IS, 183 |
| CORNELL UNIVERSITY/DEPARTMENT OF EDUCATION | P&D, 129 |
| CUTTING EDGE PRODUCTIONS, INC. | P&D, 130 |
| DEVLIN PRODUCTIONS, INC. | P&D, 132 |
| EDITEL-NEW YORK | P&D, 133; H/S/IS, 188 |
| EDUCATIONAL BROADCASTING CORPORATION (WNET/13) | P&D, 134 |
| EXHIBIT TECHNOLOGY, INC. | P&D, 135; H/S/IS, 188 |
| FORRESTER PRODUCTIONS | P&D, 135 |
| FROST & SULLIVAN, INC. | P&D, 136 |
| FUSION MEDIA, INC. | P&D, 136 |
| GERSTENMAIER, WILLIAM | P&D, 137 |
| GLYN GROUP, INC. | P&D, 138 |
| GRUMMAN AEROSPACE CORPORATION | P&D, 138; H/S/IS, 190 |
| INDIS INTERNATIONAL | P&D, 141 |
| INTERACTIVE AUTHORING, INC. | P&D, 146 |
| JAM, INC. | P&D, 148; H/S/IS, 194 |
| LEARNINGWARE CORP. | P&D, 149 |
| LEO BEISER INC. | P&D, 150 |
| M.J. ZINK PRODUCTIONS, INC. | P&D, 151 |
| MULTI IMAGE INTERNATIONAL | H/S/IS, 197 |
| NORTH AMERICAN PHILIPS CORPORATION | H/S/IS, 198 |
| OMNICOM ASSOCIATES | P&D, 156; H/S/IS, 198 |
| PARALLEL COMMUNICATIONS | P&D, 157 |
| POSITRON | H/S/IS, 200 |
| RCA VIDEODISCS | P&D, 160; H/S/IS, 202 |
| SAPAN | P&D, 162 |
| SCHECTER, ELLEN | P&D, 162 |
| SCOTT DOBBINS PRODUCTIONS | P&D, 162 |
| WINDELOTT CO. | H/S/IS, 210 |

## NORTH CAROLINA

| | |
|---|---|
| PREF, THE PLACEMENT REFERENCE NETWORK | P&D, 158 |

## OHIO

| | |
|---|---|
| COMPUTER CENTER/WARNER AMEX QUBE | H/S/IS, 183 |
| HEARTLAND COMMUNICATIONS | P&D, 138 |

## OKLAHOMA

| | |
|---|---|
| BCD ASSOCIATES, INC. | H/S/IS, 180 |

## OREGON

| | |
|---|---|
| EUGENE PUBLIC SCHOOLS | P&D, 134 |

P&D = Production and Development Resources Section
H/S/IS = Hardware/Software/Integrated Systems Section
M&R = Mastering and Replication Resources Section

## PENNSYLVANIA

| | |
|---|---|
| ACTRONICS, INC. | P&D, 119; H/S/IS, 179 |
| DBS FILMS, INC. | P&D, 132 |
| EDUCATIONAL COMPUTER CORPORATION | P&D, 134; H/S/IS, 188 |
| FRANKLIN RESEARCH CENTER | P&D, 136; H/S/IS, 189 |
| INTERACTIVE VIDEO CONCEPTS, INC. | P&D, 145; H/S/IS, 193 |
| JOSEPH KEYERLEBER PRODUCTIONS | P&D, 148 |
| MEDIA CONCEPTS, INC. | P&D, 154 |

## TENNESSEE

| | |
|---|---|
| BALL TELEVISION GROUP | P&D, 122 |

## TEXAS

| | |
|---|---|
| HAYES PRODUCTIONS, INC. | P&D, 138 |
| INTERACTIVE VIDEO CORPORATION | P&D, 146 |
| SOUTHWEST TEXAS STATE UNIVERSITY | H/S/IS, 205 |
| TEXAS INSTRUMENTS | H/S/IS, 207 |
| VIDEO ASSOCIATES LABS | H/S/IS, 209 |

## UTAH

| | |
|---|---|
| ALLEN COMMUNICATION | H/S/IS, 179 |
| DAVID O. MCKAY INSTITUTE | P&D, 130; H/S/IS, 185 |
| WICAT SYSTEMS, INC. | P&D, 176; H/S/IS, 210 |

## VIRGINIA

| | |
|---|---|
| COASTAL VIDEO COMMUNICATIONS | P&D, 127 |
| COMPUTER SCIENCES CORPORATION | P&D, 127; H/S/IS, 183 |
| FUTURE SYSTEMS, INC. | P&D, 137 |
| HAZELTINE CORPORATION/ TRAINING SYSTEMS CENTER | H/S/IS, 190 |
| HELGERSON ASSOCIATES | P&D, 139 |
| HUMAN RESOURCES RESEARCH ORGANIZATION (HUMRRO) | P&D, 139; H/S/IS, 191 |
| INTERACTIVE TELEVISION COMPANY | P&D, 145; H/S/IS, 193 |
| KELBAUGH, LARRY E. | P&D, 148 |
| PERCEPTRONICS, INC. | P&D, 157; H/S/IS, 199 |
| PERFORMANCE TECHNOLOGY, INC. | H/S/IS, 200 |
| PERGAMON INTERNATIONAL INFORMATION CORPORATION | H/S/IS, 200 |
| VIDEO SOFTWARE ASSOCIATES, INC. | P&D, 173 |

## WASHINGTON, DC

| | |
|---|---|
| CITIDISC PRODUCTIONS | P&D, 126; H/S/IS, 182 |
| CORPORATION FOR PUBLIC BROADCASTING | P&D, 129 |
| OPTICAL SOCIETY OF AMERICA | P&D, 156 |

## WASHINGTON STATE

| | |
|---|---|
| INTERMEDIA | P&D, 146; H/S/IS, 193 |
| SPIE-THE INTERNATIONAL OPTICAL ENGINEERING SOCIETY | P&D, 165 |
| SYNSOR CORPORATION | H/S/IS, 206 |
| VIDEOACTIVE, INC. | P&D, 174 |
| VIDEODISCOVERY | P&D, 175 |

## WISCONSIN

| | |
|---|---|
| BLUE LAKES COMPUTING | H/S/IS, 181 |
| BRIEN LEE & COMPANY | P&D, 123 |
| FUTUREDAY | P&D, 137 |
| UNIV. OF WISCONSIN/EDUCATIONAL TECHNOLOGY PROGRAM | P&D, 172 |

## FOREIGN COUNTRIES

### CANADA

| | |
|---|---|
| ALBERTA EDUCATIONAL COMMUNICATIONS CORP. (ACCESS) | P&D, 120 |
| ALPHATEL SYSTEMS LTD. | P&D, 121; H/S/IS, 180 |
| CABLESHARE INC. | P&D, 124; H/S/IS, 181 |
| CEDI INFORMATIQUE | P&D, 125 |
| DEPT. OF NATIONAL DEFENCE | P&D, 132; H/S/IS, 185 |
| ENERGY, MINES & RESOURCES | H/S/IS, 188 |
| INFOTECH CONSULTANTS | P&D, 142 |
| INTERACTIVE IMAGE TECHNOLOGIES INC. | P&D, 144 |
| MATROX ELECTRONIC SYSTEMS LTD. | P&D, 153; H/S/IS, 196 |
| MICHAEL J. PETRO LTD. | P&D, 154; H/S/IS, 197 |
| ONTARIO INSTITUTE FOR STUDIES IN EDUCATION | P&D, 156 |
| SIMON FRASER UNIVERSITY/INSTRUCTIONAL MEDIA CENTRE | P&D, 163 |
| SOFI INC. (INTERACTIVE TRAINING & INFO. CORP.) | P&D, 164; H/S/IS, 204 |
| SONY OF CANADA LTD. | H/S/IS, 205 |
| SOUTHERN ALBERTA INST. OF TECH./LEARNING RES. CEN. | P&D, 165 |

### ENGLAND

| | |
|---|---|
| BBC OPEN UNIVERSITY PRODUCTION CENTRE | P&D, 123 |
| CENTRE FOR REMOTE SENSING | H/S/IS, 182 |
| CLARK, DAVID R. | P&D, 126 |
| INTERACTIVE VIDEO PROGRAMMES (DRAGONFLAIR LTD) | P&D, 146 |
| OWL MICRO-COMMUNICATIONS | H/S/IS, 199 |
| RECENT PRODUCTIONS LIMITED | P&D, 160; H/S/IS, 202 |
| UNIV. COLLEGE, CARDIFF/CENTRE FOR EDUCATIONAL TECHNOLOGY | P&D, 169 |
| UNIV. OF LONDON AUDIO-VISUAL CENTRE | P&D, 171 |
| VIDEODISC TECHNOLOGY LIMITED | P&D, 174 |

**Organizations By Geographic Location**

FRANCE

LOGIVISION                          P&D, 150; H/S/IS, 195

GERMANY

INNOVISION                        P&D, 143
MAZ & MOVIE                    H/S/IS, 196
TELEMEDIA GmbH             P&D, 168; H/S/IS, 206

HOLLAND

PHILIPS INTERNATIONAL/AUDIO DIV./LASERVISION DEPT.      P&D, 158; H/S/IS, 200

ITALY

CENTRO TELEVISIVO UNIVERSITARIO (CTU)           P&D, 125
TECNETRA-TECHNOLOGY TRANSFER COMPANY       P&D, 167; H/S/IS, 206

SWITZERLAND

CAT ZURICH—COMPUTER ASSISTED TELEVIDEO AG      P&D, 124; H/S/IS, 182

P&D = Production and Development Resources Section
H/S/IS = Hardware/Software/Integrated Systems Section
M&R = Mastering and Replication Resources Section

## LEVELS 1, 2, AND 3 VIDEODISCS

| LEVEL 1 VIDEODISCS | FORMAT | *PAGE* |
|---|---|---|
| ABBA | Laser CAV | 217 |
| AMERICA LIVE IN CENTRAL PARK | Laser CAV | 224 |
| APOLLO | Laser CAV | 228 |
| ARMY COMM. TECH. OFF. (ACTO) VIDEODISC PROTOTYPES | Laser CAV | 229 |
| ASTRONOMY | Laser CAV | 231 |
| BELLY DANCING | Laser CAV | 236 |
| BOSTON | Laser CAV | 244 |
| CALEDONIAN DREAMS | Laser CAV | 249 |
| CELL, THE | Laser CAV | 253 |
| CENTERDISK | Laser CAV | 253 |
| COLORATION | Laser CAV | 262 |
| CREATIVE CAMERA, THE | Laser CAV | 265 |
| DARTMOUTH COLLEGE ARCHIVAL DISC | Laser CAV | 267 |
| DISC SUPPLEMENT TO JOURNAL, CELL MOTILITY | Laser CAV | 275 |
| DISNEYDISC OF MYSTERY AND MAGIC | CED | 276 |
| DIVE TO ANOTHER WORLD | CED, Laser CAV | 277 |
| ELEPHANT PARTS | Laser CAV | 283 |
| ENTERTAINMENT GAME, THE | CED | 285 |
| EVE, THE | Laser CAV | 287 |
| EVENING WITH RAY CHARLES, AN | Laser CAV | 287 |
| EXPLORING LANGUAGE | Laser CAV | 288 |
| FIRST NATIONAL KIDISC, THE | Laser CAV | 292 |
| FLEETWOOD MAC | Laser CAV | 293 |
| FOOTBALL FOLLIES AND SENSATIONAL SIXTIES | Laser CAV | 294 |
| FORD MOTOR VIDEO NETWORK (15 TOTAL DISC SIDES) | Laser CAV | 295 |
| FUN & GAMES | Laser CAV | 298 |
| GARDENING AT HOME | Laser CAV | 300 |
| GEOSCIENCE | Laser CAV | 301 |
| GROVER WASHINGTON, JR. IN CONCERT | Laser CAV | 308 |
| HISTORY DISQUIZ, THE | Laser CAV | 316 |
| HOW TO WATCH PRO FOOTBALL | Laser CAV | 318 |
| INTERACTIVE ARTS & SCIENCES, INC.—(CONFIDENTIAL) | CED | 325 |
| INTERACTIVE ARTS INTERNATIONAL (UNANNOUNCED) | CED, Laser CAV | 325 |
| JAZZERCISE | Laser CAV | 330 |
| JOY OF RELAXATION, THE | Laser CAV | 332 |
| KANAKO | Laser CAV | 333 |
| LANDISC I | Laser CAV | 338 |
| LANDISC II | Laser CAV | 338 |
| LANGUAGE AND LEARNING | Laser CAV | 338 |
| LIFE AND WORKS OF MICHELANGELO, THE | CED | 340 |
| LIZA | Laser CAV | 342 |
| LORETTA | Laser CAV | 343 |
| M.J. ZINK PRODUCTIONS, INC.—(NOT GIVEN) | | 345 |
| MANY ROADS TO MURDER | CED | 350 |
| MARYLAND INSTRUCTIONAL TELEVISION | Laser CAV | 351 |
| MASTER COOKING COURSE, THE | Laser CAV | 351 |
| MAZE MANIA | Laser CAV | 352 |
| MEDICAL MICROSCOPY | Laser CAV | 353 |
| MYSTERYDISC—MURDER, ANYONE? | CED, Laser CAV | 362 |
| NEIL SEDAKA IN CONCERT | Laser CAV | 363 |
| NFL SYMFUNNY AND LEGENDS OF THE FALL, THE | Laser CAV | 364 |
| OLIVIA | Laser CAV | 369 |
| OLIVIA: IN CONCERT | CED, Laser CAV | 369 |
| OLIVIA: PHYSICAL | Laser CAV | 369 |
| ORIENTAL DREAMS | Laser CAV | 371 |
| PAUL SIMON | Laser CAV | 374 |
| PBS CLOSED CAPTIONED DISC | Laser CAV | 375 |
| PERMS AND HAIRSTYLES | Laser CAV | 376 |
| POTPOURRI | Laser CAV | 382 |
| PRODUCING INTERACTIVE VIDEODISCS | Laser CAV | 384 |
| PUZZLE OF THE TACOMA NARROWS BRIDGE COLLAPSE, THE | Laser CAV | 385 |

## Levels 1, 2, And 3 Videodiscs

| | | |
|---|---|---|
| RAIDERS OF THE LOST ARK | Laser CAV | 387 |
| RAINBOW GOBLINS STORY | Laser CAV | 388 |
| RAINBOW: LIVE BETWEEN THE EYES | Laser CAV | 388 |
| RESEARCH DISC | Laser CAV | 390 |
| ROCK ADVENTURE | Laser CAV | 393 |
| SHUTTLE | Laser CAV | 403 |
| SIGHT THROUGH SOUND, AN INTERACT. INTRO MED. DIAG. | Laser CAV | 403 |
| SLEEPWALKER | Laser CAV | 405 |
| SON OF FOOTBALL FOLLIES AND BIG GAME AMERICA, THE | Laser CAV | 407 |
| SPACE AGE, THE | Laser CAV | 408 |
| SPACE SHUTTLE MISSION REPORTS: S.T.S. 5, 6 & 7 | Laser CAV | 409 |
| SPACE SHUTTLE: MISSION CONTROL | CED | 409 |
| SUPER MEMORIES OF THE SUPER BOWLS | Laser CAV | 418 |
| SUPER SEVENTIES, THE | Laser CAV | 418 |
| SWITCH ON | Laser CAV | 419 |
| TAKANAKA WORLD | Laser CAV | 421 |
| TOUCH OF LOVE ... MASSAGE, THE | Laser CAV | 430 |
| TUBES VIDEO, THE | Laser CAV | 433 |
| U.S. DEPARTMENT OF EDUCATION: WORLD OF WORK | Laser CAV | 434 |
| VIDEO MUSEUM, THE | Laser CAV | 439 |
| VIDEODISC, THE: A TOOL FOR COMMUNICATION | Laser CAV | 439 |
| VIDEODISC SERIES/APPRENTICE & JOURNEYMAN TRAINING | Laser CAV | 439 |
| VILLA ALEGRE | Laser CAV | 440 |
| VINCENT VAN GOGH: A PORTRAIT IN TWO PARTS | Laser CAV | 440 |
| VOYAGER | Laser CAV | 442 |
| WALK THROUGH THE UNIVERSE, A | CED | 442 |
| WEEK AT THE RACES, A | CED | 444 |
| WHALES | Laser CAV | 445 |
| WHITE MUSIC | Laser CAV | 446 |
| WORDS IN MOTION I | Laser CAV | 449 |
| WORDS IN MOTION II | Laser CAV | 449 |
| WORLD OF MARTIAL ARTS, THE | Laser CAV | 449 |
| WORLD ON A SILVER PLATTER, THE | Laser CAV | 450 |

### LEVEL 2 VIDEODISCS

| | | |
|---|---|---|
| ABOUT COMPUTERS | Laser CAV | 218 |
| ADVANCED VIDEO MAINTENANCE INFORMATION | Laser CAV | 220 |
| ARIS KNITWEAR | Laser CAV | 229 |
| ARMY COMM. TECH. OFF. (ACTO) VIDEODISC PROTOTYPES | Laser CAV | 229 |
| ASIA TRAVEL | Laser CAV | 230 |
| BASIC AUTOMOTIVE ELECTRONICS | | 234 |
| BASIC DIESEL PRINCIPLES | | 234 |
| BASIC TUMBLING SKILLS | Laser CAV | 235 |
| BY YOURSELF | Laser CAV | 248 |
| CELL, THE | Laser CAV | 253 |
| CHALLENGES OF GLAUCOMA, THE | Laser CAV | 254 |
| CITIBANK | Laser CAV | 258 |
| COAST GUARD | Laser CAV | 260 |
| COLLEGE USE DEMONSTRATION DISC | Laser CAV | 261 |
| CRANBERRY WORLD VISITOR'S CENTER VIDEODISC PROJECT | Laser CAV | 265 |
| DEAF AWARENESS: LET YOUR FINGERS DO THE TALKING | Laser CAV | 269 |
| DINNER FOR THE BOSS | Laser CAV | 274 |
| FAIR EMPLOYMENT PRACTICE | Laser CAV | 290 |
| FIRST SIXTY SECONDS, THE | | 292 |
| FORD MOTOR VIDEO NETWORK (15 TOTAL DISC SIDES) | Laser CAV | 295 |
| GENERAL MOTORS OF CANADA | Laser CAV | 301 |
| GO FOR THE GREEN | Laser CAV | 302 |
| GOING METRIC | Laser CAV | 303 |
| HAWAII TRAVEL | Laser CAV | 313 |
| HOW YOUR HEART AND CIRCULATORY SYSTEM WORKS | Laser CAV | 318 |
| IBM HERITAGE | Laser CAV | 321 |
| INTELLIGENT DISC, THE | Laser CAV | 324 |
| INTERACTIVE ARTS & SCIENCES, INC.—(CONFIDENTIAL) | CED | 325 |
| INTERACTIVE JULIA CHILD | Laser CAV | 325 |
| ISRAELI BOY: LIFE ON A KIBBUTZ | Laser CAV | 328 |
| KRAFT PROCESS OVERVIEW | Laser CAV | 337 |
| LRN COLLEGE DISC | Laser CAV | 344 |
| M.J. ZINK PRODUCTIONS, INC.—(NOT GIVEN) | | 345 |
| MARITZ COMMUNICATIONS CO.—(NUM. FOR BUS. & IND.) | Laser CAV | 350 |
| MEDICAL MICROSCOPY | Laser CAV | 353 |
| MEJORE SU PRONUNCIACION (IMPROVE YOUR PRONOUNCIA.) | Laser CAV | 353 |
| MERRILL LYNCH | Laser CAV | 354 |
| MILES LEARNING CENTER: APPENDECTOMY | Laser CAV | 355 |
| MILES LEARNING CENTER: ILEOCOLIC RESECTION | Laser CAV | 355 |
| MILES LEARNING CENTER: MEDICAL APPLICATIONS/HEM. | Laser CAV | 356 |
| MILES LEARNING CENTER: ORIENTATION & OPER. INSTR. | Laser CAV | 356 |
| MYSTERYDISC 2—MANY ROADS TO MURDER | Laser CAV | 362 |
| NINETY-SIX: A CATTLE RANCH IN NORTHERN NEVADA | Laser CAV | 366 |

| Title | Format | Page |
|---|---|---|
| PBS CLOSED CAPTIONED DISC | Laser CAV | 375 |
| PHILIPS VP-835 DEMONSTRATION DISC | Laser CAV | 378 |
| POTPOURRI | Laser CAV | 382 |
| PRODUCING INTERACTIVE VIDEODISCS | Laser CAV | 384 |
| PROFESSIONAL SELLING SKILLS SYSTEM III | Laser CAV | 384 |
| PUZZLE OF THE TACOMA NARROWS BRIDGE COLLAPSE, THE | Laser CAV | 385 |
| RALSTON-PURINA | Laser CAV | 388 |
| RCA ENGINEERING DEMO DISC | CED | 389 |
| READING DEVELOPMENT VIDEODISC | Laser CAV | 389 |
| REPAIR PARTS AND SPECIAL TOOLS LIST—AN UGC 74 | Laser CAV | 390 |
| RESEARCH DISC | Laser CAV | 390 |
| SIGHT THROUGH SOUND, AN INTERACT. INTRO MED. DIAG. | Laser CAV | 403 |
| SKF: DIAGNOSTIC CHALLENGES IN GASTROENTEROLOGY | Laser CAV | 404 |
| SKF: THE HIDDEN LANGUAGE OF ARTHRITIS | Laser CAV | 405 |
| SONY OVERVIEW | Laser CAV | 408 |
| SPACE SHUTTLE MISSION REPORTS: S.T.S. 5, 6 & 7 | Laser CAV | 409 |
| SPANISH MAP READING CONVERSION PROJECT | Laser CAV | 409 |
| SPORTS AND FITNESS IN ONTARIO | Laser CAV | 411 |
| STANDARD HEATS OF SOLUTION | Laser CAV | 412 |
| STAR TALK | Laser CAV | 413 |
| TEXAS INSTRUMENTS | Laser CAV | 424 |
| TOMORROW TODAY | Laser CAV | 429 |
| TRADITION AND CHANGE | Laser CAV | 431 |
| TRANSFORMING OFFICE PRODUCTIVITY (TOP) SEMINAR | Laser CAV | 431 |
| TROUBLESHOOTING HORSEPOWER LOSS | Laser CAV | 432 |
| TUMBLING (CLOSED CAPTION) | Laser CAV | 433 |
| URINARY CATHETERIZATION | Laser CAV | 435 |
| VEHICLE EMISSION CONTROL | | 437 |
| VIDEO BOOKMARK, THE | Laser CAV | 438 |
| VIDEO VERSATILITY | Laser CAV | 439 |
| VIDEODISC SERIES/APPRENTICE & JOURNEYMAN TRAINING | Laser CAV | 439 |
| VON HERRING CAPER, THE: "A CHOICE MYSTERY" | Laser CAV | 441 |
| WAR PRAYER, THE | Laser CAV | 442 |
| WEYERHAEUSER | Laser CAV | 445 |
| WHALES | Laser CAV | 445 |
| WORDS IN MOTION I | Laser CAV | 449 |
| WORDS IN MOTION II | Laser CAV | 449 |

## LEVEL 3 VIDEODISCS

| Title | Format | Page |
|---|---|---|
| A.C. SWITCHBOARD OPERATION | Laser CAV | 217 |
| ABC-NEA SCHOOLDISC | Laser CAV | 218 |
| ACTION CODE COURSEWARE—INDUSTRIAL TECH. PROGRAM | | 219 |
| ACTRONICS, INC.—(NOT GIVEN) | Laser CAV | 219 |
| ADVANCED VIDEO MAINTENANCE INFORMATION | Laser CAV | 220 |
| ADVANCED VIDEO, INC.—(NOT GIVEN) | CED, Laser CAV | 220 |
| ADVENTURE WITH DUSTY | Laser CAV | 220 |
| ALLEGHENY INTERNATIONAL INFOSOURCE I | Laser CAV | 223 |
| AMTESS: AUTOMOTIVE | Laser CAV | 225 |
| AMTESS: MISSILE | Laser CAV | 226 |
| AN/WLR-1 SYSTEM REPAIRABLE IDENTIFICATION | Laser CAV | 226 |
| ARMY COMM. TECH. OFF. (ACTO) VIDEODISC PROTOTYPES | Laser CAV | 229 |
| ART HISTORY RETRIEVAL | Thomson CSF | 230 |
| ASSESSMENT OF NEUROMOTOR DYSFUNCTION IN INFANTS | Laser CAV | 231 |
| ATTI (AMERICAN TELEPHONE & TELEGRAPH INTERNATIONAL) | Laser CAV | 231 |
| BALLET SERIES | Thomson CSF | 232 |
| BASIC SKILLS EDUCATION FOR THE U.S. ARMY | Laser CAV | 235 |
| BERLIN I | Laser CAV | 236 |
| BERLIN II | Laser CAV | 237 |
| BIO SCI VIDEODISC, THE | Laser CAV | 239 |
| BIOLOGY I & II | Laser CAV | 239 |
| BONNLAND | Laser CAV | 243 |
| BOOZ-ALLEN & HAMILTON HOME INFO SYSTEM SIMULATOR | Laser CAV | 243 |
| BOSTON DISC | Laser CAV | 244 |
| BOSTON MEDIA CONSULTANTS—(CLASS. RESEARCH DISC) | CED | 244 |
| CALL FOR FIRE | Laser CAV | 249 |
| CAPSULE, THE | Laser CAV | 250 |
| CARDIO-PULMONARY-RESUSCITATION | Laser CAV | 251 |
| CENTERDISK | Laser CAV | 253 |
| CHALLENGES OF GLAUCOMA, THE | Laser CAV | 254 |
| CHEMISTRY I & II | Laser CAV | 256 |
| CHICAGO MERCANTILE EXCHANGE | Laser CAV | 256 |
| CHRISTMAS 1983 ELECTRONIC CATALOG | Laser CAV | 257 |
| CITTA VIVA—GALLIPOLI | Laser CAV | 258 |
| CLASSROOM BEHAVIOR RECORD | Laser CAV | 259 |
| CLEAN AND INSPECT A GENERATOR | Laser CAV | 259 |
| CLEFT LIP AND PALATE CASE MANAGEMENT | Laser CAV | 260 |

| Title | Format | Page |
|---|---|---|
| COLUMBIA RIVER PROJECT | Laser CAV | 262 |
| COMPREHENSIVE CASE MANAGEMENT OF SPINA BIFIDA | Laser CAV | 263 |
| COMPREHENSIVE JOB SKILLS PRACTICE | Laser CAV | 264 |
| CONTINUING MEDICAL EDUCATION LIBRARY | Laser CAV | 264 |
| COURT TESTIMONY SKILLS FOR THE EXPERT WITNESS | Laser CAV | 265 |
| D.C. MOTOR MAINTENANCE, PT. 1 | Laser CAV | 266 |
| D.C. MOTOR MAINTENANCE, PT. 2 | Laser CAV | 266 |
| DATA DEMONS | Other | 267 |
| DECIMALS AND FRACTIONS | Laser CAV | 270 |
| DELTA CORRIDOR | Laser CAV | 270 |
| DEMONSTRATION VIDEODISC (CAT ZURICH) | Laser CAV | 271 |
| DEMONSTRATION VIDEODISC (WICAT SYSTEMS, INC.) | Laser CAV | 271 |
| DEMONSTRATION VIDEODISCS, 1982 AND 1983 (CSC) | Laser CAV | 271 |
| DENTISTRY I ("SUE") | Laser CAV | 271 |
| DENTISTRY II ("JACKIE") | Laser CAV | 271 |
| DENTISTRY III | Laser CAV | 272 |
| DENTISTRY IV | Laser CAV | 272 |
| DETERMINING FINANCIAL ELIGIBILITY | Laser CAV | 272 |
| DEVELOPMENT OF LIVING THINGS | Laser CAV | 273 |
| DIAGNOSIS & TREATMENT OF BIRTH DEFECTS | Laser CAV | 273 |
| DIAGNOSIS & TREATMENT OF SPINA BIFIDA | Laser CAV | 273 |
| DIAGNOSTIC CHALLENGES | Laser CAV | 273 |
| DIGITAL VIDEO CORPORATION—(CONFIDENTIAL) | Laser CAV | 274 |
| E-3 FLOATING DESK PULSER | Laser CAV | 281 |
| EDITING SIMULATION DISC #1 | Laser CAV | 282 |
| EEMT COMPOSITE NO. 1 | Laser CAV | 282 |
| ELECTRONIC IGNITION SYSTEM | Laser CAV | 283 |
| EPCOT CENTER VIDEO DISCS (WALT DISNEY WORLD-FL.) | Laser CAV | 285 |
| ESTABLISHING TECHNICAL ELIGIBILITY | Laser CAV | 286 |
| EVIDENCE & OBJECTIONS | Laser CAV | 288 |
| EXTRA MILE, THE | Laser CAV | 289 |
| FLIGHT 505 | Laser CAV | 293 |
| FORT SILL VIDEODISC COURSEWARE PACKAGE | Laser CAV | 295 |
| FULCRUM CENTRAL AMERICA MAPS | Laser CAV | 298 |
| GENERAL MOTORS OF CANADA | Laser CAV | 301 |
| GREENDAY VIDEO, VOL. 1 NO. 4 | Laser CAV | 308 |
| GYNECOLOGY PATIENT EDUCATION | Thomson CSF | 309 |
| HANDS-ON WORD PROCESSING | Laser CAV | 310 |
| HANSEN'S DISEASE | Laser CAV | 311 |
| HAWK TROUBLESHOOTING COURSE | Laser CAV | 313 |
| HIGH POWER ILLUMINATOR RAM RADAR DEMO | Laser CAV | 316 |
| HOW YOUR HEART AND CIRCULATORY SYSTEM WORKS | Laser CAV | 318 |
| HUMAN GENETICS TRAINING FOR NURSES: PART I | Laser CAV | 319 |
| IBM—(CONFIDENTIAL) | Laser CAV | 320 |
| IBM 1750 TELEPHONE SYSTEM | Laser CAV | 321 |
| IBM OFFICE SYSTEMS | Laser CAV | 321 |
| IBM PC EXPERIENCE | Laser CAV | 321 |
| IMPLEMENTING ADDITIONAL PROGRAMS AND POLICIES | Laser CAV | 322 |
| INTEGRATED READINESS CENTER | Laser CAV | 323 |
| INTEGRATING ADMINISTRATIVE POLICIES | Laser CAV | 324 |
| INTELLIGENT DISC, THE | Laser CAV | 324 |
| INTERACT. & INTERVEN. W/GRIEVING CLIENTS & FAM. | Laser CAV | 324 |
| INTERACTIVE ARTS & SCIENCES, INC.—(CONFIDENTIAL) | CED | 324 |
| INTERACTIVE ARTS INTERNATIONAL (UNANNOUNCED) | CED, Laser CAV | 325 |
| INTERNATIONAL HARVESTER | Laser CAV | 325 |
| INTERVENTION IN CHILD ABUSE AND NEGLECT | Laser CAV | 326 |
| INTRODUCTION TO COMPUTER LITERACY | Laser CAV | 326 |
| INTRODUCTION TO ECONOMICS | Laser CAV | 326 |
| INTRODUCTION TO THE JOIN SYSTEM | Laser CAV | 326 |
| INTRODUCTION TO THE LEARNING SYSTEM, AN | Laser CAV | 327 |
| INVESTMENT SERVICES | Laser CAV | 327 |
| JOB PERFORMANCE PACKAGE (JPP) | Laser CAV | 331 |
| JOB READINESS MASTERY TEST | Laser CAV | 332 |
| KARATE | Laser CAV | 333 |
| KENNEDY SPACE CENTER —PAD 39A | Laser CAV | 334 |
| KLAVIER IM HAUS | Laser CAV | 336 |
| KPR/WINTHROP: UNDERSTANDING CONTRAST MEDIA: | Laser CAV | 336 |
| LANDISC I | Laser CAV | 338 |
| LANDISC II | Laser CAV | 338 |
| LOCIS INTERACTIVE VIDEO-DISC, THE | Laser CAV | 342 |
| LVT-7A1 DRIVER TRAINER | Laser CAV | 345 |
| M.J. ZINK PRODUCTIONS, INC. - (NOT GIVEN) | | 345 |
| MACARIO | Laser CAV | 346 |

| Title | Format | Page |
|---|---|---|
| MANUAL SEAT/MAN SEP. FR. A MARTIN-BAKER EJECT. ST. | Laser CAV | 349 |
| MARITZ COMMUNICATIONS CO.—(NUM. FOR BUS. & IND.) | Laser CAV | 350 |
| MASTERS OF THE DRAGON | Laser CAV | 351 |
| MATERIAL SCIENCE VIDEODISC | Laser CAV | 352 |
| MATH IN BIOLOGY | Laser CAV | 353 |
| MEDICAL MICROSCOPY | Laser CAV | 353 |
| MILTON KEYNES INFORMATION TECHNOLOGY VIDEODISC | Laser CAV | 356 |
| MONTEVIDISCO (MALE VERSION) | Laser CAV | 358 |
| MORE ADVENTURES WITH DUSTY | Laser CAV | 359 |
| MOUNTED LAND NAVIGATION | Laser CAV | 359 |
| O.I.S.E. SINGLE POINT THREAD CUTTING | Laser CAV | 368 |
| OBSERVATION SCENES FOR TRAIN. IN NON-VERBAL COMM. | Philips | 368 |
| ONE THOUSAND EIGHT HUNDRED & FIFTY TWO HMPBK WHLES | Laser CAV | 371 |
| PARIS-PARIS | Laser CAV | 373 |
| PATRIOT CLS OJT PHASE 1 | Laser CAV | 374 |
| PATRIOT MISSILE SYSTEM PROTOTYPE MAINTENANCE DISC | Laser CAV | 374 |
| PEDIATRIC CARDIOVASCULAR DEFECTS | Laser CAV | 375 |
| PEDIATRIC HEMATOLOGICAL DISEASES | Laser CAV | 375 |
| PERCEPTRONICS, INC.—(NOT GIVEN) | Laser CAV | 376 |
| PERSONAL COMPUTER DIAGNOSTICS | Laser CAV | 376 |
| PERSONAL COMPUTER EXHIBIT | Laser CAV | 377 |
| PHILIPS VP-835 DEMONSTRATION DISC | Laser CAV | 378 |
| PHYSICS I & II | Laser CAV | 378 |
| POINT OF SALE DEMONSTRATION | Laser CAV | 381 |
| POTPOURRI | Laser CAV | 382 |
| POWER PLANT TOUR | Laser CAV | 382 |
| PRINCIPLES OF OPERANT CONDITIONING | Laser CAV | 383 |
| PRODUCTION ASSOCIATES, INC.—(NOT GIVEN) | Other | 384 |
| PUZZLE OF THE TACOMA NARROWS BRIDGE COLLAPSE, THE | Laser CAV | 385 |
| RADAR, MULTIPLE BEAM ACQ. & WIDE APERTURE SURVEIL. | Laser CAV | 387 |
| RALSTON-PURINA | Laser CAV | 388 |
| RENAL ANALYSIS | Laser CAV | 390 |
| REPAIR PARTS AND SPECIAL TOOLS LIST—AN UGC 74 | Laser CAV | 390 |
| RESEARCH DISC | Laser CAV | 390 |
| SALEM II | Laser CAV | 396 |
| SAN FRANCISCO | Laser CAV | 397 |
| SCE: SONGS UNITS 2 & 3 SURROGATE TRAVEL | | 398 |
| SEAFOOD DEPARTMENT OPERATIONS | Laser CAV | 398 |
| SELLING THE PSYCHOLOGICAL | Laser CAV | 399 |
| SOCIAL STYLE SALES STRATEGIES | Laser CAV | 406 |
| SONY OVERVIEW | Laser CAV | 408 |
| SPARROW MISSILE TEST SET TRAINING VIDEODISC | Laser CAV | 410 |
| SPECTRUM OF INTEGRATED LOGISTICS SUPPORT | Laser CAV | 410 |
| SPERRY DEMO | Laser CAV | 410 |
| SPINABIFIDA | Laser CAV | 410 |
| SPS-49 RADAR | Laser CAV | 411 |
| SYSCON CORPORATION, INC.—(NOT GIVEN) | Laser CAV | 420 |
| SYSTEM TEST ENHANCEMENT PROJECT (STEP) | Laser CAV | 420 |
| TELEMART | Laser CAV | 422 |
| TEST REACTOR | Laser CAV | 423 |
| THINK IT THROUGH I | Laser CAV | 426 |
| THINK IT THROUGH II | Laser CAV | 426 |
| TJS | Laser CAV | 428 |
| TPS-43 INTERACTIVE "SAMPLER" | Laser CAV | 430 |
| TRACING A GROUND | Laser CAV | 430 |
| U.S. DEPARTMENT OF EDUCATION: WORLD OF WORK | Laser CAV | 434 |
| UNIV. OF LONDON PROGRAMME DEVELOPMENT VIDEODISC | Laser CAV | 434 |
| UNIV. OF NEB./NEB. VD DESIGN/PROD. GRP—(NOT GIVEN) | Laser CAV | 435 |
| UNIV. OF NEB./NEB. VD DESIGN/PROD. GRP—(NOT GIVEN) | Laser CAV | 435 |
| URODYNAMICS IN CLINICAL PRACTICE | Laser CAV | 436 |
| UROLOGY | Thomson CSF | 436 |
| USING THE APR | Laser CAV | 436 |
| VHD INTERACTIVE VIDEODISC DEMONSTRATION | VHD | 437 |
| VIDEO ENHANCED GERMAN GATEWAY | Laser CAV | 438 |
| VIDEODISC SERIES/A&JT | Laser CAV | 439 |
| VILLA ALEGRE | Laser CAV | 440 |
| VINSON PROJECT, THE | Laser CAV | 440 |
| VISION-UNIX INTERACTIVE VIDEODISC TRAINING CURRIC. | Laser CAV | 440 |
| VISTA (VIDEODISC INTERPERS. SKILLS TRAIN. & AS.) | Laser CAV | 441 |
| WALK THROUGH THE UNIVERSE, A | CED | 442 |
| WAY WE LIVE, THE | Laser CAV | 443 |
| WESTERN ELECTRIC | Laser CAV | 444 |
| WILS-WANG INTERACTIVE LEARNING SYSTEM-WP PROGRAM | Laser CAV | 447 |
| YOU AND AFDC | Laser CAV | 451 |
| YOU CAN BUY A COMPUTER! | Laser CAV | 451 |

## VIDEODISCS BY TYPE

### ADOLESCENT

ABOUT COMPUTERS, 218
ADVENTURE WITH DUSTY, 220
CELL, THE, 253
COLLEGE USE DEMONSTRATION DISC, 261
EPCOT CENTER VIDEODISCS (WALT DISNEY WORLD-FL.), 285
HOW YOUR HEART AND CIRCULATORY SYSTEM WORKS, 318
INTERACTIVE ARTS INTERNATIONAL (UNANNOUNCED), 325
KLAVIER IM HAUS, 336
LIFE AND WORKS OF MICHELANGELO, THE, 340
MACARIO, 346
MANY ROADS TO MURDER, 350
MARYLAND INSTRUCTIONAL TELEVISION, 351
MATH IN BIOLOGY, 352
MEJORE SU PRONUNCIACION (IMPROVE YOUR PRONOUNCIATION), 353
MONTEVIDISCO (MALE VERSION), 358
MORE ADVENTURES WITH DUSTY, 359
MYSTERYDISC - MURDER, ANYONE?, 362
SLEEPWALKER, 405
SPACE SHUTTLE MISSION REPORTS: S.T.S. 5, 6 & 7, 409
SPACE SHUTTLE: MISSION CONTROL, 409
STAR TALK, 413
WAR PRAYER, THE, 443
WEEK AT THE RACES, A, 444

### ADULT

EROTICISE, 285
KANAKO, 333
ORIENTAL DREAMS, 371
PLAYBOY MAGAZINE, VOLUME 1, 380
PLAYBOY MAGAZINE, VOLUME 2, 380
PLAYBOY'S PLAYMATE REVIEW, 380
PLAYBOY VIDEO, VOLUME 1, 380
PLAYBOY VIDEO, VOLUME 2, 380
PLAYBOY VIDEO, VOLUME 3, 381
REDD FOXX - VIDEO IN A PLAIN BROWN WRAPPER, 391
TOUCH OF LOVE . . . MASSAGE, THE, 430

### ADVERTISING

ARIS KNITWEAR, 229
BOOZ-ALLEN & HAMILTON HOME INFO SYSTEM SIMULATOR, 239
CHRISTMAS 1983 ELECTRONIC CATALOG, 257
CITTA VIVA—GALLIPOLI, 258
DATA DEMONS, 267
GREENDAY VIDEO, VOL. 1 NO. 4, 308
M.J. ZINK PRODUCTIONS, INC.—(NOT GIVEN), 345
MARITZ COMMUNICATIONS CO.—(NUM. FOR BUS. & IND.), 350
PARIS-PARIS, 373
TELEMART, 422

### ARCHIVAL

ADVANCED VIDEO MAINTENANCE INFORMATION, 220
APOLLO, 228
ART HISTORY RETRIEVAL, 230
ASTRONOMY, 231
BIO SCI VIDEODISC, THE, 239
BOOZ-ALLEN & HAMILTON HOME INFO SYSTEM SIMULATOR, 239
BOSTON DISC, 244
CAPSULE, THE, 250
CELL, THE, 253
CENTERDISK, 253
COLLEGE USE DEMONSTRATION DISC, 261
DARTMOUTH COLLEGE ARCHIVAL DISC, 267
DEMONSTRATION VIDEODISC (CAT ZURICH), 270
FORD MOTOR VIDEO NETWORK (15 TOTAL DISC SIDES), 295
GENERAL MOTORS OF CANADA, 300
GEOSCIENCE, 301
GO FOR THE GREEN, 302
HANSEN'S DISEASE, 311
IBM HERITAGE, 321
INTERACTIVE ARTS INTERNATIONAL (UNANNOUNCED), 325
LIFE AND WORKS OF MICHELANGELO, THE, 340
M.J. ZINK PRODUCTIONS, INC.—(NOT GIVEN), 345
MARITZ COMMUNICATIONS CO.—(NUM. FOR BUS. & IND.), 350

MEDICAL MICROSCOPY, 353
ONE THOUSAND EIGHT HUNDRED & FIFTY TWO
  HUMPBACK WHALES, 371
PERCEPTRONICS, INC.—(NOT GIVEN), 376
RCA ENGINEERING DEMO DISC, 389
SCE: SONGS UNITS 2 & 3 SURROGATE TRAVEL, 398
SHUTTLE, 403
SPACE AGE, THE, 408
SPACE SHUTTLE MISSION REPORTS: S.T.S. 5, 6 & 7, 409
SPACE SHUTTLE: MISSION CONTROL, 409
STAR TALK, 413
UNIV. OF LONDON PROGRAMME DEVELOPMENT
  VIDEODISC, 434
UROLOGY, 436
VOYAGER, 442


## ART

ART HISTORY RETRIEVAL, 230
CENTERDISK, 253
FULCRUM CENTRAL AMERICA MAPS, 298
HERITAGE OF THE BIBLE: THE LAW AND THE PROPHETS,
  315
LIFE AND WORKS OF MICHELANGELO, THE, 340
PARIS-PARIS, 373
PHILIPS VP-835 DEMONSTRATION DISC, 378
SAN FRANCISCO, 397
SLEEPWALKER, 405
TOMORROW TODAY, 429
TUT, THE BOY KING/THE LOUVRE, 434
VINCENT VAN GOGH: A PORTRAIT IN TWO PARTS, 440


## CATALOGS

ADVANCED VIDEO MAINTENANCE INFORMATION, 220
AN/WLR-1 SYSTEM REPAIRABLE IDENTIFICATION, 226
ARIS KNITWEAR, 229
ARMY COMM. TECH. OFF. (ACTO) VIDEODISC PROTOTYPES,
  229
BOOZ-ALLEN & HAMILTON HOME INFO SYSTEM
  SIMULATOR, 243
CAPSULE, THE, 250
CELL, THE, 253
CENTERDISK, 253
CHRISTMAS 1983 ELECTRONIC CATALOG, 257
COAST GUARD, 260
COLLEGE USE DEMONSTRATION DISC, 261
GREENDAY VIDEO, VOL. 1 NO. 4, 308
LRN COLLEGE DISC, 344
MARITZ COMMUNICATIONS CO.—(NUM. FOR BUS. & IND.),
  350
MARYLAND INSTRUCTIONAL TELEVISION, 351
MERRILL LYNCH, 354
ONE THOUSAND EIGHT HUNDRED & FIFTY TWO
  HUMPBACK WHALES, 371
PERCEPTRONICS, INC.—(NOT GIVEN), 376
PHILIPS VP-835 DEMONSTRATION DISC, 378
RCA ENGINEERING DEMO DISC, 389
REPAIR PARTS AND SPECIAL TOOLS LIST—AN UGC 74, 390
TEXAS INSTRUMENTS, 424

UNIV. OF LONDON PROGRAMME DEVELOPMENT
  VIDEODISC, 434
VINCENT VAN GOGH: A PORTRAIT IN TWO PARTS, 440
WEYERHAEUSER, 445


## CHILDREN'S DISCS

ABOUT COMPUTERS, 218
BASIC TUMBLING SKILLS, 235
BIG BLUE MARBLE PRESENTS MY SEVENTEENTH SUMMER,
  237
CELL, THE, 253
DECIMALS AND FRACTIONS, 270
DISNEYDISC OF MYSTERY AND MAGIC, 276
EPCOT CENTER VIDEODISCS (WALT DISNEY WORLD-FL.),
  285
FIRST NATIONAL KIDISC, THE, 292
FREE TO BE YOU AND ME/MANDY'S GRANDMOTHER, 297
FUN & GAMES, 298
INTERACTIVE ARTS INTERNATIONAL (UNANNOUNCED), 325
LIFE AND WORKS OF MICHELANGELO, THE, 340
MARYLAND INSTRUCTIONAL TELEVISION, 351
MR. ROGERS—HELPING CHILDREN UNDERSTAND, 360
PADDINGTON BEAR, VOL. 2, 372
SCHOLASTIC PRODUCTIONS: AS WE GROW, 398
SPACE SHUTTLE MISSION REPORTS: S.T.S. 5, 6 & 7, 409
SPACE SHUTTLE: MISSION CONTROL, 409
THINK IT THROUGH I, 426
THINK IT THROUGH II, 426
TUMBLING (CLOSED CAPTION), 433
VILLA ALEGRE, 440
WEEK AT THE RACES, A, 444
WHALES, 445
WORDS IN MOTION I, 449
WORDS IN MOTION II, 449


## COMEDY

AIRPLANE, 222
AIRPLANE II, 222
ANIMAL CRACKERS, 227
ANIMAL HOUSE, 227
ANNIE HALL, 227
APARTMENT, THE, 228
APPLE DUMPLING GANG, THE, 228
ARTHUR, 230
BAD NEWS BEARS, THE, 232
BANANAS, 233
BAREFOOT IN THE PARK, 234
BEACH GIRLS, THE, 235
BEST FRIENDS, 237
BEST LITTLE WHOREHOUSE IN TEXAS, THE, 237
BLAZING SADDLES, 240
BLUES BROTHERS, THE, 242
BON VOYAGE CHARLIE BROWN (AND DON'T COME
  BACK), 243
BREAKFAST AT TIFFANY'S, 245
CADDYSHACK, 248
CAGE AUX FOLLES, LA, 248
CALIFORNIA SUITE, 249
CAT BALLOU, 252

**Videodiscs By Type**

CHEECH AND CHONG'S NEXT MOVIE, 256
CHEECH AND CHONG'S NICE DREAMS, 256
CHITTY CHITTY BANG BANG, 257
CITY LIGHTS, 258
DAY AT THE RACES, A, 268
DIVINE MADNESS, 277
DR. STRANGELOVE, 279
DUCK SOUP, 280
EASY MONEY, 281
EDDIE MURPHY—DELIRIOUS, 282
ELEPHANT PARTS, 283
EVENING WITH ROBIN WILLIAMS, AN, 287
EVERYTHING YOU ALWAYS WANTED TO KNOW ABOUT SEX . . . , 287
FAST TIMES AT RIDGEMONT HIGH, 291
FOUL PLAY, 296
FUN & GAMES, 298
FUNNY THING HAPPENED ON THE WAY TO THE FORUM, A, 299
GALLAGHER—SOME UNCENSORED EVENING, 299
GAS PUMP GIRLS, 300
GEORGE CARLIN & STEVE MARTIN HOST SATURDAY NIGHT LIVE, VOL. 1, 301
GILDA, 302
GOODBYE GIRL, THE, 305
GOODBYE, COLUMBUS, 305
GRADUATE, THE, 305
GREAT MUPPET CAPER, THE, 307
HANKY PANKY, 311
HAROLD AND MAUDE, 312
HELLO DOLLY, 314
HERE IT IS, BURLESQUE, 315
HISTORY OF THE WORLD, PART 1, 316
IMPROPER CHANNELS, 322
IN-LAWS, THE, 322
IT CAME FROM HOLLYWOOD, 328
IT'S A MAD, MAD, MAD, MAD WORLD, 328
JERK, THE, 330
LAST OF THE RED HOT LOVERS, THE, 339
LENNY BRUCE PERFORMANCE, 340
LOVE AND DEATH, 344
LOVE AT FIRST BITE, 344
MAN WITH TWO BRAINS, THE, 348
MARY TYLER MOORE SHOW, VOL. 1, THE, 350
MISSIONARY, THE, 357
MODERN TIMES, 357
MONTY PYTHON - LIVE AT THE HOLLYWOOD BOWL, 358
MONTY PYTHON AND THE HOLY GRAIL, 358
MONTY PYTHON'S LIFE OF BRIAN, 358
MOVIE MOVIE, 359
MUPPET MOVIE, THE, 360
MURDER BY DEATH, 360
MY LITTLE CHICKADEE, 361
NATIONAL LAMPOON'S ANIMAL HOUSE, 363
NEIGHBORS, 363
NEVER GIVE A SUCKER AN EVEN BREAK, 364
NIGHT SHIFT, 365
NINE TO FIVE, 365
1941, 366
NUTTY PROFESSOR, THE, 367
ODD COUPLE, THE, 368
OH, GOD!, 369
PAPER MOON, 372
PARIS-PARIS, 373
PINK PANTHER, THE, 378
PINK PANTHER STRIKES AGAIN, THE, 379
PIPPIN, 379

PLAY IT AGAIN, SAM, 380
POPEYE, 381
PORKY'S, 382
PRIVATE EYES, THE, 383
PRIVATE LESSONS, 383
PRODUCERS, THE, 394
REDD FOXX—VIDEO IN A PLAIN BROWN WRAPPER, 391
RETURN OF THE PINK PANTHER, THE, 391
REVENGE OF THE PINK PANTHER, 391
RICHARD PRYOR & STEVE MARTIN HOST SATURDAY NIGHT LIVE, VOL. 2, 391
RICHARD PRYOR LIVE IN CONCERT, 392
RICHARD PRYOR LIVE ON THE SUNSET STRIP, 392
RUSSIANS ARE COMING, THE RUSSIANS ARE COMING, THE, 393
SEEMS LIKE OLD TIMES, 399
SERIAL, 400
SHOT IN THE DARK, A, 402
SILENT PARTNER, THE, 403
SLEEPER, 405
SOME LIKE IT HOT, 407
STARTING OVER, 415
STIR CRAZY, 415
STRIPES, 416
SURVIVORS, THE, 419
TAKE THE MONEY AND RUN, 421
"10", 423
THREE STOOGES VIDEODISC, VOL. 1, THE, 427
THREE STOOGES VIDEODISC, VOL. 2, THE, 427
TIME BANDITS, 427
TOM JONES, 429
TOY, THE, 430
TRAIL OF THE PINK PANTHER, 431
UP IN SMOKE, 435
USED CARS, 436
WHAT'S NEW PUSSYCAT?, 445
WHAT'S UP DOC?, 445
WHAT'S UP TIGER LILY?, 445
WHOLLY MOSES!, 447
YELLOWBEARD, 450
YOUNG DOCTORS IN LOVE, 452
YOUNG FRANKENSTEIN, 452
ZAPPED!, 452


DANCE

AEROBICISE, 220
AEROBICISE: THE BEGINNING WORKOUT, 221
AEROBICISE: THE ULTIMATE WORKOUT, 221
AEROBICS—NEW VIDEO, 221
ALL THAT JAZZ, 223
AMERICAN IN PARIS, AN, 225
BALLET SERIES, 232
BELLY DANCING, 236
BOSTON MEDIA CONSULTANTS—(CLASS. RESEARCH DISC), 244
BRIGADOON, 246
EVENING WITH THE ROYAL BALLET - FONTEYN & NUREYEV, 287
FAME, 290
FILLE MAL GARDEE, LA, 291
42ND STREET, 296
FUN & GAMES, 298
GISELLE—NUREYEV, 302

HAIR, 310
HERE IT IS, BURLESQUE, 315
JANE FONDA'S WORKOUT, 329
JANE FONDA'S WORKOUT FOR PREG., BIRTH AND RECOVERY, 329
JAZZERCISE, 330
JOHN CURRY'S ICE DANCIN', 332
MANON, 349
NUTCRACKER SUITE, 367
NUTCRACKER, THE, 367
NUTCRACKER/BARYSHNIKOV, 367
PARIS-PARIS, 373
PETER ALLEN & THE ROCKETTES AT RADIO CITY MUSIC HALL, 377
PHILIPS VP-835 DEMONSTRATION DISC, 378
RED SHOES, THE, 391
SATURDAY NIGHT FEVER, 397
SWAN LAKE, 419
SWING TIME, 419
TOMORROW TODAY, 429
WEST SIDE STORY, 444
WIZ, THE, 448
YANKEE DOODLE DANDY, 450

GAME

ADVANCED VIDEO, INC.—(NOT GIVEN), 221
BASEBALL FUN AND GAMES, 234
BASIC SKILLS EDUCATION FOR THE U.S. ARMY, 234
BINDU VISUALIZATION OF MUSIC, 238
BOSTON MEDIA CONSULTANTS—(CLASS. RESEARCH DISC), 244
DATA DEMONS, 267
DEMONSTRATION VIDEODISC (CAT ZURICH), 270
DISNEYDISC OF MYSTERY AND MAGIC, 276
ENTERTAINMENT GAME, THE, 285
EPCOT CENTER VIDEODISCS (WALT DISNEY WORLD -FL.), 285
FIRST NATIONAL KIDISC, THE, 292
FUN & GAMES, 298
GO FOR THE GREEN, 302
INTERACTIVE ARTS & SCIENCES, INC.—(CONFIDENTIAL), 324
INTERACTIVE ARTS INTERNATIONAL (UNANNOUNCED), 325
MANY ROADS TO MURDER, 350
MARITZ COMMUNICATIONS CO.—(NUM. FOR BUS. & IND.), 350
MASTERS OF THE DRAGON, 351
MAZE MANIA: THE AMAZING MAZE GAME, 352
MYSTERYDISC—MURDER, ANYONE?, 362
MYSTERYDISC 2—MANY ROADS TO MURDER, 362
QUARTERHORSE, 385
STUNT DOG, 417
THINK IT THROUGH I, 426
THINK IT THROUGH II, 426
VON HERRING CAPER, THE: "A CHOICE MYSTERY", 441
WEEK AT THE RACES, A, 444

IN-HOUSE TRAINING

ACTION CODE COURSEWARE—INDUSTRIAL TECH. PROGRAM, 219
AN/WLR-1 SYSTEM REPAIRABLE IDENTIFICATION, 226
ARMY COMM. TECH. OFF. (ACTO) VIDEODISC PROTOTYPES, 229
BASIC AUTOMOTIVE ELECTRONICS, 234
BASIC DIESEL PRINCIPLES, 234
BERLIN I, 236
BERLIN II, 237
BONNLAND, 243
BOSTON DISC, 244
CAPSULE, THE, 250
COMPREHENSIVE JOB SKILLS PRACTICE, 263
COURT TESTIMONY SKILLS FOR THE EXPERT WITNESS, 265
DELTA CORRIDOR, 270
DETERMINING FINANCIAL ELIGIBILITY, 272
DIAGNOSIS & TREATMENT OF BIRTH DEFECTS, 273
DIAGNOSIS & TREATMENT OF SPINA BIFIDA, 273
DISCOVERING AMERICA, 275
E-3 FLOATING DESK PULSER, 281
EEMT COMPOSITE NO. 1, 283
ESTABLISHING TECHNICAL ELIGIBILITY, 286
EXTRA MILE, THE, 289
FIRST SIXTY SECONDS, THE, 292
FLIGHT 505, 293
FORD MOTOR VIDEO NETWORK (15 TOTAL DISC SIDES), 295
GENERAL MOTORS OF CANADA, 300
GO FOR THE GREEN, 302
HANSEN'S DISEASE, 311
HOW YOUR HEART AND CIRCULATORY SYSTEM WORKS, 318
IBM—(CONFIDENTIAL), 320
IBM 1750 TELEPHONE SYSTEM, 320
IBM HERITAGE, 321
IBM OFFICE SYSTEMS, 321
IBM PC EXPERIENCE, 321
IMPLEMENTING ADDITIONAL PROGRAMS AND POLICIES, 322
INTEGRATED READINESS CENTER, 323
INTEGRATING ADMINISTRATIVE POLICIES, 324
INTERACTION & INTERVENTION W/GRIEVING CLIENTS & FAMILIES, 324
INTRODUCTION TO THE LEARNING SYSTEM, AN, 327
JOB READINESS MASTERY TEST, 332
KENNEDY SPACE CENTER—PAD 39A, 334
KRAFT PROCESS OVERVIEW, 337
MARITZ COMMUNICATIONS CO.—(NUM. FOR BUS. & IND.), 350
MATERIAL SCIENCE VIDEODISC, 351
MILTON KEYNES INFORMATION TECHNOLOGY VIDEODISC, 356
MOUNTED LAND NAVIGATION, 359
NAVY MINE NEUTRALIZATION SYSTEM TRAINER, 363
OBSERVATION SCENES FOR TRAIN. IN NON-VERBAL COMM., 368
PEDIATRIC HEMATOLOGICAL DISEASES, 375
PROFESSIONAL SELLING SKILLS SYSTEM III, 384
RCA ENGINEERING DEMO DISC, 389
SALEM II, 396
SCE: SONGS UNITS 2 & 3 SURROGATE TRAVEL, 398
SEAFOOD DEPARTMENT OPERATIONS, 398
SIGHT THROUGH SOUND, AN INTERACT. INTRO MED. DIAG., 403
SONY OVERVIEW, 407
SPECTRUM OF INTEGRATED LOGISTICS SUPPORT, 410
STANDARD HEATS OF SOLUTION, 412
SWITCH ON, 419
SYSTEM TEST ENHANCEMENT PROJECT (STEP), 420
TPS-43 INTERACTIVE "SAMPLER", 430

**Videodiscs By Type**

UNIV. OF LONDON PROGRAMME DEVELOPMENT VIDEODISC, 434
USING THE APR, 436
VEHICLE EMISSION CONTROL, 437
VISION-UNIX INTERACTIVE VIDEODISC TRAINING CURRICULUM, 440
VISTA (VIDEODISC INTERPERS. SKILLS TRAIN. & AS.), 441
WILS-WANG INTERACTIVE LEARNING SYSTEM-WP PROGRAM, 447
WORLD ON A SILVER PLATTER, THE, 449
YOU AND AFDC, 451
YOU CAN BUY A COMPUTER!, 451

INSTRUCTIONAL

A.C. SWITCHBOARD OPERATION, 217
ABC-NEA SCHOOLDISC, 218
ABOUT COMPUTERS, 218
ACTION CODE COURSEWARE—INDUSTRIAL TECH. PROGRAM, 219
ACTRONICS, INC.—(NOT GIVEN), 219
ADVANCED MAP INTERPRETATION AND TERRAIN ANALYSIS, 219
ADVANCED VIDEO MAINTENANCE INFORMATION, 220
ADVENTURE WITH DUSTY, 220
AEROBICISE, 220
AEROBICISE: THE BEGINNING WORKOUT, 221
AEROBICISE: THE ULTIMATE WORKOUT, 221
AEROBICS—NEW VIDEO, 221
AMTESS: AUTOMOTIVE, 226
AMTESS: MISSILE, 226
AN/WLR-1 SYSTEM REPAIRABLE IDENTIFICATION, 228
APOLLO, 228
ARMY COMM. TECH. OFF. (ACTO) VIDEODISC PROTOTYPES, 229
ASSESSMENT OF NEUROMOTOR DYSFUNCTION IN INFANTS, 231
ASTRONOMY, 231
BALLET SERIES, 232
BASIC AUTOMOTIVE ELECTRONICS, 234
BASIC DIESEL PRINCIPLES, 234
BASIC KARATE AND SELF DEFENSE, 234
BASIC SKILLS EDUCATION FOR THE U.S. ARMY, 234
BASIC TUMBLING SKILLS, 235
BELLY DANCING, 236
BIG BLUE MARBLE PRESENTS MY SEVENTEENTH SUMMER, 237
BIO SCI VIDEODISC, THE, 239
BIOLOGY I & II, 239
BOOZ-ALLEN & HAMILTON HOME INFO SYSTEM SIMULATOR, 243
BOSTON DISC, 244
BY YOURSELF, 248
CALL FOR FIRE, 249
CAPSULE, THE, 250
CARDIO-PULMONARY-RESUSCITATION, 251
CELL, THE, 253
CHALLENGES OF GLAUCOMA, THE, 254
CHEMISTRY I & II, 256
CHICAGO MERCANTILE EXCHANGE, 256
CLASSROOM BEHAVIOR RECORD, 259
CLEAN AND INSPECT A GENERATOR, 259
CLEFT LIP AND PALATE CASE MANAGEMENT, 260
COAST GUARD, 260
COLLEGE USE DEMONSTRATION DISC, 261
COLORATION, 261
COMP. TENNIS FROM THE PROS, VOL.1: STROKES & TECH., 263
COMPREHENSIVE CASE MANAGEMENT OF SPINA BIFIDA, 263
COMPREHENSIVE JOB SKILLS PRACTICE, 263
CONTINUING MEDICAL EDUCATION LIBRARY, 264
COURT TESTIMONY SKILLS FOR THE EXPERT WITNESS, 265
CRANBERRY WORLD VISITOR'S CENTER VIDEODISC PROJECT, 265
CREATIVE CAMERA, THE, 265
D.C. MOTOR MAINTENANCE, PT. 1, 266
D.C. MOTOR MAINTENANCE, PT. 2, 266
DATA DEMONS, 267
DEAF AWARENESS: LET YOUR FINGERS DO THE TALKING, 269
DECIMALS AND FRACTIONS, 270
DEMONSTRATION VIDEODISC (CAT ZURICH), 270
DEMONSTRATION VIDEODISCS, 1982 AND 1983 (CSC), 271
DENTISTRY I ("SUE"), 271
DENTISTRY II ("JACKIE"), 271
DENTISTRY III, 272
DENTISTRY IV, 272
DETERMINING FINANCIAL ELIGIBILITY, 272
DEVELOPMENT OF LIVING THINGS, 273
DIAGNOSTIC CHALLENGES, 273
DINNER FOR THE BOSS, 274
DISNEYDISC OF MYSTERY AND MAGIC, 276
DIVE TO ANOTHER WORLD, 276
DR. SPOCK—CARING FOR YOUR NEWBORN, 279
E-3 FLOATING DESK PULSER, 281
EDITING SIMULATION DISC #1, 282
EEMT COMPOSITE NO. 1, 282
ELECTRONIC IGNITION SYSTEM, 283
EPCOT CENTER VIDEODISCS (WALT DISNEY WORLD-FL.), 285
ESTABLISHING TECHNICAL ELIGIBILITY, 286
EVIDENCE & OBJECTIONS, 288
EXPLORING LANGUAGE, 288
EXTRA MILE, THE, 289
FAIR EMPLOYMENT PRACTICE, 290
FIRST NATIONAL KIDISC, THE, 292
FIRST SIXTY SECONDS, THE, 292
FLIGHT 505, 293
FORD MOTOR VIDEO NETWORK (15 TOTAL DISC SIDES), 295
FORT SILL VIDEODISC COURSEWARE PACKAGE, 295
FUN & GAMES, 298
GARDENING AT HOME, 300
GENERAL MOTORS OF CANADA, 301
GEOSCIENCE, 301
GOING METRIC, 303
GREAT CITIES: LONDON, ROME, DUBLIN, ATHENS, 306
GREENDAY VIDEO, VOL. 1 NO. 4, 308
GYNECOLOGY PATIENT EDUCATION, 309
HANSEN'S DISEASE, 311
HAWK TROUBLESHOOTING COURSE, 313
HIGH POWER ILLUMINATOR RAM RADAR DEMO, 316
HOW TO COPE WITH . . . , 318
HOW YOUR HEART AND CIRCULATORY SYSTEM WORKS, 318
HUMAN GENETICS TRAINING FOR NURSES: PART I, 319
IBM—(CONFIDENTIAL), 320
IBM 1750 TELEPHONE SYSTEM, 320
IBM HERITAGE, 321
IBM OFFICE SYSTEMS, 321

IBM PC EXPERIENCE, 321
IMPLEMENTING ADDITIONAL PROGRAMS AND POLICIES, 322
INTEGRATED READINESS CENTER, 323
INTEGRATING ADMINISTRATIVE POLICIES, 324
INTELLIGENT DISC, THE, 324
INTERACTIVE JULIA CHILD, 325
INTERNATIONAL HARVESTER, 325
INTERVENTION IN CHILD ABUSE AND NEGLECT, 326
INTRODUCTION TO COMPUTER LITERACY, 326
INTRODUCTION TO ECONOMICS, 326
INTRODUCTION TO THE JOIN SYSTEM, 326
INTRODUCTION TO THE LEARNING SYSTEM, AN, 327
JANE FONDA'S WORKOUT, 329
JANE FONDA'S WORKOUT FOR PREG., BIRTH AND RECOVERY, 329
JIM FIXX ON RUNNING, 331
JOB PERFORMANCE PACKAGE (JPP), 331
JOB READINESS MASTERY TEST, 332
JOY OF FAMILY, 332
JOY OF RELAXATION, THE, 332
JULIA CHILD - THE FRENCH CHEF, VOL. 1, 332
KARATE, 333
KLAVIER IM HAUS, 333
KPR/WINTHROP: UNDERSTANDING CONTRAST MEDIA:, 336
KRAFT PROCESS OVERVIEW, 337
LANGUAGE AND LEARNING, 338
LIFE AND WORKS OF MICHELANGELO, THE, 340
LOCIS INTERACTIVE VIDEODISC, THE, 342
LRN COLLEGE DISC, 344
LVT-7A1 DRIVER TRAINER, 345
MACARIO, 346
MANUAL SEAT/MAN SEP. FR. A MARTIN-BAKER EJECT. ST., 349
MARITZ COMMUNICATIONS CO.—(NUM. FOR BUS. & IND.), 350
MARYLAND INSTRUCTIONAL TELEVISION, 351
MASTER COOKING COURSE, THE, 351
MATERIAL SCIENCE VIDEODISC, 351
MATH IN BIOLOGY, 352
MEET MR. WASHINGTON/MEET MR. LINCOLN, 353
MEJORE SU PRONUNCIACION (IMPROVE YOUR PRONOUNCIA.), 353
MERRILL LYNCH, 354
MILES LEARNING CENTER: APPENDECTOMY, 355
MILES LEARNING CENTER: ILEOCOLIC RESECTION, 355
MILES LEARNING CENTER: MEDICAL APPLICATIONS/ HEM., 355
MILES LEARNING CENTER: ORIENTATION & OPER. INSTR., 356
MILTON KEYNES INFORMATION TECHNOLOGY VIDEODISC, 356
MONTEVIDISCO (MALE VERSION), 358
MORE ADVENTURES WITH DUSTY, 359
NAT. GEO.: GREAT WHALES/SHARKS, 362
NAT. GEO.: THE INCRED. MACH./MYST. OF THE MIND, 363
NAVY MINE NEUTRALIZATION SYSTEM TRAINER, 363
NINETY-SIX: A CATTLE RANCH IN NORTHERN NEVADA, 366
O.I.S.E. SINGLE POINT THREAD CUTTING, 368
OBSERVATION SCENES FOR TRAIN. IN NON-VERBAL COMM., 368
PATRIOT CLS OJT PHASE 1, 374
PEDIATRIC CARDIOVASCULAR DEFECTS, 374
PEDIATRIC HEMATOLOGICAL DISEASES, 375
PERCEPTRONICS, INC.—(NOT GIVEN), 375
PERMS AND HAIRSTYLES, 376
PERSONAL COMPUTER DIAGNOSTICS, 376
PERSONAL COMPUTER EXHIBIT, 376
PHILIPS VP-835 DEMONSTRATION DISC, 377
PHYSICS I & II, 378
POWER PLANT TOUR, 378
PRINCIPLES OF OPERANT CONDITIONING, 382
PRODUCING INTERACTIVE VIDEODISCS, 383
PRODUCTION ASSOCIATES, INC.—(NOT GIVEN), 384
PROFESSIONAL SELLING SKILLS SYSTEM III, 384
PUZZLE OF THE TACOMA NARROWS BRIDGE COLLAPSE, THE, 384
RADAR, MULTIPLE BEAM ACQ. & WIDE APERTURE SURVEIL., 385
RALSTON-PURINA, 386
RCA ENGINEERING DEMO DISC, 388
READING DEVELOPMENT VIDEODISC, 389
RED ON ROUNDBALL, 389
RENAL ANALYSIS, 389
SCE: SONGS UNITS 2 & 3 SURROGATE TRAVEL, 398
SCHOLASTIC PRODUCTIONS: AS WE GROW, 398
SELLING THE PSYCHOLOGICAL, 399
SHUTTLE, 403
SIGHT THROUGH SOUND, AN INTERACT. INTRO MED. DIAG., 403
SKF: DIAGNOSTIC CHALLENGES IN GASTROENTEROLOGY, 404
SKF: THE HIDDEN LANGUAGE OF ARTHRITIS, 405
SMOKING: HOW TO STOP, 406
SOCIAL STYLE SALES STRATEGIES, 406
SONY OVERVIEW, 407
SPACE AGE, THE, 408
SPACE SHUTTLE MISSION REPORTS: S.T.S. 5, 6 & 7, 409
SPACE SHUTTLE: MISSION CONTROL, 409
SPANISH MAP READING CONVERSION PROJECT, 409
SPARROW MISSILE TEST SET TRAINING VIDEODISC, 410
SPECTRUM OF INTEGRATED LOGISTICS SUPPORT, 410
SPINABIFIDA, 410
SPS-49 RADAR, 411
STANDARD HEATS OF SOLUTION, 412
STAR TALK, 413
SURVIVAL ANGLIA'S WORLD OF WILDLIFE, VOL. 1, 418
SURVIVAL ANGLIA'S WORLD OF WILDLIFE, VOL. 2, 418
SYSCON CORPORATION, INC.—(NOT GIVEN), 420
SYSTEM TEST ENHANCEMENT PROJECT (STEP), 420
TEXAS INSTRUMENTS, 424
THINK IT THROUGH I, 426
THINK IT THROUGH II, 426
35MM PHOTOGRAPHY, 426
TJS, 428
TOMORROW TODAY, 429
TPS-43 INTERACTIVE "SAMPLER", 430
TRACING A GROUND, 430
TRANSFORMING OFFICE PRODUCTIVITY (TOP) SEMINAR, 431
TROUBLESHOOTING HORSEPOWER LOSS, 432
TUMBLING (CLOSED CAPTION), 433
U.S. DEPARTMENT OF EDUCATION: WORLD OF WORK, 434
UNDERSEA WORLD OF JACQUES COUSTEAU, VOL. 1, 434
UNIV. OF LONDON PROGRAMME DEVELOPMENT VIDEODISC, 434
URINARY CATHETERIZATION, 435
URODYNAMICS IN CLINICAL PRACTICE, 436
USING THE APR, 436
VEHICLE EMISSION CONTROL, 437
VIDEO ENHANCED GERMAN GATEWAY, 438
VIDEODISC SERIES/APPRENTICE & JOURNEYMAN TRAINING, 439
VIDEODISC, THE: A TOOL FOR COMMUNICATION, 439
VILLA ALEGRE, 440
VINCENT VAN GOGH: A PORTRAIT IN TWO PARTS, 440

VINSON PROJECT, THE, 440
VISION-UNIX INTERACTIVE VIDEODISC TRAINING CURRIC., 440
VISTA (VIDEODISC INTERPERS. SKILLS TRAIN. & AS.), 441
VOYAGER, 442
WALK THROUGH THE UNIVERSE, A, 442
WAR PRAYER, THE, 443
WEYERHAEUSER, 445
WHALES, 445
WILS-WANG INTERACTIVE LEARNING SYSTEM-WP PROGRAM, 447
WORDS IN MOTION I, 449
WORDS IN MOTION II, 449
WORLD OF MARTIAL ARTS, THE, 449
WORLD ON A SILVER PLATTER, THE, 449
YOU AND AFDC, 451
YOU CAN BUY A COMPUTER!, 451

## MEDICAL

ACTRONICS, INC.—(NOT GIVEN), 219
ASSESSMENT OF NEUROMOTOR DYSFUNCTION IN INFANTS, 231
CHALLENGES OF GLAUCOMA, THE, 253
CLEFT LIP AND PALATE CASE MANAGEMENT, 260
COMPREHENSIVE CASE MANAGEMENT OF SPINA BIFIDA, 263
CONTINUING MEDICAL EDUCATION LIBRARY, 264
COURT TESTIMONY SKILLS FOR THE EXPERT WITNESS, 265
DEMONSTRATION VIDEODISC (CAT ZURICH), 270
DENTISTRY I ("SUE"), 271
DENTISTRY II ("JACKIE"), 271
DENTISTRY III, 272
DENTISTRY IV, 272
DIAGNOSIS & TREATMENT OF BIRTH DEFECTS, 273
DIAGNOSIS & TREATMENT OF SPINA BIFIDA, 273
DIAGNOSTIC CHALLENGES, 273
GYNECOLOGY PATIENT EDUCATION, 309
HANSEN'S DISEASE, 311
HUMAN GENETICS TRAINING FOR NURSES: PART I, 319
INTELLIGENT DISC, THE, 324
INTERACTION & INTERVENTION W/GRIEVING CLIENTS & FAMILIES, 324
INTERVENTION IN CHILD ABUSE AND NEGLECT, 326
KPR/WINTHROP: UNDERSTANDING CONTRAST MEDIA:, 336
M.J. ZINK PRODUCTIONS, INC. - (NOT GIVEN), 345
MANUAL SEAT/MAN SEP. FR. A MARTIN-BAKER EJECT. ST., 349
MEDICAL MICROSCOPY, 353
MILES LEARNING CENTER: APPENDECTOMY, 355
MILES LEARNING CENTER: ILEOCOLIC RESECTION, 355
MILES LEARNING CENTER: MEDICAL APPLICATIONS/HEM., 355
MILES LEARNING CENTER: ORIENTATION & OPER. INSTR., 356
PEDIATRIC CARDIOVASCULAR DEFECTS, 375
PEDIATRIC HEMATOLOGICAL DISEASES, 375
RENAL ANALYSIS, 390
SIGHT THROUGH SOUND, AN INTERACT. INTRO MED. DIAG., 403
SKF: DIAGNOSTIC CHALLENGES IN GASTROENTEROLOGY, 404
SKF: THE HIDDEN LANGUAGE OF ARTHRITIS, 405
SPINABIFIDA, 410

UNIV. OF LONDON PROGRAMME DEVELOPMENT VIDEODISC, 434
UROLOGY, 436
WAY WE LIVE, THE, 443

## MOTION PICTURES

ABSENCE OF MALICE, 218
ABSENT-MINDED PROFESSOR, THE, 218
ADVENTURES OF ROBIN HOOD, THE, 220
ADVENTURES OF TOM SAWYER, THE, 220
AFRICA SCREAMS, 221
AFRICAN QUEEN, THE, 221
AGENCY, 221
AIR FORCE, 222
AIRPLANE, 222
AIRPLANE II, 222
AIRPORT, 222
ALAMO, THE, 222
ALICE DOESN'T LIVE HERE ANYMORE, 222
ALICE IN WONDERLAND, 222
ALIEN, 223
ALL THAT JAZZ, 223
ALL THE MARBLES, 223
ALL THE PRESIDENT'S MEN, 223
ALONE IN THE DARK, 224
ALTERED STATES, 224
AMARCORD, 224
AMATEUR, THE, 224
AMAZING SPIDERMAN, THE, 224
AMERICAN ALCOHOLIC/READING, WRITING AND REEFER, 224
AMERICAN GIGOLO, 224
AMERICAN GRAFFITI, 224
AMERICAN HOT WAX, 225
AMERICAN IN PARIS, AN, 225
AMERICAN WEREWOLF IN LONDON, AN, 225
AMITYVILLE HORROR, THE, 225
AMITYVILLE 2: THE POSSESSION, 225
AND GOD CREATED WOMAN, 226
AND JUSTICE FOR ALL, 226
ANDERSON TAPES, THE, 226
ANGEL OF H.E.A.T., 227
ANGELS WITH DIRTY FACES, 227
ANIMAL CRACKERS, 227
ANIMAL HOUSE, 227
ANNIE, 227
ANNIE HALL, 227
ANY WHICH WAY YOU CAN, 228
APACHE, 228
APARTMENT, THE, 228
APOCALPYSE NOW, 228
APPLE DUMPLING GANG, THE, 228
ARSENIC AND OLD LACE, 229
ARTHUR, 230
ATLANTIC CITY, 231
AUTHOR, AUTHOR, 232
AUTOBIOGRAPHY OF MISS JANE PITTMAN, THE, 232
AUTUMN SONATA, 232
BACK ROADS, 232
BAD NEWS BEARS, THE, 232
BAD NEWS BEARS IN BREAKING TRAINING, THE, 232
BAFFLED, 232
BANANAS, 233
BANG THE DRUM SLOWLY, 233

BARBARELLA, 233
BARBAROSA, 233
BAREFOOT CONTESSA, 233
BAREFOOT IN THE PARK, 234
BASEBALL FUN AND GAMES, 234
BASEBALL'S HALL OF FAME, 234
BATTLESTAR GALACTICA, 235
BEACH GIRLS, THE, 235
BEARS AND I, THE, 235
BEAST WITHIN, THE, 235
BEASTMASTER, THE, 235
BEDKNOBS & BROOMSTICKS, 236
BEING THERE, 236
BELL, BOOK & CANDLE, 236
BELLS ARE RINGING, 236
BEN HUR, 236
BENJI, 236
BEST FRIENDS, 236
BEST LITTLE WHOREHOUSE IN TEXAS, THE, 237
BETWEEN THE LINES, 237
BIG BAD MAMA, 237
BIG FIGHTS, VOL.1—MUHAMMAD ALI'S GREATEST FIGHTS, 237
BIG FIGHTS, VOL.2—HEAVYWGHT CHAMPIONS' GREATEST, 237
BIG FIGHTS, VOL.3—SUGAR RAY ROBINSON'S GREATEST, 238
BIG RED ONE, THE, 238
BIG SLEEP, THE, 238
BILLION DOLLAR HOBO, THE, 238
BILLY JACK, 238
BIRDMAN OF ALCATRAZ, 239
BIRDS, THE, 239
BLACK HOLE, THE, 239
BLACK STALLION RETURNS, THE, 240
BLACK STALLION, THE, 240
BLACK SUNDAY, 240
BLADE RUNNER, 240
BLAZING SADDLES, 240
BLOW OUT, 241
BLUE HAWAII, 241
BLUE LAGOON, THE, 241
BLUE MAX, THE, 241
BLUE THUNDER, 241
BLUES BROTHERS, THE, 242
BOAT, THE, 242
BODY AND SOUL, 242
BODY HEAT, 242
BOLERO, 242
BON VOYAGE CHARLIE BROWN (AND DON'T COME BACK), 243
BONNIE & CLYDE, 243
BOOGEYMAN, THE, 243
BORDER, THE, 243
BORN FREE, 244
BOSTON STRANGLER, THE, 244
BOXCAR BERTHA, 244
BOYS FROM BRAZIL, THE, 244
BOYS IN THE BAND, THE, 245
BOYS OF SUMMER, THE, 245
BRANNIGAN, 245
BREAKFAST AT TIFFANY'S, 245
BREAKHEART PASS, 245
BREAKING AWAY, 245
BREAKOUT, 245
BREATHLESS, 245
BRIAN'S SONG, 245
BRIDGE ON THE RIVER KWAI, THE, 246

BRIDGE TOO FAR, A, 246
BRIGADOON, 246
BRIMSTONE AND TREACLE, 246
BRONCO BILLY, 246
BRUBAKER, 246
BUCK ROGERS IN THE 25TH CENTURY, 246
BUDDY BUDDY, 247
BUGS BUNNY/ROAD RUNNER MOVIE, THE, 247
BUGSY MALONE, 247
BULLITT, 247
BULLWINKLE AND ROCKY AND FRIENDS, VOL. 1, 247
BULLWINKLE AND ROCKY AND FRIENDS, VOL. 2, 247
BUS STOP, 247
BUSTIN' LOOSE, 247
BUTCH CASSIDY AND THE SUNDANCE KID, 248
BUTTERFLY, 248
CABARET, 248
CADDYSHACK, 248
CAGE AUX FOLLES, LA, 248
CAGE AUX FOLLES 2, LA, 249
CALIFORNIA SUITE, 249
CANDID CANDID CAMERA, 249
CANDIDATE, THE, 249
CANDLESHOE, 249
CANNERY ROW, 250
CANNONBALL RUN, 250
CAPRICORN ONE, 250
CAPTAIN BLOOD, 250
CARBON COPY, 250
CARNAL KNOWLEDGE, 251
CARNY, 251
CARRIE, 251
CARTOON CLASSICS VOLUME #1—CHIP-N-DALE WITH DONALD DUCK, 251
CARTOON CLASSICS VOLUME #4—SPORT GOOFY, 252
CARTOON CLASSICS VOLUME #5—BEST OF 1931/48, 252
CASABLANCA, 252
CAT BALLOU, 252
CAT ON A HOT TIN ROOF, 252
CAT PEOPLE, 252
CATCH-22, 252
CAVEMAN, 253
CBS NEWS JOHN F. KENNEDY, 253
CHALLENGE, THE, 253
CHAMP, THE, 254
CHANGE OF SEASONS, A, 254
CHANGELING, THE, 254
CHAPTER TWO, 254
CHARGE OF THE LIGHT BRIGADE, THE, 254
CHARIOTS OF FIRE, 255
CHARLIE BROWN FESTIVAL, A, 255
CHARLIE BROWN FESTIVAL, VOL. 2, A, 255
CHARLIE BROWN FESTIVAL, VOL. 3, A, 255
CHARLIE BROWN FESTIVAL, VOL. 4, A, 255
CHARLOTTE'S WEB, 255
CHARLY, 256
CHATTERBOX, 256
CHEECH AND CHONG'S NEXT MOVIE, 256
CHEECH AND CHONG'S NICE DREAMS, 256
CHINA SYNDROME, THE, 257
CHINATOWN, 257
CHIP 'N DALE WITH DONALD DUCK, 257
CHITTY CHITTY BANG BANG, 257
CHU CHU AND THE PHILLY FLASH, 257
CID, EL, 258
CINCINNATI KID, THE, 258
CITIZEN KANE, 258
CITY LIGHTS, 258

CLARENCE DARROW, 258
CLASH OF THE TITANS, 259
CLASS, 259
CLASS OF 1984, 259
CLOCKWORK ORANGE, A, 260
CLOSE ENCOUNTERS OF THE THIRD KIND, 260
COAL MINER'S DAUGHTER, 260
COAST TO COAST, 261
COLLECTOR, THE, 261
COLLEGE FOOTBALL CLASSICS, VOL. 1, 261
COLUMBIA PICTURES CARTOONS, VOL. 2, 262
COMA, 262
COMANCHER, THE, 262
COME BACK TO THE FIVE & DIME, JIMMY DEAN, JIMMY DEAN, 262
COMEDY TONIGHT, 262
COMES A HORSEMAN, 263
COMING HOME, 263
CONAN THE BARBARIAN, 264
CONVERSATION, THE, 264
COUNT OF MONTE-CRISTO, THE, 264
COUNTRY GIRL, THE, 264
COURT JESTER, THE, 265
COUSIN, COUSINE, 265
CREEPSHOW, 265
CROSBY, STILLS AND NASH, 266
CRUISING, 266
CUTTER'S WAY, 266
DAMIEN-OMEN II, 267
DARBY O'GILL AND THE LITTLE PEOPLE, 267
DARK VICTORY, 267
DAVY CROCKETT & THE RIVER PIRATES, 268
DAY AT THE RACES, A, 268
DAY OF THE DOLPHIN, THE, 268
DAY OF THE LOCUST, 268
DAYS OF HEAVEN, 268
DEAD MEN DON'T WEAR PLAID, 268
DEADLY BLESSING, 269
DEATH HUNT, 269
DEATH RACE 2000, 269
DEATH WISH, 269
DEATH WISH 2, 269
DEEP, THE, 270
DEER HUNTER, THE, 270
DEFIANCE, 270
DELIVERANCE, 270
DETECTIVE, THE, 272
DIAL "M" FOR MURDER, 273
DIAMONDS ARE FOREVER, 274
DIARY OF ANN FRANK, 274
DICK CAVETT'S HOCUS POCUS, IT'S MAGIC, 274
DINER, 274
DIRTY DOZEN, THE, 275
DIRTY HARRY, 275
DISNEY CARTOON PARADE, VOL. 1, 276
DISNEY CARTOON PARADE, VOL. 2, 276
DISNEY CARTOON PARADE, VOL. 3, 276
DISNEY CARTOON PARADE, VOL. 4, 276
DISNEY CARTOON PARADE, VOL. 5, 276
DIVA, 276
DIVINE MADNESS, 277
DOCTOR ZHIVAGO, 277
DODGE CITY, 277
DOG DAY AFTERNOON, 277
DOGS OF WAR, THE, 277
DOLL'S HOUSE, A, 277
DOMINO PRINCIPLE, THE, 278
DON'T LOOK NOW, 278

DORADO, EL, 278
DOT AND THE KANGAROO, 278
DOWNHILL RACER, 278
DR. DETROIT, 278
DR. DOOLITTLE, 279
DR. NO, 279
DR. SEUSS VIDEO FESTIVAL, 279
DR. STRANGELOVE, 279
DRACULA, 279
DRAGONSLAYER, 279
DRESSED TO KILL, 280
DUCHESS AND THE DIRTWATER FOX, 280
DUCK SOUP, 280
DUMBO, 280
DUNDER KLUMPEN, 280
EARTHLING, THE, 281
EAST OF EDEN, 281
EASY MONEY, 281
EASY RIDER, 281
EDDIE MACON'S RUN, 282
ELECTRIC HORSEMAN, THE, 282
ELEPHANT MAN, THE, 283
ELMER GANTRY, 283
EMILY, 284
EMMANUELLE IN BANGKOK, 284
EMMANUELLE, THE JOYS OF A WOMAN, 284
END, THE, 284
ENDANGERED SPECIES, 284
ENDLESS LOVE, 284
ENFORCER, THE, 285
ENTER THE DRAGON, 285
ENTER THE NINJA, 285
ESCAPE FROM ALCATRAZ, 286
ESCAPE FROM NEW YORK, 286
ESCAPE TO ATHENA, 286
ESCAPE TO WITCH MOUNTAIN, 286
EUBIE!, 287
EVERY WHICH WAY BUT LOOSE, 287
EVERYTHING YOU ALWAYS WANTED TO KNOW ABOUT SEX . . . , 287
EVILSPEAK, 288
EXCALIBUR, 288
EXODUS, 288
EXORCIST, THE, 288
EXTERMINATOR, THE, 289
EYE FOR AN EYE, AN, 289
EYE OF THE NEEDLE, 289
EYES OF LAURA MARS, THE, 289
EYEWITNESS, 289
F.I.S.T., 289
FAME, 290
FAMILY ENTERTAINMENT PLAYHOUSE, VOL. 2, 290
FAN, THE, 290
FANTASTIC VOYAGE, 290
FAREWELL, MY LOVELY, 290
FAST TIMES AT RIDGEMONT HIGH, 291
FIDDLER ON THE ROOF, 291
FINAL CONFLICT, THE, 291
FINAL COUNTDOWN, THE, 291
FINAL EXAM, 291
FIREFOX, 291
FIRST BARRY MANILOW SPECIAL, THE, 292
FIRST BLOOD, 292
FIRST MONDAY IN OCTOBER, 292
FISTFUL OF DOLLARS, A, 292
FLAMING STAR, 292
FLASH GORDON, 293
FLASHDANCE, 293

FLYING DEUCES, 294
FOG, THE, 294
FOOTBALL FOLLIES AND SENSATIONAL SIXTIES, 294
FOOTLIGHT PARADE, 294
FOR A FEW DOLLARS MORE, 294
FOR YOUR EYES ONLY, 294
FORBIDDEN PLANET, 294
FORCE 10 FROM NAVARONNE, 295
FORCED VENGEANCE, 295
FOREPLAY, 295
FORMULA, THE, 295
FORT APACHE, THE BRONX, 295
42ND STREET, 296
48 HOURS, 296
FOUL PLAY, 296
FOUR FRIENDS, 296
FOUR MUSKETEERS, THE, 296
FOUR SEASONS, THE, 296
FRANKENSTEIN, 296
FRENCH CONNECTION, THE, 297
FRENCH LIEUTENANT'S WOMAN, THE, 297
FRIDAY THE 13TH, 297
FRIDAY THE 13TH, PART II, 297
FRIDAY THE 13TH, PART III, 297
FROG PRINCE, THE, 297
FROM RUSSIA WITH LOVE, 298
FUGITIVE: THE FINAL EPISODE, THE, 298
FUN IN ACAPULCO, 298
FUNNY GIRL, 298
FUNNY THING HAPPENED ON THE WAY TO THE FORUM, A, 299
FUTUREWORLD, 299
G.I. BLUES, 299
GALAXINA, 299
GALLIPOLI, 299
GAMBLER, THE, 299
GAME OF DEATH, THE, 299
GANDHI, 300
GAS PUMP GIRLS, 300
GATOR, 300
GAUNTLET, THE, 300
GENTLEMAN JIM, 301
GENTLEMEN PREFER BLONDS, 301
GETTING OF WISDOM, THE, 301
GHOST STORY, 302
GIGI, 302
GILDA, 302
GLORIA, 302
GO TELL THE SPARTANS, 303
GODFATHER, PART II, THE, 303
GODFATHER, THE, 303
GODZILLA, 303
GOLD BUG, THE/RODEO RED AND THE RUNAWAY, 304
GOLD DIGGERS OF 1933, 304
GOLDEN AGE OF COLLEGE FOOTBALL, THE, 304
GOLDEN DECADE OF COLLEGE FOOTBALL, 1970-1979, A, 304
GOLDFINGER, 304
GOLDILOCKS AND THE THREE BEARS, 304
GOOD GUYS WEAR BLACK, 304
GOOD, THE BAD AND THE UGLY, THE, 304
GOODBYE GIRL, THE, 305
GOODBYE, COLUMBUS, 305
GOSPEL, 305
GRADUATE, THE, 305
GRAPES OF WRATH, THE, 305
GREASE, 306
GREASE 2, 306

GREAT CARUSO, THE, 306
GREAT DICTATOR, THE, 306
GREAT ESCAPE, THE, 306
GREAT GATSBY, THE, 307
GREAT LOCOMOTIVE CHASE, THE, 307
GREAT MUPPET CAPER, THE, 307
GREAT SANTINI, THE, 307
GREAT SCOUT & CATHOUSE THURSDAY, 307
GREAT SPACE COASTER SUPERSHOW, 307
GREAT TRAIN ROBBERY, THE, 307
GREATEST ADVENTURE, THE, 308
GREATEST FIGHTS OF THE '70'S, THE, 308
GREATEST SHOW ON EARTH, THE, 308
GREEN BERETS, THE, 308
GULLIVER'S TRAVELS, 308
GUMBY ADVENTURE, A, 309
GUNFIGHT AT THE O.K. CORRAL, 309
GUNS OF NAVARONE, THE, 309
GUS, 309
GUYS AND DOLLS, 309
HAIR, 309
HALLOWEEN II, 310
HALLOWEEN III, 310
HAMLET, 310
HANG 'EM HIGH, 310
HANKY PANKY, 311
HANSEL & GRETEL, 311
HAPPY BIRTHDAY TO ME, 311
HAPPY HOOKER GOES HOLLYWOOD, THE, 312
HAPPY HOOKER GOES TO WASHINGTON, THE, 312
HAPPY HOOKER, THE, 312
HARD COUNTRY, THE, 312
HARDER THEY COME, THE, 312
HAROLD AND MAUDE, 312
HARPER VALLEY P.T.A., 312
HARRY CHAPIN IN CONCERT, 312
HATARI!, 313
HAWAII, 313
HE KNOWS YOU'RE ALONE, 313
HE-MAN AND THE MASTERS OF THE UNIVERSE, VOL. 1, 314
HE-MAN AND THE MASTERS OF THE UNIVERSE, VOL. 2, 314
HEARTACHES, 314
HEARTBREAK KID, THE, 314
HEAVEN CAN WAIT, 314
HEIDI, 314
HELL IN THE PACIFIC, 314
HELLO DOLLY, 314
HENRY V, 315
HERBIE RIDES AGAIN, 315
HERCULES, 315
HERE IT IS, BURLESQUE, 315
HERITAGE OF THE BIBLE: THE LAW AND THE PROPHETS, 315
HEROES, 315
HEY, CINDERELLA, 315
HIGH ANXIETY, 315
HIGH COUNTRY, 315
HIGH NOON, 316
HIGH ROAD TO CHINA, 316
HIGH SIERRA, 316
HISTORY OF THE WORLD, PART 1, 316
HOBBIT, THE, 316
HOCUS POCUS IT'S MAGIC, 316
HOLOCAUST, 317
HOMBRE, 317
HOMEWORK, 317
HONEYSUCKLE ROSE, 317
HOOPER, 317

HOPSCOTCH, 317
HORSE SOLDIERS, THE, 318
HOUSE CALLS, 318
HOUSE ON SORORITY ROW, 318
HOW TO BEAT THE HIGH COST OF LIVING, 318
HOW TO MARRY A MILLIONAIRE, 318
HOWLING, THE, 319
HUCKLEBERRY FINN, 319
HUD, 319
HUNCHBACK OF NOTRE DAME, THE, 319
HUNTER, THE, 319
HUSTLER, THE, 319
I AM A FUGITIVE FROM A CHAIN GANG, 320
I LOVE YOU, 320
I OUGHT TO BE IN PICTURES, 320
I SPIT ON YOUR GRAVE, 320
I'M DANCING AS FAST AS I CAN, 320
I, THE JURY, 320
IF YOU COULD SEE WHAT I HEAR, 322
IMPROPER CHANNELS, 322
IN PRAISE OF OLDER WOMEN, 322
IN THE HEAT OF THE NIGHT, 322
IN-LAWS, THE, 322
INCUBUS, THE, 323
INHERIT THE WIND, 323
INN OF THE SIXTH HAPPINESS, 323
INVASION OF THE BODY SNATCHERS—ORIGINAL 327
INVITATION TO THE DANCE, 327
IRMA LA DOUCE, 327
ISLAND OF DR. MOREAU, 328
IT CAME FROM HOLLYWOOD, 328
IT'S A MAD, MAD, MAD, MAD WORLD, 328
IT'S MY TURN, 328
JACK AND THE BEANSTALK, 329
JAILHOUSE ROCK, 329
JASON & THE ARGONAUTS, 329
JAWS, 329
JAWS 2, 330
JAZZ SINGER, THE, 330
JEREMIAH JOHNSON, 330
JERK, THE, 330
JESUS CHRIST, SUPERSTAR, 331
JESUS OF NAZARETH, 331
JEZEBEL, 331
JINXED, 331
JOAN OF ARC, 331
JOE, 332
JULIA, 333
JUNIOR BONNER, 333
KELLY'S HEROES, 333
KENTUCKIAN, THE, 333
KEY LARGO, 334
KIDNAPPED, 334
KIDS ARE ALRIGHT, THE, 334
KIDS FROM FAME, THE—LIVE, 334
KILLER FORCE, 334
KING AND I, THE, 335
KING CREOLE, 335
KING KONG, 335
KING KONG—ORIGINAL VERSION, 335
KING OF COMEDY, THE, 335
KING OF HEARTS, THE, 335
KISS ME GOODBYE, 335
KLAVIER IM HAUS, 336
KLUTE, 336
KNOCK ON ANY DOOR, 336
KOTCH, 336

KRAMER VS. KRAMER, 337
LADY CHATTERLEY'S LOVER, 337
LADY SINGS THE BLUES, 337
LAST AMERICAN VIRGIN, THE, 338
LAST CHASE, THE, 338
LAST OF THE RED HOT LOVERS, THE, 339
LAST TANGO IN PARIS, 339
LAST UNICORN, THE, 339
LAST WALTZ, THE, 339
LAURA, 339
LAWRENCE OF ARABIA, 339
LCA PRESENTS FAMILY ENTERTAINMENT PLAYHOUSE, 339
LEGEND OF THE LONE RANGER, THE, 339
LENNY, 340
LET IT BE, 340
LET'S BOWL, 340
LET'S SPEND THE NIGHT TOGETHER, 340
LIAR'S MOON, 340
LION IN WINTER, THE, 341
LIPSTICK, 341
LITTLE CAESAR, 341
LITTLE DARLINGS, 341
LITTLE HOUSE ON THE PRAIRIE, 341
LITTLE PRINCE, THE, 341
LIVE AND LET DIE, 342
LIVE INFIDELITY: REO SPEEDWAGON IN CONCERT, 342
LOGAN'S RUN, 342
LOLITA, 342
LONE WOLF MCQUADE, 343
LONG RIDERS, THE, 343
LONGEST DAY, THE, 343
LONGEST YARD, THE, 343
LOOKING FOR MR. GOODBAR, 343
LOONEY, LOONEY, LOONEY BUGS BUNNY MOVIE, 343
LORD OF THE RINGS, THE, 343
LOSIN' IT, 344
LOVE AND DEATH, 344
LOVE AT FIRST BITE, 344
LOVE BUG, THE, 344
LOVE ME TENDER, 344
LOVE STORY, 344
LOVING COUPLES, 344
M*A*S*H, 345
M*A*S*H, FAREWELL, GOODBYE, AMEN, 345
MACARIO, 346
MAD MAX, 346
MAGIC, 346
MAGIC PONY, THE, 346
MAGNIFICENT SEVEN, THE, 346
MAGNUM FORCE, 347
MAHOGANY, 347
MAIN EVENT, THE, 347
MAKING LOVE, 347
MAKING OF RAIDERS OF THE LOST ARK, THE, 347
MAKING OF STAR WARS & THE EMPIRE STRIKES BACK, THE, 347
MALTESE FALCON, THE, 347
MAN FOR ALL SEASONS, A, 348
MAN OF LA MANCHA, 348
MAN ON THE MOON, 348
MAN WHO CAME TO DINNER, 348
MAN WHO SHOT LIBERTY VALANCE, THE, 348
MAN WITH THE GOLDEN GUN, THE, 348
MAN WITH TWO BRAINS, THE, 348
MANDINGO, 349
MANY ADVENTURES OF WINNIE THE POOH, THE, 349
MARATHON MAN, 350
MARTY, 350

MARY POPPINS, 350
MATILDA, 352
MAUSOLEUM, 352
MEATBALLS, 352
MEET ME IN ST. LOUIS, 353
MEGAFORCE, 353
MELVIN AND HOWARD, 353
MEMORIES OF SUPER BOWL, 353
MICKEY MOUSE AND DONALD DUCK CARTOONS, 354
MICKEY MOUSE AND DONALD DUCK COLLECTION #1, 354
MICKEY MOUSE AND DONALD DUCK COLLECTION #2, 354
MICKEY MOUSE AND DONALD DUCK COLLECTION #3, 354
MIDNIGHT COWBOY, 355
MIDNIGHT EXPRESS, 355
MILDRED PIERCE, 355
MIRACLE OF LAKE PLACID—WINTER OLYMPICS 1980, THE, 356
MISFITS, THE, 356
MISS PEACH & KELLY SCHOOL GIRLS, 356
MISSING, 357
MISSION GALACTICA: THE CYLON ATTACK, 357
MISSIONARY, THE, 357
MISSOURI BREAKS, 357
MODERN PROBLEMS, 357
MODERN TIMES, 357
MOMMIE DEAREST, 357
MONSIGNOR, 357
MONTY PYTHON - LIVE AT THE HOLLYWOOD BOWL, 358
MONTY PYTHON AND THE HOLY GRAIL, 358
MONTY PYTHON'S LIFE OF BRIAN, 358
MOON IS BLUE, THE, 358
MOONLIGHTING, 358
MOONRAKER, 359
MOSES, 359
MOVIE MOVIE, 359
MR. MAGOO IN SHERWOOD FOREST, 359
MR. MAGOO, VOL. 1, 359
MR. MAGOO'S CHRISTMAS, 360
MR. ROBERTS, 360
MR. ROGERS GOES TO SCHOOL, 360
MUPPET MOVIE, THE, 360
MUPPET MUSICIANS OF BREMEN, 360
MURDER BY DEATH, 360
MURDER BY DECREE, 360
MURDER ON THE ORIENT EXPRESS, 361
MY BLOODY VALENTINE, 361
MY BODYGUARD, 361
MY FAIR LADY, 361
MY FAVORITE YEAR, 361
MY LITTLE CHICKADEE, 361
MY TUTOR, 362
NASHVILLE, 362
NATIONAL LAMPOON'S CLASS REUNION, 363
NEIGHBORS, 363
NETWORK, 363
NEVER GIVE A SUCKER AN EVEN BREAK, 364
NEW YORK YANKEES MIRACLE YEAR—1978, THE, 364
NEW YORK, NEW YORK, 364
NFL '81 OFFICIAL SEASON YEARBOOK, 364
NFL SYMFUNNY AND LEGENDS OF THE FALL, THE, 364
NICE DREAMS, 365
NIGHT GAMES, 365
NIGHT PORTER, THE, 365
NIGHT SHIFT, 365
NIGHTHAWKS, 365
NINE TO FIVE, 365
1941, 366
NO NUKES, 366
NORMA RAE, 366
NORSEMAN, THE, 366
NORTH BY NORTHWEST, 366
NORTH DALLAS FORTY, 366
NOTORIOUS, 367
NOW AND FOREVER, 367
NOW VOYAGER, 367
NUTTY PROFESSOR, THE, 367
OCTOPUSSY, 368
ODD COUPLE, THE, 368
OFFICER AND A GENTLEMAN, AN, 368
OH, GOD!, 369
OKLAHOMA, 369
OLD YELLER, 369
OMEN, THE, 369
ON GOLDEN POND, 370
ON HER MAJESTY'S SECRET SERVICE, 370
ON THE BEACH, 370
ON THE TOWN, 370
ONE & ONLY, GENUINE ORIGINAL FAMILY BAND, THE, 370
ONE FLEW OVER THE CUCKOO'S NEST, 370
ONE FROM THE HEART, 370
ONLY WHEN I LAUGH, 371
ORCA, THE KILLER WHALE, 371
ORDINARY PEOPLE, 371
OSCAR, THE, 371
OUR TOWN, 372
OUTLAND, 372
OWL AND THE PUSSYCAT, THE, 372
PAINT YOUR WAGON, 372
PAPER CHASE, THE, 372
PAPER MOON, 372
PAPILLON, 373
PARADISE, 373
PARASITE, 373
PASSION OF LOVE, 373
PATERNITY, 373
PATHS OF GLORY, 374
PATTON, 374
PAUL McCARTNEY AND WINGS—ROCKSHOW, 374
PENNIES FROM HEAVEN, 376
PETE'S DRAGON, 377
PETRIFIED FOREST, THE, 377
PHANTASM, 377
PHANTOM TOLLBOOTH, 378
PINK FLOYD—THE WALL, 378
PINK PANTHER, THE, 378
PINK PANTHER STRIKES AGAIN, THE, 379
PIPPIN, 379
PIRATE MOVIE, THE, 379
PIRATE, THE, 379
PIRATES OF PENZANCE, THE, 379
PLACE IN THE SUN, A, 379
PLANET OF THE APES, 380
PLAY IT AGAIN, SAM, 380
PLAY MISTY FOR ME, 380
PLAYBOY MAGAZINE, VOLUME 1, 380
PLAYBOY MAGAZINE, VOLUME 2, 380
PLAYBOY'S PLAYMATE REVIEW, 381
POLLYANNA, 381
POLTERGEIST, 381
POPEYE, 381
PORKY'S, 382
POSEIDON ADVENTURE, 382
POSTMAN ALWAYS RINGS TWICE, THE, 382
PRETTY BABY, 382
PRIDE OF THE YANKEES, 383
PRINCE AND THE PAUPER, 383

### Videodiscs By Type

PRINCE OF THE CITY, 383
PRIVATE BENJAMIN, 383
PRIVATE EYES, THE, 383
PRIVATE LESSONS, 383
PRIVATE POPSICLE, 384
PRODUCERS, THE, 384
PROM NIGHT, 385
PSYCHO, 385
PSYCHO 2, 385
PUBLIC ENEMY, 385
PUMPING IRON, 385
QUADROPHENIA, 385
QUEST FOR FIRE, 386
QUICK DOG TRAINING, 386
QUIET MAN, THE, 386
RACE FOR YOUR LIFE, CHARLIE BROWN, 386
RAGGEDY ANN & ANDY, 387
RAGGEDY MAN, 387
RAGING BULL, 387
RAGTIME, 387
RAIDERS OF THE LOST ARK, 387
RAISE THE TITANIC, 388
RAPUNZEL, 388
RAVEN, THE, 388
REBEL WITHOUT A CAUSE, 389
RED RIVER, 389
RED SHOES, THE, 390
REDS, 390
RETURN OF A MAN CALLED HORSE, 391
RETURN OF THE KING, 391
RETURN OF THE PINK PANTHER, THE, 391
RETURN OF THE STREET FIGHTER, 391
RETURN TO MACON COUNTY, 391
REVENGE OF THE PINK PANTHER, 391
RICH AND FAMOUS, 391
RICHARD PRYOR LIVE IN CONCERT, 392
RING OF BRIGHT WATER, 392
RIO BRAVO, 392
RIO LOBO, 392
ROAD GAMES, 392
ROAD WARRIOR, THE, 392
ROARING TWENTIES, THE, 392
ROBE, THE, 393
ROCKY, 393
ROCKY 2, 393
ROCKY 3, 393
RODAN, 394
ROLLERBALL, 394
ROLLING THUNDER, 394
ROMAN HOLIDAY, 394
ROMEO AND JULIET, 394
ROSE, THE, 394
ROSEMARY'S BABY, 394
ROUGH CUT, 395
ROYAL WEDDING, 395
RUN SILENT, RUN DEEP, 395
RUSSIANS ARE COMING, THE RUSSIANS ARE COMING, THE, 395
RUST NEVER SLEEPS, 396
S.O.B., 396
SAILOR WHO FELL FROM GRACE WITH THE SEA, THE, 396
SAMSON & DELILAH, 396
SANDS OF IWO JIMA, 397
SATURDAY NIGHT FEVER, 397
SATURN 3, 397
SAVANNAH SMILES, 397
SAVE THE TIGER, 397
SCANNERS, 397
SEA WOLVES, 398
SEARCHERS, THE, 398
SECRET OF N-I-M-H, THE, 398
SECRET POLICEMAN'S OTHER BALL, 399
SEDUCTION OF JOE TYNAN, THE, 399
SEDUCTION, THE, 399
SEEMS LIKE OLD TIMES, 399
SEMI-TOUGH, 399
SENDER, THE, 399
SENIORS, 400
SEPARATE WAYS, 400
SERGEANT YORK, 400
SERIAL, 400
SERPICO, 400
SEVEN BRIDES FOR SEVEN BROTHERS, 400
SEVEN YEAR ITCH, THE, 400
SEX ON THE RUN, 400
SGT. PEPPER'S LONELY HEARTS CLUB BAND, 401
SHAFT, 401
SHAGGY DOG, THE, 401
SHANE, 401
SHARKEY'S MACHINE, 401
SHINING, THE, 402
SHOGUN, 402
SHOGUN ASSASSIN, 402
SHOOT THE MOON, 402
SHOOTIST, THE, 402
SHOT IN THE DARK, A, 402
SHOWDOWN: SUGAR RAY LEONARD VS. THOMAS HEARNS, 402
SILENT PARTNER, THE, 403
STREAK, 404
SINBAD AND THE EYE OF THE TIGER, 404
SINGIN' IN THE RAIN, 404
SIX PACK, 404
SIX WEEKS, 404
SLAP SHOT, 405
SLAVE OF THE CANNIBAL GODS, 405
SLEEPER, 405
SLEEPING BEAUTY, 405
SLUMBER PARTY '57, 406
SMALL TOWN IN TEXAS, A, 406
SMOKEY AND THE BANDIT, 406
SMOKEY AND THE BANDIT II, 406
SOLDIER, THE, 407
SOME KIND OF HERO, 407
SOME LIKE IT HOT, 407
SOMETHING WICKED THIS WAY COMES, 407
SOMEWHERE IN TIME, 407
SON OF FOOTBALL FOLLIES AND BIG GAME AMERICA, THE, 407
SONS OF KATIE ELDER, THE, 407
SOPHIE'S CHOICE, 408
SOUND OF MUSIC, THE, 408
SOUTH PACIFIC, 408
SOUTHERN COMFORT, 408
SOYLENT GREEN, 408
SPACE AGE, THE, 408
SPACEHUNTER: ADVENTURES IN THE FORBIDDEN ZONE, 409
SPELLBOUND, 410
SPIRAL STAIRCASE, 411
SPLIT IMAGE, 411
SPY WHO LOVED ME, THE, 412
SQUIRM, 412
STAGECOACH, 412
STALAG 17, 412
STAR IS BORN, A, 412

STAR TREK—THE MOTION PICTURE, 413
STAR TREK 2—THE WRATH OF KHAN, 413
STAR TREK 6: AMOK TIME/JOURNEY TO BABEL, 413
STAR TREK, VOL. 1, 414
STAR TREK, VOL. 2, 414
STAR TREK, VOL. 3, 414
STAR TREK, VOL. 4, 414
STAR TREK, VOL. 5, 414
STAR WARS, 414
STARDUST MEMORIES, 414
START TO FINISH—THE GRAND PRIX, 414
STARTING OVER, 415
STILL OF THE NIGHT, 415
STING, THE, 415
STING 2, THE, 415
STIR CRAZY, 415
STORY OF O, THE, 416
STRANGE BEHAVIOR, 416
STRANGER IS WATCHING, A, 416
STRANGERS ON A TRAIN, 416
STRAW DOGS, 416
STREET FIGHTER, THE, 416
STREETCAR NAMED DESIRE, A, 416
STRIPES, 416
STUDENT BODIES, 417
STUNT MAN, THE, 417
SUMMER LOVERS, 417
SUMMER OF '42, 417
SUNSET BOULEVARD, 417
SUNSHINE BOYS, THE, 417
SUPER BOWL XIV: STEELERS VS. RAMS PLUS SEASON'S, 417
SUPER BOWL XV: RAIDERS VS. EAGLES PLUS SEASON'S, 418
SUPER MEMORIES OF THE SUPER BOWLS, 418
SUPER SEVENTIES, THE, 418
SUPERMAN—THE MOVIE, 418
SUPERMAN 2, 418
SURVIVAL ANGLIA'S WORLD OF WILDLIFE, VOL. 1, 418
SURVIVORS, THE, 419
SWAMP THING, 419
SWEPT AWAY, 419
SWING TIME, 419
SWISS FAMILY ROBINSON, 419
SWORD AND THE SORCERER, THE, 420
SYBIL, 420
TABLE FOR FIVE, 420
TAKE THE MONEY AND RUN, 421
TAKING OF PELHAM 1-2-3, THE, 421
TALE OF THE FROG PRINCE, 421
TALES FROM MUPPETLAND, 421
ALES FROM MUPPETLAND 2, 421
TAMING OF THE SHREW, THE, 421
TAPS, 422
TARZAN, THE APE MAN, 422
TATTOO, 422
TAXI DRIVER, 422
"10", 423
TEN COMMANDMENTS, THE, 423
TERRYTOONS, VOL. 1 FEATURING MIGHTY MOUSE, 423
TESS, 423
TEX, 423
TEXAS CHAINSAW MASSACRE, 424
THAT CHAMPIONSHIP SEASON, 424
THAT'S ENTERTAINMENT, 424
THERE'S NO BUSINESS LIKE SHOW BUSINESS, 424
THEY ALL LAUGHED, 425
THEY CALL ME BRUCE?, 425
THEY DRIVE BY NIGHT, 425
THEY SHOOT HORSES, DON'T THEY?, 425

THIEF, 425
THING, THE, 425
THING, THE (ORIGINAL 1951 VERSION), 425
THIRTY-NINE STEPS, THE, 426
THOMAS CROWN AFFAIR, THE, 426
THOUSAND CLOWNS, A, 426
THREE DAYS OF THE CONDOR, 426
THREE MUSKETEERS, THE, 427
THREE STOOGES VIDEODISC, VOL. 1, THE, 427
THREE STOOGES VIDEODISC, VOL. 2, THE, 427
THUNDERBALL, 427
THUNDERBIRDS ARE GO, 427
THUNDERBOLT AND LIGHTFOOT, 427
TICKET TO HEAVEN, 427
TILL MARRIAGE DO US PART, 427
TIME BANDITS, 427
TIME MACHINE, THE, 428
TO CATCH A THIEF, 428
TOM AND JERRY, 428
TOM AND JERRY CARTOON FESTIVAL, VOL. 1, 428
TOM AND JERRY, VOL. 2, 429
TOM JONES, 429
TOMMY, 429
TOOTSIE, 429
TORA! TORA! TORA!, 429
TOWERING INFERNO, 430
TOY, THE, 430
TRAIL OF THE PINK PANTHER, 431
TRAPEZE, 431
TREASURE ISLAND, 432
TREASURE ISLAND (WITH WALLACE BEERY), 432
TREASURE OF SIERRA MADRE, 432
TRIBUTE, 432
TRON, 432
TRUE CONFESSIONS, 433
TRUE GRIT, 433
TURNING POINT, THE, 433
TWELVE ANGRY MEN, 433
TWELVE O'CLOCK HIGH, 434
20,000 LEAGUES UNDER THE SEA, 434
2001: A SPACE ODYSSEY, 434
UP IN SMOKE, 435
URBAN COWBOY, 435
USED CARS, 436
VALLEY GIRLS, 437
VENOM, 437
VERDICT, THE, 437
VICE SQUAD, 438
VICTOR VICTORIA, 438
VICTORY, 438
VICTORY AT SEA, 438
VIDEODROME, 439
VILLAGE OF THE DAMNED, 440
VISITING HOURS, 441
VIVA LAS VEGAS, 441
WAR GAMES, 442
WAR OF THE WORLDS, THE, 442
WAR PRAYER, THE, 443
WARRIORS, THE, 443
WASHINGTON AFFAIR, THE, 443
WASN'T THAT A TIME, 443
WATERSHIP DOWN, 443
WAY WE WERE, THE, 444
WE'RE NO ANGELS, 444
WEST SIDE STORY, 444
WESTWORLD, 444
WHAT'S NEW PUSSYCAT?, 445
WHAT'S UP DOC?, 445

### Videodiscs By Type

WHAT'S UP TIGER LILY?, 445
WHEN A STRANGER CALLS, 445
WHEN WORLDS COLLIDE, 445
WHITE HEAT, 446
WHITE LIGHTNING, 446
WHOLLY MOSES!, 447
WHOSE LIFE IS IT ANYWAY?, 447
WIFEMISTRESS, 447
WILD BUNCH, THE, 447
WILD IN THE COUNTRY, 447
WIMBLEDON 1981/WIMBLEDON: A CENTURY OF GREATNESS, 448
WIMBLEDON: 1979-1980 FEATURING BORG AND MCENROE, 448
WINNIE THE POOH, 448
WINTER KILLS, 448
WIZ, THE, 448
WIZARD OF OZ, THE, 448
WOMEN IN LOVE, 448
WOODSTOCK, 448
WOODY WOODPECKER AND HIS FRIENDS, 449
WORLD ACCORDING TO GARP, THE, 449
WORLD SERIES 1980—PHIL. PHILLIES VS. KANSAS CITY ROYALS, 450
WORLD'S GREATEST LOVER, THE, 450
XANADU, 450
YANKEE DOODLE DANDY, 450
YEAR OF LIVING DANGEROUSLY, THE, 450
YELLOWBEARD, 450
YES, GEORGIO, 451
YOU ONLY LIVE TWICE, 452
YOUNG DOCTORS IN LOVE, 452
YOUNG FRANKENSTEIN, 452
YUM-YUM GIRLS, 452
Z, 452
ZAPPED!, 452
ZIEGFELD FOLLIES, 453
ZORRO, THE GAY BLADE, 453

MUSIC

ABBA, 217
ABBA IN CONCERT, 218
AIDA, 221
ALL THAT JAZZ, 223
ALLMAN BROTHERS BAND, THE, 223
AMERICA LIVE IN CENTRAL PARK, 224
AMERICAN HOT WAX, 225
AMERICAN IN PARIS, AN, 225
ANNIE, 227
APRIL WINE, 229
ASHFORD & SIMPSON, 230
BEST LITTLE WHOREHOUSE IN TEXAS, THE, 237
BILLY SQUIER LIVE IN THE DARK, 238
BINDU VISUALIZATION OF MUSIC, 238
BLONDIE: EAT TO THE BEAT, 241
BLUE HAWAII, 241
BLUES ALIVE, 241
BLUES BROTHERS, THE, 242
BOB WELCH & FRIENDS: LIVE AT THE ROXY, 242
BOHEME, LA, 242
BOSTON MEDIA CONSULTANTS—(CLASS. RESEARCH DISC), 244
BRIGADOON, 246
CALEDONIAN DREAMS, 249
CAROLE KING, 251
CHARLIE DANIELS BAND, THE, 255
CHARLOTTE'S WEB, 255
CHICK COREA/GARY BURTON-LIVE IN TOKYO, 258
CLAUDE BOLLING: CONCERTO FOR CLASSIC GUITAR AND, 259
COAL MINER'S DAUGHTER, 260
COMPLEAT BEATLES, THE, 263
DAVE MASON LIVE AT PERKINS PALACE, 268
DEXY'S MIDNIGHT RUNNERS, 273
DIANA ROSS IN CONCERT, 274
DIRT BAND, THE, 275
DIVINE MADNESS, 277
DOLLY PARTON, 278
DON KIRSHNER'S ROCK CONCERT, VOL. 1, 278
DOOBIE BROTHERS LIVE, THE, 278
DURAN DURAN, 280
DVORAK 9, 280
DVORAK'S SLAVIC DANCE, 280
EARTH, WIND AND FIRE, 281
ELECTRIC LIGHT ORCHESTRA, 283
ELEPHANT PARTS, 283
ELTON JOHN: VISIONS, 283
ELVIS—ALOHA FROM HAWAII, 284
ELVIS—HIS 1968 COMEBACK SPECIAL, 284
ELVIS ON TOUR, 284
EPCOT CENTER VIDEODISCS (WALT DISNEY WORLD-FL.), 285
EUBIE!, 287
EURYTHMICS: SWEET DREAMS (THE VIDEO ALBUM), 287
EVE, THE, 287
EVENING WITH RAY CHARLES, AN, 287
FAME, 290
FIDDLER ON THE ROOF, 291
FILLE MAL GARDEE, LA, 291
FIRST BARRY MANILOW SPECIAL, THE, 292
FLASHDANCE, 293
FLEETWOOD MAC, 293
FLEETWOOD MAC IN CONCERT MIRAGE TOUR '82, 293
FOOTLIGHT PARADE, 294
42ND STREET, 296
FROM THE NEW WORLD SYMPHONY NO. 9 IN E MINOR, OP.95, 298
FUN & GAMES, 298
FUN IN ACAPULCO, 298
FUNNY GIRL, 298
FUNNY THING HAPPENED ON THE WAY TO THE FORUM, A, 299
G.I. BLUES, 299
GIMME SHELTER, 302
GIRL GROUPS—THE STORY OF A SOUND, 302
GOLD DIGGERS OF 1933, 304
GOSPEL, 305
GRACE JONES: ONE MAN SHOW, 305
GRATEFUL DEAD IN CONCERT, 305
GRATEFUL DEAD/DEAD AHEAD, 306
GREASE, 306
GREASE 2, 306
GROVER WASHINGTON, JR. IN CONCERT, 308
HAIR, 310
HALL & OATES ROCK AND SOUL LIVE, 310
HARDER THEY COME, THE, 312
HARRY CHAPIN, THE FINAL CONCERT, 312
HELLO DOLLY, 314
HERE IT IS, BURLESQUE, 315
HONEYSUCKLE ROSE, 317
HOROWITZ, 317
ITZHAK PERLMAN, 328

JAMES TAYLOR IN CONCERT, 329
JAZZ IN AMERICA, 330
JAZZ SINGER, THE, 330
JEFFERSON STARSHIP, 330
JESUS CHRIST, SUPERSTAR, 331
JUDY GARLAND, 333
KANAKO, 333
KENNY LOGGINS ALIVE, 334
KIDS ARE ALRIGHT, THE, 334
KIDS FROM FAME, THE—LIVE, 334
KING AND I, THE, 335
KING CREOLE, 335
KINGSTON TRIO/ETC., 335
KNACK, THE—LIVE AT CARNEGIE HALL, 336
LADY SINGS THE BLUES, 337
LAST WALTZ, THE, 339
LET IT BE, 340
LET'S SPEND THE NIGHT TOGETHER, 340
LITTLE PRINCE, THE, 341
LITTLE RIVER BAND LIVE EXPOSURE, 341
LIVE FROM THE SANTA BARBARA BOWL—BOB MARLEY AND THE WALERS, 342
LIVE INFIDELITY: REO SPEEDWAGON IN CONCERT, 342
LIZA, 342
LORETTA, 343
LUCIANO PAVAROTTI IN CONCERT, 345
M. J. ZINK PRODUCTIONS, INC.—(NOT GIVEN), 345
MA VLAST (MY FATHERLAND), 346
MANHATTAN TRANSFER, 349
MANON, 349
MARY POPPINS, 350
MAZE FEATURING FRANKIE BEVERLY. HAPPY FEELIN'S— LIVE IN NEW ORLEANS, 352
MEL TORME & DELLA REESE IN CONCERT, 353
MICK FLEETWOOD—THE VISITOR, 354
MUSIC OF MELISSA MANCHESTER, THE, 361
NEIL SEDAKA IN CONCERT, 363
NEW YORK, NEW YORK, 364
NIGHT WITH LOU REED, A, 365
NUTCRACKER SUITE, 367
NUTCRACKER, THE, 367
NUTCRACKER/BARYSHNIKOV, 367
OLIVIA, 369
OLIVIA IN CONCERT, 369
OLIVIA: PHYSICAL, 369
ONE & ONLY, GENUINE ORIGINAL FAMILY BAND, THE, 370
ONE FROM THE HEART, 370
ONE NIGHT STAND: A KEYBOARD EVENT, 370
ORIENTAL DREAMS, 371
OWL AND THE PUSSYCAT, THE, 372
PAINT YOUR WAGON, 372
PARIS-PARIS, 373
PAUL MCCARTNEY AND WINGS—ROCKSHOW, 374
PAUL SIMON, 374
PAUL SIMON IN CONCERT, 375
PAVAROTTI IN LONDON, 375
PENNIES FROM HEAVEN, 376
PETE'S DRAGON, 377
PETER ALLEN & THE ROCKETTES AT RADIO CITY MUSIC HALL, 377
PETER GRIMES, 377
PHILIPS VP-835 DEMONSTRATION DISC, 378
PIAF, 378
PINK FLOYD AT POMPEII, 378
PINK FLOYD—THE WALL, 378
PIPPIN, 379
PIRATES OF PENZANCE, THE, 379
POPEYE, 381

PRESLEY—KING OF ROCK 'N' ROLL, 382
QUADROPHENIA, 385
QUEEN—GREATEST FLIX, 386
RAINBOW GOBLINS STORY, 387
RAINBOW: LIVE BETWEEN THE EYES, 388
RAY CHARLES CONCERT, 388
RCA'S ALL-STAR COUNTRY MUSIC FAIR, 389
ROCK ADVENTURE, 393
ROD STEWART—TONIGHT HE'S YOURS, 393
ROD STEWART LIVE AT THE LOS ANGELES FORUM, 394
ROSE, THE, 394
ROSTROPOVICH: DVORAK CELLO CONCERTO/ SAINT-SAENS, 395
ROXY MUSIC, 395
RUSH—EXIT STAGE LEFT, 395
RUST NEVER SLEEPS, 396
SACRED MUSIC OF DUKE ELLINGTON, 396
SAMSON ET DALILA, 396
SATURDAY NIGHT FEVER, 397
SGT. PEPPER'S LONELY HEARTS CLUB BAND, 401
SHADOWS AND LIGHT, 401
SHEENA EASTON LIVE AT THE PALACE, 401
SIMON & GARFUNKEL IN CONCERT, 404
SIMON & GARFUNKEL: THE CONCERT IN CENTRAL PARK, 404
SLIPSTREAM—STARRING JETHRO TULL, 406
SOUND OF MUSIC, THE, 408
STAR IS BORN, A, 412
STEVE MILLER BAND LIVE, 415
STEVIE NICKS IN CONCERT, 415
SWAN LAKE, 419
SWING TIME, 419
TAKANAKA WORLD, 421
TALES OF HOFFMANN, THE, 421
TEDDY PENDERGRASS IN CONCERT, 422
THAT'S ENTERTAINMENT, 424
THERE'S A MEETIN' HERE TONIGHT, 424
TO RUSSIA . . . WITH ELTON JOHN, 428
TOMMY, 429
TONY BENNETT SONGBOOK—RECORDED LIVE IN NEW YORK, 429
TOTALLY GO-GOS, 430
TUBES VIDEO, THE, 433
URBAN COWBOY, 435
VICTORY AT SEA, 438
VLADIMIR HOROWITZ IN LONDON, 441
WEST SIDE STORY, 444
WHITE MUSIC, 446
WHO—LAST AMERICAN TOUR, THE, 446
WHO ROCKS AMERICA, THE, 446
WHO'S AFRAID OF OPERA? VOL. 1, 446
WHO'S AFRAID OF OPERA? VOL. 2, 446
WHO'S AFRAID OF OPERA? VOL. 3, 446
WIZ, THE, 448
WIZARD OF OZ, THE, 448
WOODSTOCK, 448
XANADU, 450
YANKEE DOODLE DANDY, 450


**POINT-OF-PURCHASE**

ALLEGHENY INTERNATIONAL INFOSOURCE I, 223
ARIS KNITWEAR, 229
BOOZ-ALLEN & HAMILTON HOME INFO SYSTEM SIMULATOR, 243

## Videodiscs By Type

CHRISTMAS 1983 ELECTRONIC CATALOG, 257
COLORATION, 261
DATA DEMONS, 267
DEMONSTRATION VIDEODISC (CAT ZURICH), 270
GENERAL MOTORS OF CANADA, 300
GREENDAY VIDEO, VOL. 1 NO. 4, 308
HANDS-ON WORD PROCESSING, 310
IBM—(CONFIDENTIAL), 320
INVESTMENT SERVICES, 327
M. J. ZINK PRODUCTIONS, INC.—(NOT GIVEN), 345
MARITZ COMMUNICATIONS CO.—(NUM. FOR BUS. & IND.), 350
PERMS AND HAIRSTYLES, 376

### PROMOTIONAL

ARIS KNITWEAR, 229
ASIA TRAVEL, 230
ATTI (AMERICAN TELEPHONE & TELEGRAPH INTERNATIONAL), 231
BOOZ-ALLEN & HAMILTON HOME INFO SYSTEM SIMULATOR, 243
BOSTON, 244
CHICAGO MERCANTILE EXCHANGE, 256
CITTA VIVA—GALLIPOLI, 258
COAST GUARD, 260
COLLEGE USE DEMONSTRATION DISC, 261
COLORATION, 261
CRANBERRY WORLD VISITOR'S CENTER VIDEODISC PROJECT, 265
DATA DEMONS, 267
DEMONSTRATION VIDEODISC (CAT ZURICH), 270
EPCOT CENTER VIDEODISCS (WALT DISNEY WORLD-FL.), 285
GREENDAY VIDEO, VOL. 1 NO. 4, 308
HANDS-ON WORD PROCESSING, 310
HAWAII TRAVEL, 313
HIGH POWER ILLUMINATOR RAM RADAR DEMO, 316
INTERNATIONAL HARVESTER, 325
LRN COLLEGE DISC, 344
MARITZ COMMUNICATIONS CO. - (NUM. FOR BUS. & IND.), 350
MERRILL LYNCH, 354
O.I.S.E. SINGLE POINT THREAD CUTTING, 368
PARIS-PARIS, 373
PATRIOT CLS OJT PHASE 1, 374
PERMS AND HAIRSTYLES, 376
PERSONAL COMPUTER EXHIBIT, 376
PHILIPS VP-835 DEMONSTRATION DISC, 377
POTPOURRI, 382
RADAR, MULTIPLE BEAM ACQ. & WIDE APERTURE SURVEIL., 385
RALSTON-PURINA, 386
SONY OVERVIEW, 407
SPARROW MISSILE TEST SET TRAINING VIDEODISC, 410
SPERRY DEMO, 410
SPORTS AND FITNESS IN ONTARIO, 411
SPS-49 RADAR, 411
TEXAS INSTRUMENTS, 424
TOMORROW TODAY, 429
TRADITION AND CHANGE, 431
VHD INTERACTIVE VIDEODISC DEMONSTRATION, 437
VIDEO BOOKMARK, THE, 438
VIDEO MUSEUM, THE, 439
VIDEO VERSATILITY, 439
VIDEODISC, THE: A TOOL FOR COMMUNICATION, 439
WALK THROUGH THE UNIVERSE, A, 442
WESTERN ELECTRIC, 444
WEYERHAEUSER, 445
YOU CAN BUY A COMPUTER!, 451

### RESEARCH AND DEVELOPMENT

ABC-NEA SCHOOLDISC, 218
ADVANCED VIDEO MAINTENANCE INFORMATION, 219
ARMY COMM. TECH. OFF. (ACTO) VIDEODISC PROTOTYPES, 229
ART HISTORY RETRIEVAL, 230
BALLET SERIES, 232
BASIC SKILLS EDUCATION FOR THE U.S. ARMY, 234
BOSTON MEDIA CONSULTANTS - (CLASS. RESEARCH DISC), 244
CAPSULE, THE, 250
CITIBANK, 258
COURT TESTIMONY SKILLS FOR THE EXPERT WITNESS, 265
DECIMALS AND FRACTIONS, 270
DEMONSTRATION VIDEODISC (CAT ZURICH), 270
DEMONSTRATION VIDEODISC (WICAT SYSTEMS, INC.), 271
DEMONSTRATION VIDEODISCS, 1982 AND 1983 (CSC), 271
DEVELOPMENT OF LIVING THINGS, 273
DIGITAL VIDEO CORPORATION—(CONFIDENTIAL), 274
E-3 FLOATING DESK PULSER, 281
EEMT COMPOSITE NO. 1, 282
GATHERING AND VERIFYING INFORMATION, 300
GENERAL MOTORS OF CANADA, 300
GREENDAY VIDEO, VOL. 1 NO. 4, 308
GYNECOLOGY PATIENT EDUCATION, 309
HANSEN'S DISEASE, 311
INTEGRATED READINESS CENTER, 323
INTERACTIVE ARTS & SCIENCES, INC.—(CONFIDENTIAL), 324
ISRAELI BOY: LIFE ON A KIBBUTZ, 328
KLAVIER IM HAUS, 336
KRAFT PROCESS OVERVIEW, 337
MACARIO, 346
MARITZ COMMUNICATIONS CO.—(NUM. FOR BUS. & IND.), 350
MONTEVIDISCO (MALE VERSION), 358
NAVY MINE NEUTRALIZATION SYSTEM TRAINER, 363
O.I.S.E. SINGLE POINT THREAD CUTTING, 368
ONE THOUSAND EIGHT HUNDRED & FIFTY TWO HUMPBACK WHALES, 371
PATRIOT MISSILE SYSTEM PROTOTYPE MAINTENANCE DISC, 374
PBS CLOSED CAPTIONED DISC, 375
POINT OF SALE DEMONSTRATION, 381
RCA ENGINEERING DEMO DISC, 389
REPAIR PARTS AND SPECIAL TOOLS LIST—AN UGC 74, 390
RESEARCH DISC, 390
SPANISH MAP READING CONVERSION PROJECT, 409
SPECTRUM OF INTEGRATED LOGISTICS SUPPORT, 410
STANDARD HEATS OF SOLUTION, 412
SYSTEM TEST ENHANCEMENT PROJECT (STEP), 420
TEST REACTOR, 423
TPS-43 INTERACTIVE "SAMPLER", 430
UNIV. OF LONDON PROGRAMME DEVELOPMENT VIDEODISC, 434
UNIV. OF NEB./NEB. VIDEODISC DESIGN/PROD. GRP-(NOT GIVEN), 435

UNIV. OF NEB./NEB. VIDEODISC DESIGN/PROD. GRP-(NOT GIVEN), 435
UROLOGY, 436
VIDEO ENHANCED GERMAN GATEWAY, 438
VISTA (VIDEODISC INTERPERS. SKILLS TRAIN. & AS.), 441
WALK THROUGH THE UNIVERSE, A, 442

## TECHNICAL

A.C. SWITCHBOARD OPERATION, 217
ABC-NEA SCHOOLDISC, 218
ACTION CODE COURSEWARE—INDUSTRIAL TECH. PROGRAM, 219
ADVANCED VIDEO MAINTENANCE INFORMATION, 219
AMTESS: AUTOMOTIVE, 225
AMTESS: MISSILE, 226
AN/WLR-1 SYSTEM REPAIRABLE IDENTIFICATION, 226
APOLLO, 228
ARMY COMMUNICATIVE TECH. OFFICE (ACTO) VIDEODISC PROTOTYPES, 229
ASTRONOMY, 231
BASIC AUTOMOTIVE ELECTRONICS, 234
BASIC DIESEL PRINCIPLES, 234
BASIC SKILLS EDUCATION FOR THE U.S. ARMY, 234
BOSTON DISC, 244
BOSTON MEDIA CONSULTANTS—(CLASS. RESEARCH DISC), 244
CALL FOR FIRE, 249
CAPSULE, THE, 250
CARDIO-PULMONARY-RESUSCITATION, 251
CLEAN AND INSPECT A GENERATOR, 259
COLUMBIA RIVER PROJECT, 262
D.C. MOTOR MAINTENANCE, PT. 1, 266
D.C. MOTOR MAINTENANCE, PT. 2, 266
DATA DEMONS, 267
DEMONSTRATION VIDEODISC (WICAT SYSTEMS, INC.), 271
DEMONSTRATION VIDEODISCS, 1982 AND 1983 (CSC), 271
DISC SUPPLEMENT TO JOURNAL, CELL MOTILITY, 275
EEMT COMPOSITE NO. 1, 282
ELECTRONIC IGNITION SYSTEM, 283
EXTRA MILE, THE, 289
FORD MOTOR VIDEO NETWORK (15 TOTAL DISC SIDES), 295
FORT SILL VIDEODISC COURSEWARE PACKAGE, 295
GENERAL MOTORS OF CANADA, 300
GEOSCIENCE, 301
HANSEN'S DISEASE, 311
HAWK TROUBLESHOOTING COURSE, 313
HIGH POWER ILLUMINATOR RAM RADAR DEMO, 316
IBM—(CONFIDENTIAL), 320
IBM 1750 TELEPHONE SYSTEM, 320
IBM OFFICE SYSTEMS, 321
INTELLIGENT DISC, THE, 324
INTRODUCTION TO THE JOIN SYSTEM, 326
JOB PERFORMANCE PACKAGE (JPP), 331
KRAFT PROCESS OVERVIEW, 337
LVT-7A1 DRIVER TRAINER, 345
M. J. ZINK PRODUCTIONS, INC.—(NOT GIVEN), 345
MANUAL SEAT/MAN SEP. FR. A MARTIN-BAKER EJECT. ST., 349
MARITZ COMMUNICATIONS CO.—(NUM. FOR BUS. & IND.), 350
MATERIAL SCIENCE VIDEODISC, 351
MILES LEARNING CENTER: ORIENTATION & OPER. INSTR., 356
MILTON KEYNES INFORMATION TECHNOLOGY VIDEODISC, 35
NAVY MINE NEUTRALIZATION SYSTEM TRAINER, 363
NINETY-SIX: A CATTLE RANCH IN NORTHERN NEVADA, 366
O.I.S.E. SINGLE POINT THREAD CUTTING, 368
ONE THOUSAND EIGHT HUNDRED & FIFTY TWO HUMPBACK WHALES, 371
PATRIOT CLS OJT PHASE 1, 374
PATRIOT MISSILE SYSTEM PROTOTYPE MAINTENANCE DISC, 374
PERCEPTRONICS, INC.—(NOT GIVEN), 376
PERSONAL COMPUTER DIAGNOSTICS, 377
PHILIPS VP-835 DEMONSTRATION DISC, 378
RADAR, MULTIPLE BEAM ACQ. & WIDE APERTURE SURVEIL., 386
RCA ENGINEERING DEMO DISC, 389
REPAIR PARTS AND SPECIAL TOOLS LIST—AN UGC 74, 390
SCE: SONGS UNITS 2 & 3 SURROGATE TRAVEL, 398
SHUTTLE, 402
SIGHT THROUGH SOUND, AN INTERACT. INTRO MED. DIAG., 403
SONY OVERVIEW, 407
SPACE AGE, THE, 408
SPACE SHUTTLE MISSION REPORTS: S.T.S. 5, 6 & 7, 409
SPANISH MAP READING CONVERSION PROJECT, 409
SPARROW MISSILE TEST SET TRAINING VIDEODISC, 410
SPS-49 RADAR, 411
SYSCON CORPORATION, INC.—(NOT GIVEN), 420
TRACING A GROUND, 430
TROUBLESHOOTING HORSEPOWER LOSS, 432
UNIV. OF LONDON PROGRAMME DEVELOPMENT VIDEODISC, 434
VEHICLE EMISSION CONTROL, 437
VIDEODISC SERIES/APPRENTICE & JOURNEYMAN TRAINING, 439
VINSON PROJECT, THE, 440
VISION-UNIX INTERACTIVE VIDEODISC TRAINING CURRIC., 440
VOYAGER, 442
WILS-WANG INTERACTIVE LEARNING SYSTEM-WP PROGRAM, 447